Theoretical Systems Ecology

ADVANCES AND CASE STUDIES

Efraim Halfon

Basin Investigation and Modeling Section
National Water Research Institute
Canada Centre for Inland Waters
Burlington, Ontario, Canada

ACADEMIC PRESS
New York   San Francisco   London   1979
A Subsidiary of Harcourt Brace Jovanovich, Publishers

COPYRIGHT © 1979, BY ACADEMIC PRESS, INC.
ALL RIGHTS RESERVED.
NO PART OF THIS PUBLICATION MAY BE REPRODUCED OR
TRANSMITTED IN ANY FORM OR BY ANY MEANS, ELECTRONIC
OR MECHANICAL, INCLUDING PHOTOCOPY, RECORDING, OR ANY
INFORMATION STORAGE AND RETRIEVAL SYSTEM, WITHOUT
PERMISSION IN WRITING FROM THE PUBLISHER.

ACADEMIC PRESS, INC.
111 Fifth Avenue, New York, New York 10003

United Kingdom Edition published by
ACADEMIC PRESS, INC. (LONDON) LTD.
24/28 Oval Road, London NW1 7DX

Library of Congress Cataloging in Publication Data
Main entry under title:

Theoretical systems ecology.

    Includes bibliographies.
    1. Ecology--Mathematical models. I. Halfon,
Efraim. II. Title.
QH541.15.MdT43   574.5'01'84   78–27014
ISBN 0–12–318750–8

PRINTED IN THE UNITED STATES OF AMERICA

79 80 81 82 83 84 9 8 7 6 5 4 3 2 1

Theoretical Systems Ecology

ADVANCES AND CASE STUDIES

CONTRIBUTORS

M. B. Beck
John W. Brewer
William G. Cale, Jr.
C. Cobelli
Michael Conrad
John T. Finn
B. S. Goh
Efraim Halfon
John Harte
Aleksej G. Ivakhnenko
Clark Jeffries
George J. Klir
Georgij I. Krotov
A. Lepschy

Dennis P. Lettenmaier
Orie L. Loucks
Patrick L. Odell
P. Jussi Orava
Bernard C. Patten
Jeffrey E. Richey
G. Romanin-Jacur
D. D. Šiljak
Madan G. Singh
Robert E. Ulanowicz
Vladimir N. Visotsky
Vicki Watson
Jackson R. Webster
Bernard P. Zeigler

Contents

List of Contributors	xiii
Preface	xv

Preview: Theory in Ecosystem Analysis
Efraim Halfon

1.	Introduction	1
2.	Aggregation and Organization	3
3.	Model Structures, Formalisms, and Theory of Modeling	4
4.	System Identification	6
5.	Model Analysis, Control Theory, and Stability	8
6.	Outlook	11
	References	12

Part I Aggregation and Organization

Chapter 1 Multilevel Multiformalism Modeling: An Ecosystem Example
Bernard P. Zeigler

1.	Introduction	18
2.	The Ecosystem: Questions of Interest and Models	18
3.	Organization of Questions and Models	20
4.	Experimental Frames	21
5.	Constructed Models	29
6.	Organization of Models	40
7.	Applicability of Frames to Models	43
8.	Summary	44

v

9. Discussion 45
   Appendix: Some Results on Estimated Parameters
   and Model Cross Comparison 49
   References 54

Chapter 2 Concerning Aggregation in Ecosystem Modeling

William G. Cale, Jr. and Patrick L. Odell

1. Introduction 55
2. Modeling: General 59
3. Modeling: Specific 61
4. The Aggregation Model 63
5. Concluding Remarks 69
   Appendix 74
   References 76

Chapter 3 Use of First-Order Analysis in Estimating Mass Balance Errors and Planning Sampling Activities

Dennis P. Lettenmaier and Jeffrey E. Richey

1. Introduction 80
2. First-Order Analysis 82
3. Results 92
4. Applications to Experimental Design 94
5. Summary 101
   Appendix: Notation 102
   References 103

Part II Model Structures, Formalisms, and Theory of Modeling

Chapter 4 Prediction, Chaos, and Ecological Perspective

Robert E. Ulanowicz

1. Introduction 107
2. Determinism and Chaos 109
3. Ecological Perspective 113

4.	Summary	115
	References	116

Chapter 5 Hierarchical Organization of Ecosystems
Jackson R. Webster

1.	Introduction	119
2.	Definition of Hierarchy	120
3.	The Hierarchical Organization of Nature	122
4.	Hierarchical Levels of Ecological Interest	125
5.	Application of Hierarchy Theory to Ecosystems	126
	References	128

Chapter 6 Hierarchical Adaptability Theory and Its Cross-Correlation with Dynamical Ecological Models
Michael Conrad

1.	Introduction	132
2.	Review of Adaptability Theory	133
3.	Hierarchical Adaptability Theory	137
4.	Cross-Correlation with Dynamical Ecological Models	143
5.	Conclusion	148
	References	149

Chapter 7 Structure and Stability of Model Ecosystems
D. D. Šiljak

1.	Introduction	151
2.	System Structure	154
3.	Partitions and Condensations	160
4.	Vulnerability of Structure	164
5.	Vulnerability of Stability	169
6.	Conclusion	179
	References	179

Chapter 8 Systems Approach to Continental Shelf Ecosystems

Bernard C. Patten and John T. Finn

1.	Introduction	184
2.	Causal Theory of Environment	185
3.	Causal Analysis of Ecosystems	192
4.	Flow Analysis of the Ross Sea Pelagic Ecosystem	199
5.	Summary	209
	References	210

Chapter 9 A Framework for Dynamical System Models: Cause–Effect Relationships and State Representations

P. Jussi Orava

1.	Introduction	214
2.	Input–Output System	215
3.	Causality	224
4.	State	227
5.	Discussion of Ecological Examples	231
	References	232

Part III System Identification

Chapter 10 Structural Identifiability of Linear Compartmental Models

C. Cobelli, A. Lepschy, and G. Romanin-Jacur

1.	Introduction	237
2.	Linear Time-Invariant Compartmental Models	239
3.	The Problem of Structural Identifiability	241
4.	Structural Properties Related to Identifiability	243
5.	The Analysis of Structural Identifiability	246
6.	Examples and Conclusions	251
	References	257

Chapter 11 Model Structure Identification from Experimental Data

M. B. Beck

1.	Introduction	260
2.	System Identification: A Brief Review	261
3.	Model Structure Identification: Black Box Models	270
4.	Model Structure Identification: Internally Descriptive Models	273
5.	Conclusions	287
	References	287

Chapter 12 Computer-Aided Systems Modeling

George J. Klir

1.	Introduction	291
2.	Relevant Concepts	295
3.	Systems Modeling	302
4.	Examples of Systems Modeling in Ecology	312
5.	Conclusions	320
	References	322

Chapter 13 Identification of the Mathematical Model of a Complex System by the Self-Organization Method

Aleksej G. Ivakhnenko, Georgij I. Krotov, and Vladimir N. Visotsky

1.	Introduction	326
2.	Present State of the Theory of Computer-Aided Self-Organization of Mathematical Models	326
3.	Computer-Aided Self-Organization of Models	328
4.	Discovery of Laws with the Aid of GMDH	336
5.	Application of GMDH to Environmental Problems	339
6.	Conclusions	350
	References	352

Part IV Model Analysis, Control Theory, and Stability

Chapter 14 An Analysis of Turnover Times in a Lake Ecosystem and Some Implications for System Properties

Vicki Watson and Orie L. Loucks

1.	Introduction	356
2.	Methods	361
3.	Results	368
4.	Discussion	374
5.	Summary and Conclusions	380
	References	381

Chapter 15 The Usefulness of Optimal Control Theory to Ecological Problems

B. S. Goh

1.	Introduction	385
2.	Discrete-Time Optimal Control	387
3.	Continuous-Time Optimal Control	394
4.	Conclusions	398
	References	398

Chapter 16 Toward Optimal Impulsive Control of Agroecosystems

John W. Brewer

1.	Introduction	401
2.	Optimal Single-Impulse Control of **S**-Shaped Growth	403
3.	Numerical Experiments	412
4.	Comments on the Optimal Impulse Control of **J**-Shaped Growth	414
5.	Concluding Remarks	416
	References	416

Chapter 17 Hierarchical Methods in River Pollution Control

Madan G. Singh

1.	Introduction	420
2.	Problem Formulation	421
3.	The Three-Level Method of Tamura	426
4.	The Time Delay Algorithm of Tamura	430
5.	The Interaction Prediction Approach	434
6.	River Pollution Control	437
7.	Hierarchical Feedback Control for Linear Quadratic Problems	444
8.	Extension to the Servomechanism Case	448
9.	Conclusions	450
	References	451

Chapter 18 Ecosystem Stability and the Distribution of Community Matrix Eigenvalues

John Harte

1.	Introduction	453
2.	A Practical Measure of Stability	455
3.	Analysis of the Stability Measure	456
4.	Discussion	462
	Appendix	463
	References	465

Chapter 19 Robust Stability Concepts for Ecosystems Models

B. S. Goh

1.	Introduction	467
2.	Global and Finite Stability	469
3.	Nonvulnerability	478
4.	Sector Stability	482
5.	Conclusion	486
	References	486

Chapter 20 Stability of Holistic Ecosystem Models

Clark Jeffries

1.	Introduction	489
2.	Holistic Ecosystem Modeling	494
3.	Discussion	499
4.	Conclusion	501
	Appendix	502
	References	503

Index 505

List of Contributors

Numbers in parentheses indicate the pages on which the authors' contributions begin.

M. B. Beck (259), International Institute for Applied Systems Analysis, Schloss Laxenburg, 2361 Laxenburg, Austria

John W. Brewer (401), Department of Mechanical Engineering, University of California, Davis, California 95616

William G. Cale, Jr. (55), Environmental Sciences Program, University of Texas at Dallas, Richardson, Texas 75080

C. Cobelli (237), Laboratorio per Ricerche di Dinamica dei Sistemi e di Elettronica Biomedica, Consiglio Nazionale delle Ricerche, C.P. 1075, 35100 Padova, Italy

Michael Conrad (131), Department of Computer and Communication Sciences, University of Michigan, Ann Arbor, Michigan 48104

John T. Finn* (183), Department of Zoology, Institute of Ecology, University of Georgia, Athens, Georgia 30602

B. S. Goh (385, 467), Mathematics Department, University of Western Australia, Nedlands, W.A. 6009, Australia

Efraim Halfon (1), Basin Investigation and Modeling Section, National Water Research Institute, Canada Centre for Inland Waters, Burlington, Ontario, Canada L7R 4A6

John Harte (453), Energy and Environment Division, Lawrence Berkeley Laboratory, Berkeley, California 94720

Aleksej G. Ivakhnenko (325), Vladimirskaya 51/53 k. 14, Kiev-3, 252003 U.S.S.R.

Clark Jeffries (489), General Delivery, La Ronge, Saskatchewan, Canada S0J 1L0

George J. Klir (291), Department of Systems Science, School of Advanced Technology, State University of New York at Binghamton, Binghamton, New York 13901

Georgij I. Krotov (325), Vladimirskaya 51/53 k. 14, Kiev-3, 252003 U.S.S.R.

A. Lepschy (237), Instituto di Elettrotecnica e di Elettronica, Universita di Padova, 35100 Padova, Italy

Dennis P. Lettenmaier (79), Department of Civil Engineering and College of Fisheries, University of Washington, Seattle, Washington 98195

Orie L. Loucks (355), The Institute of Ecology, Holcomb Research Institute, Butler University, Indianapolis, Indiana 46208

Patrick L. Odell (55), Environmental Sciences Program, University of Texas at Dallas, Richardson, Texas 75080

P. Jussi Orava (213), Systems Theory Laboratory, Department of Electrical Engineering, Helsinki University of Technology, SF-02150 Otaniemi, Finland

**Present address:* Ecosystems Center, Marine Biological Laboratory, Woods Hole, Massachusetts 02543.

Bernard C. Patten (183), Department of Zoology, Institute of Ecology, University of Georgia, Athens, Georgia 30602

Jeffrey E. Richey (79), College of Fisheries, University of Washington, Seattle, Washington 98195

G. Romanin-Jacur (237), Laboratorio per Ricerche di Dinamica dei Sistemi e di Elettronica Biomedica, Consiglio Nazionale delle Ricerche, C.P. 1075, 35100 Padova, Italy

D. D. Šiljak (151), Department of Electrical Engineering and Computer Science, University of Santa Clara, Santa Clara, California 95953

Madan G. Singh (419), Laboratoire d'Automatique et d'Analyse des Systemes du CNRS 7, 31400 Toulouse, France

Robert E. Ulanowicz (107), Center for Environmental and Estuarine Studies, University of Maryland, Solomons, Maryland 20688

Vladimir N. Visotsky (325), Vladimirskaya 51/53 k. 14, Kiev-3, 252003 U.S.S.R.

Vicki Watson (355), Department of Botany and Institute for Environmental Studies, University of Wisconsin, Madison, Wisconsin 53706

Jackson R. Webster (119), Department of Biology, Virginia Polytechnic Institute and State University, Blacksburg, Virginia 24061

Bernard P. Zeigler (17), Department of Applied Mathematics, The Weizmann Institute of Science, Rehovot, Israel

Preface

After many years of development, systems ecology is having a large impact upon all aspects of environmental research. The system approach with its body of concepts and techniques has broadened the ecologists' perspectives and has attracted students into ecology from other disciplines. Their common ground is systems science and ecological theory. Their goals have been the establishment of a sound theoretical basis and the development of mathematical models. Unfortunately, there has been a lack of communication between theoreticians, and modelers and field ecologists. Modelers have dealt almost exclusively with difference and differential equations to model ecosystems and to produce simulations, computer solutions of the equations, that could be used for forecasting. They approached the problems of model development, simplification, identification, and analysis on an *ad hoc* basis. Theoreticians worked on a theory of modeling, but their concepts seldom were used in complex modeling exercises. Perhaps the language of system theory was too mathematical for many ecologists to understand and apply.

The purpose of this book is to try to bridge this gap. It is to present to theoretical systems ecologists and other theoreticians in systems science recent advances in the field. Since the language of system theory is mathematics, many chapters are mathematically sophisticated. It is also a purpose of the book to present to scientists, who do not have the background to follow mathematical concepts, some aspects of systems ecology with which they are not familiar in a way they can understand and apply. The examples at the end of each chapter have this function. They show how theory can be used successfully and fruitfully to improve the development and analysis of models.

Notation has been kept uniform as much as possible, given the difference in topics included in the book. References have been spelled out to allow easy access to information complementary to that presented here. To demonstrate that a set of data contains information that can be extracted with system techniques and used at different stages of model construction and usage, three authors (Beck, Ivakhnenko, and Singh), have used data from the river Cam in England. To show how systems ecology has evolved in places other than North America, authors from nine countries contributed to this effort.

Three classes of problems are analyzed in the book: (1) Selection of components comprising the system model. The first two sections deal with theory of modeling, formalisms, classes, and properties of models. (2) Definition of the relationships and interactions between the system variables. The section on Identification deals with the problem of extracting information from data for the purpose of deriving the model structure. (3) Model analysis. Several sections cover this aspect. To represent current trends, several chapters on stability and control theory are included.

This book was conceived in the stimulating research environment of the Canada Centre for Inland Waters, and I am grateful to Floyd C. Elder and Theodore J. Simons who provided a sheltered climate for unencumbered research. I am grateful to Lawrence R. Pomeroy, Bernard C. Patten, and Rolf E. Bargmann, who directed me and helped to bring out and develop my true research interests. I have come to appreciate the theoretical aspects of systems ecology during the several years I spent with the systems ecology group at the University of Georgia. Together with Bernard C. Patten, Jack B. Waide, Jack R. Webster, and William G. Cale provided me with guidance in the difficult art of ecological modeling.

I thank all the authors for writing, and rewriting the chapters until the referees and myself were satisfied. It was a pleasure to act as their editor. Many persons contributed their ideas, talent, and time. Most of the following also acted as referees: L. J. Bledsoe, W. G. Cale, M. Conrad, J. J. Duffy, J. T. Finn, B. S. Goh, J. Harte, A. G. Ivakhnenko, C. Jeffries, D. P. Lettenmaier, S. H. Levine, M. McLean, J. Orava, B. C. Patten, D. Sahal, D. Siljak, R. V. Thomann, R. E. Ulanowicz, V. Watson, J. R. Webster, and B. P. Zeigler. Ms. J. Fleet, Ms. V. Hamilton, and Ms. N. Snelling helped with the typing and other clerical tasks.

My wife Silvia gracefully put up with all the pressures during the organization of the book.

Efraim Halfon

PREVIEW: THEORY IN ECOSYSTEM ANALYSIS

Efraim Halfon

1. Introduction	1
2. Aggregation and Organization	3
3. Model Structures, Formalisms, and Theory of Modeling	4
4. System Identification	6
5. Model Analysis, Control Theory, and Stability	8
6. Outlook	11
References	12

1. INTRODUCTION

A system may be defined as a set of elements standing in an interrelation among themselves and with the environment. It is generally agreed that a "system" is a model of general nature, that is, a conceptual analog of certain rather universal traits of observed entities. In other words, system-theoretical arguments pertain to, and have predictive value, inasmuch as general structures are concerned (von Bertalanffy, 1972). As researchers in other disciplines have done before, ecologists have turned to the system approach (e.g., Van Dyne, 1969; Patten, 1971; Odum, 1971).

The system approach is based on the evidence that certain system properties do not depend on the specific nature of the individual system, that is, they are valid for systems of different nature as far as the traditional classification of science (physical, biological, social) is concerned (Klir, 1972). Therefore, sophisticated procedures developed for the analysis of complex systems, mainly electrical, can now be applied to ecological systems where analytical methodology is far less advanced. When quantitative formalisms such as algebraic or differential equations are used, the similarity of different systems becomes a subject of interest. A model is a conceptualization of the real system; they are similar. Usually a model cannot be considered a unique representation of the real system. Indeed an infinite number of models may be conceptualized. When a model is built, the similarity relation must be understood and quantified.

Compartmental analysis is a phenomenological and macroscopic approach for modeling physicochemical process. A compartment (or state variable, object, element, etc., according to the different terminologies common to systems theory) is a basic unit of functional interest. It may be a species of algae in a lake, all plankton, or the whole lake itself, depending on the study goals. The choice of the compartment may be arbitrary, but any decision made at the early stages of model development will influence all of the other results. This choice of the compartment defines the relation between models and systems. This is the *aggregation problem*.

The second main topic of interest in model development is the *system identification problem*; the study of relations among compartments. The formulation of a correct model structure is as important as the solution to the aggregation problem. Most of the volume (13 out of 20 chapters) is dedicated to these problems of model development. It is my belief that a solid theoretical foundation is important if we want to continue to progress in the systems analysis of ecosystems. The concept of similarity is crucial to any form of general systems theory and thus crucial to the understanding of a theory of modeling of all systems and, particularly in this instance, of ecological systems.

This volume contains four major parts and their sequence follows the usual course of thinking in system science.

Part I discusses some fundamental system problems and focuses on the aggregation problem and its relation to sampling activities. Here some theoretical foundations are laid. Problems related to model development are analyzed.

Part II includes information on modeling approaches and philosophy. The emphasis here is on model structure and includes formalisms, classes, and model properties. Three main topics in this part are hierarchical models, structure properties, and the relation of causality to model structures.

Part III introduces methodologies and computer techniques of system identification. These methods, however, cannot be separated from the modeling philosophies of their originators: Klir and Ivakhnenko present their inductive approaches to general systems theory and their identification methods reflect their respective beliefs.

Part IV contains studies on model analysis, and the focus is on structural properties, such as stability, flow analysis, and general systems properties. The other topic of interest in this part is the applicability of control theory to ecological models. Goh presents the basics in Chapter 15, followed by more sophisticated applications (Chapters 16 and 17).

2. AGGREGATION AND ORGANIZATION

"A model which must be capable of accounting for all the input–output behavior of a real system and be valid in all allowable experimental frames can never be fully known" (Zeigler, 1976). This model, which Zeigler calls the *base* model, would be very complex and require such great computational resources that it would be almost impossible to simulate. For ecosystems, the base model can never be fully known because of the complexity of the system and the impossibility of observing all possible states. However, given an experimental frame of current interest, a modeler is likely to find it possible to construct a relatively simple model that will be valid in that frame. This is a *lumped* model. It is the experimenter's image of the real system with components lumped together and interactions simplified (Zeigler, 1976).

Modeling an ecosystem requires knowledge of the real system, obtained with experiments, and its abstraction within a mathematical framework. Systems methods can be used effectively in the latter phase of model development. Indeed, when coupled with experimental work, system-theoretic concepts can help in the development of an ecologically realistic mathematical model. The *state space approach* is the most widely used in modern systems analysis because it allows description of both observable and unobservable variables. The models are memoryless and nonanticipatory and the state of the system is predicted using information on the present state and inputs to the system.

How do we choose the state variables and what kind of errors originate when we develop a homomorphic model? The problem is that ecosystem models have been developed *a priori* as aggregations without regard to the consequences of that aggregation. According to Zeigler (Chapter 1), the organization of simulation models is accomplished by considering several elements and their relationships. These elements are a collection of experimental frames, the real system, and the domain of possible models. The experimental frames specify the restrictions on experimental access to the real system. The models are assumed to be transition systems which are specifiable at various levels of structure and behavior and within short-hand conventions (e.g., sequential machines, discrete event, and differential equation formalisms). Thus, the models are the various distributed and lumped models which can be postulated to account for the observed data and to predict the results of future experiments.

In Chapter 2, Cale and Odell analyze errors introduced by using a lumped (homomorphic) model instead of an additive (isomorphic) model. With the exception of the case of identical time constants, Cale and Odell believe that error is inherent in any dynamic aggregate regardless of how that aggregate was created. By assuming that the population level of organization is conceptually analogous to the base model of Zeigler, Cale and Odell introduce methods for quantifying aggregation error and indicate that at least four general time domain error responses exist for linear aggregates of causally independent (see Patten and Finn, Chapter 8) components. These two first chapters complement each other: Cale and Odell prove that lumped models have errors associated with them, while Zeigler points out that complex models can be employed to validate simpler ones. Thus several models can be developed, and if the correspondence between models is known, a more refined (less aggregate) model can be used to gain confidence in a more abstract but general lumped model. This problem is also dealt with at a more abstract level in Chapters 5, 6, and 8.

Chapter 3 deals with the errors that originate during sampling activities and in estimating the mass balance. First-order analysis is employed to solve this problem. The results show that this technique can be used to place confidence intervals around the components of a mass balance and to assess the accuracy of the overall mass balance. In the investigation of sampling strategies, the method allows studying the effects of sampling frequencies and station locations.

3. MODEL STRUCTURES, FORMALISMS, AND THEORY OF MODELING

In this part, we are introduced to general abstract ideas, formal approaches, and some expected trends in this area. Hierarchical theory plays an important role here. Structural properties related to reachability, causality, and stability are analyzed in detail. The relevant role of these ideas is reflected in their general applicability.

Ulanowicz (Chapter 4) explores possible trends in ecosystem analysis. His view is that current efforts at total ecosystem model prediction have obtained meager success, and, therefore, he questions whether current methods can ever achieve reasonable predictions. His approach is highly empirical and macroscopic but this empiricism is embodied in a formal theory. He analyzes several *a posteriori* methods, spectral analysis and ensemble theory, and concludes that these are potential tools to realize better models.

Webster (Chapter 5) is interested in the hierarchical organization of ecosystems and shows that even if we do not know whether nature is truly organized hierarchically, man's perception of nature is hierarchical. Each element in the hierarchy consists of systems of the next lower level and is characterized by behaviors occurring more slowly than behaviors at the next lower level. These levels, however, are coupled and interact with one another; some relationships however are stronger than others. He then describes how different philosophies of science have related different level behaviors. Webster's view is that at each level there are behaviors which cannot be predicted from the laws of lower levels; that is, they are more than the sum of the lower level parts. These behaviors are entirely consistent with lower level laws, so that the hierarchical structure constrains all level behaviors. When ecosystems are viewed as hierarchies, its levels are often given as organisms, populations, communities, and ecosystems. Webster, however, argues that populations and communities do not interact more within themselves than with other populations and communities, respectively, ideas which Goh (Chapter 19) rejects. This view is also different from that of Cale and Odell (Chapter 2) who argue that populations are a "reasonable unit of functional interest..." whereas Webster states that "identification of populations as subsystems of an ecosystem is...questionable." Rather, other physical subunits with a high level of internal interactions should be chosen.

Conrad (Chapter 6) adds to the hypotheses on hierarchical organization of ecosystems his views on the adaptability of biological systems. Adaptability is defined operationally in terms of the maximum tolerable uncertainty of the environment and entails behavioral uncertainty, ability to anticipate the environment, and indifference to the environment. His main conclusion is that the dynamics of any particular level in an ecosystem appear autonomous (i.e., are more amenable to description by a model incorporating fewer variables associated with other levels), but the contribution of adaptabilities at other levels can be critical for maintaining these dynamics.

Chapter 7 introduces a mathematical framework for considerations of structural properties of ecosystem models. This framework is based upon directed graphs (digraphs) and the object of the work is to associate them with dynamic systems. The states, inputs, and outputs of the system are identified by state, input, and output vertices of the corresponding digraph. These concepts will surface again in Chapter 20. Other concepts that are of interest for the analysis of model structure are input and output reachability. They specify how inputs affect the states of the system, and how the states can be observed at the outputs. Again here we are introduced to some concepts that will be used later in Chapter 8, 9, 10, and 14. Structure properties are then related to questions regarding the system stability, and Šiljak shows how a large class of models are stable despite arbitrary structural perturbations.

Chapter 8 deals with a particular class of problems related to the system structure: causality—or how the system elements interact with one another. The purpose of this chapter is to outline the causal theory of the organism/environment relation and to illustrate application of the theory by causal analysis of a continental shelf ecosystem model. As pointed out by Webster (Chapter 5), the conceptual subdivision of an ecosystem into functional units needs close scrutiny. Following one of the notations in general system theory (Zadeh and Desoer, 1963), these elements are also called *objects*. Patten and Finn use the term *holon* (Koestler, 1967) to describe a general system object. This object is a causal link whose action converts cause into effect. In a theory of Patten *et al.* (1976), the holon is oriented, that is, its attribute set is partitioned into inputs and outputs. The modeling philosophy of Patten and Finn follows from the ideas of systems theorists, namely Zadeh, Desoer, and Mesarovic. These approaches are of a *deductive* nature, whereas Klir's and Ivakhnenko's (Chapters 12 and 13) might be called *inductive*. The causal holon is the building block of causal systems in a modeling paradigm in which systems (wholes) represent a synthesis of objects (parts).

Chapter 9 is the most mathematical chapter in the book and perhaps one of the most difficult to follow for the uninitiated. The subject is again the study of cause–effect relationships and it follows Patten and Finn's chapter which should be read first. As in Chapter 8, the formalism of this chapter belongs to the deductive development and includes many conventional notions presented in some established works bearing on the subject (e.g., Windeknecht, 1967; Zadeh, 1969; Kalman *et al.*, 1969; Mesarovic, 1972). This chapter does not contain the description of any methods which can be used to establish cause–effect relationships in a mathematical model, but it is an introduction and discussion of a new formalism, that is, the principles which govern model construction, and is included here because it may be valuable in developing common languages for specialists in various fields of applications such as ecology.

4. SYSTEM IDENTIFICATION

System identification is a field of system theory that can be formally defined as the process of determining coefficients, parameters, and structure of a mathematical model in such a way as to describe a physical process in accordance with some predetermined criteria. In the development of ecosystem models the solution to the identification problem is important because it permits objective determination of the existence of interactions among components. Time series observations help in the development of the model structure and quantification of the parameter values.

Chapter 10 introduces the problem of structural identifiability. Cobelli *et al.* discuss whether identification methods can give acceptable solutions given the possibility of doing some experiments with the system. This analysis is an *a priori* process. Some structural model properties which are strictly related to identifiability are used, in particular, controllability, observability, and connectability. Three models of nutrient cycling are analyzed. The identifiability tests are performed directly with information extracted from the diagram of the compartmental models. These models are (a) a four-compartment model of the calcium cycle in a watershed ecosystem; (b) a four-compartment model describing the phosphorus dynamics in fresh water; and (c) a six-compartment model of the magnesium cycle in a tropical forest ecosystem.

Chapter 11 introduces some theoretical techniques which can be used for the solution of the identification problem. Two kinds of models are investigated: black-box (or input–output) models and internally descriptive (or mechanistic) models. These two model representations reflect two opposite, yet complementary, approaches to modeling. Their identification needs different methods and Beck describes them in detail. There are two specific identification problems that need to be solved: (a) to determine which of the several inputs are related in a significant way to the outputs, and (b) to define the time dependence of the relationships between inputs and outputs. The identification of black-box models is accomplished by computing the sample cross-correlation functions for the data. The identification of internally descriptive models is accomplished by the Extended Kalman Filter. The filter theory and an algorithm for its implementation are also presented. Since its application depends on the particular problem, a few guidelines are offered on the mechanics of implementing the filter for any given system. An example deals with a river quality model, that is, DO–BOD–algae interactions. The experimental data are taken from a field study of the river Cam in eastern England. This river is also a source of data for Ivakhnenko *et al.* who use it as an example for their identification scheme.

Chapter 12 deals with the problem of finding the model structure. Klir describes the fundamentals of his general systems theory and shows how by following an appropriate strategy one can derive information about the structure of the system from the given data. The structure identification procedure is assumption free. As such, it does not give any *a priori* meaning to the variables. The interpretation of the identified structure systems is done after they have been identified. Data collected from Lake Ontario during a 1-year period are used. Another example is the identification of hydrology oriented variables collected in a watershed of the Andrews Experimental Forest in Oregon (U.S.A.). The study found that the identified structure conformed to the one recognized by ecologists. This indicates that information about the structure is implicitly included in the data and can be utilized for developing models in areas where knowledge regarding the structure is not available.

Chapter 13 introduces the Group Method of Data Handling or Self-Organization Method. Its theory has been developed over a decade at the Institute of Cybernetics of the Academy of Sciences of the Ukrainian SSR. This theory is known only to a few specialists outside the USSR, Tamura and Ikeda in Japan, and Duffy and Franklin in the United States; they have applied it to environmental problems. Ivakhnenko's idea is that if the data are not too variable, the computer itself can find the best unique model for prediction or the best one exhibiting cause and effect relationships. By application of the self-organization method, the computer should be able to objectively discover the natural law that exists in the object under study. The self-organization of models can be regarded as a specific algorithm of computer artificial intelligence. The set of data is from the river Cam in England and from a reservoir near Kiev, Ukrainian SSR. The algorithm is used for the identification of both spatial and temporal patterns to determine the best model for prediction and for structural realism.

In a way, these last two chapters complement Zeigler's and Patten and Finn's articles by presenting alternative modeling theories. The former two follow Mesarovic's and Wymore's deductive approaches where goal-seeking description of behavior, together with the consistency and completeness of axiomatic theories, plays an important role. The latter two identify system traits rather than define the concept of a system axiomatically, and the identification is based on information derived from time series observations.

5. MODEL ANALYSIS, CONTROL THEORY, AND STABILITY

This last part is directed to readers who at this point may ask: We understand how we should go about choosing the state variables, deciding what kind of structure we should use, and identifying the principles at the basis of some system theories employed by ecologists for their purposes; now what do we do with the model? Some answers to this question were already given earlier (e.g., Chapters 4, 7, 8, and 11). but here some authors try to give a more concrete answer. As the reader can notice, not one chapter in this part deals with the simulation *per se*. This subject has already been the topic of numerous books (Patten, 1972; Canale, 1976; Hall and Day, 1977; Scavia, 1978). Instead, the emphasis here is on the information that can be extracted from a model once it has been developed and run on a computer. The first chapter of this part (14) is dedicated to the analysis of one well-known model "CLEANER" developed in the early 1970s for the IPB project. The following three chapters deal with control theory. The last three are dedicated to stability analysis; however, since this is a diverse field of study, only a few approaches could be included.

Watson and Loucks (Chapter 14) take the *holistic* view that the system properties are more numerous than and different from what can be found by analysis of the system parts. Since CLEANER is a large-scale model of a lake, it is interesting to study its system properties. One of the features which has been the subject of long investigations and modeling exercises is the system feedback or closed-loop properties. In ecosystems this is exemplified by nutrient cycling, and, therefore, this property is analyzed to reveal information on the system couplings. Moreover, the use of a model allows the elucidation of system properties from a dynamic point of view which emphasizes rates and time constants of processes acting on the static structure of the system. One method of flow analysis used here is equivalent to that proposed by Finn (1976) and also described in this volume (Chapter 8). This is an index of recycling and can be used to establish conditions that lead to bioregulation in the ecosystem studied.

Chapter 15 introduces the basic concepts of Optimal Control Theory. A more advanced treatment is presented in Chapter 17. These chapters show that there are two great difficulties in applying optimal control theory to the management of ecosystems. It requires an adequate model of the ecosystem and large computational resources. As we saw in Chapter 14, ecological models tend to be large scale and complex, and their reduction to lumped models, following Zeigler's advice, is usually not accomplished. Thus, most methods in control theory tend to reduce the computational requirements to allow the usage of complex models. Goh, in his introduction, presents some historical developments and shows how the classical calculus of variations was used to compute optimal rocket trajectories. It is interesting to note how these space applications are now being used for ecological problems. Brewer (Chapter 16) uses impulsive control (originally used to minimize fuel consumption of artificial satellites) to manage an agro-ecosystem, and Beck (Chapter 11) employs the Kalman filter, also used to compute trajectories of satellites, for the system identification of his river Cam model.

Optimal control can be used with both discrete-time and continuous-time models, for example, difference and differential equations. Discrete-time optimal control is nothing more than a nonlinear programming problem with a special structure. However, the large number of variables can cause problems. This mathematical tool has been successfully applied to a number of ecological problems and Goh describes some applications in fishery management.

Another application is presented by Brewer who uses optimal impulsive control to study the movement, transformation, and impact of biocides within the environment. He observes that biocides sometimes quickly disperse from the area where they have been applied. Their application can, therefore, be idealized and the biocide application can be described, in a highly ideal manner, by the Dirac delta function. Brewer's desideratum is to minimize the amount of biocide put into an ecosystem to obtain a given result.

Finally, in Chapter 17, Singh presents hierarchical methods applied to river pollution control. As such, this chapter is closely related to several others in the book—to Chapter 5 for the hierarchical approach, and to Chapters 11 and 13 for the use of data from a river in England. The chapter is a review of some recent results in hierarchical optimization and control for linearly interconnected dynamical systems. Both open- and closed-loop control methods are covered. For the open-loop case the Goal Coordination method for both discrete and continuous dynamical systems, the method of Tamura, and the prediction approach are described in particular reference to the river problem. For the closed loop case, the prediction approaches of Singh *et al.* (1976) are outlined. These methods are quite general and can be applied to other problems in ecology. This chapter is the second most difficult to read because of the necessity of formal mathematics. Several examples, however, are included in the theoretical treatment to make the subject more understandable.

The last three chapters deal with one of the central problems in ecology, that is, how to predict what will happen to a disturbed ecosystem. As is the case when one has to decide which state variable should be chosen to represent an ecosystem, it is important in the stability problem to establish a few properties to measure so that, on the basis of measurements it should be possible to predict the system reactions to perturbations. This problem has been introduced by Šiljak (Chapter 7) for structural perturbations. The approach in Chapter 18 is to introduce a definition of stability which offers a useful quantitative measure of ecosystem response to stress. This is motivated by the need of bringing theoretical stability analysis more into alignment with experimental constraints and practical needs. Harte thus deals with the problem of relative stability rather than of absolute stability. The assumption is that ecosystems are usually stable (see also Waide *et al.*, 1974; Patten, 1975; Halfon, 1976). Harte then proceeds, in a rather model-independent fashion, to relate relative stability of a system to certain combinations of possibly measureable system variables.

Chapter 19 introduces one standard test for stability, the *Liapunov method*. After presenting the method, Goh modifies it to make it applicable to ecological problems. He uses a "two-sided energy principle" to construct a Liapunov function. The main conclusion of the paper is that an ecosystem model is robust relative to all types of perturbations if it is a collection of subsystems for which self-regulating interactions dominate all other interactions. This view is different from that proposed by Webster in Chapter 5.

Finally, Jeffries (Chapter 20) explores the influence of one stable system on another. Liapunov functions are used here for the purpose of analyzing hierarchies of communities. A mathematical model of energy transfer is derived. This model allows for two basic types of energy transfer between compartments: predation and detritus donation. The stability of coupled systems depends then on the food web structure of the communities.

6. OUTLOOK

This volume includes original contributions in theoretical systems ecology which I think are important as a basis for future developments. Not all existing trends could be included (see Innis, 1977), but these look most promising. Some scientists are interested in general systems theory and in developing a theory for ecological systems; others are more interested in systems methods which can be applied to simulation models. Others approach more abstract problems. All these aspects of systems ecology are important and need careful attention. For many ecologists, systems ecology is only simulation, that is, the representation of an ecosystem as a set of equations which can then be used for prediction. The other face of systems ecology is at least as important, because this is where real progress will be generated. In fact, there is a difference, between mathematical model building and the system approach. Modelers want to construct an adequate mathematical representation of a given system. For the theoretician, the analysis itself, rather than the real system, is the focus of interest. The theoretical investigations are a study of the abstract system of relations (Rapoport, 1972). Since these properties are common to all mathematical systems isomorphic to that under study, the conclusions are expected to be valid for all real systems of which the mathematical systems are adequate representations. This is why some topics such as stability, causality, and hierarchical organizations relate not only to ecological systems but to all systems. It is this generality that is important and interesting.

One of the problems that needs to be solved is terminology. In this volume terms are used as uniformly as possible. Frequently different names are used for the same concept or the same name is used for different concepts. This problem is typical not only of systems ecology but also of general systems theory and other sciences. The task of unifying these concepts is not easy and will probably take a long time to accomplish. For example, the term "system" is used as meaning "model" or "real system" or "object." This confusion derives from the fact that several authors follow the terminology of different general systems theories (e.g., Klir, Mesarovic, Zadeh, von Bertalanffy, etc.) whose concepts do not often overlap. For example, Patten and Finn use the term object for "a set of variables together with a set of relations between them" (Zadeh and Desoer, 1963), which in Klir's terminology is quite close to (but not identical with) the concept of behavior. A reader interested in understanding this terminological "jungle" is advised to read Klir's (1972) book.

Where do we go from here? The contributors indicate that it is important to work toward a formal theory of ecosystems, since a working definition of this system is essential to make progress in ecosystem ecology. To this end, Patten and Finn, Webster, Ulanowicz, and Šiljak provide some answers and approaches. System methods which are appropriate for the analysis of ecological systems should be developed. These methods should also be of general nature so that other disciplines can use them. We are now looking at ecosystems in all their complexity, and this feature has become the focus of interest. Most authors seem inclined toward a holistic rather than a reductionistic approach. New possibilities have been discovered by approaching the problems at the general level: For example, the modeling theories of Zeigler, Patten, and Klir should be carefully studied. Finally, modelers should understand the theoretical aspects of systems ecology. Much can be gained from the interaction of the two groups.

I have tried to outline some important trends at this time. The reader is now invited to proceed and read what the authors have to say.

REFERENCES

Canale, R. P., ed. (1976). "Modeling Biochemical Processes in Aquatic Ecosystems." Ann Arbor Sci. Publ., Ann Arbor, Michigan.
Finn, J. T. (1976). Measures of ecosystem structure and function derived from analysis of flows. *J. Theor. Biol.* **36**, 363–380.
Halfon, E. (1976). Relative stability of ecosystem linear models. *Ecol. Modell.* **2**, 279–286.
Hall, C. A. S., and Day, J. W. (1977). "Ecosystem Modelling in Theory and Practice: An Introduction with Case Studies." Wiley (Interscience), New York.
Innis, G. S., ed. (1977). "New Directions in the Analysis of Ecological Systems," Parts 1 and 2, Simul. Counc. Proc. Ser., Vol. 5. Soc. Comput. Simul., La Jolla, California.

Kalman, R. E., Falb, P. L., and Arbib, N. A. (1969). "Topics in Mathematical System Theory." McGraw-Hill, New York.
Klir, G. J., ed. (1972). "Trends in General Systems Theory." Wiley (Interscience), New York.
Koestler, A. (1967). "The Ghost in the Machine." Macmillan, New York.
Mesarovic, M. D. (1972). A mathematical theory of general systems. *In* "Trends in General Systems Theory" (G. J. Klir, ed.), pp. 251–269. Wiley (Interscience), New York.
Odum, E. P. (1971). "Fundamentals of Ecology," 3rd ed. Saunders, Philadelphia, Pennsylvania.
Patten, B. C., ed. (1971). "Systems Analysis and Simulation in Ecology," Vol. 1. Academic Press, New York.
Patten, B. C., ed. (1972). "Systems Analysis and Simulation in Ecology," Vol. 2. Academic Press, New York.
Patten, B. C. (1975). Ecosystem linearization: An evolutionary design problem. *Am. Nat.* **109**, 529–539.
Patten, B. C., Bosserman, R. W., Finn, J. T., and Cale, W. G. (1976). Propagation of cause in ecosystems. *In* "Systems Analysis and Simulation in Ecology" (B. C. Patten, ed.), Vol. 4, pp. 457–549. Academic Press, New York.
Rapoport, A. (1972). The uses of mathematical isomorphism in general system theory. *In* "Trends in General Systems Theory" (G. J. Klir, ed.), pp. 42–77. Wiley (Interscience), New York.
Scavia, D., ed. (1978). "Perspectives in Aquatic Ecosystem Modelling." Ann Arbor Sci. Publ., Ann Arbor, Michigan (in press).
Singh, M. G., Hassan, M., and Titli, A. (1976). A multi level controller for interconnected dynamical systems using the prediction principle. *IEEE Trans. Systems, Man, and Cybernetics* **SMC-6**, 233–239.
Van Dyne, G. M., ed. (1969). "The Ecosystem Concept in Natural Resource Management." Academic Press, New York.
von Bertalanffy, L. (1972). The history and status of general systems theory. *In* "Trends in General Systems Theory" (G. J. Klir, ed.), pp. 21–41. Wiley (Interscience), New York.
Waide, J. B., Krebs, J. E., Clarkson, S. P., and Setzler, E. M. (1974). A linear systems analysis of the calcium cycle in a forested watershed ecosystem. *Prog. Theor. Biol.* **3**, 261–345.
Windeknecht, T. G. (1967). Mathematical systems theory: Causality. *Math. Syst. Theory* **1**, 279–288.
Zadeh, L. A. (1969). The concepts of system, aggregate, and state in system theory. *In* "System Theory" (L. A. Zadeh and E. Polak, eds.), pp. 3–42. McGraw-Hill, New York.
Zadeh, L. A., and Desoer, C. A. (1963). Linear systems theory. The state space approach. McGraw-Hill, New York.
Zeigler, B. P. (1976). "Theory of Modelling and Simulation." Wiley (Interscience), New York.

Part **I**

AGGREGATION AND ORGANIZATION

1. INTRODUCTION

Ecosystems, as examples of large-scale multifaceted systems, require that a multiplicity of models be developed since a single all-encompassing model, however desirable as a conceptual goal, is not a practical object. By decomposing questions and modeling objectives into an ordered structure of elements called experimental frames (Zeigler, 1976a), useful partial models may be constructed, validated, and employed, each one attuned to a particular experimental frame. Concomitant with the pluralism of such partial models is the recognition that models are expressible in different formalisms, each offering conceptual and computational advantages within its domain of application (Zeigler and Barto, 1977). But now, in addition to the familiar activities involving construction and validation of individual models, a host of organizational activities aimed at integrating the collection of models into a synergistic whole is required. The computer can be programmed to aid in executing these activities to a much greater degree than it is doing today.

A theoretical basis for structuring the organization of partial models has been developed (Zeigler, 1978). In this paper, we illustrate our approach and discuss its application to the patch-structured universes employed by Huffaker (1958; Huffaker *et al.*, 1963) to study predator–prey coexistence. We show how the approach facilitates the development of mutually supportive detailed and abstract models that in conjunction, provide both accurate ecological realism at one extreme and general insight into the essential mechanisms at work, at the other.

2. THE ECOSYSTEM: QUESTIONS OF INTEREST AND MODELS

The problem of predator–prey coexistence in patchy environments has received much theoretical attention of late (Levin, 1976; Hassel and May, 1974; Maynard Smith, 1974). Most of this work has employed the conventional differential and difference equation formalism, but following on suggestions stressing the importance of discrete processes (Maynard Smith, 1974), Zeigler (1977a) has shown that the discrete event formalism and associated simulation languages can provide effective, comprehensible explanations of predator–prey coexistence. Also, there have been very few attempts to fit the theoretical models to laboratory or field data. In contrast, our general approach is illustrated by the development of four related models, expressed in either the differential equation or discrete event formalism and constructed at different levels of abstraction, ranging from the most detailed level where close comparison of model behavior with experimental data is possible, to the most abstract where overall properties are discernible in a relatively simple manner.

The real system to which the modeling is directly addressed is that of the controlled universes constructed by Huffaker (1958; Huffaker et al., 1963). Basically, these consist of spatial arrays of oranges (patches) of controllable nutritional value, inhabitable by prey and predator mites, and interconnected by migration pathways of controllable difficulty. Many other discrete food unit environments fit this general form.

The most detailed model, *the base model*, is of the stochastic differential equation type. In it the local state (situation on each patch) is determined by a Lotka–Volterra type of differential equation governing the joint food, prey, and predator dynamics; the impetus for emigration and the effect of immigration are logically determined from the local state (food, prey, predator); and the migration process is of the stochastic random walk variety. This model permits the identification of the parameters of the local Lotka–Volterra dynamics from data for single patches. Because of its size and detail, it is not feasible for computer simulation, however. This motivates the construction of a second model which is both feasibly simulateable and amenable to validation against data collected from universes in which the effect of migration is at issue.

This (second) model is of the stochastic discrete event type and is simulated in SIMSCRIPT, a well-known discrete event language. The model keeps track of the same state variables as its predecessor but updates them only at "event times." The tables required for scheduling events and executing the updates were derived by appropriately partitioning the local state space and summarizing the Lotka–Volterra trajectories between partition boundaries. (This required a once-and-for-all simulation of the Lotka–Volterra equations.) This technique for representing differential equation models in summary form as discrete event models is quite general (Zeigler, 1977b).

The third model in the hierarchy is also of the discrete event type. It is an abstraction of its predecessor, in which the local situation is represented by a small number of discrete states (empty, some prey, etc.) and the migration processes are also suitably simplified. Since actual population counts have been discarded, this model cannot make quantitative global population predictions. The model lends itself, however, to convenient parameter study of persistence and the development of patterned interaction.

The last lumped model is of the deterministic differential equation type. It describes the global behavior of its predecessor operating under the so-called "random phase–random space" mode (Zeigler, 1977a). The model yields simple algebraic expressions for the equilibrium distribution of patch states and thus explains the form of the dependence of persistence on migration parameters.

3. ORGANIZATION OF QUESTIONS AND MODELS

The integration and organization of the above models are achieved within the formal system suggested by Zeigler (1976a, 1977a). The following is an informal review of the concepts involved.

We distinguish the following elements:

$\mathscr{E}$—a collection of *experimental frames*. A frame $E \in \mathscr{E}$ represents a restricted set of questions by specifying the restrictions on experimental access to the real system sufficient to answer them. Such a frame E determines a collection of data sets $\mathscr{D}(E)$, such that each $D \in \mathscr{D}(E)$ is an *a priori* possible result of complete data acquisition within frame E.

$\mathscr{R}$—the *real system*, comprises the specific data that have been, or would be, collected by experimenting with the system. Thus $\mathscr{R}$ associates with each experimental frame E a unique data set $\mathscr{R}(E) \in \mathscr{D}(E)$.

$\mathscr{M}$—*the domain of possible models*. These are assumed to be transition systems which are specifiable at various levels of structure and behavior and within various short-hand conventions such as the sequential machines, discrete event, and differential equation formalisms.

A full description of these basic elements and the concepts they embody may be found in Zeigler (1976b, Chapters 2 and 11). In Section 4, these elements are formulated in the context of patch-structured universes and the experiments of Huffaker (1958; Huffaker *et al.*, 1963) in particular. Roughly, the *experimental frames* will encode the various choices of observables (species counts in patches) and conditions (initial stocking of species, structure of universe) under which experiments were run. The *real system* is the data collectible by making the implied observations under the given conditions. Finally, the *models* are the various distributed and lumped models which can be postulated to account for the observed data and to predict the results of future experiments.

We can imagine an ideal state of affairs in which for each frame $E \in \mathscr{E}$ there is a known model $M \in \mathscr{M}$ which *best* answers the questions posable in E. By *best* we mean that the model can reproduce without error the data set $\mathscr{D}(E)$ in a manner which requires the least consumption of computer resources. Realistically, this ideal is not realizable after a necessarily finite span of data aquisition. The dynamics of modeling concern successive approximations to the ideal.

The problem is as follows: At any time t, the data already acquired in frame E will be some subset $\mathscr{R}^t(E) \subseteq \mathscr{R}(E)$. Many, perhaps most frames, will not even have been considered. Those that have, form the subset

$$\mathscr{E}^t = \{E | \mathscr{R}^t(E) \neq \varnothing\}.$$

$\varnothing$ is the empty set. Similarly, only a small subset $\mathscr{M}^t$ of the possible models $\mathscr{M}$ will have been constructed as potential model candidates. Thus the state of affairs at any time t is reflected in the triple $\mathscr{E}^t$, $\mathscr{R}^t$, $\mathscr{M}^t$.

4. EXPERIMENTAL FRAMES

Let us examine the elements of the triple at time $t =$ June 1961, the date of the last observation recorded by Huffaker et al. (1963).

There are four main types of experimental frames. As displayed in Table I, these types are distinguished by the descriptors "global," "local," "total," and "occupancy." The global descriptor refers to the fact that all cells (locations where an orange or a substitute rubber ball may be placed) in the universe are being observed. In contrast, in the local condition, only some subset of the cells are of interest. The total descriptor refers to the fact the quantities of interest in a frame have been totaled to produce aggregate quantities, so that only these aggregates are observable in the frame. Finally, the occupancy descriptor refers to the fact that a frame permits only the observation of discrete occupancy states, such as whether or not a cell is empty, whether or not a prey colony has been established on the cell, and whether or not there are no, few, or many predators present.

Table I also summarizes the kinds of questions associated with each frame. The occupancy frames are the most restricted. Nonetheless they permit consideration of the persistence of predator–prey relations since to determine whether or not there are any prey or predators requires only a binary categorization (present/not present) for each cell. At the other extreme, the global frames permit observation of detailed spatial distribution of species. The total frames correspond to classical populations in which spatial structure has been averaged out. It is evident that certain frames are potentially more informative than others. In a moment, we shall formally characterize this fact in terms of the "derivability" relation (Zeigler, 1978).

The experimental frames $\mathscr{E}$ are defined in Table II.[1] Each frame names a set of variables of interest, called the *compare* variables, and a set of

---

[1] Tables I, II, and III are rather complete listings of the elements involved. They are given here in such complete form as evidence that the decomposition of data advocated in this paper is feasible in practice. The reader need not be familiar with the detailed contents of the tables to understand the rest of the chapter.

Table I Experimental Frames and Their Posable Questions

Plane of frame (representative frames) given	Description	Posable questions concern
E^{global} (degree of abstraction = 0)	Food amount and population variables for each cell	Spatial characteristics of predator–prey, prey–food interaction
E^{local}	Food amount and population variables for a subset of cells keeping all others zero	Predator–prey, prey–food, interaction in local patch
$E^{\text{global, total}}$ (degree of abstraction = 1)	Food amount and population variables totaled over all cells	Space-averaged population sizes in predator–prey, balance of prey–food, interaction (classical lumped populations)
$E^{\text{occupancy, global}}$ (degree of abstraction = 1)	Discrete food and population states for each cell (empty, some prey, many prey, some predators, etc.)	Persistence of predator–prey, balance of prey–food, interactions; effect of cell geometry
$E^{\text{occupancy, global, total}}$ (degree of abstraction = 2)	Totals of cells in the various states as given in $E^{\text{occupancy, global}}$	Persistence of predator–prey, balance of prey–food, interactions under random phase–random space conditions

variables determining the conditions under which experiments are to be performed, called the *control* variables. The most inclusive frame, $E_{\text{food, prey, pred}}^{\text{global}}$, specifies as compare variables: food amount, prey, and predator population counts in each cell. This constitutes a total of $3N$ variables, where N is the number of cells in the universe. There are no control variables for this frame. An example of a frame which has a nonempty set of control variables is $E_{\text{food, prey/pred}}^{\text{global}}$ whose compare variables are all $2N$ food and prey variables and whose control variables are all N predator variables.

In a frame with no control variables, compare variables readings are recorded against time for the duration of any particular experiment (see Fig. 1). This yields a time function also called a segment, or a trajectory which is

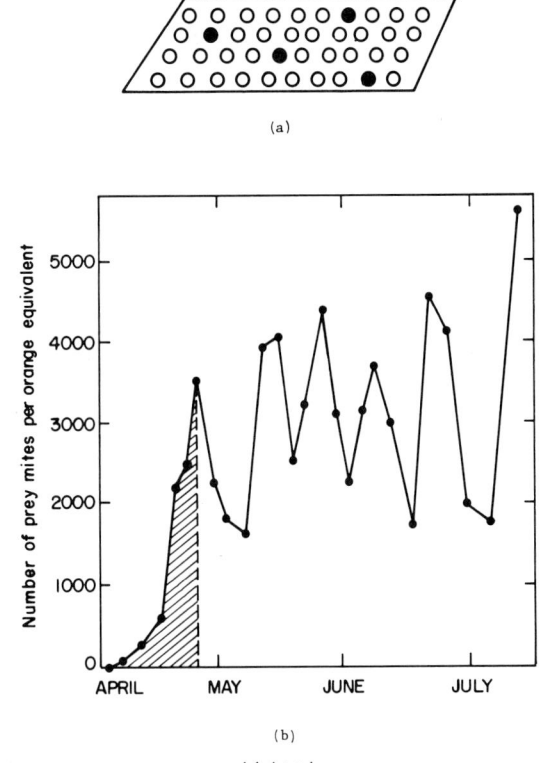

Figure 1. A data element of a frame $E_{\text{food, prey/pred}}^{\text{global, total}}$ (b) (redrawn from Fig. 8, Huffaker, 1958). The universe consists of four oranges embedded in an array of rubber balls [oranges are the solid circles in (a)]. The hatched initial portion of (b) is the data element belonging to the frame $E_{\text{food, prey/pred}}^{\text{local, total}}$ where the subset referred to in the "local" designation is the set of oranges indicated in (a).

Table II Experimental Frames and Relevant Data Sets in Huffaker Universe

Frame (E)	Description	Associated data sets $\mathscr{N}^t$ (E) (Table III)
$E^{\text{global}}_{\text{food, prey, pred}}$	Food, prey, predator variables for each cell	Missing
$E^{\text{global}}_{\text{food, prey, pred}}$	Food, prey variables for each cell in absence of predator	Missing
$E^{\text{global}}_{\text{food, prey, pred}}$	Food variable for each cell in absence of prey and predator	Missing (but partial descriptions of the orange replenishment schedules used are given)
$E^{\text{global, total}}_{\text{food, prey, pred}}$	Totals of food, prey, and predator over all cells	58, II(A-I), Figs. 9–18[a] 63, II(3, 4), Figs. 3, 4 63, I-(4, 5, 6), Fig. 5
$E^{\text{global, total}}_{\text{food, prey, pred}}$	Totals of food, prey over all cells in absence of predator	58, I(A, B, C) Figs. 6, 7, 8 63, E-2, Fig. 2
$E^{\text{global, total}}_{\text{food, prey, pred}}$	Total of food over all cells in the absence of prey and predator	Missing
$E^{\text{local}}_{\text{food, prey, pred}}$	Food, prey, predator variables for a subset of cells keeping all others zero	Missing

$E_{\text{food, prey/pred}}^{\text{local}}$	Food, prey variables for a subset of cells in absence of predator	Missing
$E_{\text{food, prey, pred}}^{\text{local}}$	Food variable for a subset of cells in absence of prey and predator	Missing (some very incomplete description of orange quality and spoilage rates given)
$E_{\text{food, prey, pred}}^{\text{local, total}}$	Totals of food, prey, predator variables for a subset of cells keeping all others zero	58, II A, Fig. 9 58, II B, Fig. 10 58, II C, Fig. 11
$E_{\text{food, prey/pred}}^{\text{local, total}}$	Totals of food, prey variables for a subset of cells in absence of predator, keeping all others zero	58, IA, Fig. 6; initial parts of 58, IB, Fig. 7 58, IC, Fig. 8 63, E2, Fig. 2
$E_{\text{prey, pred}}^{\text{occupancy, global}}$	Occupancy states of prey and predator for each cell	58, II I, Fig. 18 63, II–(3, 4), Figs. 3, 4
$E_{\text{prey, pred}}^{\text{occupancy, global, total}}$	Totals of cells in prey and predator occupancy states	Computable from 58, II I, Fig. 18 63, II–(3, 4), Figs. 3, 4

[a] 58, II(A–I), Figs. 9–18 denotes that the data set is presented in Huffaker (1958) Sections II(A–I) and Figs. 9–18, for example; 63 is Huffaker *et al.* (1963).

26   Bernard P. Zeigler

Table III Data Elements For Huffaker Universes $\mathscr{R}^{t=1961}$

Data element key[a]	Description (as given by Huffaker)
58, IA, Fig. 6	Predators absent, simplest universe, four large areas of food, grouped at adjacent, joined positions
58, IB, Fig. 7	Predators absent, four large areas of food widely dispersed
58, IC, Fig. 8	Predators absent, 20 small areas of food alternating with 20 positions with no food
58, IIA, Fig. 9	Predators present, simplest universe, four large areas of food, grouped at adjacent joined positions
58, IIB, Fig. 10	Predators present, eight large areas of food, grouped at adjacent joined positions
58, IIC, Fig. 11	Predators present, six whole oranges as food, grouped at adjacent joined positions
58, IID, Fig. 12	Predators present, four large areas of food widely dispersed
58, IIE, Fig. 13	Predators present, eight large areas of food widely dispersed
58, IIF, Figs. 14, 15	Predators present, 20 small areas of food alternating with 20 foodless positions
58, IIG, Fig. 16	Predators present, 40 small areas of food occupying all positions
58, IIH, Fig. 17	Predators present, 120 small areas of food occupying all 120 positions (barriers to migration added)
58, II I, Fig. 18	Predator–prey oscillations, 120 small areas of food occupying all 120 positions (barriers to migration added)
63, E-2, Fig. 2	Predators absent, complex 3-shelf universe, 210 small areas of food
63, II-(3, 4), Figs. 3, 4	Predators present, 3-shelf universe, 252 small areas of food
63, I-(4, 5, 6), Fig. 5	Predators present, complex 3-shelf universe, 252 larger areas of food

[a] 58, IA, Fig. 6 denotes that the data set is presented in Huffaker (1958), Section IA, and Fig. 6; 63 is Huffaker *et al.* (1963).

referred to as a *data element* belonging to the frame. The set of all such data elements observable in a particular Huffaker universe is the data set $\mathscr{R}(E)$ assigned by such a real system to frame E. The set of all such possible data sets assignable by the possible Huffaker universes is $\mathscr{D}(E)$.

In Table III, the data elements recorded by Huffaker (1958; Huffaker *et al.*, 1963) are listed. The collection of these data elements constitutes $\mathscr{R}^t$, the real system data until time $t =$ June 1961. In Table II, these data elements are distributed among the experimental frames. The data elements associated with a frame E in Table II constitute the subset $\mathscr{R}^t(E)$ of $\mathscr{R}(E)$, namely, the data acquired until time t in frame E.

Table II displays some frames as having "missing" data sets. A frame such as $E^{\text{global}}_{\text{food, prey, pred}}$ for which this is true is conceivable, i.e., it is an element of $\mathscr{E}$, but up to time $t = 1961$, no data have been collected for it, that is, it is not an element of $\mathscr{E}^{(t=1961)}$. $E^{\text{global}}_{\text{food, prey, pred}}$ is marked as "missing" because the food amount at each *individual* orange is not recorded in the Huffaker experiments, even though certain aggregate utilizations are.

Although many of the conceivable frames were actually realized in the Huffaker experiments, in the current modeling effort we found that the missing frames often contained information which could have been extremely helpful. One of the benefits of representation of experimentation in the experimental frame formalism is that certain experiments may be suggested by the logical structure of frame organization which may turn out to be crucial in later modeling. These might not have been thought of in an unstructured experimental approach.

In a frame with control variables specified, the above concepts hold except for the fact that the compare variable readings are only recorded for as long as the control variables remain zero. Thus, for example, in the *class* of frames denoted by $E^{\text{local}}_{\text{food, prey, pred}}$, each frame specifies a subset of cells S such that the food amount, prey, and predator counts of cells in S are the *compare* variables, and all other prey and predator densities are *control* variables. Data are collectible within such a frame as long as no prey or predators establish themselves on oranges other than those in S. When S consists of a single orange, such data give a picture of the local interaction of food, prey, and predators uncontaminated by colony establishments on other oranges, or subsequent remigration from these colonies. While Huffaker performed no experiments with single oranges, the same principle holds when S is taken as the subset of initially seeded oranges, given Huffaker's observation that migration occurs only *due to food depletion or overpopulation* (indeed, this is the basis of our discrete event models). For example, see Fig. 1.

Note that "keeping control variables zero" is a special case of the "range of validity" specification given by Zeigler (1976b, Chapter 11).

4.1. Organization of Frames

The frames in $\mathscr{E}$ are partially ordered[2] by a relation "is derivable from" or in short "$\leqslant$." $E' \leqslant E$ means that the restrictions on data acquisition imposed in frame E' are over and above those in frame E. As a consequence, data collectible in E' can also be deduced from data collectible in E, and questions posable in E' are posable in E as well.

[2] The relation $\leqslant$ is a *partial order* if it is reflexive and transitive. The reader unfamiliar with this concept may think in terms of the numerical relation $\leqslant$ (less than or equal to) which is a total order. In a partial order, some pairs may not be related.

28   Bernard P. Zeigler

Formalizing one step further, we require an *onto* mapping from $\mathscr{D}(E)$ to $\mathscr{D}(E')$, where mapping D to D' has the interpretation that data set D' is derivable from data set D (we write $D' \leqslant D$) by employing the unique set of operating specified by the pair (E', E). Such operations will in general be information destroying in nature, so that, questions answerable given D cannot be answered given D'.

Figure 2 displays the "is derivable from" relation in our current example. Three types of operations are employed in this case to reduce data

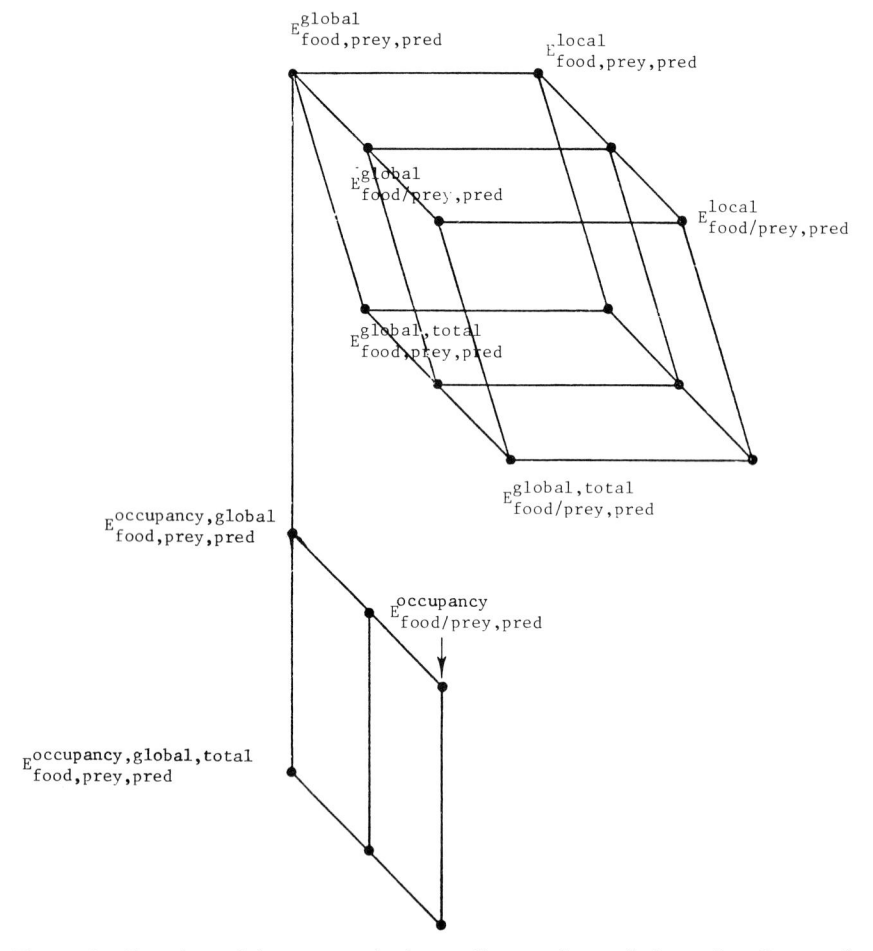

Figure 2. Experimental frames organized according to planes of abstraction. Degree of abstraction increases from top to bottom. Nodes represent frames and lines (implicitly directed from top to bottom) represent the derivability relation.

sets, one to another. This is apparent in the following definition given first for frames having no control variables: $E \leqslant E'$ if

(1) the *compare* variables of E are a subset of *compare* variables of E',

or

(2) the *compare* variables of E are simple sums of the compare variables of E', or

(3) the *compare* variables of E are obtainable by discretizing the *compare* variables of E' (for example, the variable with range {empty, some prey, maximum prey} is obtainable from the variable "prey count"), or

(4) any composition of the above.

The operation types associated with (1), (2), (3) above are *selection*, *aggregation*, and *coarsening*, respectively.

For frames having control variables the definition is: $E \leqslant E'$ if

(a) the *compare* variables of E may be computed from the *compare* variables of E' employing selection, aggregation, or coarsening, in any composition (as above).

(b) the *control* variables of E include those of E', and if there are any additional control variables in E, they must be computed from the compare variables of E' employing selection, aggregation, and coarsening, in any composition.

In Fig. 2 and Table II, the frames are further organized into *planes*. Frames in the same plane are relatable using only the selection operation. If $E \leqslant E'$, then we place E on a lower plane than E' if at least one aggregation or coarsening operation must be used to derive the variables of E from those of E'. In fact, the minimal number of such operations necessary to make this derivation is a measure of the distance between planes. In particular, the distance from the base, or most inclusive plane, to a given plane is a measure of the *degree of abstraction* embodied by the latter plane.

Note, however, that the $\leqslant$ relation is a partial order so that there may be more than one distinct plane with the same degree of abstraction.

Table I shows the naturalness of the plane notion in the ecosystem context by relating the plane of a frame to the kind of question posable within it. As expected, the higher the degree of abstraction, the more restricted the question posable. But note that the global/local distinction does not involve a change in degree of abstraction.

5. CONSTRUCTED MODELS

Table IV provides summary descriptions of the models constructed to date, $\mathcal{M}^{t=1977}$. The conceptual basis for the occupancy and random phase-

Table IV Models and Brief Discription

Model (formalism)	Description
Base (combined differential equation-discrete event; combined stochastic-deterministic)	Local state (situation on each orange) determined by Lotka–Volterra type differential equation governing joint food, prey, and predator dynamics; impetus for emigration and the effect of immigration are logically determined from local state (food, prey, predator); migration process is of stochastic random walk variety; orange replacement schedule simulates that employed by Huffaker.
Lumped (discrete event; combined stochastic-deterministic)	Keeps track of same state variables as base model but updates them only at "event" times. The tables required for scheduling events and executing the updates were derived from base model local interaction as discussed in text.
Occupancy (discrete event; combined stochastic-deterministic)	The local situation is represented by a small number of discrete states (empty, some prey, etc.) and the migration processes are simplified to the Bernoulli trial type.
Random Phase-Space (differential equation; deterministic)	Derived from occupancy model under random phase–random space hypothesis. Describes the dynamics of the occupancy probabilities of the discrete states when occupancy model is operating in random phase-random space mode.

space model has been fully described by Zeigler (1977a). We now describe the base model and the discrete event lumped model derived from it.

5.1. Base Model

The base model postulates food–prey–predator interaction on an orange in isolation to be specified by the differential equation:

$$dr/dt = -ux\,\text{pos}(r), \tag{1a}$$

$$dx/dt = (b\,\text{pos}(r) - d)x - cxy, \tag{1b}$$

$$dy/dt = -d'y + c'xy, \tag{1c}$$

where r is the food amount (measured in fraction of unused orange surface), x, y are prey and predator population sizes, and $\text{pos}(r)$ is 1 if r is positive and 0 otherwise. The meaning attached to the six parameters involved is given in Table V. The underlying time unit is 1 day.

Table V Use of Experimental Frames in Identifying Model Parameters

Model component	Parameter	Description	Identified in experimental frame
Local interaction (large population model)			
	b	Prey birth rate	$E_{\text{food, prey/pred}}^{\text{local, total}}$
	d	Prey death rate	$E_{\text{food, prey/pred}}^{\text{local, total}}$
	u	Prey food utilization	$E_{\text{food, prey/pred}}^{\text{local, total}}$
	d'	Predator death rate	$E_{\text{food, prey, prey}}^{\text{local, total}}$
	c	Predation rate	$E_{\text{food, prey, prey}}^{\text{local, total}}$
	c'	Predation efficiency	$E_{\text{food, prey, prey}}^{\text{local, total}}$
Food replenishment	Threshold	Threshold on prey population below which orange is replaced	$E_{\text{food, prey/pred}}^{\text{local, total}}$
Prey migration	$pyrem$	Prey fraction remaining after emigration	$E_{\text{food, prey/pred}}^{\text{global, total}}$
	$pysurvive$	Probability of migrating prey finding a cell	$E_{\text{food, prey/pred}}^{\text{global, total}}$
	$meanpysearch$	Mean search time for prey finding a cell	$E_{\text{food, prey/pred}}^{\text{global, total}}$
	$pydif_1, pydif_2$	Prey diffusivities in horizontal and vertical directions	$E_{\text{food, prey/pred}}^{\text{global, total}}$
Pred migration	$pdrem$	Analogous	$E_{\text{food, prey, pred}}^{\text{global, total}}$
	$pdsurvive$	to	$E_{\text{food, prey, pred}}^{\text{global, total}}$
	$meanpdsearch$	prey	$E_{\text{food, prey, pred}}^{\text{global, total}}$
	$pddif_1, pddif_2$		$E_{\text{food, prey, pred}}^{\text{global, total}}$
Local interaction (small population model)	d''	Predator death probability at low prey size	$E_{\text{food, prey, pred}}^{\text{global, total}}$
	c''	Minimum prey required to initiate predator reproduction	$E_{\text{food, prey, pred}}^{\text{global, total}}$
	$\bar{c}$	Fraction of prey used to create 1 predator	$E_{\text{food, prey, pred}}^{\text{global, total}}$

Equation (1a) asserts that food utilization is proportional to prey density (recipient limited interaction) as long as food remains. Equations (1a) and (1b) are Lotka–Volterra relations without self-limitation.

We assume, following Huffaker's observations, that prey migrate only when food is exhausted (r first becomes 0). The migration is effected as follows:

(1) When (and if) food is exhausted, a fraction *pyrem* of the current prey remain on the orange [and are subject to the dynamics of Eq. (1)].

(2) Of the migrating prey [(1-*pyrem*) times current population], a fraction, *pysurvive*, are assumed to actually reach a cell (the rest are assumed to die).

(3) For each of the migrating individuals, a search time T_s is sampled from an exponential distribution with mean *meanpysearch*.

(4) The cell assigned to the individual is computed by quantizing spatial coordinates derived from normal distributions (independent for each dimension) with mean, the current cell location and standard deviation $pydif_1 \cdot T_s^{1/2}$ (in case of horizontal dimensions) or $pydif_2 \cdot T_s^{1/2}$ (in case of vertical dimension).

(5) After time T_s has elapsed, the individual is added to the population of his assigned cell if food remains there; otherwise, with probability *pysurvive*, he is sent to step (3) for further migration (with probability, 1-*pysurvive*, he dies).

The migration thus implemented is a random walk with constant probability of stopping. We postulate our mites to search blindly and "bump into" orange locations. It is important to note that emigration is not continuous but occurs only at certain points in the local cycle. We have shown (Zeigler, 1977a) that continuous migration is unlikely to stabilize a locally unstable system such as Huffaker's.

Predator migration is carried out exactly as prey migration with respective parameters *pdrem*, *pdsurvive*, $pddif_1$, $pddif_2$, with the following exceptions:

(1′) Predator migration is initiated when a local maximum in predator density is reached,

(5′) After time T_s has elapsed, the individual is added to the population of his assigned cell unless the prey population is below *eqprey*, the equilibrium prey level computed from Eqs. (1b) and (1c).

If the prey population is below *eqprey*, a small-population stochastic model takes effect. With probability *pdrem′*, the invading predator remains; otherwise, it is migrated as in step (3). A predator that remains dies with probability d''. If the predator lives, it creates another predator if there are at least c'' prey and $\bar{c}x$ are used up as a result (where x is the current prey population size).

Note that although the small-population submodel is a stochastic version of the deterministic Lotka–Volterra model used for large numbers, it may be a crude summary of the local interaction in these circumstances, thus optimal settings of the primed parameters may bear little relation to their deterministic counterparts.

Indeed, this methodology suggests that a second-level spatial characterization of each orange could be built. Such a model would be tuned to more finely structured local observations and a simplified version would replace (or perhaps turn out to be identical with) our current base submodel. In the general terminology, no such experimental frame currently belongs to $\mathscr{E}^t$, although one might be forced to create such a frame, if the current models (in which the small-population submodel participates) prove unable to match the data gathered within existing frames (see Appendix).

5.2. Discrete Event Lumped Model

The first lumped model (illustrated in Fig. 3) is a discrete event version of our base model. The migration is unchanged but the local interaction is described in summary transition function form obviating the necessity for step-by-step simulation of the differential equations.

For example, consider the food–prey submodel [Eqs. (1a) and (1b) with $y = 0$]. For positive $r(0)$, it is possible to solve analytically to find

$$r(t) = r(0) - (u/a)(x(t) - x(0)), \tag{2a}$$
$$x(t) = x(0) e^{at}, \tag{2b}$$

where $a = b - d$ (net prey growth rate).

The time for r to reach 0 is given from Eq. (2) by

$$\tau = \tfrac{1}{a} \ln\{[ar(0)/ux(0)] + 1\} \tag{3a}$$

and the prey population at that time is

$$x(\tau) = [ar(0)/u] + x(0). \tag{3b}$$

The discrete event model keeps track of the values r, x, and y for each cell. If at some time t, a prey individual migrates to an empty cell with food amount r, Eq. (3a), with $x(0) = 1$ and $r(0) = r$ is used to schedule the subsequent emigration (to occur at $t + \tau$). When the emigration event occurs, Eq. (3) is used to update the prey population, and of course, the food amount is set to zero. The consequent prey die-out is scheduled to occur in elapsed time $(1/d) \cdot \ln[x(\tau) \cdot pyrem]$.

At every subsequent immigration the cell state is updated. Suppose that a time e has elapsed since the last immigration. Then using Eq. (2) with $t = e$, and $r(0)$, $x(0)$ being the values at the last update, the correct state pertaining just before the immigration is computed. To the prey number so computed we add 1 to account for the immigrating individual and then use Eq. (3) to reschedule the emigration event.

It can be shown that this discrete event algorithm exactly reproduces the behavior of the base model prey–food interaction. The addition of the

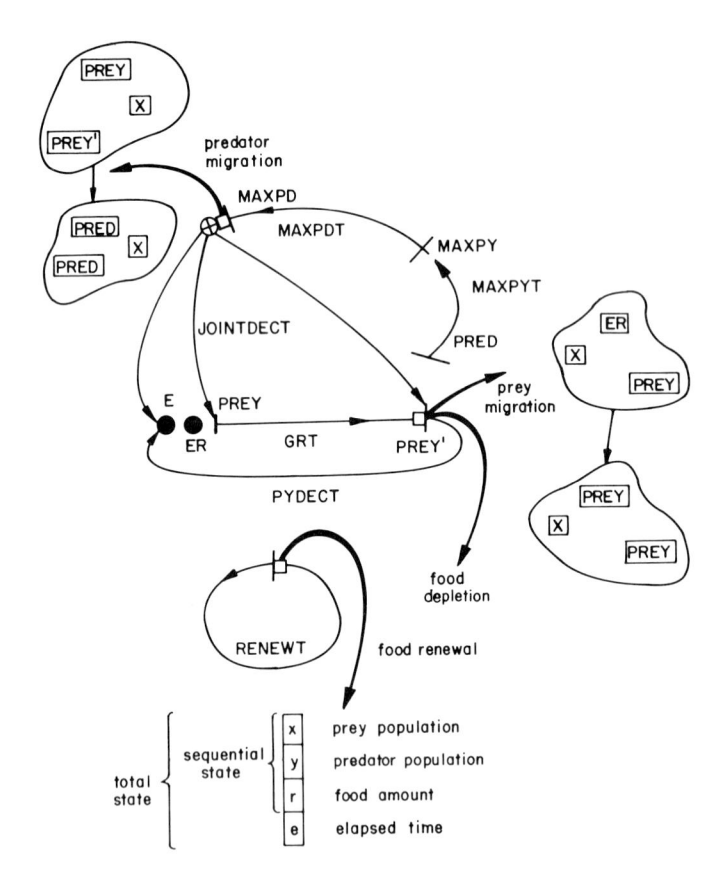

Figure 3. Discrete event representation of the base model. The discrete states are E (empty, no food), ER (empty, food replenished), $PREY$ (prey colony established), $PREY'$ (prey colony at maximum size), $PRED$ (predators invaded), $MAXPD$ (predator colony at maximum size). Scheduling times are: GRT (growth time of prey colony), $PYDECT$ (decay time of prey colony), $MAXPYT$ (time to reach maximum prey size after predator invasion), $MAXPDT$ (time to reach maximum predator population from maximum prey population), and $JOINTDECT$ (time from maximum predator to end of cycle).

predator is handled in principle in the same way, except that the scheduling and update functions cannot be obtained analytically but can be approximated with a once-and-for-all sampling of the trajectories generated by Eq. (1) (Zeigler, 1977b). We shall provide a brief description of this process.

Example. Consider the case where prey have colonized an orange but have not yet exhausted the food. Then Eqs. (1b) and (1c) reduce to the

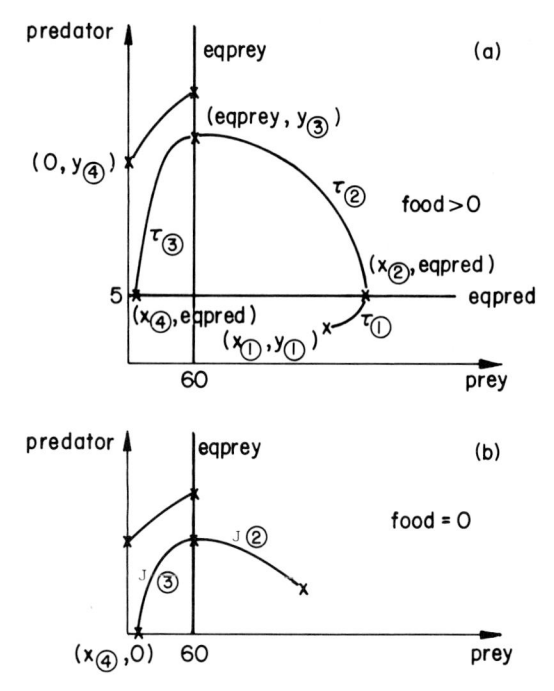

1. Multilevel Multiformalism Modeling: An Ecosystem Example

Figure 4. Typical trajectories in the prey–predator (x, y) plane with food (a) $r > 0$ and (b) $r = 0$.

Lotka–Volterra dynamics and one easily obtains the equilibrium prey and predator isoclines, namely

$$x^* = eqprey = d'/c', \qquad y^* = eqpred = c/a.$$

These isoclines divide the plane into four regions as shown in Fig. 4a. A typical trajectory initiated by a predator immigration is segmented into an initial joint growth phase $\tau_{(1)}$, a predator growth phase $\tau_{(2)}$, and a joint decline phase $\tau_{(3)}$. Generalizing Huffaker's observation to predators, we postulate that predators emigrate at maximum predator population at the end of the $\tau_{(2)}$ phase on the *eqprey* isocline and at the end of the $\tau_{(3)}$ phase when the prey minimum is reached (either a crash occurs if predators are numerous enough or the *eqpred* isocline is reached). Any prey left are assumed to remain on orange and subsequently take part in the standard food–prey interaction.

Scheduling and updating for each of these boundary crossings were done by use of tables generated from a CSMP simulation of Eq. (1) and shown in Fig. 5. These tables are interesting in themselves; to our knowledge, they represent the first such global study of Lotka–Volterra

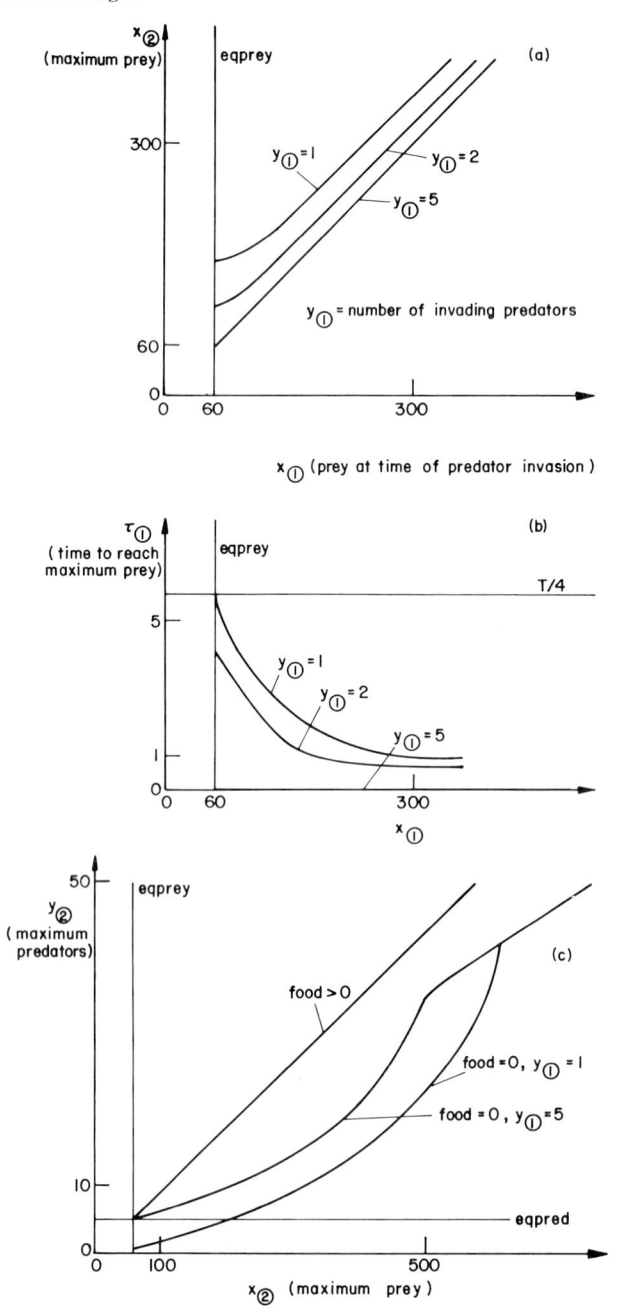

Figure 5. (a–f). State update and scheduling curves obtained by simulation of Eq. (1). Symbols shown are keyed to Fig. 4.

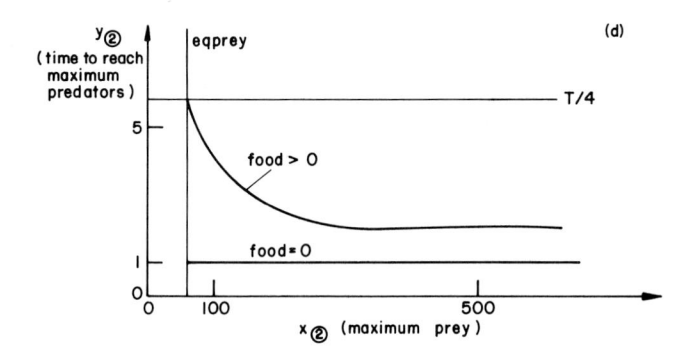

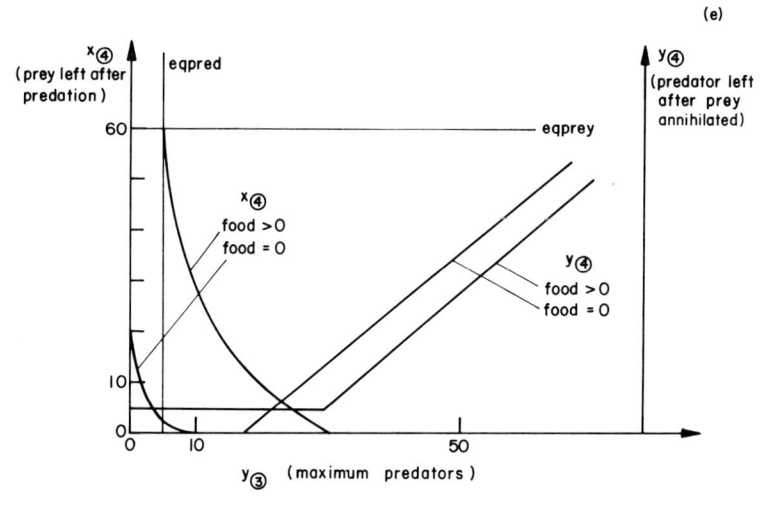

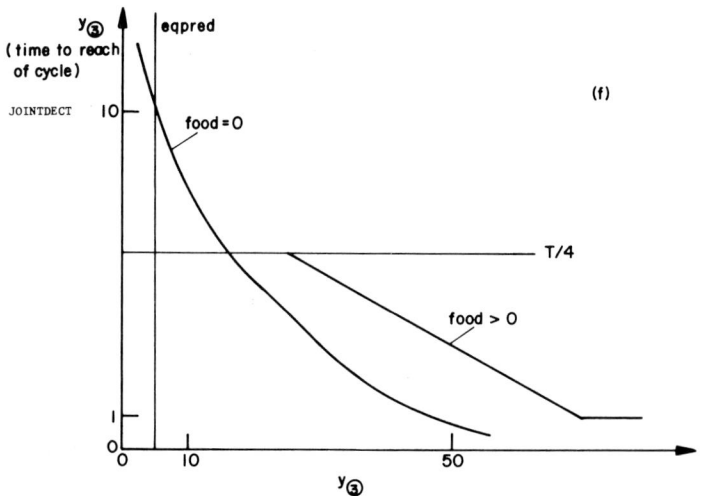

1. Multilevel Multiformalism Modeling: An Ecosystem Example

dynamics. The parameter T shown is the period of the cycle obtained by linearization around the equilibrium point $T = 2\pi/ad^{1/2}$. The time to cross from one boundary to the next is approximately $T/4$ near equilibrium but declines rapidly as initial populations increase.

A disadvantage of generating tables by simulation is that it must be done potentially anew for each set of parameters. This makes it important to be able to identify the local interaction parameters before all others (Section 7.1). The theoretical basis for discrete event representation of systems is given by Zeigler (1977b).

5.3. Occupancy Models

The overall occupancy model is described in Fig. 6. As can be seen, the local description is reduced to a small number of states (empty, patch

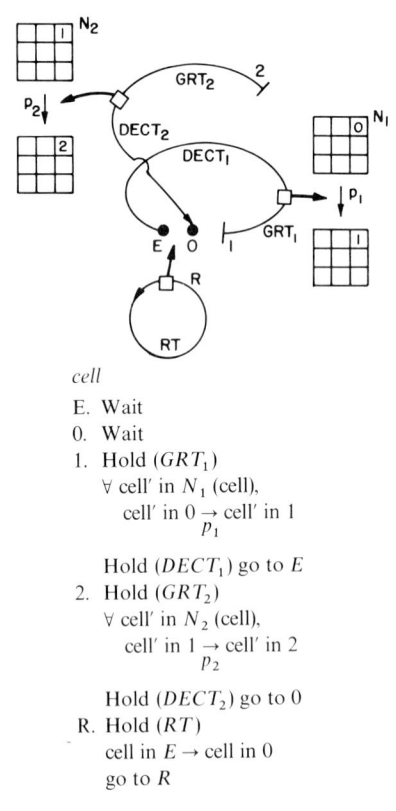

cell
E. Wait
0. Wait
1. Hold (GRT_1)
   $\forall$ cell' in N_1 (cell),
      cell' in $0 \underset{p_1}{\rightarrow}$ cell' in 1

   Hold $(DECT_1)$ go to E
2. Hold (GRT_2)
   $\forall$ cell' in N_2 (cell),
      cell' in $1 \underset{p_2}{\rightarrow}$ cell' in 2

   Hold $(DECT_2)$ go to 0
R. Hold (RT)
   cell in $E \rightarrow$ cell in 0
   go to R

Figure 6. The occupancy model. Discrete states are 0 (empty, food replenished), 1 (some prey), 2 (some predator), E (empty, no food).

occupied by prey only, etc.), and the local dynamics are reduced to timed transitions from state to state. The migration component is also simplified by specifying neighborhoods for each species (for each cell) and migration is effected by means of independent Bernoulli trials governed by specified probabilities and conditions at the cells in the neighborhood of a migration-active cell. (See Zeigler, 1977a, for a full explanation.) Cell spaces up to quite large sizes [we have commonly investigated 30×30 arrays (900 cells)] can readily be simulated in discrete event languages such as SIMSCRIPT. We are able to study by this means the spatial patterns which are associated with persistence and extinction as they are governed by the geometry of the space, the characteristics of the neighborhoods, and the settings of the other parameters.

5.4. Random Phase-Space (RPS) Models

The RPS hypothesis assumes that the cells in a given state are uniformly distributed in both space and phase (elapsed time in the state) at all times. On the basis of this hypothesis we may derive the differential equation system shown in Table VI. These equations are simple enough to be solved for equilibrium isoclines and thus give qualitative information about how persistence is governed by the various parameters. Here persistence is judged relative to prespecified extinction levels such that if the prey- and predator-occupied cell fraction fall below these levels, the system

Table VI   Random Phase-Space Model[a]

[a] Let x = fraction of cells in state 1 (some prey),
let y = fraction of cells in state 2 (some prey, some predator),
let z = fraction of cells in state 0 (empty, some food),
and let u = fraction of cells in state E (empty, no food).
Then $dx/dt = -(x/T_1) + zx(p_1 N_1/T_1) - yx p_2 N_2/T_2$, $dy/dt = -(y/T_2) + yx p_2 N_2/T_2$, $dz/dt = (u/RT) - zx p_1 N_1/T_1 + (y/T_2)$, with $u = 1 - (x+y+z)$, where $T_1 = GRT_1 + DECT_1$, $T_2 = GRT_2 + DECT_2$.

Equilibrium fraction	No predator (food, prey/pred)	Some predator (food, prey, pred)
x^* (avg. prey cell)	$1-(p_1 N_1)^{-1}/1+RT \cdot T_1^{-1}$	$(p_2 N_2)^{-1}$
y^* (avg. pred. cell)	0	$\dfrac{1-(p_1 N_1)^{-1}-(p_2 N_2)^{-1}(1-RT \cdot T_1^{-1})}{1+p_2 N_2 T_2^{-1}(p_1 N_1)^{-1} T_1}$
z^* (avg. food cell)	$(p_1 N_1)^{-1}$	$(p_1 N_1)^{-1} + y^* p_2 N_2 T_2^{-1}(p_1 N_1)^{-1} T_1$
u^* (avg. utilization)	$1-(x^*+z^*)$	$1-(x^*+y^*+z^*)$

is assumed to go extinct. Moreover, the equations can be easily simulated to generate the associated dynamic behavior. The predictions thus made can be matched against the behavior of the simulated occupancy models, to the advantage of both model types (see p. 52).

6. ORGANIZATION OF MODELS

The hierarchy of models is displayed in Fig. 7. Models are organized in a manner parallel to that of experimental frames. Model M is on the same plane with model M' if M is a subcomponent of M'. Models of greater degrees of abstraction (lower level planes) are derived from more refined models by simplification procedures based on aggregation, coarsening, and discrete-eventization (discussed in Section 3).

The criterion which justifies the simplification is that the mappings involved be homomorphisms. We have given extensive expositions of such model relations (Zeigler, 1976a) and their use in organizing models (Zeigler, 1977b). For an exposition and example in the compartmental ecosystem context, see Zeigler (1976b).

6.1. Relation of Parameters

Basically a homomorphism between models is a correspondence between their state spaces in which corresponding states transit to corresponding states and yield corresponding outputs. Such a relation usually implies a correspondence between parameter values as well, so that parameter settings of a lumped model may be completely determined by those of a more refined morphic preimage.

Indeed, an example of such a parameter correspondence was given in the derivation of the discrete event lumped model (Section 5.2). The scheduling and update tables of the discrete event model are parameters— one can treat them as entities to be arbitrarily adjusted until the desired behavior is achieved. On the other hand, the morphism, which underlies the construction of the lumped model, uniquely prescribes these tables for each setting of the local interaction parameters (b, d, u, d', c, c').

The same concept of parameter correspondence is illustrated in the parameter complexes appearing in the RPS equations, as determined by the parameters of the occupancy model (Table VI).

To complete the chain, it should be indicated how the parameters of the occupancy model are related to those of the more refined discrete event lumped model. In this case, however, a morphism cannot be established to hold strictly over the complete state spaces of the two models, and we must

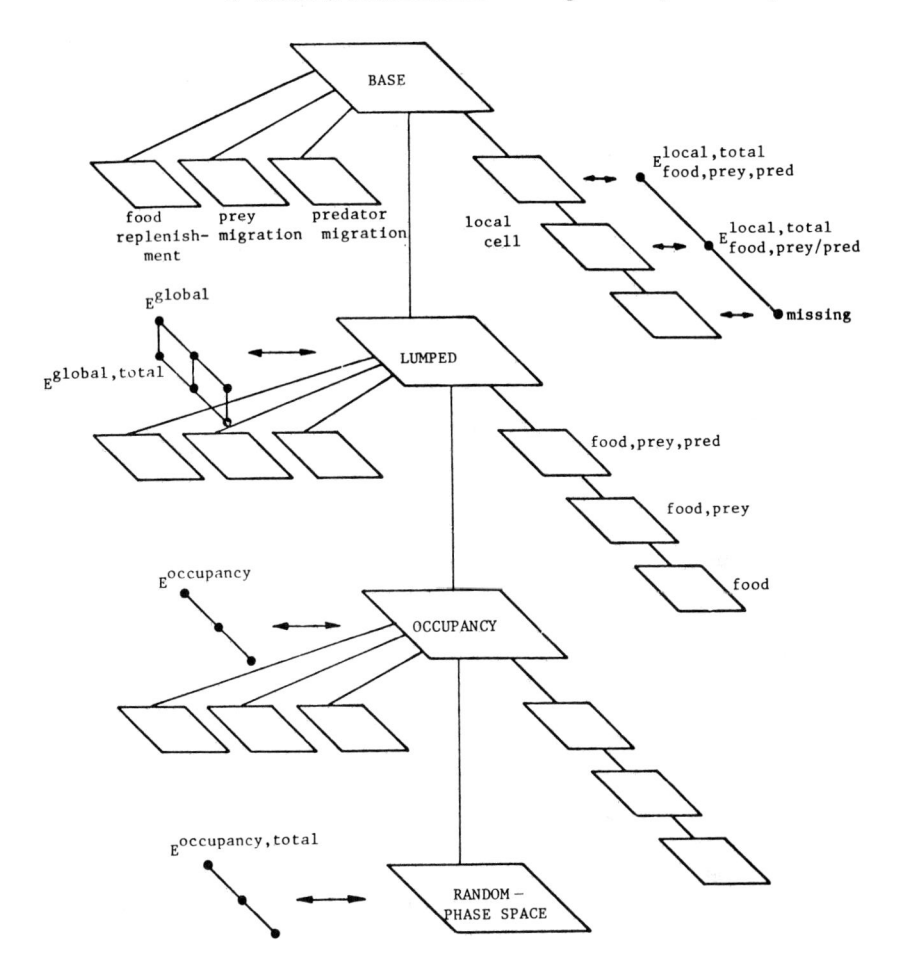

1. Multilevel Multiformalism Modeling: An Ecosystem Example 41

Figure 7. The organization of models $\mathcal{M}_0^{t=1977}$. Also shown are the experimental frames applicable to the various models.

be content with estimating average parameter value settings, or at least, ranges to which they can be bounded. We now outline how this may be done.

Table VII provides ranges for the patch life cycle parameters GRT_i, DCT_i, $i = 1, 2$. The derivation is straightforward from Section 5.2. Although the migration mechanisms in the lumped and occupancy models are not directly comparable, they can be matched through the notion of *effective neighborhood*. The effective neighborhood of a species is the expected number of cells colonized in a migration episode, given that all cells in the

Table VII Estimation of Occupancy Model Parameters

Parameter	Range of values in terms of lumped discrete event model parameter values
GRT_1 (Growth time of prey colony)	$[0, (1/a)\ln ar_0/u]$
DCT_1 (Decay time of prey colony)	$[0, (1/d)\ln ar_0/u]$
GRT_2 (Time to maximum predator population)	$[0, T/2] (T = 2\pi/ad^{1/2})$
DCT_2 (Time to extinction of predator population measured from maximum population point)	$[0, T/4]$
$p_1 N_1$ (Effective prey colonization neighborhood)	From Fig. 8 with a number of samples $= (ar_0/u)(1-pyrem) \cdot pysurvive$ and random walk parameters $meanpysearch, pydif_1, pydif_2$.
$p_2 N_2$ (Effective predator colonization neighborhood)	From Fig. 8 with a number of samples $= (c/c') \cdot (ar_0/u) \cdot (1-pdrem) \cdot pdsurvive$ and random walk parameters $meanpdsearch, pddif_1, pddif_2$.

space are colonizable. In the occupancy model this is just $p_i N_i$ for species i. In the more refined model, the effective neighborhood is the expected number of distinct cells accessed in a migration episode. Figure 8 plots the number of distinct cells accessed versus the number of samples taken from the random walk distribution (with parameters typical in the case of extended coexistence). The maximum number of distinct cells accessed can be estimated by noting that the random walk distribution with mean search time T_s and diffusivity d in one dimension appears (from simulation) to be normally distributed with standard derivation $\sigma = dT_s^{1/2}$. Thus, the number of distinct cells rises at first in proportion to the number of samples; then it approximates the number contained within a radius 3σ of the active cell as the number of samples increases to moderate values. (Theoretically it continues to rise very slowly beyond this point.) Now the number of samples in the migration episode is just the number of migrants and can be bounded above as indicated in Table VII. Combining this number with Fig. 8 yields the effective neighborhood bound.

Finally, we note that yet more refined models can be postulated which would place constraints on the parameters of the base model. Thus in our base model, the prey death rate (in patch) d, the probability of survival *pysurvive*, and the search time parameter *meanpysearch* are independently

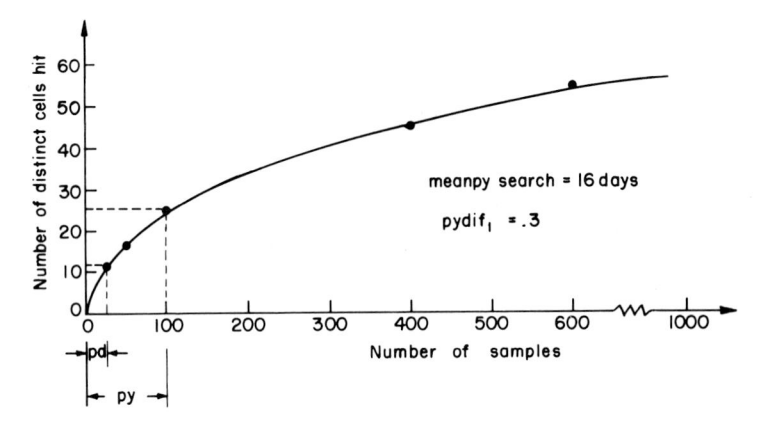

1. Multilevel Multiformalism Modeling: An Ecosystem Example   43

Figure 8. The cumulative number of cells hit versus the number of samples from the random walk distribution with parameters typical in the case of extended predator–prey persistence. The numbers *pd* and *py* indicate upper bounds on the numbers of predators and prey emigrating in a migration episode as estimated in Table VII.

adjustable. If we postulate that a migrant survives only if his lifetime exceeds his search time, where these random variables are independent and exponentially distributed, we derive the relation:

$$pysurvive = (1 + d \cdot mean\ pysearch)^{-1}. \qquad (4)$$

Parameter values obtained after adjustment can be checked against this relation. Large discrepancy might indicate dependence of the variables or given cause to reexamine the model structure and/or parameter settings.

7. APPLICABILITY OF FRAMES TO MODELS

A frame E is *applicable to a model* M if the compare variables specified by E are (1) included among the descriptive variables of M, or (2) may be obtained from the descriptive variables of M by aggregation or coarsening.

Figure 7 depicts the *core* of the applicability relation (see also Tables I and V). *Core* is used here because one can infer applicability to higher plane models of frames applicable to lower plane models (Zeigler, 1978; Axiom 8).

Roughly, if E is applicable to M, then M can potentially answer the questions of interest in E. M is *valid for* $\mathscr{R}^t$ in E if M can reproduce $\mathscr{R}^t(E)$, the data collected up to time t in E. Viewed another way, E applicable to M means that the data $\mathscr{R}^t(E)$ may be employed to identify the parameters of M, i.e., the parameters may be adjusted until a best fit with the data $\mathscr{R}^t(E)$ is obtained.

7.1. Parameter Identification

As shown in Table V, the experimental frame and model organizations made it possible to identify the parameters in a sequential manner, thus greatly reducing the search space at each stage. The parameters relating to local food–prey–predator interaction, prey migration, predator migration and finally predator–prey (small-population) interaction were adjusted in this order.

The test of such a procedure is that reasonable fits to the data are obtainable at later stages by holding fixed the parameters identified at earlier stages. When acceptable agreement at later stages cannot be obtained, this may indicate that the prerequisite independence assumed for earlier adjusted parameters does not hold. *In terms of experimental frames, the control conditions of a frame may not in fact hold. Indeed, it often is implicitly assumed by modelers that certain global interactions can be ignored in certain circumstances, and this may turn out to be unjustifiable.* In our case, $E^{\text{local, total}}$ frames assume that migration effects have been nullified, the justification for which lies in Huffaker's verbal account of migration episodes accompanying the time series data.

If there is reason to doubt that the control conditions of an experimental frame are not satisfied, a readjustment of parameters may be attempted. To the extent that such a readjustment is small, the decomposition into experimental frames will have been beneficial.

In the Appendix, we report on the parameters identified in some of the key experiments and cross compare models in this regard.

8. SUMMARY

The concepts discussed in the paper are summarized as follows.

$\mathcal{E}, \mathcal{R}, \mathcal{M}$ is the triple of conceivable experimental frames, real system data, and possible models, respectively, underlying the modeling study of the Huffaker Universes.

$\mathcal{E}^{t=1961}$ is the subset of frames for which data had been collected until 1961.

$\mathcal{R}^{t=1961}(E)$ denotes the data collected within frame E until 1961.

$\mathcal{M}^{t=1977}$ is the subset of models considered until 1977 (the current time) by the present modeler.

An experimental frame E in $\mathcal{E}$ specifies a (compare, control) variable pair. A data element of a frame E is a time series of compare variable values obtained under conditions where control variables are kept at zero levels.

Frames are partially ordered by the derivability relation; $E \leqslant E'$

means that data elements of E are derivable from those of E' by employing selection, aggregation, and coarsening operations. Each frame may be assigned a degree of abstraction equal to the minimum number of operations required to derive it form a fixed most inclusive frame.

Models may similarly be partially ordered by use of morphism relations. A homomorphism is a mapping from a refined model to a coarse one which preserves the transition structures. A homomorphism induces a mapping from the parameter assignments of the finer model to those of the coarser one.

If a frame E is applicable to a model M, this means that the behavior generated by M can be interpreted as data within frame E. One interpretation of this fact is that the real system data collected within E can be employed to identify the parameters of M.

9. DISCUSSION

There are a number of levels at which the integrated approach to modeling illustrated here may be discussed. We briefly consider some of them.

9.1. Large-Scale Multifaceted System Modeling

Our formalism has been constructed from a general starting point—the theory of systems and its specialization to modeling and simulation. Thus, it is aimed for application to "large-scale" systems in general. We have placed the large scale in quotation marks to signify our belief that "large scaleness" is *a matter of approach rather than of fact*. Indeed, a system is called large scale precisely when one recognizes that to deal with it successfully requires the consideration of many factors and aspects. There are some systems which strikingly have this characteristic—environmental systems, urban systems, and so on, that are indeed large scale. But *microscale* systems such as the biological cell are equally complex, when examined in all their facets. Thus we propose the term *multifaceted* to connote the systems (viewpoint) we are addressing.

In this chapter, our approach is illustrated in a particular ecosystem context. But some general points clearly emerge.

Simpler models can give qualitative and sometimes quantitatively accurate predictions.

The RPS model gives good estimates of an average cell occupancy fractions when its underlying conditions hold. More generally it may give correct qualitative relationships (effect of parameter settings) even when its quantitative predictions are inaccurate.

Simpler models can be employed to check more complex ones.

If the correspondence between models is known, behaviors of the models may be compared. This can be employed at:

(a) the development stage; if the simpler model is known to be correctly implemented (or does not require simulation), then the logic of the more complex model can be verified by comparison of model behavior (this is an important special case of redundancy use for program verification, Bosworth, 1975).

(b) the prediction stage; the more the predictions of various models agree, the greater may be the confidence in the predictions. Where serious disagreements occur, confidence considerations may determine the choice of which to believe, or lead to the conclusion that more development is necessary.

Complex models can be employed to validate simpler ones.

Conversely, if the correspondence between models is known, a more refined model whose details are tied to a particular real system can be used to gain confidence in a more abstract but general model. Thus by validating the lumped discrete event model against Huffaker's data, and finding that our corresponding occupancy and RPS models produce matching behavior, we gain confidence in the abstractions employed to derive the simpler models, that is, that patches, rather than individuals, are sufficient entities for analysis of persistence. Holling (1974) and others have employed a simulation model at the level of detail of our lumped discrete event model to check out the wider consequences of optimal control policies derived from a simpler analytic model.

Models may be introduced independently or derived from existing ones.

It may be sometimes advantageous to construct a model from "phenomenological considerations" rather than from "first principles." However, when a homomorphism can be established between a more refined model and such an *ad hoc* model, additional advantages of the kind indicated above accrue. In addition, if a base model is available on which to base construction of a lumped model, constructs may be suggested which would not have come to mind in a phenomenological approach (Whitehead, 1978).

Needed experiments may be implied by the experimental frame organization.

The logical structure of the experimental frame organization may suggest conceivable frames in $\mathscr{E}$ that have not yet been realized to date (are not in $\mathscr{E}^t$), and might not be thought of in an unstructured experimental approach. For example, data on the orange spoilage and prey–food interaction suggested by frames of the form $E_{food/prey, pred}$ and $E_{food, prey/pred}$ are missing and would be helpful to model construction and validation.

Model and experimental frame organizations may be extended at all levels.

The multifaceted system approach explicitly recognizes that model construction and validation is a never-ending process. For example, as accuracy demands in some frame increase, it may be found that the current stock of models is inadequate to meet these demands. This may spur the formulation of new experimental frames, data acquisition within them, and construction and validation of models which would guide the refinement of the original models so as to meet the increased accuracy requirements. This paradigm is illustrated in our finding that small-population interaction on a patch may play a more important role in determining average population levels than was suspected originally. Development of a credible small-population submodel could be based on a spatial model of the predator–prey interaction on a patch developed from data acquired in an appropriately defined experimental frame. Complexity constraints would prohibit incorporating such a spatial model directly into our local interaction model and thus simplifications would be sought, perhaps resulting in refinements of the classical Lotka–Volterra model along lines developed by Hassel *et al.* (1976).

An example of refinement at the other extreme of abstraction is given by the incorporation by Gurney and Nisbet (1978) of fluctuation terms in the RPS model which enables it to predict equilibrium fluctuation magnitudes from steady-state population levels.

9.2. Ecosystem Modeling

In this paper we have illustrated our large-scale multifaceted approach in a highly restricted ecosystem context. Having dealt only with two species and three trophic levels, we have only scratched the surface of the possibilities and problems that would arise in dealing with a natural ecosystem. Yet extension of the experimental frames on the same plane of abstraction to many species would simply involve the specification of frames by pairs (A, B), where A is the subset of species to be observed (whose descriptive variables are the compare varibles), and B is the subset of

At the above settings, the maximum prey size on an orange is $u/a \approx 2 \times 10^4$ mites per orange equivalent (so, for example, a 1/10 exposed orange area can support 2000 preys at the perigee of growth).

A.2. Local Food–Predator–Prey Interaction

The parameter settings $d' = 0.30\,\text{day}^{-1}$, $c = 0.05\,\text{day}^{-1}$, and $c' = 0.006\,\text{day}^{-1}$ were estimated employing data element 58, IIA, Fig. 9 (in frame $E^{\text{local, total}}_{\text{food, prey, pred}}$). Employing a CSMP simulation of Eqs. (1a)–(1c) we adjusted the parameters d', c, and c' so as to fit as closely as possible the prey and predator curves in Fig. 9 Huffaker (1958).

With the estimated parameters, we have the equilibrium prey and predator levels as 50 and 5 mites on a patch, respectively.

A.3. Prey Migration

The data necessary for identifying prey migration parameters in the absence of predators are available only for the hazard-free universes in the Huffaker 1958 study and the complex universe of the 1963 study (Huffaker et al., 1963), but not for the 1958 universe in which prey–predator coexistence was achieved. Employing hazard-free universe data sets 58, IB, Fig. 7 and 58, IC, Fig. 8 (in frame $E^{\text{global, total}}_{\text{food, prey/pred}}$) we adjusted the parameters *pyrem*, *pysurvive*, *meanpysearch*, and *pydif*$_1$ of our discrete event lumped model so as to have the SIMSCRIPT generated curves match the data curves as closely as possible in average prey produced and number of prey maxim produced in the experimental interval. Estimated obtained were *pyrem* = 0.9, *pysurvive* = 0.9, *meanpysearch* = 0.1 days, *pydif*$_1$ = 20. Employing the data set 63 E-2, Fig. 2 for the prey–food interaction in the complex 1963 universe, we estimated in the same manner that *pyrem* = 0.9, *pysurvive* = 0.5, *meanpysearch* = 13 days, *pydif*$_1$ = 0.3, and *pydif*$_2$ = 0.15. Thus, as expected, prey mites in the complex universe take much longer on the average (13 versus 0.1 days) to cover much less distance ($13^{1/2} \times 0.3 \approx 1.0$ versus $0.1^{1/2} \times 20 \approx 6.0$, see Section 7) than they do in the hazard-free cases.

A.4. Predator Migration

Predator migration parameters *pdrem*, *pdsurvive*, *meanpdsearch*, and *pddif*$_1$ were adjusted in the discrete event lumped model so as to fit as closely as possible the data element 58, III, Fig. 8 representing the 1958 universe in which coexistence was established. The settings of the prey migration parameters were those determined from the complex 1963

Table A.I Comparison of Data and Lumped Discrete Event Behavior in Frame $E^{\text{global, total}}_{\text{food, prey, pred}}$ [a]

Density	Data	Model[b]
Maximum prey (predators absent)	Missing	4600[c]
Average prey (predators absent)	Missing	2400
Maximum prey	2000	3500
Average prey	900	730
Maximum predators	50	46
Average predators	12	13

[a] The case of predator–prey coexistence (Huffaker, 1958, Section III, Fig. 8).
[b] Parameter assignments are $b = 0.55 \,\text{day}^{-1}$, $d = 0.30 \,\text{day}^{-1}$, $u = 1.3 \times 10^{-5}$, $d' = 0.30 \,\text{day}^{-1}$, $c = 0.05 \,\text{day}^{-1}$, $c' = 0.005 \,\text{day}^{-1}$; $pyrem = 0.9$, $pysurvive = 0.35$, $meanpysearch = 13 \,\text{days}$, $pydif_1 = 0.3$; $pdrem = 0.6$, $pdsurvive = 0.5$, $meanpdsearch = 14 \,\text{days}$, $pddif_1 = 0.2$; $pdrem' = 0.3$, $d'' = 0.0$, $c'' = 10.0$, $\bar{c} = 0.0$.
[c] All densities quoted in mites per orange equivalent (Huffaker, 1958).

universe just described. (Subsequent trials with deviations from these settings did not significantly improve the results.) The predator migration parameters were initially set equal to those of the prey and a fairly broad neighbourhood of parameter assignments centered on the initial settings was investigated.

It was found that coexistence is robust in this neighborhood in that most simulation runs ended with both predators and prey still around. However, it did not seem possible to achieve very close quantitative agreement. We noticed that the predator-occupied cell fraction was too small and this seemed to be due to the fact that in our original model, predators invading patches of low prey density (less than *eqprey*) were always returned immediately for continued migration. It thus appeared that predator invasion of low density patches was a significant process and we accordingly modified our small-population submodel to its current form. With this modification we were able to bring the statistics shown in Table A.I generated by the simulation quite close to those of the data. Although the averages agree quite well, the model overestimates the prey maximum considerably, which may point to a further needed modification. (In analogy with the predictions of a Lotka–Volterra model, the overshoot could be the sensitive result of too low an initial predator population, and thus not an intrinsic model shortcoming.) The best-fit parameter settings are indicated in Table A.I.

Table A.II Comparison of Data and Lumped Discrete Event Model Behavior in Frame $E^{\text{occupancy, total}}_{\text{food, prey, pred}}$

Cell occupancy	Data	Model
Average prey cell (state 1)[a]	17	28
Average pred. cell (state 2)[b]	11	15
Standard deviation/prey cell[c]	16	19
Standard deviation/pred. cell	11	11

[a] A prey cell is a cell occupied by at least some prey but no predator.
[b] A predator cell is a cell occupied by at least some predator.
[c] Measures the amplitude of oscillation considered as a fluctuation about the average (Gurney and Nisbet, 1978).

In Table A.II, the same data are analyzed from the cell occupancy point of view (frame $E^{\text{occupancy, total}}_{\text{food, prey, pred}}$). [Note the model in question is the lumped discrete event model not the occupancy model; the occupancy states can be computed from the finer population count information.]

As can be seen, the statistics from model and data are remarkably close, save for considerable overestimation in the average prey cell count. This is understandable in view of the maximum prey population overestimation.

It should be noted that the average occupancy counts are not necessarily correlated with the average population counts. As we have noted, the somewhat independent occupancy perspective was useful in diagnosing a shortcoming of the model.

A.5. Occupancy and RPS Models

Employing the parameter values of Table A.I, we can determine corresponding parameter values for the occupancy model, using the relations of Table VII. In order to explore the behavior of the occupancy model in this space, we fixed all but the migration parameters at the extremes of their ranges and sampled the model behavior for allowable assignments of the latter parameters. Employing the equilibrium relations in Table VI, we can uniquely determine the effective neighborhoods $p_1 N_1$ and $p_2 N_2$ of the RPS model required to reproduce the occupancy averages of the data (58, II I, Fig. 8) shown in Table A.II. As shown in Table A.III, these are within but at the lower end of the ranges computed from Table VII. However, simulation of the occupancy model with these parameter settings resulted in quick elimination of the prey. Only when the effective neighborhood was considerably increased and the effective predator neighborhood considerably decreased was coexistence obtained in 10×10

Table A.III Cross Comparison of Model Behavior in Frame $E^{\text{occupancy, total}}_{\text{food, prey, pred}}$

Effective prey neighborhood, $P_1 N_1$	Effective predator neighborhood, $P_2 N_2$	Occupancy model[a] Avg. prey cell	Occupancy model[a] Avg. pred. cell	RPS model[a] Avg. prey cell	RPS model[a] Avg. pred. cell
$\varepsilon[0, 25]$ 4	$\varepsilon[0, 12]$ 8	10 × 10 extinct		17	11
4	8	30 × 30 extinct			
4	4	10 × 10 extinct			
4	4	30 × 30 44	9	30	18
24	24	10 × 10 60	28	48	50
Data[b]		17	11	17	11
Lumped discrete event[b]		28	15	28	15

[a] Other parameter values (in days): $GRT_1 = 20$, $DCT_1 = 20$, $GRT_2 = 5$, $DCT_2 = 2$, $RT = 44$.
[b] From Table A.II.

cell array. (Halving the predator neighborhood was sufficient for coexistence in a 30 × 30 cell array. The 100-cell array is more representative of the 120-cell 1958 universe.) The effective neighborhoods obtained in this way are still within the ranges computed from Table VII. However, the occupancy averages obtained from the occupancy model for both predator and prey in these cases tend to exceed those of the lumped discrete event model and the real system data.

In sum, this between-model comparison seems to indicate that the random-phase condition is only approximately being satisfied in the lumped discrete event model and the real system. While the occupancy and RPS models predict that coexistence is possible within the allowed parameter space, they do not do very well in predicting the observed occupancy cell averages unless the number of cells is considerably increased.

ACKNOWLEDGMENTS

The assistance of Ramon Guardons in many phases of the present project is gratefully acknowledged.
This research was supported in part by the National Science Foundation Grant No. MCS76-04297.

REFERENCES

Bosworth, J. L. (1975). Software reliability by redundancy. Ph.D. Thesis, Univ. of Michigan, Ann Arbor.
Gurney, W. S. C., and Nisbet, R. M. (1978). Fluctuations in predator–prey populations in patchy environments. *J. Anim. Ecol.* **47**, 85–102.
Hassel, M. P., and May, R. M. (1974). Aggregation in predators and insect parasites and its effect on stability. *J. Anim. Ecol.* **41**, 567–594.
Hassel, M. P., Lawton, J. H., and Beddington, J. R. (1976). The components of arthropod predation. *J. Anim. Ecol.* **45**, 135–186.
Holling, C. S. (1974). "Project Status Report: Ecology and Environmental Project," IIASA SR-74-2-EC. Laxenburg, Austria.
Huffaker, C. B. (1958). Experimental studies in predation: Dispersion factors and predator prey oscillations. *Hilgardia* **27**, 243–383.
Huffaker, C. B., Shea, E. P., and Herman, S. G. (1963). Experimental studies on predation: Complex dispersion and level of food in an acarine predator prey interaction. *Hilgardia* **54**, 305–329.
Levin, S. A. (1976). Spatial patterning and the structure of ecological communities. *In* "Some Mathematical Questions in Biology" (S. H. Levine, ed.), Vol. VII, A.M.S. Publ.
Maynard Smith, J. (1974). "Models in Ecology." Cambridge Univ. Press, London and New York.
Whitehead, B. (1978). A neural network model of human pattern recognition. Ph.D. Thesis, Univ. of Michigan, Ann Arbor.
Zeigler, B. P. (1976a). "Theory of Modelling and Simulation." Wiley, New York.
Zeigler, B. P. (1976b). The aggregation problem. *In* "Systems Analysis and Simulation in Ecology" (B. C. Patten, ed.), Vol. 4, pp. 299–311. Academic Press, New York.
Zeigler, B. P. (1977a). Persistence and patchiness of predator-prey systems induced by discrete event population exchange mechanisms. *J. Theor. Biol.* **67**, 687–713.
Zeigler, B. P. (1977b). "Systems Simulateable by the Digital Computer: Discrete Event Representable Models," Logic of Computers, Tech. Rep., Univ. of Michigan, Ann Arbor.
Zeigler, B. P., and Barto, A. G. (1977). Alternative formalisms for bio- and eco-system modelling. *In* "New Directions in the Analysis of Ecological Systems" (G. S. Innis, ed.), Part II, Simul. Counc. Proc. Ser., Vol. 5, pp. 167–178. Soc. Comput. Simul., La Jolla, California.
Zeigler, B. P. (1978). Structuring the organization of partial models. *Int. J. Gen. Syst.* **4**, 81–88.

Chapter 2

CONCERNING AGGREGATION IN ECOSYSTEM MODELING

William G. Cale, Jr. and Patrick L. Odell

1. Introduction	55
2. Modeling: General	59
2.1 Identification Problem	59
2.2 Inference Problem	59
2.3 Parameter Selection Problem	61
3. Modeling: Specific	61
4. The Aggregation Model	63
5. Concluding Remarks	69
Appendix	74
References	76

1. INTRODUCTION

The aggregation problem is well known in economics. Economic models are characteristically complex and not directly applicable to real data since sufficient observations for parameterization are unavailable (Chipman, 1975). To solve this dilemma aggregate models are developed which group "true model" variables into sums or weighted averages, these aggregate variables then being put into relationships which mimic corresponding relations in the "true model" (Chipman, 1975). As pointed out by Fisher (1969) aggregate functions are used at various levels of economic inquiry (entire economies, manufacturing, welfare systems, etc.) and their use has resulted in both empirical and theoretical inferences regarding such things as technical change, investment rate, and economic growth. Once an aggregation strategy is decided, methods are available to optimize the resultant model (Chipman, 1976). Aggregate functions in economics thus derive from complex theoretical models and their successful application results from a well-understood methodology for construction and optimization.

Ecological systems are similar to economic systems in that they are complex and therefore difficult to describe by simple models. Historically, ecosystem models have been developed a priori as aggregations without regard to the consequences of that aggregation. This situation stems in part from lack of a unified theory against which to characterize the aggregates and in part from lack of data to model species separately. Recently both empirical and system theoretic work on aggregation in ecological models has begun. Wiegert (1975) studied a model of algae–fly interactions in a hot springs ecosystem. Using five different aggregation schemes, he concluded that the disaggregate model produced the most reliable output. Zeigler (1976) defines a base model to be a complete conceptualization of a system at a particular level of resolution. Working from a theoretical perspective he shows that if the time constants of aggregated species are identical, homomorphic models may be developed which preserve the state transitions of the base model. Using the data of Huffaker (1958; Huffaker *et al.*, 1963), Zeigler (this volume) develops a base model and derives three increasingly less complex homomorphic models from this. He shows that collectively ecological realism is maintained at one extreme and general insight into mechanisms of operation is realized at the other. The importance of maintaining state transitions is emphasized by discussing techniques for deriving the homomorphic models. Halfon and Reggiani (1978) discuss criteria for selecting an optimal model complexity which describes the system's behavior and also maintains ecological reliability. The purpose of this paper is to define in a relative precise manner the aggregation problem in ecology. The paper will include an analysis of those errors introduced by using an *aggregate model* instead of a simple *additive* (isomorphic) *model*. We assume that the latter representation—analogous to the Zeigler (1976) base model and the "true model" of Chipman (1975)—is unattainable below the population level (except perhaps in very simple systems), that the population level is a reasonable unit of functional interest, and that the additive model, if constructed, would simulate behavior of the real system at the highest level of accuracy possible under the modeling strategy employed.

2. Concerning Aggregation in Ecosystem Modeling

It is not our intent to argue the merits and limitations of various modeling strategies (e.g., linear versus nonlinear versus finite difference). These issues have been discussed at length in the literature (Patten et al., 1975; Waide and Webster, 1976; Waide et al., 1974; Watt, 1975; Bledsoe, 1976). Rather, we will examine one common methodology—first-order linear differential equations—with the goal of analyzing error[1] resulting from aggregating several first-order components into a single aggregate, the aggregate itself being described by a first-order equation. We believe that error is inherent in any dynamic aggregate regardless of how that aggregate was created. Restricting the analysis to linear models serves only to give a frame of reference with respect to analyzing error. It is hoped that future work will not only expand on what is presented here but also will begin to examine more complex models.

Definition 1. Let S denote an ecosystem composed of p distinct subsystems S_i, $i = 1, 2, \ldots, p$. That is,

$$S = \bigcup_{i=1}^{p} S_i. \tag{1}$$

Note that S is a single ecosystem which may or may not be a member of a collection of ecosystems. If the purpose of the analyst is to make an inference about a collection of similar ecosystems, or a population of ecosystems, then one needs to consider characteristics of the population of such ecosystems.

Definition 2. A set $Ⓢ = \{S\}$ of ecosystems will denote a collection of ecosystems. Each member S of the collection will necessarily contain the same distinct subsystems in order to be in $Ⓢ$.

In the real world it is almost always true that each subsystem S_i is also composed of a set of distinct subsystems, say S_{ij} where $j = 1, 2, \ldots, m$. That is,

$$S_i = \bigcup_{j=1}^{m} S_{ij}, \quad i = 1, 2, \ldots, p. \tag{2}$$

However, for our purposes we will assume (1) is an acceptable

---

[1] The concept of error will be developed in the paper. Informally, it is the deviation at time t of the state of an aggregate component from the sum of the states at t of the n separate components composing the aggregate. It is thus the deviation of the aggregate model from the isomorphic model for the n components. We assume throughout that an isomorphic model at the population level would yield the best possible approximation to the true state of the system and thus would measure error from this best possible approximation.

representation of the system and that each S_i, for biotic components, represents a single population (species).

The purpose in modeling an ecosystem is to facilitate predicting the state of an ecosystem at any time, t, with respect to some well-defined property.

Definition 3. Let P be a property of an ecosystem S which varies continuously with time, then $x(t)$ is said to be the state of the system S with respect to the property P if

(a) $x(t) \geqslant 0$ for all t, $0 \leqslant t < \infty$,
(b) $x(t)$ is continuous, $0 \leqslant t < \infty$, and
(c) $x(t)$ has meaning in a ecological sense and is a meaningful measure of P.

We will simply say that $x(t)$ is the state of the ecosystem S.

Clearly, one can never know $x(t)$ exactly. It is assumed that the best approximation to it would result from an isomorphic model constructed at a high degree of resolution, say the atomic level. Owing to the complexity of ecological systems it seems that the best isomorphic model attainable might be at the species level. Obviously, there is no clear distinction between isomorphic and homomorphic (aggregate) models. Thus, let us for present purposes draw it here, at the population level. There are then two sources of error, (a) the difference between $x(t)$ and the state predicted by the isomorphic model and (b) the difference between the state predicted by the isomorphic model and that predicted by the homomorphic model. We shall assume that the first error is sufficiently small that it may be ignored. That is, for practical purposes,

$$|x(t)_{\text{IM}} - x(t)| = 0 \quad \text{for all} \quad t.$$

It is assumed that we do know the form of the isomorphic model, that is, p linear first-order differential equations. Thus, our analysis seeks to examine the nature of the error that results by aggregating p first-order equations into m first-order equations, where $m < p$. That is to say, we are interested in the error resulting from aggregation. In the discussion that follows $x(t)$ is referred to as the true state of the system. However, for our purpose in this paper it is the state that would be predicted by an isomorphic model.

A basic goal of a modeler is to select a function $x^*(t, \beta)$, where β represents a finite number of parameters, so that $x^*(t, \beta)$ is a "good" approximation to the true state $x(t)$. In general, $x^*(t, \beta)$ will be a vector function $\mathbf{x}^*$, some of whose elements are already aggregates.

2. MODELING: GENERAL

2.1. Identification Problem

Several problems arise immediately, and quite naturally, when one attempts to model a real system. Let S' denote the system we wish to model, then from (1)

$$S' = \bigcup_{i=1}^{p'} S_i', \tag{3}$$

where p' denotes the number of subsystems S_i' which makes up S'. In his ignorance the modeler may not be able to *identify* all the subsystems which make up S', hence he will necessarily model S' by S as defined in (1). Clearly, p may not be equal to p' in the model, nor will the selection of subsystems be identical. The modeler hopes that since

$$S' = \left[\bigcup_{i=1}^{p} S_i\right] \cup \left[\bigcup_{S_i \not\subset S} S_i'\right] - \left[\bigcup_{S_i \not\subset S} S_i\right] \tag{4}$$

the $\bigcup_{S_i \not\subset S} S_i'$ and $\bigcup_{S_i \not\subset S} S_i$ are small and/or relatively unimportant in his analysis when S is used instead of S'. Note that

$$\bigcup_{S_i' \not\subset S} S_i' \tag{5}$$

and

$$\bigcup_{S_i \not\subset S'} S_i \tag{6}$$

denote those subsystems which were there but were not modeled (5) and those subsystems which were modeled but not there (6). The magnitude of these errors will be expected to appear when one compares model output with observations from the system.

2.2. Inference Problem

A very important concept one must be clear on is to what population of systems one is going to claim when applying the results generated by the model. There are three possible cases which are admissible: (a) a single ecosystem, (b) a collection of ecosystems of which the one in (a) is but a single member, and (c) a single ecosystem, but an unknown member from the collection of ecosystems in (b). This chapter is concerned primarily with situation (a); yet it is important that a clear understanding of the difference between (a), (b), and (c) be attained.

Our purpose in (a) is to *approximate* $x(t)$, the true state for a single ecosystem S'. The results generated by our model are valid for that ecosystem and not for any other one unless we *assume* that *all* ecosystems in the collection of ecosystems are not significantly different from the one we have modeled.

In order to carry out (b) and/or (c) one must be able to perform (a) well since to do (b) and (c) well requires that (a) be done well.

Definition 4. Let $x^*(t)$ be a model for $x(t)$, then $x^*(t)$ approximates $x(t)$ on the interval $0 \leq t < \infty$ with accuracy $\varepsilon > 0$, if

$$\max_{0 \leq t < \infty} |x^*(t) - x(t)| \leq \varepsilon. \tag{7}$$

Other measures of accuracy may prove easier to manipulate mathematically than the one posed in (7). We will list two for reference only. These are

$$\int_0^\infty (x^*(t) - x(t))^2 \, dt \leq \varepsilon \tag{8}$$

and

$$\int_0^\infty x^*(t) - x(t) \, dt \leq \varepsilon. \tag{9}$$

Definition 5. Let Ⓢ denote a collection of a single type of ecological system; for each $S \subset$ Ⓢ, let there be an $x_s(t)$, then the probability density function (p.d.f.)

$$f(x_s(t)) \tag{10}$$

of $x_s(t)$ is said to be a description of the state of the population Ⓢ.

Definition 6. If $x^*(t)$ denotes an approximation of $x(t)$ for some system $S \subset$ Ⓢ, then the p.d.f.

$$f^*(x_s(t)) \tag{11}$$

is the description of the approximate state of Ⓢ.

In most real-world situations our goal is to know $x_s(t)$ for every $S \subset$ Ⓢ, but due to our ignorance or our inability to model we usually try to select $x^*(t)$ so that (7) holds for some preassigned value of ε, and then derive (11) based on selected information from observations made on a randomly selected sample of systems in Ⓢ.

Finally, when studying a specified system $S \subset$ Ⓢ but not knowing which one it is, we wish to *predict* $x_s(t)$ using information contained in (11). This can be thought of as a Bayes estimation problem (Mood and Graybill, 1963) and will not be discussed here.

Again, our purpose in this chapter is to approximate $x(t)$ by a model $x^*(t)$. The description of a population of specified systems, and the prediction of one in the population are important questions which can and should be formulated so that when data are available the models will exist.

2.3. Parameter Selection Problem

When one approximates an unknown function $x(t)$ with a known function $x^*(t)$, he restricts his choice to a family of functions usually defined by a set of parameters which we shall denote by β. These questions arise naturally:

(a) What family of functions should be chosen? (Polynomial, Fourier series, sets of linear differential equations, etc.)

(b) Upon selecting the family, how many parameters should one select?

(c) And, what parameters should be selected as relevant and/or significant?

Definition 7. Let $\circledX(t) = \{x^*(t, \beta); \beta \in \Omega\}$ be a set of functions admissible for approximating $x(t)$ where β is $q \times 1$ vector value element of a well-defined parameter vector space Ω. The parameter selection problem is selecting q and the q properties which define the set Ω.

Note that one can now interpret that the problem of approximating $x(t)$ becomes one of selecting β so that

$$\max_t |x^*(t, \beta) - x(t)| < \varepsilon \qquad (12)$$

and

$$x^*(t, \beta) \in \circledX(t)$$

defined in Definition 7. Note that for a specified ε there may be more than one β which satisfies.

3. MODELING: SPECIFIC

Let S be a specified ecological system defined with respect to a property P. Let there be k trophic levels in the system,

$$S_i = \bigcup_{j=1}^{m} S_{ij}, \quad i = 1, 2, \ldots, k,$$

with each trophic level possessing m species.

Definition 8. The state of the system is

$$x(t) = \sum_{i=1}^{k} \sum_{j=1}^{m} x_{ij}(t)$$

and the system state vector is the $\sum_{i=1}^{k} m_i \times 1$ column vector

$$\mathbf{x}(t) = [x_{11}(t), \ldots x_{j_1 1}(t); x_{12}(t), \ldots x_{j_2 2}(t); \ldots; x_{1k}(t), \ldots x_{j_k k}]^T. \quad (13)$$

$(\cdot)^T$ means the transpose of $(\cdot)$.

Definition 9. The general state vector linear model is defined by the vector differential equation

$$\dot{\mathbf{x}}(t) = M(t)\mathbf{x}(t) + \mathbf{u}(t), \quad (14)$$

where $M(t)$ is $\sum_{i=1}^{k} m_i \times \sum_{i=1}^{k} m_i$ matrix and $u(t)$ is a $\sum_{i=1}^{k} m_i \times 1$ vector of input values.

Definition 10. If $M(t) = M$ and $\mathbf{u}(t) = \mathbf{u}$ for all t, then we say the state vector is generated by a linear model with constant coefficients and inputs.

Note that the matrix

$$M(t) = \begin{bmatrix} M_{11}(t) & M_{12}(t) & \cdots & M_{1k}(t) \\ M_{21}(t) & M_{22}(t) & \cdots & M_{2k}(t) \\ \vdots & \vdots & \ddots & \vdots \\ M_{k1}(t) & M_{k2}(t) & \cdots & M_{kk}(t) \end{bmatrix}$$

can be partitioned into $m_i \times m_j$ submatrices with respect to sizes of trophic levels. In most cases, $M_{ii}(t)$ is diagonal and $M_{ij}(t)$ may or may not be null.

Through our analysis in this paper we will consider a single trophic level and assume M_{ii} is diagonal. For convenience then we will concern ourselves with a model of a single trophic level made up of p distinct species and return to (1) as our system model. The state in Definition 8 becomes

$$x(t) = \sum_{i=1}^{p} x_i(t) \quad (15)$$

and the state vector is

$$\mathbf{x}(t) = (x_1(t), \ldots, x_p(t)). \quad (16)$$

2. Concerning Aggregation in Ecosystem Modeling

Definition 11. The additive model of the state of an ecosystem $x(t)$ is $x(t)$ in (15), where $x(t)$ is given by solving (14) for $x_i(t)$, $i = 1, 2, \ldots, p$.

Since we are going to further assume that the linear model has constant coefficients, then M is diagonal and defined by

$$M = \begin{bmatrix} -l_1 & 0 & \cdots & 0 \\ 0 & -l_2 & \cdots & 0 \\ \vdots & & \ddots & \vdots \\ 0 & 0 & \cdots & -l_p \end{bmatrix}, \quad (17)$$

where each l_i, $i = 1, 2, \ldots, p$, denotes the turnover rate of one species. One can write

$$x_i(t) = \frac{u_i}{l_i} + \left(x_i(0) - \frac{u_i}{l_i}\right) e^{-l_i t} \quad (18)$$

for $i = 1, 2, \ldots, p$; the state vector is known as well as the state of the system by substituting (18) into (14).

Definition 12. The equilibrium turnover rate (or simply the turnover rate) is defined by

$$l_i = u_i/x_i^e(0), \quad (19)$$

where

$$x_i^e(0) = x_i(\infty), \quad (20)$$

the equilibrium state for $i = 1, 2, \ldots, p$.

Note that

$$l_i = u_i/x_i(\infty), \quad (21)$$

$$l_i/u_i = 1/x_i(\infty), \quad (22)$$

$$\sum_{i=1}^{p} \frac{u_i}{l_i} = \sum_{i=1}^{p} x_i(\infty), \quad \text{and} \quad (23)$$

$$x_i(t) = x_i(\infty) + (x_i(0) - x_i(\infty)) e^{-l_i t}. \quad (24)$$

4. THE AGGREGATION MODEL

In this section we will define what we mean by the aggregation problem and develop a methodology for evaluating errors when using an aggregate model when the state of an ecosystem is given by the additive model defined in (14), (18), and (24).

Definition 13. Let $x_A(t)$ be an approximation model for $x(t)$, where

$$x_A(t) = \frac{u_A}{l_A} + \left(x_A(0) - \frac{u_A}{l_A}\right)e^{-l_A t}, \tag{25}$$

where

(a) $u_A = \sum_{i=1}^{p} u_i$,   (b) $l_A = \sum_{i=1}^{p} \beta_i l_i$,

(c) $\beta_i = \dfrac{x_i^e(0)}{\sum_{i=1}^{p} x_i^e(0)} = \dfrac{x_i(\infty)}{\sum_{i=1}^{p} x_i(\infty)}$,   (d) $x_A(0) = \sum_{i=1}^{p} x_i(0)$.

Using (19), (20), (21), and (23) we find that

$$\frac{u_A}{l_A} = \sum_{i=1}^{p} x_i(\infty), \qquad l_A = \sum_{i=1}^{p} u_i \Big/ \sum_{i=1}^{p} x_i(\infty),$$

$$x_A(t) = \sum_{i=1}^{p} x_i(\infty) + \left(\sum_{i=1}^{p} x_i(0) - \sum_{i=1}^{p} x_i(\infty)\right) e^{-l_A t}. \tag{26}$$

Suppose that the true state of the system equals $x(t)$ as given by the additive model; then how well $x_A(t)$ approximates $x(t)$ is the basic question posed in this paper.

Definition 14. The relative error in using $x_A(t)$ to approximate $x(t)$ is defined by $q(t)$ where

$$q(t) = [x(t) - x_A(t)]/x(t). \tag{27}$$

The bias is given by the numeration, that is,

$$b(t) = x(t) - x_A(t). \tag{28}$$

Theorem 1. (i) $q(0) = \lim_{t \to 0} q(t) = 0$.

(ii) $q(\infty) = \lim_{t \to \infty} q(t) = 0$.

(iii) $x(0) = x_A(0)$.

(iv) $x(\infty) = x_A(\infty)$.

(v) $\dot{x}(t) = \sum_{i=1}^{p} [x_i(0) - x_i(\infty)] e^{-l_i t}(-l_i). \tag{29}$

(vi) $\dot{x}_A(t) = \left[\sum_{i=1}^{p} x_i(0) - x_A(\infty)\right] e^{-l_A t}(-l_A). \tag{30}$

(vii) $l_A = \sum_{i=1}^{p} u_i \Big/ \sum_{i=1}^{p} x_i(\infty)$.

(viii) $b(t) = \sum_{i=1}^{p} (x_i(0) - x_i(\infty))(e^{-l_i t} - e^{-l_A t})$. (31)

(ix) If $l_1 = l_2 = \ldots = l_p$, then $b(t) = q(t) = 0$ for all t.

(x) If $x_i(0) = x_i(\infty)$ for every $i = 1, 2, \ldots, p$, then $b(t) = q(t) = 0$ for all t.

Proof. If $t = 0$, from (15), (24), and (25) (a), (b), and (d),

$$x_A(0) = \sum_{i=1}^{p} x_i(0) = x(0)$$

and $q(0) = 0$.

If $t \to \infty$,

$$x_A(\infty) = \lim_{t \to \infty} x_A(t) = \sum_{i=1}^{p} \lim_{t \to \infty} x_i(t) = \sum_{i=1}^{p} x_i(\infty) = x(\infty).$$

Finally $q(\infty) = 0$. Hence (i), (ii), (iii), and (iv) are true.

The results (v) and (vi) are obtained by direct differentiation of $x(t)$ and $x_A(t)$ as defined by (15), (24), and (26). The results (vii), (viii), (ix), and (x) are obtained by substitution and resolving terms.

Clearly if $q(t)$ is nonzero, then $q(t)$ is maximal for at least one t such that $\dot{q}(t) = 0$. Figure 1 shows the four possible error responses that may occur when aggregating linear components where the M_{ij} ($i \neq j$) elements in the system matrix are zero. In Table Ia we show a three-compartment additive model and its associated aggregate. In Table Ib we show four sets of initial conditions for this model, each set yielding a different error type. Note that there are infinitely many sets of $x_i(0)$ and that each set produces one of the four error types.

Note that one might wish to select a set of β_i's so that $q(t)$ or $b(t)$ is minimized. From (31) we know that

$$b(t) = \sum_{i=1}^{p} (x_i(0) - x_i(\infty))(e^{-l_i t} - \exp(-\sum_{i=1}^{p} \beta_i l_i t)) \quad (32)$$

$\sum \beta_i = 1, \qquad \beta_i \geq 0 \qquad \text{for} \quad i = 1, 2, \ldots, p.$

Clearly selecting the β_i's so that $b(t)$ is minimal is an optimization problem with linear and inequality constraints. The optimal β_i's can be found numerically. This however is a pedagogical problem since in practice one aggregates because one feels less confident in estimating the value of l_i, $i = 1, 2, \ldots, p$, than in estimating l_A, the aggregate turnover rate.

The additive model is recursive in the following sense:

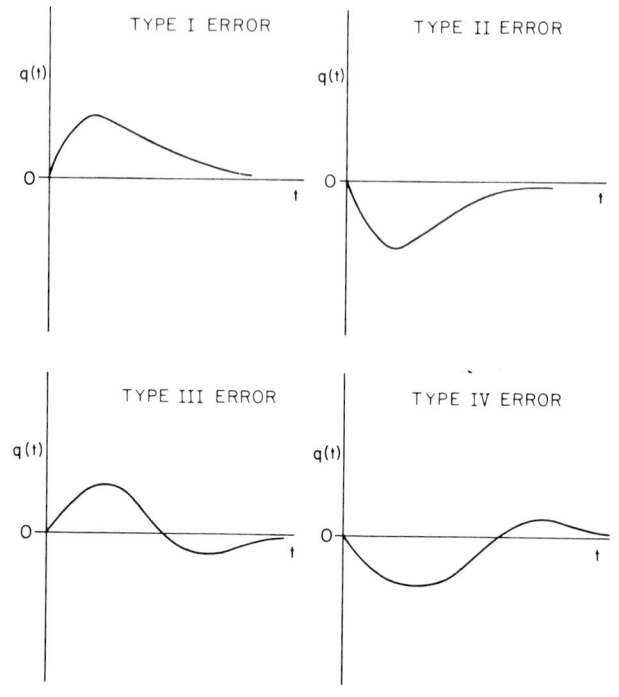

Figure 1. Four general types of error observed by altering the initial condition of a linear aggregate.

$$x^p(t) = \sum_{i=1}^{p} x_i(t) = x^{p-1}(t) + x_p(t). \qquad (33)$$

The aggregate model is also recursive in a more complicated form.

$$\begin{aligned}
x_A^p(t) &= \sum_{i=1}^{p} x_i(\infty) + \left[\sum_{i=1}^{p} x_i(0) - \sum_{i=1}^{p} x_i(\infty)\right] e^{-l_A^p t} \\
&= \sum_{i=1}^{p-1} x_i(\infty) + x_p(\infty) + \left[\sum_{i=1}^{p-1} x_i(0) - \sum_{i=1}^{p-1} x_i(\infty)\right] e^{-l_A^p t} \\
&\quad + [x_p(0) - x_p(\infty)] e^{-l_A^p t}.
\end{aligned}$$

But

$$\begin{aligned}
l_A^p &= l_A^{p-1} - l_A^{p-1} + l_A^p \\
&= l_A^{p-1} + \sum_{i=1}^{p-1} \beta_i^p l_i + \beta_p^p l_p - \sum_{i=1}^{p-1} \beta_i^{p-1} l_i \\
&= l_A^{p-1} + \beta_p^p l_p + \sum_{i=1}^{p-1} (\beta_i^p - \beta_i^{p-1}) l_i
\end{aligned}$$

2. Concerning Aggregation in Ecosystem Modeling

$$= l_A^{p-1} + \beta_p{}^p l_p - \sum_{i=1}^{p-1} \beta_i^{p-1} l_i \left(1 - \frac{\beta_i^p}{\beta_i^{p-1}}\right)$$

$$= l_A^{p-1} + \beta_p{}^p l_p - \sum_{i=1}^{p-1} \beta_i^{p-1} l_i \left(1 - \frac{\sum_{i=1}^{p-1} x_i(\infty)}{\sum_{i=1}^{p} x_i(\infty)}\right)$$

$$= l_A^{p-1} + \beta_p{}^p l_p - \beta_p{}^p \sum_{i=1}^{p-1} \beta_i^{p-1} l_i$$

$$= l_A^{p-1} + \beta_p{}^p l_p - \beta_p{}^p l_A^{p-1} = [1 - \beta_p{}^p] l_A^{p-1} + \beta_p{}^p l_p. \tag{34}$$

Then one can write the recursive form

$$x_A{}^p(t) = x_A^{p-1}(t) + \left[\sum_{i=1}^{p-1} x_i(0) - \sum_{i=1}^{p-1} x_i(\infty)\right] e^{-l_A{}^p t}$$
$$+ [x_p(0) - x_p(\infty)] e^{-l_A{}^p t} + x_p(\infty) \tag{35}$$

where $l_A{}^p$ can be replaced by (34), not a parsimonious form.

In order to determine an upper bound on the bias $b(t)$, the following analysis applies. Using (31) we know that

$$b(t) = \sum_{i=1}^{p} (x_i(0) - x_i(\infty))(e^{-l_i t} - e^{-l_A t}).$$

$$\leqslant p \max_i \{|x_i(0) - x_i(\infty)|\} \max_i \{|e^{-l_i t} - e^{-l_A t}|\}$$

$$\leqslant p \max_i \{|x_i(0) - x_i(\infty)|\} \max_i \{D_i(t)\},$$

where

$$D_i(t) = e^{-l_i t} - e^{-l_A t}.$$

But

$$l_A = (\sum \beta_i l_i) \quad \text{and} \quad \sum \beta_i = 1$$

or

$$\sum \beta_i \min\{l_i\} \leqslant l_A \leqslant \sum \beta_i \max\{l_i\}$$
$$\min\{l_i\} \leqslant l_A \leqslant \max\{l_i\}$$
$$e^{-\max\{l_i\}t} \leqslant e^{-l_A t} \leqslant e^{-\min\{l_i\}t}$$
$$e^{-l_i t} - e^{-\min\{l_i\}t} \leqslant e^{-l_i t} - e^{-l_A t} \leqslant e^{-l_i t} - e^{-\max\{l_i\}t}$$
$$e^{-\max\{l_i\}t} - e^{-\min\{l_i\}t} \leqslant e^{-l_i t} - e^{-l_A t} \leqslant e^{-\min\{l_i\}t} - e^{-\max\{l_i\}t}$$

or

$$|e^{-l_i t} - e^{-l_A t}| \leq |e^{-\min\{l_i\}t} - e^{-\max\{l_i\}t}|$$

for all $i = 1, 2, \ldots, p$.

Finally,

$$b(t) \leq p \max_i \{|x_i(0) - x_i(\infty)|\} |e^{-\min\{l_i\}t} - e^{-\max\{l_i\}t}| \tag{36}$$

$|e^{-\min\{l_i\}t} - e^{-\max\{l_i\}t}|$ is max when

$$t = \left[\frac{\ln(\min\{l_i\}/\max\{l_i\})}{-(\max\{l_i\} - \min\{l_i\})} \right].$$

The time of maximum bias does not necessarily correspond to the time of maximum relative error. We solve for that time as follows:

$$q(t) = \frac{x(t) - x_A(t)}{x(t)} = 1 - \frac{x_A(t)}{x(t)}$$

$$= 1 - \frac{1}{R(t)} \quad \left(\text{where } R(t) = \frac{x(t)}{x_A(t)}\right),$$

$$\frac{dq(t)}{dt} = \frac{d}{dt}\left[1 - \frac{1}{R(t)}\right] = \frac{d}{dt}[1 - R(t)^{-1}]$$

$$= [R(t)]^{-2} \frac{dR(t)}{dt}.$$

Setting $dq(t)/dt$ to zero,

$$0 = [R(t)]^{-2} \frac{dR(t)}{dt}.$$

Since $[R(t)]^{-2}$ is always nonzero, the above is only true when $dR(t)/dt = 0$. We find a t when this is true.

$$0 = \frac{dR(t)}{dt},$$

$$0 = \frac{d}{dt}\left[\frac{x(t)}{x_A(t)}\right],$$

$$0 = \frac{x_A(t)[dx(t)/dt] - x(t)[dx_A(t)/dt]}{[x_A(t)]^2},$$

$$0 = x_A(t)\frac{dx(t)}{dt} - x(t)\frac{dx_A(t)}{dt},$$

2. Concerning Aggregation in Ecosystem Modeling

$$0 = \left[\frac{u_A}{l_A} + \left(x_A(0) - \frac{u_A}{l_A}\right)e^{-l_A(t)}\right]\left\{\sum_{i=1}^{p}\left[-l_i x_i(0)e^{-l_i t} + u_i e^{-l_i t}\right]\right\}$$

$$-\sum_{i=1}^{p}\left[\frac{u_i}{l_i} + \left(x_i(0) - \frac{u_i}{l_i}\right)e^{-l_i t}\right]\left[-l_A x_A(0)e^{-l_A t} + u_A e^{-l_A t}\right],$$

$$0 = [H + A e^{-l_A t}]\sum_{i=1}^{p} B_i e^{-l_i t} - \sum_{i=1}^{p}[C_i + D_i e^{-l_i t}]F e^{-l_A t}, \tag{37}$$

where

$$A = x_A(0) - (u_A/l_A), \qquad B_i = -l_i x_i(0) + u_i,$$
$$C_i = u_i/l_i, \qquad D_i = x_i(0) - (u_i/l_i),$$
$$F = u_A - l_A x_A(0), \qquad H = u_A/l_A.$$

Equation (37) reduces to

$$H \sum_{i=1}^{p} B_i e^{-l_i t} + e^{-l_A t}\left(\sum_{i=1}^{p} G_i e^{-l_i t} - F \sum_{i=1}^{p} C_i\right) = 0, \tag{38}$$

where $G_i = AB_i - FD_i$. Equation (38) is a linear combination of exponentials which may be solved for t by numerical approximation. We used Newton's formula (Scheid, 1968) for the example given in the Appendix of this chapter.

5. CONCLUDING REMARKS

Systems ecologists face three types of modeling situations: (a) modeling a known system using data only from that system, (b) modeling a known system using data from that system and a group of closely related systems, and (c) modeling or describing a collection of systems of a particular type. Understanding behavior in models in (a) is essential to understanding behavior in (b) and (c). To this end we have examined one aspect of that modeling problem, namely, the introduction of error resulting from aggregating causally independent linear components. The following conclusions are drawn:

1. Except in two restricted situations aggregate models as described have error associated with them. Zero error results whenever the turnover rates of the separate aggregate components are identical (see also Zeigler, 1976) or (the trivial case) whenever the initial conditions of the separate components are respectively equal to their equilibrium value. In lieu of actually computing $b(t)_{max}$ or $q(t)_{max}$ for a particular aggregate, a reasonable modeling strategy would be to restrict aggregations to components with nearly identical l_i's. Since the aggregate turnover rate is a weighted sum of

Table Ia A Simple Three Compartment Additive Model and Its Associated Aggregate

Compartment	l_i	u_i	$x_i(\infty)$	β_i	Differential equation
$x_1(t)$	2.0	20.0	10.0	10/45	$\dot{x}_1(t) = 20 - 2x_1(t)$
$x_2(t)$	2.0	30.0	15.0	15/45	$\dot{x}_2(t) = 30 - 2x_2(t)$
$x_3(t)$	0.5	10.0	20.0	20/45	$\dot{x}_3(t) = 10 - 0.5x_3(t)$
					$x(t) = x_1(t) + x_2(t) + x_3(t)$
$x_A(t)$	$\frac{4}{3}$	60.0	45.0	NA	$\dot{x}_A(t) = 60.0 - \frac{4}{3}x_A(t)$

Table Ib Four Sets of Initial Conditions for the Model in Table Ia, Each of Which Yields a Different Error Type

i	$x_i(0)$	$x_A(0)$	Error type
1	15.0		
2	20.0	95.0	Type I
3	60.0		
1	0.0		
2	0.0	0.0	Type II
3	0.0		
1	1.0		
2	1.0	19.0	Type III
3	17.0		
1	15.0		
2	20.0	60.0	Type IV
3	25.0		

the l_i's ($l_A = \sum_{i=1}^{p} l_i \beta_i$) consideration should also be given to selecting from the class of similar l_i's those which also have similar equilibrium states. This conclusion is similar to the Lange–Hicks condition discussed by Simon and Ando (1961). The condition states that two or more state variables which respond similarly to a given set of conditions may be aggregated together into a single variable, this aggregate being a properly weighted average of the original variables.

By treating only aggregates within trophic levels we have assumed the matrix $M(t)$ in the general state equation to be nearly decomposable. That is, interspecific relationships between certain groups in a single trophic level are considered unimportant except to the extent that their aggregate state influences other system components. A complete discussion of nearly decomposable matrices in the analysis of linear economic models may be found in Simon and Ando (1961).

2. Concerning Aggregation in Ecosystem Modeling

Table II Four Examples of Additive Models Which Yield the Same Resultant Aggregate[a]

Case	i	l_i	$x_i(\infty)$	β_i	u_i/u_A	u_i	Additive model	Forced response
I	1	4	60	$\frac{1}{3}$	$\frac{2}{3}$	240	$\dot{x}_1(t) = 240 - 4x_1(t)$	
($p = 2$)	2	1	120	$\frac{2}{3}$	$\frac{1}{3}$	120	$\dot{x}_2(t) = 120 - x_2(t)$	Fig. 2
							$x(t) = x_1(t) + x_2(t)$	
II	1	6	30	1/6	3/6	180	$\dot{x}_1(t) = 180 - 6x_1(t)$	
($p = 3$)	2	2	60	2/6	2/6	120	$\dot{x}_2(t) = 120 - 2x_2(t)$	Fig. 3
	3	$\frac{2}{3}$	90	3/6	1/6	60	$\dot{x}_3(t) = 60 - \frac{2}{3}x_3(t)$	
							$x(t) = x_1(t) + x_2(t) + x_3(t)$	
III	1	8	18	1/10	4/10	144	$\dot{x}_1(t) = 144 - 8x_1(t)$	
($p = 4$)	2	3	36	2/10	3/10	108	$\dot{x}_2(t) = 108 - 3x_2(t)$	
	3	$\frac{4}{3}$	54	3/10	2/10	72	$\dot{x}_3(t) = 72 - \frac{4}{3}x_3(t)$	Fig. 4
	4	$\frac{1}{2}$	72	4/10	1/10	36	$\dot{x}_4(t) = 36 - \frac{1}{2}x_4(t)$	
							$x(t) = \sum_{i=1}^{4} x_i(t)$	
IV	1	16	5	1/36	8/36	80	$\dot{x}_1(t) = 80 - 16x_1(t)$	
($p = 8$)	2	7	10	2/36	7/36	70	$\dot{x}_2(t) = 70 - 7x_2(t)$	
	3	4	15	3/36	6/36	60	$\dot{x}_3(t) = 60 - 4x_3(t)$	
	4	5/2	20	4/36	5/36	50	$\dot{x}_4(t) = 50 - (5/2)x_4(t)$	
	5	8/5	25	5/36	4/36	40	$\dot{x}_5(t) = 40 - (8/5)x_5(t)$	Fig. 5
	6	1	30	6/36	3/36	30	$\dot{x}_6(t) = 30 - x_6(t)$	
	7	4/7	35	7/36	2/36	20	$\dot{x}_7(t) = 20 - (4/7)x_7(t)$	
	8	$\frac{1}{4}$	40	8/36	1/36	10	$\dot{x}_8(t) = 10 - \frac{1}{4}x_8(t)$	
							$x(t) = \sum_{i=1}^{8} x_i(t)$	

[a] Here $u_A = 360$, $x_A(\infty) = 180$, $l_A = 2$; thus $\dot{x}_A(t) = 360 - 2x_A(t)$. In each case the β_i's were taken to be i/Σ_i. Input fractions u_i/u_A were found by subtracting the associated β_i from $(p + 1)/\Sigma_i$. Thus, $\Sigma \beta_i = \Sigma (u_i/u_A) = 1.0$ for each case.

2. Aggregations of causally independent linear components produce at least four general error types (Fig. 1). Empirical work indicates that the particular type is dependent upon selection of initial conditions. In simulations of many different aggregate models (including the example in Tables Ia, b) we observed that forced responses show Type II, free responses show Type I, and Types III and IV may be produced by trial and error manipulation of the initial conditions. We suspect there is a simple algorithm based on l_i's and β_i's for predicting the error type a priori.

3. The time of occurrence of maximum relative bias and maximum relative error do not necessarily correspond. Methods are given for computing each. These times are particularly important in modeling applications where real-world decisions are to be based, at least in part, on model output. It is quite possible that the simulation interval of interest will occur during the time of $b(t)_{\max}$ or $q(t)_{\max}$. The modeler must be aware of

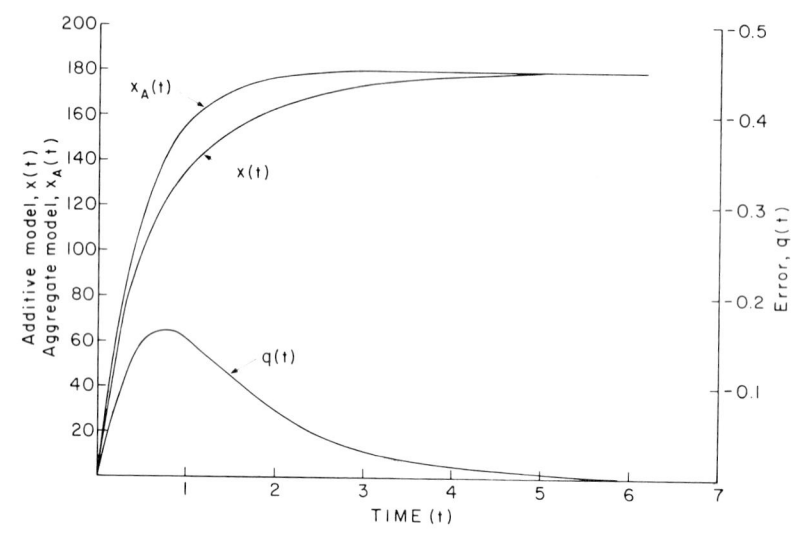

Figure 2. Time domain response of Case I in Table II with associated relative error $q(t)$.

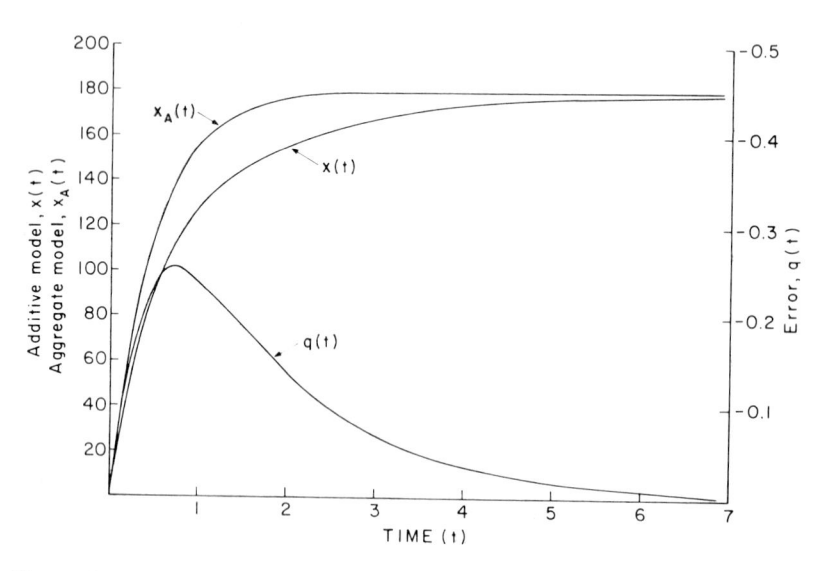

Figure 3. Time domain response of Case II in Table II with associated relative error $q(t)$.

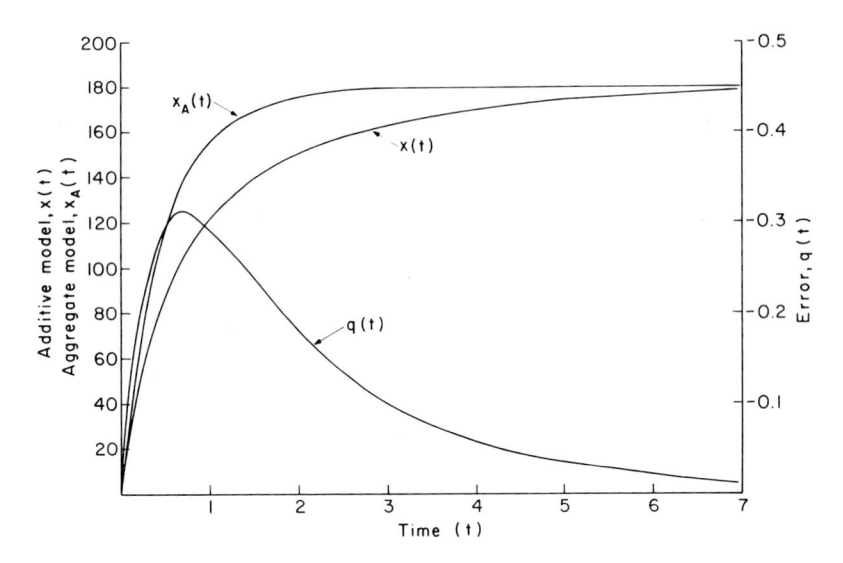

2. Concerning Aggregation in Ecosystem Modeling 73

Figure 4. Time domain response of Case III in Table II with associated relative error $q(t)$.

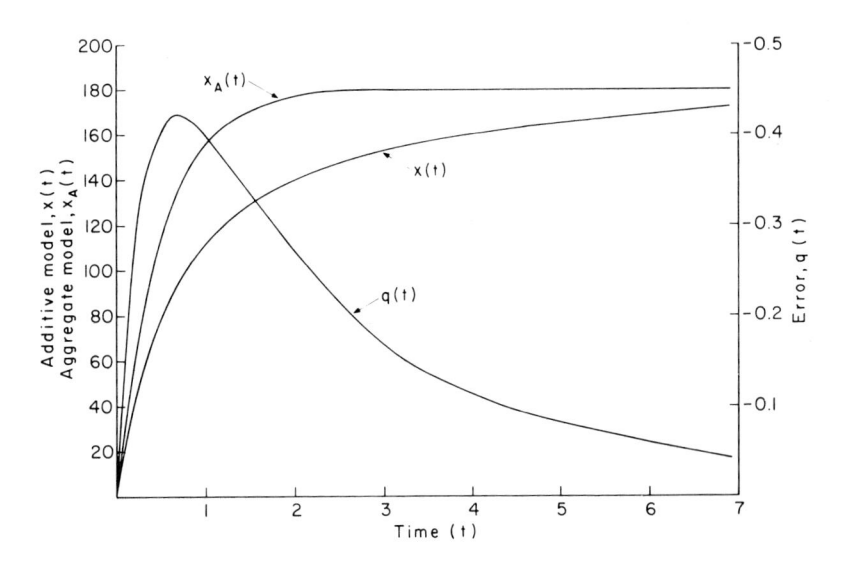

Figure 5. Time domain response of Case IV in Table II with associated relative error $q(t)$.

this time and the magnitude of the error to properly interpret the simulations.

4. Relative error and bias both approach zero as $t \to 0$ and as $t \to \infty$. Thus, there are intervals in a simulation which yield more accurate results than others.

5. Additive models are recursive in a simple manner. Aggregate models are also recursive but the recursion is more complex.

6. In Table II we give four additive models, each of which has the same associated aggregate model. By setting $x_i(0) = x_A(0) = 0$ for each case, forced response simulations were performed. Time behavior of $q(t)$, $x(t)$, and $x_A(t)$ for each case is shown in Figs. 2, 3, 4, and 5. Note that as the number of components increases, $q(t)_{max}$ for the Type II error response also increases. Since the shapes of the $q(t)$ curves are the same in each case, a relatively high $q(t)_{max}$ implies a longer period of high error in the aggregate model. It appears then that error tends to increase as the number of components in the aggregate increases if the modeler is insensitive to the values of l_i's and β_i's in the aggregate. In this example, the time of occurrence of $q(t)_{max}$ was hastened by the addition of more components to the aggregate.

It is clear that work is required in the general area of aggregation in ecological models. The approach suggested here of examining errors should be expanded systematically by analyzing errors for aggregations of components with direct causal linkage (Patten et al., 1976) with and without feedback. A study of the behavior of causally linked aggregates should be followed by an analysis of whole-system behavior. Similar studies using other than linear models should be valuable.

One goal of systems ecologists is to develop ecosystem models which accurately simulate dynamic behavior over a wide range of inputs. Error analysis provides a measure of this accuracy. While the intent is to ultimately eliminate errors from ecosystem models, an immediate advantage of error analysis is that output from models in current usage may be more properly interpreted. This is important not only in the scientific community but also in the political sector where model output is being used as part of the decision-making process.

APPENDIX

As an example for computing time of maximum relative error, consider Case I in Table II. We have

$$\dot{x}_1(t) = 240 - 4x_1(t),$$

2. Concerning Aggregation in Ecosystem Modeling

$$\dot{x}_2(t) = 120 - x_2(t),$$
$$\dot{x}_A(t) = 360 - 2x_A(t),$$

with $x_1(0) = x_2(0) = x_A(0) = 0$. Using (37) and (38) with

$A = x_A(0) - u_A/l_A = 0 - \frac{360}{2} = -180,$
$B_1 = -l_1 x_1(0) + u_1 = -4 \cdot 0 + 240 = 240,$
$B_2 = -l_2 x_2(0) + u_2 = -1 \cdot 0 + 120 = 120,$
$C_1 = u_1/l_1 = \frac{240}{4} = 60,$
$C_2 = u_2/l_2 = \frac{120}{1} = 120,$
$D_1 = x_1(0) - u_1/l_1 = 0 - \frac{240}{4} = -60,$
$D_2 = x_2(0) - u_2/l_2 = 0 - \frac{120}{1} = -120,$
$F = u_A - l_A x_A(0) = 360 - 2 \cdot 0 = 360,$
$H = u_A/l_A = \frac{360}{2} = 180,$
$G_1 = AB_1 - FD_1 = (-180)(240) - (360)(-60) = -21{,}600,$
$G_2 = AB_2 - FD_2 = (-180)(120) - (360)(-120) = 21{,}600,$

and

$$0 = f(t) = H \sum_{i=1}^{p} B_i e^{-l_i t} + e^{-l_A t} \left(\sum_{i=1}^{p} G_i e^{-l_i t} - F \sum_{i=1}^{p} C_i \right),$$

we have

$f(t) = 180 \, (240 \, e^{-4t} + 120 \, e^{-t}) + e^{-2t} \, (-21{,}600 \, e^{-4t}$
$\qquad + 21{,}600 \, e^{-t} - 360 \cdot 60 - 360 \cdot 120)$
$\quad = 2 e^{-4t} + e^{-t} - e^{-6t} + e^{-3t} - 3 e^{-2t},$
$f'(t) = -8 e^{-4t} - e^{-t} + 6 e^{-6t} - 3 e^{-3t} + 6 e^{-2t}.$

Substituting into Newton's formula,

$$t_{n+1} = t_n - (f(t)/f'(t)),$$

$$t_{n+1} = t_n - \frac{2 e^{-4t} + e^{-t} - e^{-6t} + e^{-3t} - 3 e^{-2t}}{-8 e^{-4t} - e^{-t} + 6 e^{-6t} - 3 e^{-3t} + 6 e^{-2t}}.$$

Select a value for t_n, say 0.6. Then by iteration of the above;

t_n	t_{n+1}
0.6	0.7762219
0.7762219	0.7644570
0.7644570	0.7644902
0.7644902	0.7644902

Use this value to compute $q(t)_{max}$ from (27).

$$q(t) = (x(t) - x_A(t))/x(t).$$

$$x(0.76449) = x_1(0.76449) + x_2(0.76449) = 57.18099 + 64.13145$$
$$= 121.31244$$
$$x_A(0.76449) = 140.98382.$$

$$q(t)_{max} = (121.31244 - 140.98382)/121.31244 = -0.16215.$$

In this example the aggregate model at worst underestimates the true value by 16.2%.

REFERENCES

Bledsoe, L. J. (1976). Linear and nonlinear approaches for ecosystem dynamic modeling. In "Systems Analysis and Simulation in Ecology" (B. C. Patten, ed.), Vol. 4, pp. 283–298. Academic Press, New York.

Chipman, J. S. (1975). Optimal aggregation in large scale econometric models. *Sankhya, Ser. C* **37**, 121–159.

Chipman, J. S. (1976). Estimation and aggregation in econometrics: An application of the theory of generalized inverses. In "Generalized Inverses and Applications" (M. Z. Nashed, ed.), pp. 549–769. Academic Press, New York.

Fisher, F. M. (1969). The existence of aggregate production functions. *Econometrica* **37**, 553–577.

Halfon, E., and Reggiani, M. G. (1978). Adequacy of ecosystem models. *Ecol. Modell.* **4**, 41–50.

Huffaker, C. B. (1958). Experimental studies on predation: Dispersion factors and predator prey oscillations. *Hilgardia* **27**, 243–383.

Huffaker, C. B., Shea, E. P., and Herman, S. G. (1963). Experimental studies on predation: Complex dispersion and level of food in an acarine predator-prey interaction. *Hilgardia* **54**, 305–329.

Mood, M. M., and Graybill, F. A. (1963). "Introduction to the Theory of Statistics," pp. 187–189. McGraw–Hill, New York.

Patten, B. C., Egloff, D. A., Richardson, T. H., and 38 co-authors (1975). Total ecosystem model for a cove in Lake Texoma. In "Systems Analysis and Simulation in Ecology" (B. C. Patten, ed.), Vol. 3, pp. 205–421. Academic Press, New York.

Patten, B. C., Bosserman, R. W., Finn, J. T., and Cale, W. G., Jr. (1976). Propagation of cause in ecosystems. In "Systems Analysis and Simulation in Ecology" (B. C. Patten, ed.), Vol. 4, pp. 457–579. Academic Press, New York.

Scheid, F. (1968). "Theory and Problems of Numerical Analysis." McGraw–Hill, New York.

Simon, H. A., and Ando, A. (1961). Aggregation of variables in dynamic systems. *Econometrica* **29**, 111–138.

Waide, J. B., and Webster, J. R. (1976). Engineering systems analysis: Applicability to ecosystems. In "Systems Analysis and Simulation in Ecology" (B. C. Patten, ed.), Vol. 4, pp. 330–371. Academic Press, New York.

Waide, J. B., Krebs, J. E., Clarkson, S. P., and Setzler, E. M. (1974). A linear systems analysis of the calcium cycle in a forested watershed ecosystem. *Prog. Theor. Biol.* **3**, 261–345.

Watt, K. E. F. (1975). Critique and comparison of biome ecosystem modeling. *In* "Systems Analysis and Simulation in Ecology" (B. C. Patten, ed.), Vol. 3, pp. 139–152. Academic Press, New York.

Wiegert, R. G. (1975). Simulation modeling of the algal-fly components of a thermal ecosystem: effects of spatial heterogeneity, time delays, and model condensation. *In* "Systems Analysis and Simulation in Ecology" (B. C. Patten, ed.), Vol. 3, pp. 157–181. Academic Press, New York.

Zeigler, B. P. (1976). The aggregation problem. *In* "Systems Analysis and Simulation in Ecology" (B. C. Patten, ed.), Vol. 4, pp. 299–311. Academic Press, New York.

Chapter 3

USE OF FIRST-ORDER ANALYSIS IN ESTIMATING MASS BALANCE ERRORS AND PLANNING SAMPLING ACTIVITIES

Dennis P. Lettenmaier and Jeffrey E. Richey

1. Introduction	80
2. First-Order Analysis	82
2.1 Findley Lake Mass Balance	85
2.2 Spatial Variation	87
2.3 Application to Lake Mass Balance	90
3. Results	92
3.1 Estimated Mass Balance Errors	92
3.2 Implications for Systems Analysis	94
4. Applications to Experimental Design	94
4.1 Example Application	95
4.2 Implications for Experimental Design	97
4.3 Error Sensitivity to Stage–Discharge Parameter Estimation	100
5. Summary	101
Appendix: Notation	102
References	103

1. INTRODUCTION

Efforts to understand and subsequently to model ecosystem behavior are predicated on an assumed knowledge of the dynamics of mass or energy flow from one compartment to another. Mass balance techniques, in which there is an accounting of mass entering, leaving, being produced, and being stored within an ecosystem, are useful tools in systems ecology. These techniques are essentially derivatives of compartment theory, where system structure is elucidated through a mathematical treatment of the observed patterns of flow of a tracer substance (see, for example, Zilversmit *et al.*, 1942; Solomon, 1960; Conover and Francis, 1973; Richey, 1974). The mass balance itself is thus simply an effort to conduct a complete mass flow accounting for an ecosystem, based on observed data. However, since observed data collected from natural systems are almost always probabilistic rather than deterministic in both space and time, any mass balance has some associated error due to imprecision of the estimates made. Often this error, when expressed as a fraction of the quantities being estimated, is substantial, for example, in the range 0.25–1.0. Past efforts at quantifying this error have met with little success, however. Commonly the results of applying classical error analysis formulas (Beers, 1953) have been error estimates so large that most analysts have been unwilling to admit to the indicated error levels, and subsequently quantitative error analysis has been dropped altogether. The difficulties encountered are usually associated with improper treatment of the correlation between elements in the mass balance. For instance, in a lake ecosystem, inflow and outflow volume estimates are often not uncorrelated, as is assumed in the simplest (and most widely used) form of the error analysis formulas.

Despite the almost total absence of attempts at the analysis of mass balance errors in past ecosystem studies, the assessment of error magnitudes could and should play an important part in ecosystem studies. For instance, consider the gross mass balance for a lake ecosystem nutrient:

$$\int_t^{t+\Delta t} m_i(t)\,dt + \int_t^{t+\Delta t} m_{ip}(t)\,dt - \int_t^{t+\Delta t} m_o(t)\,dt = s_{t+\Delta t} - s_t, \qquad (1)$$

where m_i is the inflowing mass from all sources, m_{ip} is the mass internally produced, m_o is the mass outflowing, and $s_{t+\Delta t} - s_t$ is the mass storage change in the lake. If, for instance, m represents dissolved inorganic phosphorus, inflows may be from streams, groundwater, and direct precipitation; internal production may come from zooplankton excretion (positive) and phytoplankton uptake (negative), and the storage change would simply represent the change in dissolved mass over the accounting period. Hence, for a given accounting period τ the mass balance would be written

$$m_s + m_g + m_p + m_{ip} - m_o - \Delta s_\tau = r \tag{2}$$

with the terms on the left-hand side of Eq. (2) representing surface, net groundwater, direct precipitation mass inflows, internal production, surface outflow, and storage change, respectively. The residual r is the imbalance between the difference of all assumed sources less all sinks minus storage change. If the balance were exact, r would of course be zero; however, it is possible that not all the sources and sinks have been accounted for by Eq. (2), so there may be a nonzero residual.

Unfortunately, the terms on the left-hand side of Eq. (2) are not available directly, but must be estimated from imprecise data, so an estimated mass balance

$$\hat{r} = \hat{m}_s + \hat{m}_g + \hat{m}_p + \hat{m}_{ip} - \hat{m}_o - \hat{\Delta}s_\tau \tag{3}$$

must be made, where $\hat{}$ denotes sample estimates made from real data. Hence, $\hat{r}$ will almost certainly not be zero, but the magnitude of $\hat{r}$ may be attributable to either model inadequacy (i.e., $r \neq 0$ in Eq. 2) or random sampling error. Error analysis can serve a very useful purpose in distinguishing between these two sources, for if a probabilistic statement of the error of the estimate $\hat{r}$ of r can be made, a confidence interval

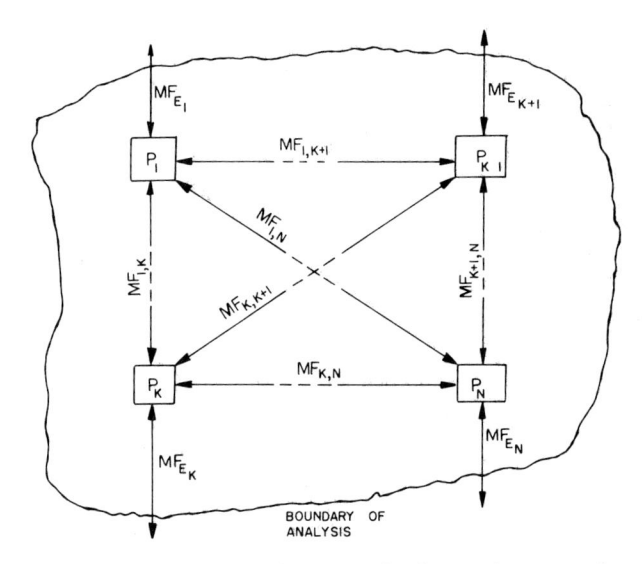

Figure 1. Pool/Flow ecosystem model conceptualization. P denotes pool mass, MF_E is external mass flow, MF is internal mass flow between pools.

$\hat{r} - \delta_1 \leqslant r \leqslant \hat{r} + \delta_2$ can be constructed. If the confidence interval does not include zero, $r \neq 0$ is implied at the given confidence level. Hence, error analysis can serve a very useful purpose in establishing the adequacy of the mass balance accounting. Also, the width of the confidence interval $\delta_1 + \delta_2$ is related to the manner in which the data from which $\hat{r}$ is estimated are collected; hence, error analysis can serve as a guide to sample program design.

Finally, error analysis can prove extremely useful in the construction, calibration, and verification of dynamic (deterministic) ecosystem models. Most such models ultimately depend on a compartment-flow structure as illustrated in Fig. 1. Based on estimates of the pool volumes (masses) and flows at different points in time, estimates of the type of model (e.g., difference equations) most suitable to describe the flows between the pools are made. Pool flows are essentially described by the terms in the left-hand side of Eq. (1); the right-hand side gives the change in pool volume. In the model calibration stage, model parameters are adjusted to make the model best "fit" the data-based estimates of pool volumes and flows. This procedure is complicated by the random nature of the estimated pool volumes and flows described above. If the randomness of these estimates is not properly accounted for, the calibration procedure is simply chasing random errors. Assessment of mass balance estimation errors, then, opens the door to a number of quantitative estimates of goodness of fit of the model to the data-based mass balance estimate.

2. FIRST-ORDER ANALYSIS

Cornell (1972) and Benjamin and Cornell (1970) have pointed out that in many water resource applications, knowledge of the mean and variance of a quantity in lieu of its complete probability density function may be acceptable, particularly in cases where derivation of the probability density function is difficult or impossible. Such cases occur frequently when probabilistic statements regarding $y = f(\mathbf{x})$ are desired, and only the joint probability density function of (the vector) $\mathbf{x}$ is known. Unless f has a very simple form, it is very difficult to arrive at a closed form solution for the probability density function of y. In many cases, too, only the mean and covariance matrix of $\mathbf{x}$, $Q_x = E(\mathbf{x} - \boldsymbol{\mu}_x)(\mathbf{x} - \boldsymbol{\mu}_x)^T$ are known (where $E(\cdot)$ denotes the statistical expectation) or can be estimated accurately, so the analysis can only be carried to estimation of the second central moment (variance) of y. In such cases first-order analysis is an extremely useful tool.

In first-order analysis, a Taylor series expansion is taken for the mean and deviation from the mean of y as a function of the mean and covariance of $\mathbf{x}$:

3. Use of First-Order Analysis   83

$$\mu_y \simeq f(\mu_x), \qquad y - \mu_y \simeq \left. \frac{\partial f}{\partial x_i} \right|_{\mu_x}^T (x - \mu_x)$$

from which

$$E(y - \mu_y)^2 \simeq E\left\{ \left.\frac{\partial f}{\partial x_i}\right|_{\mu_x}^T (x - \mu_x)(x - \mu_x)^T \left.\frac{\partial f}{\partial x_i}\right|_{\mu_x} \right\}$$

$$= \left.\frac{\partial f}{\partial x_i}\right|_{\mu_x}^T Q_x \left.\frac{\partial f}{\partial x_i}\right|_{\mu_x} = b^T Q_x b \simeq \sigma_y^2. \qquad (4)$$

Each of the terms on the right-hand side of Eq. (3) clearly fits into this format since each is functionally dependent on several variables. Using again the dissolved inorganic phosphorus example, we estimate the mass fluxes as the trapezoid rule integration of discrete measurements of concentrations and flow volumes taken over time. For instance, the surface stream outflow mass flux might be estimated as

$$m_o = cd_1 \sum_{t_o}^{t_1} fo_t + \sum_{i=1}^{n-1} \frac{cd_i + cd_{i+1}}{2} \sum_{t_i}^{t_{i+1}-1} fo_t + cd_n \sum_{t_n}^{t_f} fo_t, \qquad (5)$$

where cd_i is the discharge concentration at the ith sample date, t_i is the time of the ith sample, and fo_t is the daily average outflow on day t. The right-hand side of Eq. (5) gives the functional form of f in this example, where the elements of x are the measured concentrations and the outflow values between measurements. In the case where the outflow is measured by a continuous stage recorder and converted to flow by means of a stage–discharge relationship, $fo_t \simeq ah_t^b$, with h_t the daily average stage height at day t; the parameters a and b and the stage heights may be treated as random variables in addition to the concentration measurements.

Stage heights, however, can usually be measured quite accurately; most of the random variability in flow estimates results from error in the estimation of the parameters a and b. The computation is greatly simplified if the daily average stages are treated as constants. In this case, the independent variable vector for the outflow mass flux becomes

$$x = \begin{vmatrix} a \\ b \\ cd_1 \\ cd_2 \\ \vdots \\ cd_n \end{vmatrix}$$

and the linearization vector

$$\mathbf{b} = \begin{vmatrix} \dfrac{\partial \hat{m}_o}{\partial a} \\ \dfrac{\partial \hat{m}_o}{\partial b} \\ \dfrac{\partial \hat{m}_o}{\partial cd_1} \\ \vdots \\ \dfrac{\partial \hat{m}_o}{\partial cd_n} \end{vmatrix} = \begin{vmatrix} b_1 \\ b_2 \\ \vdots \\ b_{n+2} \end{vmatrix}$$

has elements

$$b_1 = cd_1 \sum_{t_o}^{t_1 - 1} h_t^b + \sum_{i=1}^{n-1} \frac{cd_i + cd_{i+1}}{2} \sum_{t_i}^{t_{i+1}-1} h_t^b + cd_n \sum_{t_n}^{t_f} h_t^b,$$

$$b_2 = ab \left\{ cd_1 \sum_{t_o}^{t_1 - 1} h_t^{b-1} + \sum_{i=1}^{n-1} \frac{cd_i + cd_{i+1}}{2} \sum_{t_i}^{t_{i+1}-1} h_t^{b-1} + cd_n \sum_{t_n}^{t_f} h_t^{b-1} \right\},$$

$$b_3 = \sum_{t_o}^{t_1 - 1} ah_t^b + \tfrac{1}{2} \sum_{t_1}^{t_2 - 1} ah_t^b,$$

$$b_j = \tfrac{1}{2} \left\{ \sum_{t_{i-1}}^{t_i - 1} ah_t^b + \sum_{t_i}^{t_{i+1}-1} ah_t^b \right\} \quad (4 \leq j \leq n+1)(i = j-2),$$

$$b_{n+2} = \tfrac{1}{2} \sum_{t_{n-1}}^{t_n - 1} ah_t^b + \sum_{t_n}^{t_f} ah_t^b.$$

The covariance matrix P_{xx} is defined as

$$P_{xx} = U_x^T R_{xx} U_x,$$

where U_x is the diagonal matrix with elements $U_{x_{ii}}$, the standard deviations of x_i, and $R_{xx_{ij}}$ is the correlation between x_i and x_j. Neither U_x nor R_{xx} is usually known a priori; however, estimates of the elements are often available. For instance, the variances of a and b (x_1 and x_2) result directly from the linear regression of the logarithm of flow on the logarithm of stage to estimate the stage–discharge parameters. The stage–discharge coefficient a is the exponential of the regression estimate; hence, its variance is given by $\sigma_a^2 = (\exp(\sigma_{a'}^2) - 1)(\exp(\sigma_{a'}^2 + 2a'))$, where σ_a' is the standard deviation of $a' = \ln(a)$, when it is assumed that the residuals from the linear (log–log) regression are normally distributed. The stage–discharge parameters a and b are assumed uncorrelated. The variances of the outflow concentrations may be estimated from the concentration measurements directly. The correlations between measured outflow concentrations desired are the measurement error correlations. In most cases these will be zero. It is important here to note that the quantities being estimated (e.g., inflow, outflow, etc.) are themselves random variables, usually with nonzero correlation in time. However, it is these random variables themselves, rather than their population values (e.g., long-term historic means) which the mass balance attempts to estimate, so only the measurement error need be considered, which will often be uncorrelated in time.

With sample estimates of U_x and R_{xx}, P_{xx} may be computed and **b** evaluated. Direct substitution into Eq. (4) will then yield the desired estimate of σ_m^2, the variance of the estimated outflow mass flux. Likewise, evaluation of the necessary linearization vectors, variances, and correlations will yield estimated variances for $\hat{m}_s$, $\hat{m}_g$, $\hat{m}_p$, $\hat{m}_{ip}$, and Δs_r. However, in addition to correlations within the elements of each of these terms, cross correlations must also be accounted for in order to arrive at an estimate for σ_r^2. Also, some of the terms included require assessment of spatial correlation (e.g., phytoplankton uptake measured at a few locations, often only one, must be extrapolated to an estimate for the entire lake volume). The approach to handling these difficulties is best illustrated in a specific example.

2.1. Findley Lake Mass Balance

Findley Lake is a small (11.7 ha) subalpine lake located near the crest of the Cascade Mountains about 80 km east of Seattle at an elevation of 1129 m. The lake is usually ice covered from mid-December to early July, and is extremely oligotrophic; for example, phosphate levels are usually near the lower limits of detectability. Estimation of mass balances for Findley Lake is complicated by the existence of several small inlet streams and the difficulty of access in winter months, when snow accumulations may reach 6 m. The outlet is gaged; however, inlet streamflows are estimated only with concurrent chemical sampling. It has been found that these periodic spot measurements are inadequate to estimate inflow volumes. Groundwater flow to and from the lake is composed of a shallow component representing diffuse inflow to the lake, mostly during the snowmelt season, and a deep underflow composed of seepage loss from the lake which is observed to surface as a small spring below the cirque which forms the lake basin. The magnitude of this outflow has been measured by estimating lake volume change and evaporation after a long period with no precipitation during which no outflow was recorded. Net deep groundwater flow was then estimated by difference. The deep groundwater estimate so derived has been found to be quite small relative to the annual flow volume.

Periodic measurements of shallow groundwater inflow and surface stream concentrations have suggested that the chemical composition of the diffuse, shallow groundwater inflow is very similar to that of the surface streams. Consequently, the sum of the shallow groundwater and surface inflow has been estimated by difference between the gage outflow and estimated deep groundwater flow, and the recorded surface stream concentrations have been applied to this difference to arrive at inflow mass flux.

Because of the small volume and low recorded dissolved inorganic phosphorus concentration in the deep groundwater, this term has been neglected. Hence, the Findley Lake dissolved inorganic phosphorus mass balance is assumed to be

$$r = m_s + m_p + m_{ip} - m_o - \Delta s_\tau.$$

Four time horizons, τ, were taken to correspond to the Findley Lake climate. These are (1) winter, January 1–April 30, characterized by large amounts of precipitation, mostly as snow, with moderate flow volumes and ice covering the lake surface; (2) spring, May 1–July 15, characterized by large runoff volumes, mostly from snowmelt, with ice covering the lake surface; (3) summer, July 16–September 30, characterized by low flow volumes and an ice-free lake surface; and (4) fall, October 1–December 31, a season of usually very low runoff with the lake ice-free until near the end of the season.

The terms in the budget are clearly not independent—for instance, the stage–discharge parameters occur in both the mass inflow and mass outflow estimates because of the manner in which the hydrologic balance is estimated. In order to compute the estimated error of the residual term, r, a first-order analysis may again be used, i.e., $\sigma_r^2 = \mathbf{c}^T P_w \mathbf{c}$,

$$\mathbf{w} = \begin{vmatrix} m_s \\ m_p \\ m_{ip} \\ m_o \\ \Delta s_\tau \end{vmatrix} \quad \text{and} \quad \mathbf{c} = \begin{vmatrix} \dfrac{\partial r}{\partial m_s} \\ \dfrac{\partial r}{\partial m_p} \\ \dfrac{\partial r}{\partial m_{ip}} \\ \dfrac{\partial r}{\partial m_o} \\ \dfrac{\partial r}{\partial \Delta s_\tau} \end{vmatrix} = \begin{vmatrix} 1 \\ 1 \\ 1 \\ -1 \\ -1 \end{vmatrix}.$$

P_w, the covariance matrix of $\mathbf{w}$ may be estimated using a multivariate extension of Eq. (4). Let $m_s = w_1(\mathbf{x}_i)$, $m_p = w_2(\mathbf{x}_2)$, $m_{ip} = w_3(\mathbf{x}_3)$, $m_o = w_4(\mathbf{x}_4)$, $\Delta s_t = w_5(\mathbf{x}_5)$. Similarly,

$$\mathbf{b}_1 = \frac{\partial m_s}{\partial \mathbf{x}_1}, \quad \mathbf{b}_2 = \frac{\partial m_p}{\partial \mathbf{x}_2}, \quad \text{etc.}$$

Note that

$$P_{w_{ij}} = E(w_i - \mu_i)(w_j - \mu_j) \simeq \mathbf{b}_i^T E(\mathbf{x}_i - E\mathbf{x}_i)(\mathbf{x}_j - E\mathbf{x}_j)^T \mathbf{b}_j = \mathbf{b}_i^T P_{x_i x_j} \mathbf{b}_j,$$

where $\mathbf{x}_i$ is the vector of independent variables for w_i, $\mathbf{b}_i$ is the partial derivative vector for w_i, and $P_{x_i x_j}$ is the covariance matrix of $\mathbf{x}_i$ and $\mathbf{x}_j$ defined as

$$P_{x_i x_j} = V_{x_i} R_{x_i x_j} V_{x_j},$$

with V_{x_i}, V_{x_j} the diagonal matrices of standard deviations of $\mathbf{x}_i$ and $\mathbf{x}_j$ and $R_{x_i x_j}$ the cross-correlation matrix of $\mathbf{x}_i$ with $\mathbf{x}_j$.

The elements of the covariance matrix of w, calculated in a straightforward manner, may now be used. The $\mathbf{b}_i$ vectors and the standard deviation matrices V_{x_i} must be evaluated in order to compute the diagonal terms of P_w (the variances of $w_1, w_2, \ldots$), so the only additional terms which must be evaluated are the cross-correlation matrices. Many of the cross correlations will be zero. However, some independent variables occur in more than one of the $\mathbf{x}_i$, for example, outflow stage heights are used to evaluate both the inflow and outflow masses; hence the cross correlations corresponding to these elements are 1.0. For the Findley Lake dissolved inorganic phosphorus balance, the independent variable vectors are given in Table I. Evaluation of the partial derivative vectors is straightforward.

2.2. Spatial Variation

The independent variable vectors $\mathbf{x}_i$ contain estimates of each of the time-dependent variables at each measurement time, so time variability is accounted for explicitly in the first-order formulation. However, generally, each of the time-dependent variables is also space-dependent. In the case of the inflow and outflow concentrations, spatial variability is much less significant than temporal variability because stream cross sections are quite small. For in-lake measurements—that is, internal production terms and the storage change terms—spatial variation may be significant. For instance, the variances of these terms are a function of the number of sample stations from which the areal average is computed. In Findley Lake, the use of a single sample station has historically been based on the small size of the lake and its mixing characteristics (average residence time is only about 45 days). Also, the bathymetry of the lake is such that most of the lake volume is spanned by only a relatively small fraction of the lake surface area. However, if it is desired to assess the effect of incorporating additional

Table I Functional Forms and Independent Variables for Findley Lake Dissolved Inorganic Phorphorus Budget

Dependent variable	Functional form	Independent variable vector	Dimension
Inflow $w_1 = m_S$	$ci_1 \sum\limits_{t_o}^{t_1-1}(ah_t^b - g) + \sum\limits_{i=1}^{n-1} \frac{ci_i + ci_{i+1}}{2} \sum\limits_{t_i}^{t_{i+1}-1}(ah_t^b - g)$ $+ ci_n \sum\limits_{t_n}^{t_f}(ah_t^b - g)$ n = Number of concentration observations t_i = Time (day number) of ith concentration observation g = Daily average deep groundwater inflow or outflow h_t = Outflow stage on day t ci_i = Inflow concentration on ith sample day t_o = Beginning day of season t_f = Ending day of season	$\begin{array}{\|c\|} a \\ b \\ g \\ ci_1 \\ \vdots \\ ci_n \end{array}$	$n+3$
Direct precipitation $w_2 = m_p$	$cp_1 \sum\limits_{t_o}^{t_1-1} p_t + \sum\limits_{i=2}^{m-1} \frac{cp_i + cp_{i+1}}{2} \sum\limits_{t_i}^{t_{i+1}-1} + cp_m \sum\limits_{t_m}^{t_f} p_t$ p_t = Total precipitation on day t t_i = Date of ith rainfall concentration measurement m = Number of rainfall concentration measurements	$\begin{array}{\|c\|} cp_1 \\ \vdots \\ cp_m \end{array}$	m

Internal production $w_3 = m_{ip}$	$(ze_1 - pu_1)(t_1 - t_0) + \sum_{i=1}^{n-1}(ze_i - pu_i) + (ze_{i+1} - pu_{i+1})(t_{i+1} - t_i)$ $+ (ze_n - pu_n)(t_f - t_n)$	$\begin{array}{c}ze_1\\ \vdots \\ ze_1\\ pu_1 \\ \vdots \\ pu_n\end{array}\Big\} 2n$
	ze_i = Phytoplankton volume-weighted average excretion on ith day of measurement	
	pu_i = Phytoplankton volume-weighted uptake on ith day of measurement	
Outflow $w_4 = m_o$	$cd_1 \sum_{t_o}^{t_1-1} ah_t^b + \sum_{i=1}^{n-1} \frac{cd_i + cd_{i+1}}{2} \sum_{t_i}^{t_{i+1}-1} ah_t^b + cd_n \sum_{t_n}^{t_f} ah_t^b$	$\begin{array}{c}cd_1\\ \vdots \\ cd_n\end{array}\Big\} n$
	cd_i = Inflow concentration on ith sample day h_t = Outflow stage on day t	
Storage change $w_5 = \Delta s_t$	$cl_n v_n - cl_1 v_o$	$\left.\begin{array}{c}cl_1\\ cl_n\end{array}\right\} 2$

[a] v_o = initial lake volume; v_n = final lake volume; cl_n = lake volume weighted concentration on nth (last) sample day; cl_1 = lake volume weighted concentration on first sample day.

stations, use may be made of results given by Rodriguez-Iturbe and Mejia (1974) for rainfall network design, reviewed below.

Rodriguez-Iturbe and Mejia base their work on the assumption of a process which is stationary in both space and time, and for which temporal and spatial correlations are separable, that is,

$$r(s_i, t_i; s_j, t_j) = r_1(s_i; s_j)r_2(t_i; t_j).$$

These assumptions are apparently met throughout much of the year (excluding the growing season) at Findley Lake. The assumption of spatial stationarity is justified on the basis of the well-mixed character of the lake; temporal stationarity, as observed earlier, is usually a tenable approximation except during the growing season.

Spatial correlation is assumed to be described by

$$r_1(s_i, s_j) = b_r \|s_i - s_j\| K_1(b_r \|s_i - s_j\|) \tag{6}$$

in two dimensions, where b_r is a constant, $\|s_i - s_j\|$ is the scalar distance between spatial locations s_i and s_j, and $K_1(\cdot)$ is a modified Bessel function of the second kind. It is beyond the scope of this work to discuss the basis for this form of spatial correlation; however, Whittle (1954) shows the physical justification for such a form for diffusion processes, which has obvious implications for the lake mass balance application.

Based on these assumptions, Rodriguez-Iturbe and Mejia (1974) show that the variance of the estimate $\hat{\mu}_z$ of the areal mean m_z of a process z is

$$E(m_z - \hat{\mu}_z)^2 = \sigma_p^2 g(n_s),$$

where z_k is the process measurement at location l, n_s is the number of spatial station locations, σ_p^2 is the point variance of the process z, and the variance reduction factor g is a function of the number of stations and the spatial correlation function. Note that m_z is itself a random variable; for example, the dissolved mass of a chemical constituent in the lake at any given time. Hence, the variance is reduced by a factor depending on the number of stations (and their location).

2.3. Application to Lake Mass Balance

The formulation given above is valid for a process with a two-dimensional spatial distribution. For a lake mass balance, the problem is clearly three-dimensional. When the problem is to estimate the total mass or mass flux of a dissolved substance in a lake, a two-dimensional approximation may be sustained by using discrete "slices" of the lake, and treating each slice or slab as a two-dimensional process. The approximate integration is then

$$\mathscr{S}_i = \frac{1}{n_s} \sum_{i=1}^{n_d} \mathscr{A}_i h_i \sum_{j=1}^{n_s} z_{ij},$$

with $\mathscr{A}_i$ and h_i the average slab area and depth, respectively, and where the variance of the depth summation

$$\operatorname{var}\left(\sum_{j=1}^{n_s} z_{ij}\right) = g(n_s, \mathscr{A}_i) \sigma_{p_i}^2.$$

The notation $g(n_s, \mathscr{A}_i)$ is used to indicate the dependence of g on the number of stations and spatial area in each slice, and $\sigma_{p_i}^2$ indicates the possible dependence of process point variance in each slab on the depth of the slab.

It should be noted that at each time, the lake concentration (storage change) and internal production terms are of the form of $\mathscr{S}$. The variances of $\mathscr{S}_i$, which are the variances of these terms to be used in calculating the overall mass balance error, may themselves be computed using first-order analysis, where

$$\mathbf{b} = \begin{vmatrix} \mathscr{A}_1 h_1 \\ \mathscr{A}_2 h_2 \\ \vdots \\ \mathscr{A}_{n_d} h_{n_d} \end{vmatrix},$$

$$\Sigma = \begin{vmatrix} (g_1(n_s, \mathscr{A}_1))^{1/2} \sigma_{p_1} & & 0 \\ & (g_2(n_s, \mathscr{A}_2))^{1/2} \sigma_{p_2} & \\ 0 & & (g_{n_d}(n_s, \mathscr{A}_{n_d}))^{1/2} \sigma_{p_{n_d}} \end{vmatrix}.$$

The correlations between slabs may be estimated from observed data; the lag one Markov structure with decaying exponential correlation function parametrized by ρ_z is convenient for this purpose; hence, the resulting correlation matrix is approximated as

$$R = \begin{vmatrix} 1 & \rho_z^{|h_1 - h_2|} & \rho_z^{|h_1 - h_3|} & \cdots \\ \rho_z^{|h_2 - h_1|} & 1 & \rho_z^{|h_2 - h_3|} & \cdots \\ \rho_z^{|h_3 - h_1|} & \rho_z^{|h_3 - h_2|} & 1 & \cdots \\ \cdot & \cdot & & 1 \end{vmatrix},$$

where h_i is the average depth of the ith layer. This formulation has the advantage of requiring only a single correlation parameter. The variance of $\mathscr{S}_i$ is then estimated in the usual manner as $\sigma_s^2 = \mathbf{b}^T \Sigma^T R \Sigma \mathbf{b}$. The only remaining parameters necessary to carry out this computation are the b_{r_i}, the spatial correlation parameters. These may be conveniently estimated by using as a basis the estimated correlation at any given distance on the order of one-half the square root of the average slab area (the characteristic length for each slab). Such a correlation was available at Findley Lake by using the in-lake and outflow chemistry data to compute an estimated correlation coefficient. This estimate was required in any event for computation of the correlation matrix of the mass outflow term.

With an estimate of the desired spatial correlation, the correlation parameter can be estimated directly from the functional form of the spatial correlation making use of a table of Bessel functions. Subsequently, the variance reduction factor for each depth may be taken directly from plots given by Rodriguez-Iturbe and Mejia (1974), and ultimately the variance of each areally averaged term estimated at each time. The process, while tedious, is admirably suited to automated computation, as is the entire first-order analysis error estimation procedure. It is worth noting, however, that when only a single station is used, the variance reduction factor is always 1.0, so much of the above computation is not required.

3. RESULTS

3.1. Estimated Mass Balance Errors

Table II shows the results of the error analysis for the Findley Lake dissolved inorganic phosphorus balance for calendar years 1974 and 1975. In both years, the estimated residual for seasons 1 and 2 is positive and is substantially larger than the estimated residual error. This suggests that the measured influx of dissolved inorganic phosphorus exceeds the total of the losses accounted for. The possible significance of this observation must be tempered by a consideration of bias in estimation.

Estimation bias is simply a reflection of the fact that the statistical expectation of the residual, or average over all possible outcomes, may not be equal to the true residual, i.e., $E(\hat{r}) = r + \delta$, where δ is the bias. Bias may result from the discrete character of the observations and resulting integration bias, that is, the approximation of integrals by summations both in depth for the depth-integrated variables and in time for all variables. Other possible sources of bias are the time of sampling related to possible diurnal variations, and the possible "pulse" nature of storm events—in other words, the possibility of extreme nutrient fluxes occurring during events when no samples were taken. Unfortunately, the true bias cannot be known without an essentially continuous sampling program. However, if some assumptions regarding the underlying processes are made, the bias can at least be estimated. This is done in the following section. For the purposes of the present analysis of the Findley Lake residual estimates, it need only be noted that the bias is apparently quite low for seasons 1 and 2.

Table II Estimated Findley Lake Dissolved Inorganic Phosphorus Mass Balance and Errors[a]

	Season 1	2	3	4	Year
1974					
Inflow	0.47 (0.05)	0.76 (0.07)	0.06 (0.01)	0.14 (0.01)	1.43 (0.09)
Direct precipitation	0.00 (0.00)	0.05 (0.01)	0.00 (0.00)	0.02 (0.00)	0.07 (0.01)
Internal production	0.23 (0.13)	0.00 (0.06)	−0.45 (0.31)	0.38 (0.27)	0.14 (0.44)
Outflow	0.28 (0.02)	0.47 (0.04)	0.02 (0.00)	0.15 (0.01)	0.92 (0.05)
Storage change	0.08 (0.01)	−0.01 (0.01)	−0.07 (0.01)	−0.01 (0.01)	0.03 (0.01)
Residual	0.34 (0.15)	0.35 (0.12)	−0.30 (0.31)	0.40 (0.29)	0.69 (0.47)
1975					
Inflow	0.28 (0.03)	0.33 (0.03)	0.13 (0.01)	0.22 (0.03)	0.96 (0.05)
Direct precipitation	0.00 (0.00)	0.17 (0.03)	0.00 (0.00)	0.00 (0.00)	0.17 (0.03)
Internal production	0.12 (0.07)	0.06 (0.03)	0.32 (0.47)	0.01 (0.10)	0.47 (0.49)
Outflow	0.12 (0.01)	0.17 (0.02)	0.11 (0.01)	0.35 (0.03)	0.75 (0.04)
Storage change	0.06 (0.01)	−0.03 (0.01)	−0.04 (0.01)	0.01 (0.01)	−0.08 (0.01)
Residual	0.22 (0.08)	0.42 (0.06)	0.38 (0.47)	−0.13 (0.11)	0.93 (0.49)

[a] One standard deviation in parentheses. All units in kg/hectare.

3.2. Implications for Systems Analysis

First-order analysis, by its nature, yields no information as to the underlying probability density function of the quantity being estimated (in this case, the residual of the mass balance). However, use may be made of the central limit theorem, which simply states that the sum of a large number of independent random variables, regardless of their individual marginal probability distributions, is approximately normally distributed. The application here is complicated by the fact that the residual is the sum of only five random variables, which certainly is not large, and the fact that the inflow and outflow terms are highly correlated, so the independence assumption is not met. On the other hand, many of the operations involved in computing the terms in the balance are additions or subtractions, so there is some basis for application of the central limit theorem to the five terms making up the balance. However, if all of these terms were themselves normal, the residual would also be normal by the well-known property that linear combinations of normally distributed random variables are normal, regardless of the correlation between the terms. Relying on this property and the central limit theorem for justification, we may use the normal distribution to set approximate confidence bounds on the residual estimate.

For the normal distribution, the 95% confidence bounds for a two-sided test on $\hat{r}$ are $\hat{r} \pm 1.96\sigma$. However, σ is not known, and must be estimated, so the t distribution would be more appropriate for setting confidence bounds. Unfortunately, there is no basis for approximating the appropriate degrees of freedom for the t distribution. As a rough approximation, use of approximate 95% confidence interval of $\pm 2S$, where S is the estimated standard deviation of the residual (from the first-order analysis) is reasonable. Consequently, it is expected that the estimated residual should be within two of its standard deviations of zero. This is clearly not the case for either the first or second season in 1974 or 1975. However, the residuals for seasons 3 and 4 for both years do lie within the confidence bounds. The conclusion seems to be that there is a significant sink term for dissolved inorganic phosphorus during the winter months which has not been accounted for during the season of ice cover. This is in contrast to earlier speculation that the lake ecosystem was relatively dormant during the winter, and that the most important processes occur during seasons 3 and 4.

4. APPLICATIONS TO EXPERIMENTAL DESIGN

While first-order analysis is a useful tool for estimating mass balance errors, it also is potentially of great value in applications to experimental design. Present experimental design techniques used in lake ecosystem sampling programs are usually limited to sample station location considerations, that is, where stations should be located once the number of stations to be used has been decided. However, the more basic question of selecting sample station densities (spatial) and frequencies (temporal) is rarely addressed quantitatively. The proper combination of spatial density and temporal frequency is an important design problem. This is especially true in lake ecosystem studies where considerable time must be spent on data collection and analysis, and the luxury of superfluous data is one which usually cannot be afforded.

4.1. Example Application

As an example of the use of first-order analysis to aid in experimental design, we continue consideration of the Findley Lake dissolved inorganic phosphorus mass balance. Rather than utilizing real data directly, we have used available data to formulate a surmised "base" mass balance for which the mass balance errors are estimated. The linearization vectors $\mathbf{b}_i$ are then expanded about this base mass balance, rather than the observed data as is done in the estimation of mass balance errors directly from observed data. The base mass balance is also used in evaluation of those standard deviations which are expressed as a function of the process mean level.

The terms in the mass balance itself are the same as those given in Table I. The base levels and standard deviations are given in Table IIIa. In Table IIIb the spatial correlation parameter, b_r, assumed equal for all depth slabs, is given. This parameter was evaluated by substitution in Eq. (6) on the basis of a correlation (r_1) of 0.5 for a distance $\|s_i - s_j\|$ of 200 m. On the basis of calculated average slab areas, the normalized correlation parameters $\mathscr{A}_i b_r^2$, used in estimation of spatial variance reduction factors from Rodriguez-Iturbe and Mejia (1974), were calculated. These terms are also included in Table IIIb. It should be noted that while the base mass balance and standard deviations given in Table IIIa are representative of levels estimated from real data, the values used are selected as typical, rather than on the basis of a rigorous estimation procedure.

With the parameters in Tables IIIa and b defined, it is possible to evaluate mass balance errors for any desired station density and sampling frequency. This evaluation was performed for seasons 1 and 3, which were selected to provide contrast between the biologically inactive winter season and the summer growth season. With the base mass balance selected, it is also possible to evaluate the expected or mean mass balance for each sampling strategy. The difference between the expectation and the base mass balance for any strategy can be evaluated given the bias in each term

Table IIIa Parameter Means and Standard Deviations Used in Computing Residual Error for Experimental Design[a]

Dependent variable	Independent variable	Mean	Standard deviation	Units
Inflow	a	23.3	30.0	m³/sec
	b	2.1	0.7	
	g	3000	3000	m³/day
	ci_i	1.8 S_1, S_2	1.0	μg/l
		1.2 S_3		
		1.5 S_4		
	h_t	Recorded 1975 daily averages	0.0	ft
Direct precipitation	cp_i	2.4	1.0	μg/l
	p_t	Weather service monthly historic means for Stampede Pass (10 km NE Findley Lake)	0.0	m³
Internal production	ze_i	0.012 S_1, S_2		
		Note 1[b] S_3		
		0.012 S_4	0.012	kg/day
	pu_i	0 S_1, S_2		
		Note 2[c] S_3, S_4	$0.66\,pu_i$	kg/day
Outflow	a			
	b	Same as for inflow		
	cd_i	2.5 S_1, S_2		
		1.8 S_4	1.0	μg/l
Storage change	cl_n	1.8 S_1, S_2, S_4		
		1.0 S_3	1.0	μg/l
	cl_1	1.8 S_1, S_2, S_3		
		1.0 S_4	1.0	μg/l
	v_o	8.6×10^5	0.0	m³
	v_n	8.6×10^5	0.0	m³

[a] S_i denotes season i. See Table I for notation.
[b] Note 1: For season 3,
$ze_i = 0.012 + (0.12 - 0.012)(t_i - 196)/35, \quad t_i \leq 232,$
$\quad\;\; = 0.12 - (0.12 - 0.012)(t_i - 232)/41, \quad t_i > 232.$
[c] Note 2: For season 3,
$pu_i = 0.17 \exp(-0.011(t_i - 230)), \quad t_i \geq 220,$
$\quad\;\; = 0.17(t_i - 196)/24, \quad t_i < 220.$
For season 4,
$pu_i = 0.17 \exp(-0.011(t_i - 220)).$

Table IIIb Normalized Spatial Correlation Parameters Used in Computing Residual Error for Experimental Design

Depth (m)	$\mathcal{A}_i b_r^2$
0–3.5	3.38
3.5–7.5	2.56
7.5–12.5	1.13
12.5–17.5	0.47
17.5–22.5	0.16
22.5–25.0	0.02

of the mass balance as well as the bias in the residual. For the winter season, all concentrations are assumed constant, so there is no bias in any of the terms. However, during season 3 the magnitude of the internal production term varies over the season, hence some bias is likely to be present. This bias can be reduced by concentrating sampling times near the period of maximum growth. This also should reduce the error levels themselves somewhat, since the variance of the phytoplankton uptake term is proportional to its level.

In order to investigate the effect of a nonuniform sampling strategy for season 3, mass balance errors were evaluated for a uniform sampling strategy as well as for an alternative criterion based on triangular density sampling. In the triangular density strategy, sample times were selected such that the areas under a triangle with endpoints at the initial and final days of the season and the maximum at a point halfway between the phytoplankton uptake and zooplankton excretion maxima (taken as Julian day 226) were equal. This selection rule, although arbitrary, provides a simple mechanism for selecting sampling times concentrated near the point of maximum internal production for any sampling frequency.

4.2. Implications for Experimental Design

Mass balance error levels (one standard deviation) and biases were calculated for 1, 2, 3, 5, and 10 sample stations and for 3, 5, 10, 15, 20, and 25 sample dates per season for seasons 1 and 3. For season 3, both a uniform sampling frequency and the nonuniform strategy (triangular density) detailed above were evaluated. In all cases the computed bias was negligible compared to the one standard deviation error levels. The estimated residual one standard deviation error levels are plotted in Figs. 2–5. In addition to the residual errors plotted, the errors for each term in the balance were evaluated. The principal difference between seasons 1 and

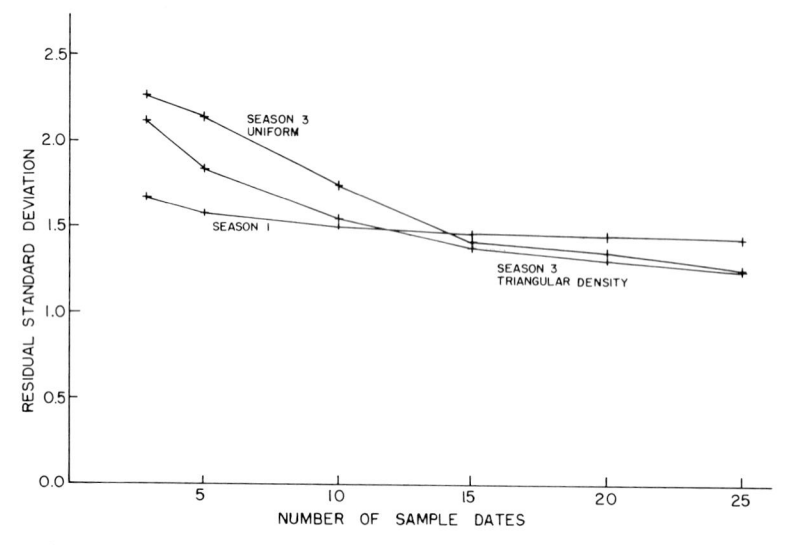

Figure 2. Residual standard deviation (in kg P) for dissolved inorganic phosphorus mass balance at Findley Lake with one sample station.

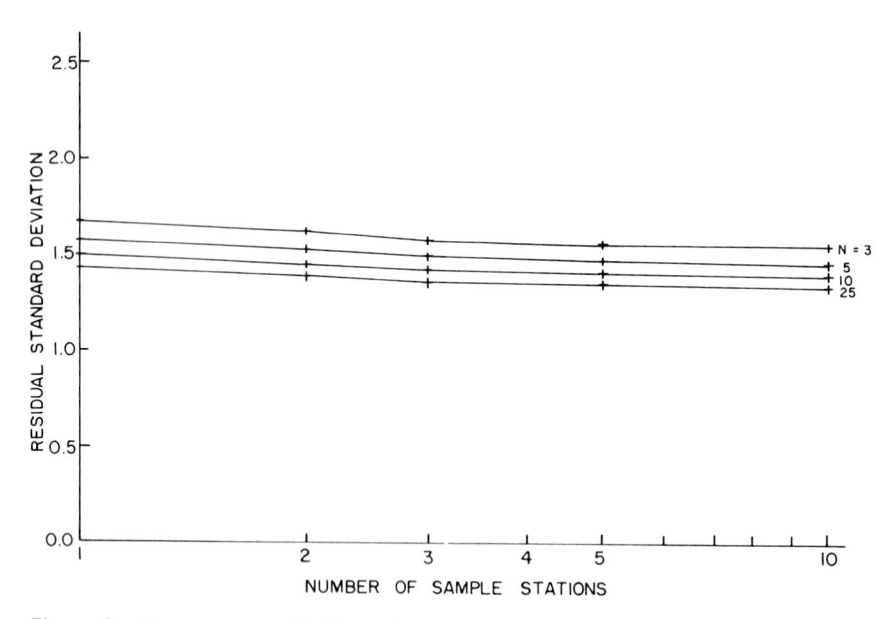

Figure 3. Season one residual standard deviation (in kg P) for dissolved inorganic phosphorus mass balance at Findley Lake. N denotes number of samples taken, sampling strategy is uniform.

3. Use of First-Order Analysis   99

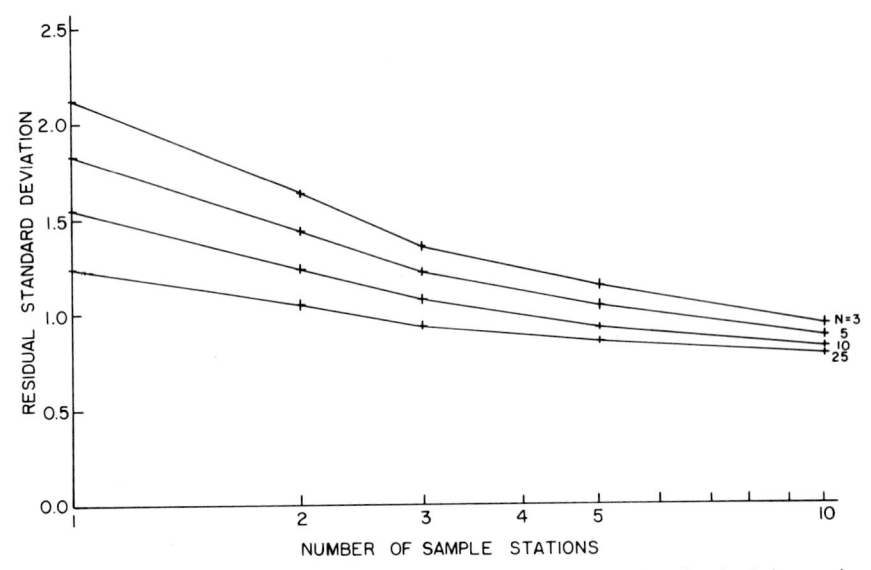

Figure 4. Season three residual standard deviation (in kg P) for dissolved inorganic phosphorus mass balance at Findely Lake. N denotes number of samples taken, sampling strategy is triangular density.

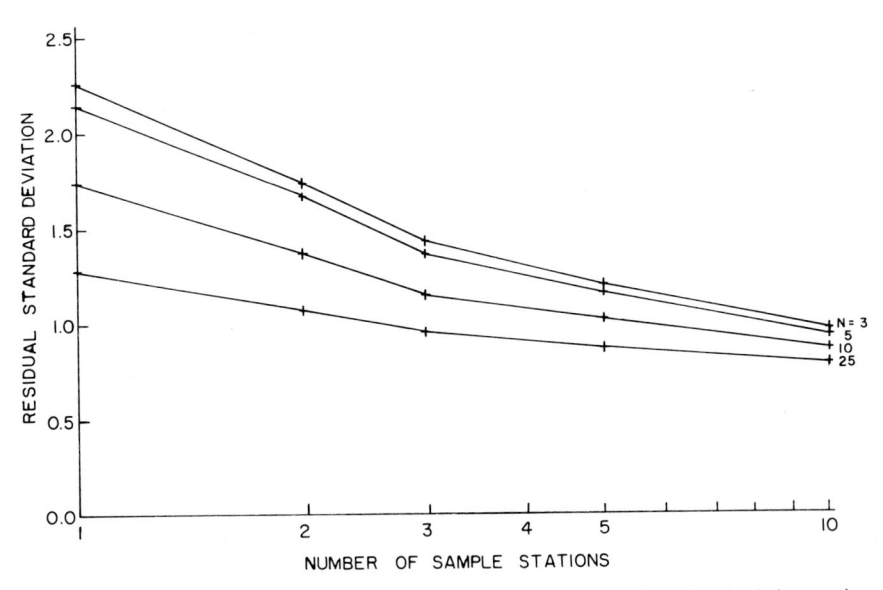

Figure 5. Season three residual standard deviation (in kg P) for dissolved inorganic phosphorus mass balance at Findley Lake. N denotes number of samples taken, sampling strategy is uniform.

3 is that inflow and outflow errors dominate in season 1, while internal production errors dominate in season 3. Also, the inflow and outflow errors are dominated by variability in the stage–discharge relationship, so that increasing sample frequency has little effect on residual error. This result has important implications, since accessibility during the winter season, particularly to the inflow creeks, is a real problem.

Figure 2 shows the lack of sensitivity of the season 1 residual error to sampling frequency. Also shown is the effect of the triangular distribution sampling frequency on the season 3 residual error level. The effect is greatest for moderate sampling frequencies; at high sampling frequencies little difference is noted between the two sampling strategies. Since the season 3 error is dominated by the internal production error, which is itself strongly dependent on sampling frequency, the residual error has a strong dependence on sampling frequency.

Figures 3–5 show residual error as a function of station density for several sampling frequencies for both seasons 1 and 3 (uniform and triangular density sampling strategies). Again, the insensitivity of season 1 residual error to experimental design is apparent. This family of curves may be used in several ways to aid in experimental design. For instance, if the total sampling effort (number of stations times sample frequency) is constrained, the optimum strategy may be selected by searching for the minimum residual error over all combinations of constant sample station density times sampling frequency. If an overview of required sampling effort as a function of error level is desired, this evaluation may be carried out over a range of minimum residual errors to give a plot of minimum error level as a function of total sampling effort (station density times frequency).

4.3. Error Sensitivity to Stage–Discharge Parameter Estimation

The result that residual error in season 1 is controlled by the variability in the stage–discharge relationship suggests the desirability of evaluating the effect that reduction of variances for the stage–discharge parameters will have on the residual error. This is especially significant because improvement in the stage–discharge relationship is possible, either by increasing the number of stage–discharge evaluations used in estimating the stage–discharge relationship or possibly by changing the channel structure at the gaging location, such as by installation of a weir. Figure 6 shows the residual error as a function of sampling frequency for stage–discharge parameter standard deviations reduced by factors of 2 and 10 from those given in Table IIIa for both seasons 1 and 3 (triangular density sampling). As expected, the effect is greatest in season 1, with the

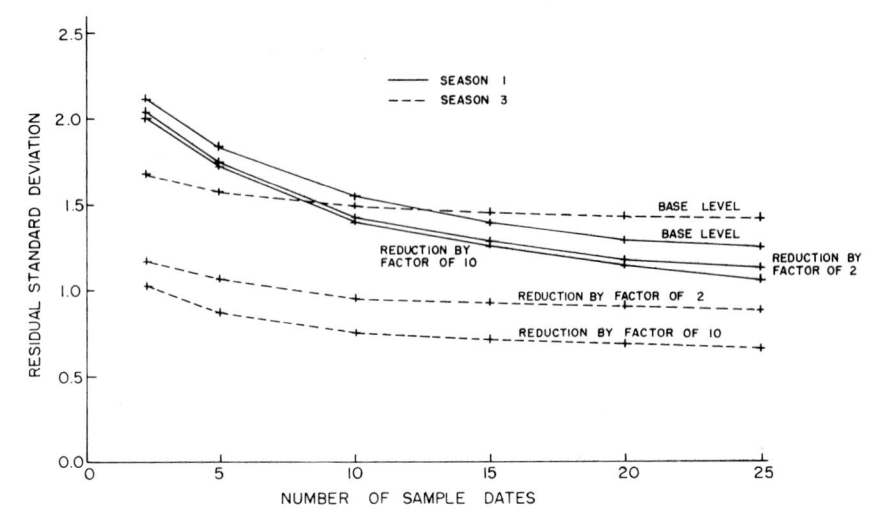

Figure 6. Sensitivity of residual standard deviation (in kg P) for Findley Lake dissolved inorganic phosphorus mass balance to stage-discharge parameter standard deviations with single sample station. Base levels are given in Table III.

greatest residual error reduction occurring as the result of a reduction of the parameter standard deviations by a factor of 2. An additional reduction by a factor of 5 (total reduction by a factor of 10) does not have nearly as large an effect in further reducing the residual error, apparently because error levels from the three terms not dependent on the stage–discharge relationship (direct precipitation, storage charge, internal production) become of greater importance as the inflow and outflow errors are reduced. Likewise, reduction of the stage–discharge error by a factor of 2 has only a small effect on the season 3 residual error, and a further reduction by a factor of 5 has almost no incremental effect.

5. SUMMARY

First-order analysis may serve at least two useful functions in mass balance studies. After the fact, it may be used both to place confidence intervals around the components of a mass balance (i.e., the integrated flows such as inflow, outflow, storage change, and the contributions of various internal sources and sinks) and to assess the accuracy of the overall mass balance, which may be measured as the difference between all internal and external sources less all internal and external sinks less storage change. If this residual is not significantly different from zero, the budget cannot be demonstrated to be out of balance, i.e., it cannot be disproved that all the sources and sinks have been accounted for. On the other hand, if a significant imbalance results, the functional form of the balance is questioned, and some important sources or sinks may have been omitted.

The magnitude of the residual error, as well as the errors or standard deviation of the balance components, is determined by the sampling program. A second function of first-order analysis may be to investigate the effect that alternate sampling strategies have on residual error and perhaps to optimize the sampling program. This may be very important in cases where the confidence bounds on the residual error of an existing mass balance are so wide that the mass balance provides little information on the system dynamics. The two most easily controlled variables in the sample design are sample station density and sampling frequency. In addition, the stage–discharge parameter error was found to contribute heavily to residual error in the Findley Lake example in those seasons where external source/sink terms dominated. Hence, in addition to the functional dependence of mass balance error on the sampling strategy, sensitivity may be tested against those additional parameters whose error levels may be controlled. In the Findley Lake case, hydrologic error has been reduced by construction of an outlet weir to obtain a stage–discharge relationship with substantially reduced parameter error.

APPENDIX: Notation

Scalars

$\mathscr{A}_i$	Average area of depth slab i
a	Stage–discharge coefficient
b	Stage–discharge exponent
b_r	Spatial correlation parameter
cd_i	Outflow concentration at measurement time i
g	Variance reduction factor for spatial sampling
fo_t	Total outflow volume on day t
h_i	Thickness of depth slab i
h_t	Daily average stage height on day t
K_1	Modified Bessel function of the second kind
m_g	Integrated groundwater mass transport
m_{ip}	Integrated internal production
m_o	Integrated outflow mass transport
m_p	Integrated precipitation mass transport
m_s	Integrated surface mass transport
m_y	Areal mean of random variable y
$m_i(t)$	Instantaneous inflow mass transport
$m_{ip}(t)$	Instantaneous internal production
$m_o(t)$	Instantaneous outflow mass transport

m_z	Spatial mean of process z
n_s	Number of sample stations
r	Mass balance residual
$r(s_i, t_i; s_j, t_j)$	Generalized spatial correlation function
r_1	Spatial correlation function
$\mathscr{S}_i$	Discrete estimate of m_z
s_t	Dissolved mass at time t
s_i	Spatial location of sample station i
t_i	Day of ith sample
t_o	Initial day of season
t_f	Final day of season
y	Dependent variable
Δs_r	Mass storage change over time Υ
δ	Estimation bias
δ_1	Estimated residual confidence interval lower half-width
δ_2	Estimated residual confidence interval upper half-width
ρ_z	Depth slab correlation parameter
σ_r^2	Residual variance
σ_y^2	Dependent variable variance
μ_y	Dependent variable mean

Vectors

b	Linearization vector
b$_i$	Linearization vector for w_i
c	Linearization vector for residual computation
w	Mass balance independent variable vector
x	General independent variable vector for univariate analysis
x$_i$	ith independent variable vector for multivariate analysis
μ_x	Independent variable mean

Matrices

P_w	Covariance matrix of mass balance residual
P_{xx}	Independent variable covariance matrix for arbitrary mass balance term
$P_{x_i x_j}$	Covariance matrix of **x**$_i$ and **x**$_j$
Q_x	Independent variable covariance matrix
R_{xx}	Correlation matrix of **x**
$R_{x_i x_j}$	Cross-correlation matrix of **x**$_i$ with **x**$_j$
U_x	Diagonal matrix of standard deviations of **x**
V_{x_i}	Diagonal matrix of standard deviations of elements of **x**$_i$
Σ	Covariance matrix for depth slab integration

REFERENCES

Beers, Y. (1953). "Introduction to the Theory of Error." Addison–Wesley, Reading, Massachusetts.

Conover, R. J., and Francis, V. (1973). The use of radioactive isotopes to measure the transfer of materials in aquatic food chains. *Mar. Biol.* **18**, 272–283.

Cornell, C. A. (1972). First order analysis of model and parameter uncertainty. *Proc. Int. Symp. Uncertainties Hydrol. Water Resour. Syst.*, Vol. III, 1972, pp. 1245–1272.

Benjamin, J. R., and Cornell, C. A. (1970). "Probability, Statistics, and Decision for Civil Engineers." McGraw-Hill, New York.

Richey, J. E. (1974). Phosphorus dynamics in Castle Lake, California. Ph.D. Thesis, University of California, Davis.

Rodriquez-Iturbe, I., and Mejia, J. M. (1974). The design of rainfall networks in time and space. *Water Resour. Res.* **10**, 713–728.

Solomon, A. K. (1960). Compartmental methods of kinetic analysis. *Miner. Metab.* **1**, Part A, 119–167.

Whittle, P. (1954). On stationary processes in the plane. *Biometrika* **41**, 434–449.

Zilversmit, D. B., Entenman, C., and Fishler, M. C. (1942). On the calculation of "turnover time" and "turnover rate" from experiments involving the use of labeling agents. *J. Gen. Physiol.* **26**, 325–331.

Part II
MODEL STRUCTURES, FORMALISMS, AND THEORY OF MODELING

Chapter **4**

PREDICTION, CHAOS, AND ECOLOGICAL PERSPECTIVE

Robert E. Ulanowicz

1. Introduction	107
1.1 The Modeling Process	107
1.2 Incomplete Results of Ecosystem Models	108
1.3 Three Views on the Adequacy of Ecosystem Modeling	108
2. Determinism and Chaos	109
3. Ecological Perspective	113
4. Summary	115
References	116

1. INTRODUCTION

1.1. The Modeling Process

A glance at the table of contents of this volume reveals that the organization of the chapters bears close resemblance to the much-discussed phases of the modeling process.

For example, the first step in a multicomponent systems analysis is usually the definition of what the compartments are and how their state may be described quantitatively. The first three chapters on the aggregation problem indicate that this task is certainly nontrivial.

The identification of the qualitative interactions between components usually follows. While one may speak in general terms about causality, in the ecological realm it is usually the transfer of matter, energy, or information that constitutes such interaction.

Choosing a mathematical statement to describe the time evolution of the interactions (usually in terms of the states of the compartments) constitutes what some (Dale, 1970) refer to as the modeling step in systems analysis. Despite the fact that this phase is the most explicitly mathematical, it generates relatively little theoretical interest as witnessed by the absence of contributions on the problem in this book.

This brings us to the final step of model verification (elsewhere referred to as model validation). A model is generally held to be valid insofar as it can reproduce the behavior of the system under conditions different from those used to create the model, i.e., to the degree that it has prediction ability. To be sure, there are occasions where invalid models provide useful insights, and the organization which the modeling scheme lends to one's thought processes or experimental program is not to be underestimated; but it may still be argued that the *sine qua non* of model validation remains—prediction ability.

1.2. Incomplete Results of Ecosystem Models

As defined, the modeling procedure is quite broad. For example, the chosen compartments may contain a single organism, a population of organisms, or an ensemble of populations. A population compartment, in turn, could be homogeneous or possess structure in size or age. Likewise, the mathematical statements may be deterministic or probabilistic in nature, linear or nonlinear, autonomous or able to possess memory. Examples of all types appear in the literature.

In practice, however, ecosystem modelers have focused upon particular types of model constructions. Most analyses seem to treat compartments which are nonsegregated (populations or trophic levels) and homogeneous. The accompanying mathematics is, for the most part, deterministic and autonomous.

Despite the enormous effort which has gone into such ecosystem modeling, there remains a paucity of models which meets the rigorous validation criteria which might be applied to a model of a mechanical system. Most ecosystem models are short on prediction ability. Even some classical examples, such as the Lotka–Volterra predator–prey scheme, have not spawned validated examples.

1.3. Three Views on the Adequacy of Ecosystem Modeling

There are three possible schools of thought concerning the apparent failure of ecosystems analysis to predict with confidence.

The first school points to our ignorance and the need for more data. The present approach is considered to be sound, but our inability to gather enough information on the components and interactions proscribes ascertaining the proper constitutive relations which will lead to valid models. In time, additional information will evolve models with greater predictive ability. Meanwhile, we can appreciate the new insights into particular ecosystems which the modeling exercise uncovers. This is a frequent argument given to funding agencies to support modeling projects. I have used it on occasion myself.

The second group is more iconoclastic (Mann, 1975). The current approach is futile because the models are poorly posed. To use populations as components is reductionistic and blinds one to emergent properties. To use trophic levels as compartments is nonsensical in light of the confused webs which characterize most ecosystems. Anyhow, stochastic influences upon model parameters yield confidence limits on ecosystems predictions which are too large to make them of any practical use. This group is not without its optimism, however. As our ability to look at ecosystems holistically advances, properties will appear along with their own laws, just as ethology emerges when we expand our scale of observation from the cellular level to the organism and then to the social unit.

Several published works of the past few years reveal that a third opinion on the ecosystem modeling problem is possible. This outlook allows that the modeling construct in vogue may be sound; however, it will not lead to useful results on the systems level because the deterministic behavior of large ecosystem models is indistinguishable from chaos. That is, prediction ability is short range at best. The only obvious emergent property of ecosystems is chaos. This does not exclude the possibility that macroscopic properties with their own laws exist. But the likelihood is that they are imposed by constraints from without (the abiotic universe) rather than emergent from within.

Because this outlook is relatively new and seemingly paradoxical, it is helpful to regard its development in more detail beginning with its theoretical origins in meteorology.

2. DETERMINISM AND CHAOS

The transition of a fluid flow field from laminar to turbulent has long intrigued theoreticians. Whether the Navier–Stokes equations (or any other form of Newton's second law) were germane to the turbulent field was a matter of much debate. To the meteorologist the issue was more than an interesting theoretical question. With the advent of computational machinery efforts were underway to apply the discrete forms of the equations of motion to meteorological flow fields for the purpose of forecasting weather. The limits to employing deterministic tools on a turbulent field thus became a matter of applied interest.

It was against this background that Lorenz (1963) published his elegant treatise on deterministic nonperiodic flow. Lorenz reduced the equations of motion for a particular two-dimensional, rotating, heated fluid to three, first-order, quadratic differential equations:

$$\dot{x} = -10x + 10y, \qquad \dot{y} = -xz + 27x - y, \qquad \dot{z} = xy. \qquad (1)$$

Here x, y, and z are complicated transforms of the stream function and excess temperature. The exciting point of this analysis is that these simple equations (mathematical cousins of the Lotka–Volterra equations) behave in such a peculiar manner.

The solutions to Eq. (1) always remain bounded. After a given period of time, however, the variables x and y go into oscillations with no finite period. This is strange behavior for a deterministic function; however, the path is uniquely determined in that starting the solution at any point on the trajectory will always result in tracing out the same pathway. Further experimentation with the system reveals that the trajectories are not stable in the strict mathematical sense. Starting the system arbitrarily close to the given initial conditions will result in a trajectory that eventually bears no coherence to the given one.

Trying to predict the future behavior of such a system is obviously perplexing. Since one can never measure initial conditions with exactitude, model and prototype evolutions are bound to become incoherent eventually. Likewise, efforts to replicate the behavior of such a system under controlled conditions is sure to meet with difficulty. In short, there is little to distinguish this deterministic entity from one that behaves chaotically.

Lorenz traced the chaotic behavior of his continuous system to discrete transitions of the trajectory between regions with qualitatively different behavior. He characterized these transitions by the numerical sequence:

$$\begin{aligned} m_{n+1} &= 2m_n & &\text{if } m_n < \tfrac{1}{2}, \\ m_{n+1} &\text{ is unidentified} & &\text{if } m_n = \tfrac{1}{2}, \\ m_{n+1} &= 2 - 2m_n & &\text{if } m_n > \tfrac{1}{2}, \\ 0 &< m_0 < 1, \end{aligned} \qquad (2)$$

and showed for a nondenumerable set of irrational m_0 that the behavior of the sequence was qualitatively similar to that of Eq. (1). At about the same time that Lorenz was making these observations, Ulam (1963) was reporting chaotic-like behavior in quadratic numerical transforms such as

$$x_{n+1} = 2x_n y_n + 2x_n z_n + 2y_n z_n,$$
$$y_{n+1} = x_n^2 + z_n^2, \qquad (3)$$
$$z_{n+1} = y_n^2.$$

Needless to say, the bizarre behavior of these numerical sequences captivated applied mathematicians such as Li and Yorke (1975), who set about elucidating the kinetics of pathological nonlinear numerical sequences. Fortunately, Yorke also appreciated the potential application of such mathematics to ecological situations. He brought his example of the discrete form of the logistic equation

$$x_{n+1} = y x_n [1 - x_n/K] \qquad (4)$$

to the attention of May (1974), who thereupon devoted much effort to making deterministic, chaotic behavior familiar to ecologists (May, 1975; Hassell et al., 1976; May and Oster, 1976).

May's analyses center about the simple, one-dimensional discrete logistic equation:

$$x_{n+1} = x_n \exp[r(1 - x_n/k)], \qquad (5)$$

which is qualitatively similar to Eq. (4). It is not too difficult to show that Eqs. (4) and (5) possess three parameter ranges, each endowed with qualitatively different behavior. In particular, Eq. (4) always possesses a stable equilibrium point when $2 > r > 0$. In the range $2.570 > r > 2.00$ the solutions possess stable cycles of period 2^n (n, integer) beginning with a two-point cycle near $r = 2.0$ and changing to 4-, 8-, 16-, ..., etc., point cycles as r increases. Finally, cycles of arbitrary period, or aperiodic (chaotic) behavior occur for $r > 2.570$.

May's emphasis upon these one-dimensional systems has the advantage that analytical methods can readily be brought to bear on the equations. It also points out the fact that even the simplest of nonlinear ecological models can give rise to chaotic behavior.

When data from real populations are applied to the discrete logistic algorithms, however, none of the naturally occurring populations possesses parameters in the chaotic range. Indeed, few naturally occurring populations show behavior more interesting than a monotonically damped return to steady state. The single example of a real population in the chaotic regime comes from a laboratory controlled population of blowflies (Nicholson, 1954).

Hence, the behavioralist or experimental ecologist might be inclined to dismiss the whole discussion about determinism and chaos as the rantings of theoreticians more concerned with their equations than with reality. To do so at this time, however, would be unfortunate and premature—for two reasons.

To begin with, the analytical work has emphasized single populations, whereas ecosystems, by definition, consist of collections of interacting populations. The little theoretical work that has been done on interacting populations indicates that chaotic behavior is *more likely* to occur with several species (May and Oster, 1976). Theoretical studies of host–parasite interactions (Beddington *et al.*, 1975) and competing species (Hassell and Comins, 1976) indicate that chaos will intercede sooner (at lower parameter values) than in the single-population case.

It is tempting to proceed by induction to the many-species problem, where interactions which deviate only slightly from linearity give rise to chaos. In fact the mechanical analog of this many-species problem has been well studied (Tuck and Menzel, 1972). The Fermi–Pasta–Ulam system of many mass points connected by springs with weakly nonlinear properties readily gives rise to apparent chaos. Hence, if these analogies and intuitions hold, the most obvious *emergent* property of ecosystems is chaos!

The second reason for not dismissing the possibility of chaotic-like determinism in ecosystems lies with the consequences of a negative result. One way to argue the nonexistence of chaos on an ecosystems scale would be to demonstrate the existence of an emergent, organizing property. Such a discovery would be a major breakthrough in ecological theory.

There remains one final possibility—total ecosystems do not become chaotic because they are inherently linear. The radical hypothesis that ecosystems are fundamentally linear was formalized by Patten (1975). Furthermore, there are empirical results which purport that linear models are somewhat more robust than their nonlinear counterparts (Ulanowicz *et al.*, 1978). It is yet to be resolved, however, whether linear models work a little better because the prototype systems are inherently linear or because, within the limited predictability possible in chaotic systems, linear models work just as well as (or better than) anything else.

Before leaving the subject of deterministic, chaotic-like behavior, it should be pointed out that the discussion in the ecological literature has dwelled upon discrete-time numerical sequences. The possibility that discrete-mass or discrete-space numerical models might also lead to chaotic-like behavior has not been fully assessed. Cohen (1976), for example, shows how certain discrete reproduction processes such as the breeding of pigs or the growth of algae may not converge to a single limit as numbers become large (i.e., they possess a nondegenerate limit). He points out how his examples share the lack of predictability and reproducibility exhibited by those of May.

Of course, one should always remember, too, that chaotic-like behavior is not confined to discrete sequences and was, in fact, first studied as resulting from continuous, ordinary differential equations with constant coefficients.

3. ECOLOGICAL PERSPECTIVE

From a pragmatic point of view, it matters little whether the holist or the chaoticist holds the proper perspective on ecosystems behavior. (The two are not necessarily mutually exclusive.) Common to both philosophies is the belief that ecosystems modeling, as it has been known for the past few decades, has reached the point of diminishing returns, and that radical progress in the understanding of ecosystems will ensue only by expanding the scales of observation in ecosystems modeling. Conceptually, this means defining new macroscopic variables and deriving meaningful relationships among them.

The search for a macroscopic ecosystems theory is likely to be a long and costly endeavor wrought with much futile effort. Nevertheless, it is one of the most intriguing contemporary issues in basic science. It should command the attention of the best theoretical ecologists and the support of every sponsoring agency.

But how do we begin? In fact, efforts are already underway—it is just that some efforts are not advertised as endeavors in macroscopic ecology and thereby escape our recognition.

A case in point is the long controversy over the relationship (if any) between diversity and stability. These variables are properly macroscopic. Diversity is an ensemble property abstracted in an ad hoc manner from physics and information theory. Much attention has been focused on the ambiguities of defining diversity. Unfortunately, much less effort has been devoted to unambiguously defining stability.

The issue is not always perceived as macroscopic, however. The proper approach would then be phenomenological in nature, with effort aimed at defining stability as a function of diversity in much the same empirical way that engineers once sought to relate the efficiency of engines to the temperature difference driving them. Instead, the literature is replete with attempts to arrive at the relationship deductively from simple model examples, that is, from the lower level in the hierarchy (see May, 1973).

Other efforts are appropriately empirical, but are so presented by conventional modeling philosophy that their potential value to macroscopic ecology is obscured. Thus Bargmann and Halfon (1977), Mobley (1973), and Ulanowicz *et al.* (1978) approach modeling in an *a posteriori* fashion, allowing the data to define interactions. The same methodologies applied to macroscopic variables could prove to be very useful tools.

One empirical approach is intentionally divorced from the mainstream of ecological modeling. Platt and Denman (1975) have advocated spectral analysis as a useful way of presenting time series data so as to evoke new hypotheses. Their strategy is to relate the Fourier spectrum of one component to that of another. In this vein Platt (1972) and Powell *et al.* (1975) compare the spectra of chlorophyll abundance with that of turbulent water motion to define regions of frequency space where the profiles are closely related or significantly different. One can envision the result of such analysis as an empirical correlation between two variables over segments of their spectra. This is in contrast to the cause–effect relation which conventional modeling assumes valid for any continuous time scale.

Platt and Denman emphasize a tendency of nonlinear systems towards "periodic (cyclic) organization in time, in space, or in both." They further argue that characteristic periods should emerge from many-species nonlinear ensembles such as ecosystems. Their speculations on this point stand in contradiction to the intuition of May and Oster. Therefore, further work with spectral analysis of whole ecosystems may help to resolve whose notion is in closer agreement with biological reality.

Of course, the argument for a macrobiology is not new. It arises by analogy to the relationship between statistical mechanics and classical thermodynamics (e.g., Kerner, 1971). It seems, therefore, that the tools of statistical thermodynamics would be the proper instruments with which to begin the development of a macrobiological theory.

Unfortunately, there seems to be little inclination among the ecological community to pursue this line of investigation (the sole exception being diversity indices which come to ecology from statistical mechanics via the intermediate discipline of information theory). The reason for this reticence seems to be twofold.

First, there exists the usual communications barrier between the physicist and ecologist, promoted by the reluctance of most of the parties to spend a significant period of study in the other's discipline.

More easily remedied, however, is the desire of both parties to draw the analogy too closely. Kerner's (1957) work provides a good example.

Without going into detail, Kerner began with a set of generalized Lotka–Volterra predator–prey differential equations written for many species. Under the assumption of antisymmetry of the interaction terms, one may derive a constant of motion for the system, invoke the ergodic hypothesis, and thereby define macroscopic variables such as ecotemperature.

In the eyes of Kerner and his critics the validity of the concept of ecotemperature rests upon the strengths or weaknesses of the derivation.

But this is drawing the analogy with statistical mechanics too tightly. After all, the laws of mechanics and thermodynamics are quite precisely defined. In ecology the models for population dynamics are usually analogies in themselves and only the vaguest notions exist for what macrobiological laws might be. Therefore, it appears premature to argue whether the ergodic hypothesis is applicable to ecosystems, when the entire argument is to connect two realms whose underpinnings are quite uncertain. Rather, Kerner's arguments can be regarded as heuristic in nature and the variable ecotemperature can be subjected to empirical scrutiny in its own right. Such an ad hoc adaptation of a variable is, after all, what occurred with species diversity.

In qualitative terms, the ecotemperature of a species is the expectation value of the square of the deviation of a species from its mean, divided by the mean. When expressed in energetic terms, the "temperature" of lower trophic species is probably greater than that of more predatory species. The difference in ecotemperature, therefore, suggests itself as a candidate for the force conjugate to the energetic flow between compartments (Ulanowicz, 1972). Of course, temperature and energy flow immediately suggest an analog to entropy and an inverse to Odum's (personal communication) much sought after "energy quality."

"Ecopressure" as a quantity which is equal throughout an ecosystem at steady state could likely be heuristically derived. The analogs from classical and irreversible thermodynamics are manifold and exciting. However, caution should be exercised so as to not blind oneself to any new phenomena peculiar to the thermodynamics of macrobiological systems. Phenomenology, not strict analogy, is what is necessary.

4. SUMMARY

In the relative inability of ecological models to provide a degree of robust prediction, ecosystems analysis has encountered its counterpart to the "Ultraviolet Catastrophe" of physics. Just as the difficulties posed by the breakdown of previous theories led to the magnificent advances of quantum physics, the search for alternate strategies of ecosystems analysis holds promise for a major breakthrough in the understanding of ecosystems function. Dilemma and chaos usually force a new perspective, and a different perspective is necessary if one is to view new wonders.

ACKNOWLEDGMENTS

I would like to thank Edward Kerner, Kenneth Mann, Trevor Platt, and Jim Yorke for reading the draft manuscript and offering their comments and encouragement.

Contribution No. 789, Center for Environmental and Estuarine Studies of the University of Maryland.

REFERENCES

Bargmann, R. E., and Halfon, E. (1977). Efficient algorithms for statistical estimation in compartmental analysis: Modelling ^{60}Co kinetics in an aquatic microcosm. *Ecol. Modell.* **3**, 211–226.

Beddington, J. R., Free, C.A., and Lawton, J. H. (1975). Dynamic complexity in predator-prey models framed in difference equations. *Nature (London)* **255**, 58–60.

Cohen, J. E. (1976). Irreproducible results and the breeding of pigs. *BioScience* **26**, 391–394.

Dale, M. B. (1970). Systems analysis and ecology. *Ecology* **51**, 2–16.

Hassell, M. P., and Comins, H. N. (1976). Discrete time models for two-species competition. *Theor. Pop. Biol.* **9**, 202–221.

Hassell, M. P., Lawton, J. H., and May, R. M. (1976). Patterns of dynamical behavior in single-species populations. *J. Anim. Ecol.* **45**, 471–486.

Kerner, E. H. (1957). A statistical mechanics of interacting biological species. *Bull. Math. Biophys.* **19**, 121–146.

Kerner, E. H. (1971). Statistical-mechanical theories in biology. *Adv. Chem. Phys.* **19**, 325–352.

Li, T.-Y., and Yorke, J. A. (1975). Period three implies chaos. *Am. Math. Mon.* **82**, 985–992.

Lorenz, E. N. (1963). Deterministic nonperiodic flow. *J. Atmos. Sci.* **20**, 130–141.

Mann, K. H. (1975). Relationship between morphometry and biological functioning in three coastal inlets of Nova Scotia. *In* "Estuarine Research" (L. E. Cronin, ed.), Vol. 1, pp. 634–644. Academic Press, New York.

May, R. M. (1973). "Stability and Complexity in Model Ecosystems." Princeton Univ. Press, Princeton, New Jersey.

May, R. M. (1974). Biological populations with nonoverlapping generations: Stable points, stable cycles, and chaos. *Science* **186**, 645–647.

May, R. M. (1975). Biological populations obeying difference equations: Stable points, stable cycles, and chaos. *J. Theor. Biol.* **51**, 511–525.

May, R. M., and Oster, G. F. (1976). Bifurcations and dynamic complexity in simple ecological models. *Am. Nat.* **110**, 573–599.

Mobley, C. D. (1973). A systematic approach to ecosystems analysis. *J. Theor. Biol.* **42**, 119–136.

Nicholson, A. J. (1954). An outline of the dynamics of animal populations. *Aust. J. Zool.* **2**, 9–65.

Patten, B. C. (1975). Ecosystem linearization: An evolutionary design problem. *Am. Nat.* **109**, 529–539.

Platt, T. (1972). Local phytoplankton abundance and turbulence. *Deep-Sea Res.* **19**, 183–187.

Platt, T., and Denman, K. L. (1975). Spectral analysis in ecology. *Annu. Rev. Ecol. Syst.* **6**, 189–210.

Powell, T. M., Richerson, P. R., Dillon, T. M., Agee, B. A., Dozier, B. J., Godden, D. A., and Myrup, L. O. (1975). Spatial scales of current speed and phytoplankton biomass fluctuations in Lake Tahoe. *Science* **189**, 1088–1090.

Tuck, J. L., and Menzel, M. T. (1972). The superperiod of the nonlinear weighted string (FPU) problem. *Adv. Math.* **9**, 399–407.
Ulam, S. M. (1963). Some properties of certain non-linear transformations. *In* "Mathematical Models in Physical Sciences" (S. Drobot and P. Viebock, eds.), pp. 85–95. Prentice-Hall, Englewood Cliffs, New Jersey.
Ulanowicz, R. E. (1972). Mass and energy transfer in closed ecosystems. *J. Theor. Biol.* **34**, 239–253.
Ulanowicz, R. E., Flemer, D. A., Heinle, D. R., and Huff, R. T. (1978). An empirical model of an estuarine ecosystem. *Ecol. Modell.* **4**, 29–40.

Chapter 5

HIERARCHICAL ORGANIZATION OF ECOSYSTEMS

Jackson R. Webster

1. Introduction 119
2. Definition of Hierarchy 120
3. The Hierarchical Organization of Nature........... 122
   3.1. Structure of the Levels of Organization Hierarchy... 122
   3.2. Function of the Levels of Organization Hierarchy... 123
4. Hierarchical Levels of Ecological Interest........... 125
5. Application of Hierarchy Theory to Ecosystems....... 126
   References............................ 128

1. INTRODUCTION

In a recent commentary, Guttman (1976) stated that the levels of organization concept of biological organization "if stated in any but the sloppiest and most general terms... is a useless and misleading concept." This conviction contrasts with the observations of other scientists. Weiss (1969) stated, "...the principle of hierarchic order in living nature reveals itself as a demonstrable descriptive fact." Von Bertalanffy (1968) observed, "Such hierarchical structure... is characteristic of reality as a whole and of fundamental importance especially in biology, psychology, and sociology."

The usefulness of the levels or hierarchy concept in biology is in pedagogical organization of biological knowledge and identification of interests and training of biologists. A hierarchical perspective also provides a mechanism for interaction among scientists with interests at different hierarchical levels. The vagueness of biological hierarchies stems in part from imprecise terminology and careless analogies (Bossort et al., 1977). But the essence of hierarchical organization is a vagueness and loss of detail in proceeding from one level to a higher level. As recognized by Poincaré (1905), understanding and explanation in science involves generalization and simplification, hence the emergence of hierarchical levels in our object of study. Whether Nature is truly organized hierarchically is moot. Man's perception of nature is hierarchical.

Hierarchy theory with its coupled debate of reductionism versus holism has been covered in several recent books (Koestler and Smythies, 1969; Whyte et al., 1969; Pattee, 1973; Ayala and Dobzhansky, 1974). In application to biological organization, these books have concentrated on the levels from organism down. Using ideas developed in these four books and elsewhere, I have focused this chapter on ecological levels of organization; organism, population, community, and ecosystem.

2. DEFINITION OF HIERARCHY

A hierarchy is formed by a partial ordering of a set (Simon, 1973). That is, a hierarchical organization of a set, U, with subsets A, B, C, ..., is formed by ordering the subsets by a relation, R, which specifies that the elements of A are higher than the elements of B which are higher than the elements of C.... The relation R is a binary relation in U such that:

(1) b higher than c and c higher than b, where b and c are elements of U, implies $b = c$ (R is antisymmetric); and

(2) b higher than c and c higher than d, where b, c, and d are elements of U, implies b higher than d (R is transitive) (Bunge, 1969).

There are many relations which satisfy these criteria. In the strictest use of the term hierarchy, R would be a ranking or dominance relationship. As an example, an army (U) is composed of personnel (the elements of U) which are ranked (R) into a hierarchy: generals...lieutenants... sergeants...privates. R thus divides the set into levels.

Simon (1973) used a Chinese box to exemplify a hierarchy. The boxes, large and small, are the elements of U. R implies containment: A box is higher than another box if it contains that box.

In each of the previous hierarchy examples, the elements of U were independent except through their relation defined by R, and all the elements of U were similar, all soldiers or all boxes. In many interesting examples of hierarchies, this is not so. In a structural hierarchy each element of U is composed of other elements of U. A book is a good example. A book is made up of chapters, pages, paragraphs, sentences, words, and letters. Each chapter, page, etc., may be considered an element of U, the set of all things which are part of the book. R may be interpreted as "consists of." A book (level 1) consists of chapters (level 2), chapters consist of pages, and so forth. R is antisymmetric—a sentence does not consist of paragraphs, and transitive—a book consists of letters.

Two important properties of structural hierarchies must be noted. First, the entire structure, the book, exists completely at all levels. The levels represent only different perspectives on the book. Consider the different ways in which a reader, publisher, printer, copy editor, and typesetter view a book. Each sees the book at a different level, yet each ultimately sees the entire book.

Second, Guttman (1976) would have that elements at each level consist entirely and exclusively of elements of the next lower level. This may be true for some structural hierarchies, but for generality it is unnecessary. Taken out of the abstract, a page of a book consists not only of words, or sentences, or paragraphs, but also of the paper on which they are printed. More restrictive hierarchies also exist, for example, the hierarchy of taxonomic categories. A class consists of orders; not some orders, some families, and a few other nondescript groups. Even if a class has only a single species, there are still appropriate family and order names into which the species is categorized.

Proceeding up this hierarchy of hierarchy examples (R might be defined as "is more complex than"), the elements of U need not be physical objects. As in systems theory (e.g., as described by Caswell *et al.*, 1972), an abstract object may be defined by its dynamic behaviors. A set of such abstract objects may then form a hierarchy by a relation defined on the behaviors. History exemplifies such a dynamic hierarchy in which R is based on behavioral frequencies. If we are interested in a very brief period of history, we might concentrate on day-to-day occurrences. However, in a comparison of political administrations, such day-to-day happenings would be glossed over in monthly or yearly generalities. At higher levels even these behaviors would be lost, such as in comparison of the Greek and Roman Empires. At a level far above this is the perspective of H. G. Wells' "The Outline of History." Each perspective represents a viable historical endeavor; however, the hostility with which "The Outline of History" was received by some historical specialists (Toynbee, 1935, in Iberall, 1972) is analogous to similar hostility in the natural sciences (Iberall, 1972).

3. THE HIERARCHICAL ORGANIZATION OF NATURE

Odum (1959) visualized the biological spectrum as protoplasm, cells, tissues, organs, organ systems, organisms, populations, communities, ecosystems, and biosphere.[1] This spectrum can be extended down into the nonbiological realm: macromolecules, molecules, atoms, subatomic particles, and upward to astronomical proportions. This hierarchy is an ordering of the set of all natural (i.e., not man-created) systems. The ordering relation is both "consists of" and "behaves at a lower frequency than." This hierarchy, known as levels of organization or levels of integration hierarchy, is both a structural and dynamic hierarchy. Each element, that is, each natural system, consists of systems of the next lower level and is characterized by behaviors occurring more slowly than behaviors at the next lower level. Behaviors of atoms and subatomic particles occur in fractions of milliseconds. Organismic behaviors occur over hours, days, and years. Ecosystem behaviors occur over much longer time periods, hundreds of years, possibly even hundreds of millions of years.

The levels of organization hierarchy was recognized by Aristotle: "Nature proceeds little by little from things lifeless to animal life in such a way that it is impossible to determine the exact line of demarcation, nor on which side thereof an intermediate from should lie" ("Historia Animalium," Book VIII, Chapter 1). However, only relatively recently has there been any formal treatment of structural and functional aspects of the levels of organization hierarchy.

3.1. Structure of the Levels of Organization Hierarchy

Simon (1962, 1973) described the structure of this hierarchy as having both a vertical separation that isolates each level from levels above and below, and a horizontal separation that segregates the components of any level into groups, thus defining the level above. Vertical separation is based on behavioral frequencies. If we focus on a single level of this hierarchy, higher level behavior occurs so slowly that it is perceived as constant. Lower level behavior occurs so rapidly that all we observe is a sampled statistical behavior. Unfortunately, we cannot observe this hierarchy from outside. We human organisms are ourselves part of the levels of organization hierarchy. We exist at the organismal level, and in our attempts to perceive various levels, we cannot extricate ourselves from our level. Unaided, we see only a narrow band of frequencies, from seconds up to a human lifespan, and a narrow band of structures. To look either up or

---

[1] As discussed below, there is disagreement over these levels.

down we must use tools. Examples of tools which allow us to look downward in space/time are microscopes, which give a finer spatial resolution, and chromatography which allows us to separate compounds based on molecular velocities. Looking upward from the organismic level requires other kinds of tools, for example, H. T. Odum's "macroscope" (1971). For spatial phenomena we have aerial photographs and LANDSAT imagery. Earth pictures taken from the moon present a striking new perspective of the biosphere. For examining higher level dynamic behaviors, we have historical records, sediment analysis, and fossil records.

The horizontal structure of the levels of organization hierarchy depends on the isolation of the systems making up any level and upon their segregation into groups which form the systems of the next higher level. A system is a set of interacting objects (in this case the objects are lower level systems). Since all but abstract systems are open systems, interacting with "outside" objects, a specific system exists only by definition. One way of isolating one system from another is by the degree of interaction. As Simon (1973) describes the situation, "Everything in the world is connected with everything else..., but some things are more connected than others." Thus at the atomic level, all atoms interact, at least indirectly, but the atoms that make up specific molecules interact more strongly with each other than they do with atoms of other molecules. The integrity of a system exists by its high degree of internal interaction. Each system is connected with other systems by the weaker connections between component objects. These systems exhibit a "loose horizontal coupling" (Simon, 1973). But the loose or weak connections are the object interactions that identify the next higher level of the hierarchy. This reveals a third ordering relation for the levels of organization hierarchy, bond strength. Nuclear particles are held together by pion fields with energies of the order of 140 MeV. Atomic bonds are much weaker. Covalent bonds between molecules involve energies of only on the order of 5 eV (Simon, 1973).

3.2. Function of the Levels of Organization Hierarchy

It is essential to recognize that from whatever level we view nature, we see exactly the same system. The observed behaviors are manifestations of behaviors at other levels. Various philosophies of science have related different level behaviors in different ways. From a strict holistic philosophy, a behavior at one level cannot be explained in terms of lower level behaviors. The higher level behavior results from a synergism. It is something more than the sum of the lower level parts.

Antithetical to this philosophy is reductionism. From a reductionistic viewpoint, higher level behavior is nothing more than a definable combination of lower level behaviors. The dichotomy of these philosophies developed during the nineteenth century and has extended into the twentieth century. For certain emergent behaviors, life and consciousness, the dichotomy may never be resolved, as neither holism nor reductionism is provable (Platt, 1969a). However, in general, a midground between mystic holism and Laplacian reductionism has been described (Novikoff, 1945; Muller, 1958; Koestler, 1967, 1969; Weiss, 1969). At each level, there are behaviors and laws relative to these behaviors which are specific to that level and not defined for other levels. These behaviors cannot be predicted from the laws of lower levels, but they are entirely consistent with lower level laws.

The best example of this relationship was presented by Rosen (1969). A gas can be considered at two separate levels, macroscopically as a continuous fluid and microscopically as a large number of individual particles. At the macroscopic level, the gas laws are applicable. The relevant behaviors are temperature, pressure, and volume. At the microscopic level, Newton's laws of motion apply. The behaviors are position and velocity. Can the transition from Newton's laws to gas behavior be made? Without invoking Heisenberg's Uncertainty Principle, it is easy to see that simply because of the number of particles involved, the transition is impossible. There is, however, a connection between levels, statistical mechanics. Gas behavior can be related to an average particle behavior.

There is also downward causation (Cambell, 1974) in the levels of organization hierarchy. Higher level organizations place constraints or boundary conditions (Polayni, 1968) on lower level behaviors. An infinite variety of nuclear particle arrangements are possible, yet in only a hundred' or so do the interacting forces produce a stable nucleus. The atoms can be abstractly arranged in an infinity of molecules, but only a relative few actually exist. Only a very few molecular associations manifest attributes that we call life (Muller, 1958). The organization at each level provides a stability to the lower level systems and in so doing constrains the possibilities of lower level behavior. This is not really a downward causation but a selection of persistent organizations. Since multilevel systems cannot form instantly (Gerard, 1969), the lower level of organization must have existed before the higher level. The higher level organization then evolved to include those lower level systems which through their interaction obtained the greatest stability (Levins, 1973).

Within the levels of organization, there are both upward and downward behavioral constraints. The systems at each level conform to lower level laws and through natural selection the behaviors have been constrained by higher level organization. Koestler (1967) proposed the term "holon" for these "Janus-faced" systems. He intended this term to supplant thinking in terms of parts and wholes, by a multilevel, stratified approach to natural science (Koestler, 1969). In the following section, the proposed ecological levels of the levels of organization hierarchy are examined as holons (see also Patten and Finn, Chapter 8, this volume).

4. HIERARCHICAL LEVELS OF ECOLOGICAL INTEREST

The ecological levels of the levels of organization hierarchy are often given as organisms, populations, communities, and ecosystems. For each of these levels to exist according to Simon's (1962, 1973) hierarchy concept, each system at each level must have greater interaction among its subsystems than interaction with other systems. That is, each system is defined by strong internal interaction and weak external interaction. Certainly there are more and stronger interactions within an organism than between organisms. Similarly, at the ecosystem level there are more interactions within an ecosystem, for example, a lake or a forest, than between ecosystems, such as interactions across the land–water interface. However, this system identification may not hold for populations and communities. Phytophagous insects of a forest canopy insect community do not interact more strongly with each other than with the forest tree community. Organisms of a phytoplankton community do not interact more among themselves than with the abiotic and zooplankton components of an aquatic ecosystem. Or, using community in the broader sense as all the organisms of a defined area, interactions among the organisms of a forest community are no stronger than their interactions with the nonliving parts of the forest. A community is not a subsystem but a conceptual part of an ecosystem (Schultz, 1967). In the third edition of his ecology textbook, E. P. Odum (1971) revised the ecological levels to organism systems, population systems, and community systems, that is, ecosystems. The ecosystem concept is based on recognition of strong interactions between living organisms and their abiotic environment. Artificial separation of the living components from an ecosystem is seldom useful, and in practice, seldom considered.

Identification of populations as subsystems of an ecosystem is also questionable. Do the organisms comprising a population interact more strongly intraspecifically or interspecifically? That is, are interactions among members of a population stronger than interspecific competition and trophic relations? Aristotle recognized that the life of animals may be divided into two acts—procreation and feeding ("Historia animalium," Book VIII, Chapter I). Recognition of one or the other of these two activities as stronger represents a schism among ecologists. Evolution depends on procreation; ecological function depends on trophic interactions. We might in fact think of two intertwined hierarchies with the ecological hierarchy in a sense cutting across the genetic hierarchy (Wright, 1959). Although much has been written to bridge the schism between evolution and ecology, or perhaps evolutionary ecology and functional ecology, the division still remains with many unanswered questions (Hutchinson, 1965; Smith, 1975).

From a functional standpoint, ecosystems are comprised of interacting organisms. Rowe (1961), from a different viewpoint, recognized that populations and communities are not part of the levels of organization hierarchy. He suggested that components at each level must have physical boundaries. For example, individual organisms and specific ecosystems can be identified by existent or defined boundaries. This is not true of populations or communities. One cannot place or define a physical boundary around a population or community and have only that community or population within the boundary. The isolate is certain to also contain some of the abiota of the ecosystem. In practice, Rowe's physical boundaries often coincide with minimum interaction surfaces. Wilson (1969) pointed out that natural boundaries may be recognized by either minimum interaction or some form of closure, either topological or temporal. The most useful boundaries for scientific study are those which coincide with boundaries for other properties (Platt, 1969b). For example, a watershed divide is not only a physical boundary but also a minimum interaction surface—water on one side of the divide does not interact with water on the other side.

Are there, then, any levels between organism and ecosystem? Trophic levels (Lindeman, 1942), functional groups (Cummins, 1974; Botkin, 1975), and guilds (Root, 1973, 1975) fail to fill this void for the same reasons that populations and communities do. The major interactions are between groups rather than within. Rowe (1961) found no intermediate levels, only larger and smaller ecosystems. The only satisfactory subdivision of an ecosystem is into smaller physical subunits which exhibit a high level of internal interaction. A forest might be divided into canopy, forest floor, and soil subsystems; a lake might be divided into littoral, pelagic, and benthic subsystems. Within these systems the interacting subsystems must be recognized as organisms.

5. APPLICATION OF HIERARCHY THEORY TO ECOSYSTEMS

MacFadyen (1975) suggested that the most obvious problems in ecology arise because ecology is concerned with at least two levels of organization. This should be turned into a strength rather than a weakness. All natural investigations should proceed at at least two levels (Feibleman, 1954; Bartholomew, 1964; Schultz, 1967). Behavior at any level is explained in terms of the level below, and its significance is found in the level above. Ecosystem behavior can be explained in terms of organism behavior. The significance of organism behavior can then be found in ecosystem behavior. But, as elaborated above, ecosystem behavior cannot be predicted from the laws of organism behavior. This is a philosophical mistake made in many ecosystem models. We cannot expect ecosystem behavior to emerge from a set of organismic equations. Fortunately the mistake is not made in practice. No one has attempted to model an ecosystem by writing a differential equation for every organism in an ecosystem. Populations, trophic levels, and functional groups or guilds are used as concepts for lumping organisms with similar traits, that is, for defining average organism behavior. Overton (1975; White and Overton, 1974) has been more specific in applying hierarchical structure to his modeling "paradigm." Behavior at a mechanistic level is translated into behavior at a higher level by a "ghost" module in the model structure.

Advances in ecosystem ecology must proceed first from an understanding of ecosystem level behaviors and laws. Next comes specification of organism-level dynamics and finally identification of the statistical formalism connecting the two. In biology, and especially ecology, this order, from higher level behavior to lower level behavior, and then the connection between the two, has been reversed (Rosen, 1969). Poincaré (1905) suggested that physical chemistry at that time was hampered by a general grasp of third and fourth decimal places. Perhaps the same is now true in ecology. We know so much about organism behavior that we have difficulty finding the larger regularities. We must search for overriding simplicity in the large-scale complexity (Odum, 1977). The start of this search is the definition of ecosystem behaviors, as we have few such concepts (Overton, 1975). Schultz (1967) listed six ecosystem properties: productivity, stability, cyclicity, diversity, trophic structure, and entropy. Odum (1977) called for measurement of ecosystem level properties in impact evaluation, but gave only two examples, P/R ratios and diversity. However, in an earlier paper (Odum, 1969) he listed 24 ecosystem attributes which deserve consideration. In our examination of these properties, our search for other properties, and our expressions of rules governing these properties (hypotheses), we should look for simplicity, not a rigorous or exact simplicity, but an approximate simplicity (Poincaré, 1905).

ACKNOWLEDGMENTS

Dr. B. C. Patten read an early version of this paper and provided many helpful suggestions.

REFERENCES

Ayala, F. J., and Dobzhansky, T., eds. (1974). "Studies in the Philosophy of Biology." Univ. of California Press, Berkeley.
Bartholomew, G. A. (1964). The roles of physiology and behavior in maintenance of homeostasis in the desert environment. *Symp. Soc. Exp. Biol.* **18**, 7–29.
Bossort, A. K., Jasieniuk, M. A., and Johnson, E. A. (1977). Levels of organization. *BioScience* **27**, 82.
Botkin, D. B. (1975). Functional groups of organisms in model ecosystems. *In* "Ecosystem Analysis and Prediction" (S. A. Levin, ed.), pp. 98–102. SIAM, Philadelphia, Pennsylvania.
Bunge, M. (1969). The metaphysics, epistemology and methodology of levels. *In* "Hierarchical Structures" (L. L. Whyte, A. G. Wilson, and D. Wilson, eds.), pp. 17–26. Am. Elsevier, New York.
Cambell, D. T. (1974). "Downward Causation" in hierarchically organized biological systems. *In* "Studies in the Philosophy of Biology" (F. J. Ayala and T. Dobzhansky, eds.), pp. 179–186. Univ. of California Press, Berkeley.
Caswell, H., Koenig, H. E., Resh, J. A., and Ross, Q. E. (1972). An introduction to systems science for ecologists. *In* "Systems Analysis and Simulation in Ecology" (B. C. Patten, ed.), Vol. 2, pp. 3–78. Academic Press, New York.
Cummins, K. W. (1974). Structure and function of stream ecosystems. *BioScience* **24**, 631–641.
Feibleman, J. K. (1954). Theory of integrative levels. *Br. J. Philos. Sci.* **5**, 59–66.
Gerard, R. W. (1969). Hierarchy, entitation, and levels. *In* "Hierarchical Structures" (L. L. Whyte, A. G. Wilson, and D. Wilson, eds.), pp. 215–228. Am. Elsevier, New York.
Guttman, B. S. (1976). Is "Levels of Organization" a useful biological concept? *BioScience* **26**, 112–113.
Hutchinson, G. E. (1965). "The Ecological Theater and the Evolutionary Play." Yale Univ. Press, New Haven, Connecticut.
Iberall, A. S. (1972). "Toward a General Science of Viable Systems." McGraw-Hill, New York.
Koestler, A. (1967). "The Ghost in the Machine." Macmillan, New York.
Koestler, A. (1969). Beyond atomism and holism—the concept of the holon. *In* "Beyond Reductionism" (A. Koestler and J. R. Smythies, eds.), pp. 192–232. Hutchinson, London.
Koestler, A., and Smythies, J. R., eds. (1969). "Beyond Reductionism." Hutchinson, London.
Levins, R. (1973). The limits of complexity. *In* "Hierarchy Theory" (H. H. Pattee, ed.), pp. 109–127. George Braziller, New York.
Lindeman, R. L. (1942). The trophic-dynamic aspect of ecology. *Ecology* **23**, 399–418.
MacFadyen, A. (1975). Some thoughts on the behavior of ecologists. *J. Ecol.* **63**, 379–391.
Muller, C. H. (1958). Science and philosophy of the community concept. *Am. Sci.* **46**, 294–308.
Novikoff, A. B. (1945). The concept of integrative levels and biology. *Science* **101**, 209–215.
Odum, E. P. (1959). "Fundamentals of Ecology," 2nd ed. Saunders, Philadelphia, Pennsylvania.
Odum, E. P. (1969). The strategy of ecosystem development. *Science* **164**, 262–276.
Odum, E. P. (1971). "Fundamentals of Ecology," 3rd ed. Saunders, Philadelphia, Pennsylvania.

Odum, E. P. (1977). The emergence of ecology as a new integrative discipline. *Science* **195**, 1289–1292.
Odum, H. T. (1971). "Environment, Power and Society." Wiley (Interscience), New York.
Overton, W. S. (1975). An ecosystem modelling approach in the Coniferous Biome. *In* "Systems Analysis and Simulation in Ecology" (B. C. Patten, ed.), Vol. 3, pp. 117–138. Academic Press, New York.
Pattee, H. H., ed. (1973). "Hierarchy Theory." George Braziller, New York.
Platt, J. R. (1969a). Commentary on the limits of reductionism. Part I. *J. Hist. Biol.* **2**, 140–147.
Platt, J. R. (1969b). Theories on boundaries in hierarchical systems. *In* "Hierarchical Structures" (L. L. Whyte, A. G. Wilson, and D. Wilson, eds.), pp. 201–213. Am. Elsevier, New York.
Poincaré, H. (1905). "Science and Hypothesis." Scott, London.
Polayni, M. (1968). Life's irreducible structure. *Science* **160**, 1308–1312.
Root, R. B. (1973). Organization of a plant-arthropod association in simple and diverse habitats: The fauna of collards (*Brassica oleracea*). *Ecol. Monogr.* **43**, 95–124.
Root, R. B. (1975). Some consequences of ecosystem texture. *In* "Ecosystem Analysis and Prediction" (S. A. Levin, ed.), pp. 83–92. SIAM, Philadelphia, Pennsylvania.
Rosen, R. (1969). Hierarchical organization in automata theoretic models of biological systems. *In* "Hierarchical Structures" (L. L. Whyte, A. G. Wilson, and D. Wilson, eds.), pp. 179–199. Am. Elsevier, New York.
Rowe, J. S. (1961). The level of integration concept and ecology. *Ecology* **42**, 420–427.
Schultz, A. M. (1967). The ecosystem as a conceptual tool in the management of natural resources. *In* "Natural Resources: Quality and Quantity," pp. 139–161. Univ. of California Press, Berkeley.
Simon, H. A. (1962). The architecture of complexity. *Proc. Am. Philos. Soc.* **106**, 467–482.
Simon, H. A. (1973). The organization of complex systems. *In* "Hierarchy Theory" (H. H. Pattee, ed.), pp. 3–27. George Braziller, New York.
Smith, F. E. (1975). Ecosystems and evolution. *Bull. Ecol. Soc. Am.* **56**, 2–6.
Toynbee, A. J. (1935). "A Study of History," Vol. I. University Press, London.
von Bertalanffy, L. (1968). "General Systems Theory." George Braziller, New York.
Weiss, P. A. (1969). The living system: Determinism stratified. *In* "Beyond Reductionism" (A. Koestler and J. R. Smythies, eds.), pp. 3–55. Hutchinson, London.
White, C., and Overton, W. S. (1974). "Users Manual for FLEX 2 and FLEX 3 Model Processors," Bull. No. 15, review draft. For. Res. Lab., Oregon State University, Corvallis.
Whyte, L. L., Wilson, A. G., and Wilson, D., eds. (1969). "Hierarchical Structures." Am. Elsevier, New York.
Wilson, A. (1969). Closure, entity and level. *In* "Hierarchical Structures" (L. L. Whyte, A. G. Wilson, and D. Wilson, eds.), pp. 54–55. Am. Elsevier, New York.
Wright, S. (1959). Genetics and the hierarchy of biological sciences. *Science* **130**, 959–965.

Chapter **6**

HIERARCHICAL ADAPTABILITY THEORY AND ITS CROSS-CORRELATION WITH DYNAMICAL ECOLOGICAL MODELS

Michael Conrad

1. Introduction	132
2. Review of Adaptability Theory	133
2.1 Measures of Adaptability	133
2.2 Relation between Adaptability and Environmental Uncertainty	135
2.3 Connection to Ecological Variables	136
3. Hierarchical Adaptability Theory	137
3.1 The Decomposition Scheme	137
3.2 Dependencies among Different Levels of Organization	139
3.3 Self-Justifiability of Hierarchical and Compartmental Descriptions	140
3.4 Dynamical Independence of Levels	141
4. Cross-Correlation with Dynamical Ecological Models	143
4.1 Dynamical Notions of Stability	143
4.2 Interpretation of Dynamical Models (Simplest Example)	145
4.3 Relevance to Construction of Dynamical Models	147
5. Conclusion	148
References	149

1. INTRODUCTION

Many present-day models of ecological systems are dynamical in nature, viz., based on differential equations of the form

$$\dot{x}_i = f(x_1, \ldots, x_n), \qquad i = 1, \ldots, n, \tag{1}$$

where the x_i are generally associated with species or different forms of a given species, but may be associated with more lumped units (such as patches or trophic groups). For such systems a key issue is stability (or, more precisely, the formulation and applicability of various stability concepts). The basic consideration is that a system incapable of either absorbing or absorbing and then dissipating disturbances will change and change again until it assumes a form which at least appears to forget these disturbances. If it assumes such a persistent form it will, by definition, stay in it much longer than any of the impersistent forms and in this sense it is the persistent forms which "earn the biggest share of the time pie." Indeed the process of ecological succession can be viewed in terms of such a sequence of changes to more and more persistent forms of organization, that is, forms more and more capable of absorbing and forgetting disturbances, either of external or internal origin.

The problem of the persistent functioning of complex biological systems in the face of environmental disturbance is fruitfully treated by an alternative method of analysis. This method (which might be called adaptability analysis) is fundamentally statistical in nature, selecting as its fundamental object of study the hierarchy of mechanisms and modes of adaptability in an ecosystem and seeking to specify relations between environmental uncertainty and the behavioral uncertainties at different levels in the hierarchy, the ability to anticipate the environment, and the indifference to the environment. An advantage of the method is that it is possible to manage hierarchical structure (e.g., community, population, organism, genome) in a natural way and therefore to directly specify the sources of absorption and dissipation of disturbance. Adaptability analysis is complementary to dynamical styles of analysis in the sense that what is sacrificed in detailed trajectorial description it provides in terms of the description of the regulative processes which make possible any given trajectorial behavior.

This chapter begins (Sections 2, 3.1, 3.2) with a review of relevant aspects of adaptability theory and its hierarchical formulation (but without considering informational processes which underlie anticipation of the environment, cf. Conrad, 1977a). Attention is then focused on the justifiability of community descriptions structured in terms of level (or compartmental) organization and on the conditions for dynamical independence of levels. This makes it possible to cross-correlate basic elements of the adaptability theory framework with concepts deriving from the dynamical systems approach, and also suggests a number of considerations relevant to the interpretation and construction of dynamical ecological models. A peculiar but only apparently paradoxical consequence of the analysis is that the sources of absorption and dissipation of disturbance (i.e., modes of adaptability) most important for supporting the dynamics at any given level are least explicit in the dynamical description of this level.

2. REVIEW OF ADAPTABILITY THEORY

2.1. Measures of Adaptability

Within the framework of the theory of adaptability reviewed here adaptability is defined as ability to continue to function in the face of environmental uncertainty (cf. Conrad, 1972a,b, 1975, 1976a,b, 1977a,b). This is a stripped-down definition, and has both the advantages and disadvantages of a stripped-down definition. It omits any reference to, for example, nature and intensity of the disturbance or the time scale of the disturbance, or even diversity of the environment (as opposed to its indeterminacy). However, it has the advantages that it can be formed into a definition which is mathematically fruitful and provides a natural framework for the addition of these extra dimensions.

The environment and also the (biotic) community are considered to consist of ensembles of distinguishable states, with the behavior of the former represented by the set of transition probabilities

$$\omega' = \{p[\beta^v(t+\tau)|\alpha^r(t), \beta^s(t)]\}, \qquad \text{(all } r, s, v), \tag{2}$$

where β^v and β^s are states v and s of the environment, α^r is state r of the community, t is the time, and τ is a time interval. The measure of environmental uncertainty is given by

$$H(\omega') = -\sum p[\alpha^r(t), \beta^s(t)]p[\beta^v(t+\tau)|\alpha^r(t), \beta^s(t)] \log p[\beta^v(t+\tau)|\alpha^r(t), \beta^s(t)], \tag{3}$$

where the sum runs over r, s, v. This is sometimes called the entropy of the transition scheme (cf. Khinchin, 1957) and is a measure of the behavioral uncertainty of the environment, that is, its statistical spread of transition probabilities weighted by the probability of the initial state. Again, it should be emphasized that behavioral uncertainty is not the same as diversity. A periodically varying environment might be quite diverse, but nevertheless statistically predictable, and conversely.

The idea of maximum allowable (or maximum tolerable) environmental uncertainty can be determined in a reasonably objective way for individual organisms or populations (using, e.g., death or extinction as the criterion). To obtain an operationally meaningful definition for an entire community requires a somewhat weaker condition. There are two possibilities. The first is to search for the most uncertain environment which does not drive the system to a qualitatively different (e.g., edaphic) climax. This defines the capability of the particular form of the community to absorb or dissipate disturbance, but it of course does not define the full adaptability of the community per se (which may include the possibility for assuming numerous qualitatively distinct forms of organization). There is, however, another approach, which is more practical in the laboratory and would seem to lend itself more naturally to field situations (where the environment is only observable, not manipulable). This is the culture system approach.

The basic idea of the culture system approach (which from the time of Koch has been the fundamental method of microbiology) is that to obtain organisms with particular characteristics one defines an environment in which these characteristics are necessary. While there may be no guarantee that any particular strain will emerge, there is certainly a guarantee that no strain will emerge which is incapable of functioning in the defined environment, and furthermore no strain is likely to emerge which is incapable of functioning competitively in this environment. In the present case the idea is extended to an entire ecosystem (i.e., a multiculture rather than a pure culture), and a defining feature of the environment is its behavioral uncertainty. Such a multiculture system must undergo successive changes in organization until it assumes a form which is stable under these prevailing conditions of uncertainty, therefore a form with a well-defined *minimum* adaptability. Furthermore, since superfluous adaptability (i.e., never used mechanisms of dissipating or absorbing environmental disturbance) inevitably involves costs (ultimately in terms of utilization of energy by organisms in the system) it is reasonable to postulate that there will be a tendency for actual adaptability to decrease in the direction of minimum allowable adaptability. It should be emphasized that this would only be a tendency since some forms of adaptability (e.g., ability to spore or latent variability locked into chromosome inversions) are not energetically very costly and could not be expected to put their carriers at a serious disadvantage, at least over successional time scales. However, it is reasonable to assume that the most expensive forms of adaptability would tend to be dispensed with on successional time scales and therefore that the actual uncertainty of the culturing environment would provide a reasonable measure of the adaptability of any particular community form emerging and stabilizing in this environment. Furthermore, it provides an effective index of community adaptability, insofar as the shift to a new community form entails disruption of community organization and instigates a new successional process.

The culture system approach to adaptability and to the relation between minimum adaptability (or uncertainty of the culturing environment) and the stability of succession is the object of current research in the author's laboratory. This aspect of the problem, however, is outside the scope of the present paper.

2.2. Relation between Adaptability and Environmental Uncertainty

One advantage of the statistical definition is that it allows for an analysis of the relationship between the statistical properties of the environment and the statistical properties of the community (taken to mean all the biotic components of the ecosystem). The transition scheme of the community can be specified by expressions analogous to (2) and the uncertainty of the transition scheme by expressions analogous to (3). It is also possible to define a conditional uncertainty for the behavior of the community given the contemporaneous behavior of the environment, viz.,

$$H(\omega|\omega') = -\sum p[\alpha^r(t), \beta^s(t), \beta^v(t+\tau)] p[\alpha^u(t+\tau)|\alpha^r(t), \beta^s(t), \beta^v(t+\tau)] \\ \cdot \log p[\alpha^u(t+\tau)|\alpha^r(t), \beta^s(t), \beta^v(t+\tau)], \quad (4)$$

where ω is the transition scheme of the community. An analogous conditional uncertainty can be defined for the behavior of the environment given the contemporaneous behavior of the community.

Among these uncertainties obtains a simple identity:

$$H(\omega) - H(\omega|\omega') + H(\omega'|\omega) = H(\omega'). \quad (5)$$

Denoting the transition schemes in the most uncertain tolerable environment by hats, it is possible to write the inequality

$$H(\hat{\omega}) - H(\hat{\omega}|\hat{\omega}') + H(\hat{\omega}'|\hat{\omega}) \geq H(\omega'), \quad (6)$$

where the equality holds when $\omega' = \hat{\omega}'$.

The term $H(\hat{\omega})$ is the potential behavioral uncertainty of the community, i.e., the size of its repertoire of behavioral modes. $H(\hat{\omega}|\hat{\omega}')$ is the potential behavioral uncertainty of the community given the behavior of the environment. It thus decreases as the ability of the community to anticipate the environment increases and increases with internal sources of uncertainty within the community reflecting, e.g., lack of integration among its various components). $H(\hat{\omega}'|\hat{\omega})$ is the potential behavioral uncertainty of the environment given the behavior of the community. Thus it increases with increasing indifference (or insensitivity) to the environment, either arising from physical mechanisms of isolation (e.g., protective structures, restriction of spatial range) or from despecializations which decrease the number of metabolic requirements or increase the number of food sources. However, this term can also reflect pathological behavior, e.g., behavior which, as a consequence of injury, provides no information about the environment.

The entire left-hand side of Eq. (6) is, by definition, the adaptability of the community. Recalling the assumption that excess components of adaptability (i.e., superfluous modes of behavior, better than necessary ability to anticipate the environment, greater than necessary indifference) are always costly, Eq. (6) can be more appropriately written

$$H(\hat{\omega}) - H(\hat{\omega}|\hat{\omega}') + H(\hat{\omega}'|\hat{\omega}) \to H(\omega'), \tag{7}$$

where the arrow expresses the evolutionary tendency for adaptability to decrease to its minimum possible value in the course of evolution or succession. According to Eq. (7) the potential behavioral uncertainty of the community, less this uncertainty as reduced by anticipation, must be greater than the actual uncertainty of the environment, but it may be smaller if the community is selectively indifferent to the environment. If the contribution of any one of these components of adaptability increases (decreases) the tendency would be for this to be compenstated by a decrease (increase) in the contribution of one or both of the other components.

A simplifying assumption in the above discussion is that the community could provide information about the contemporaneous behavior of the environment and conversely. In the strict sense this is clearly impossible, except on a time scale sufficiently coarse to obscure the time delays required for the establishment of correlations. The details of the finer picture (the information transfer picture, cf. Conrad, 1977a) will not be essential for the discussion which follows and will not be reviewed here.

2.3. Connection to Ecological Variables

To this point the only assumption, as far as state specification is concerned, is that there are ensembles of distinguishable states for both community and environment. Now it is necessary to connect these states to variables ordinarily used in ecosystem biology. For the community these variables should include species (with values representing presence or absence), foodweb structure, number of organisms in each species, location of these organisms, physiological state of each organism, pattern of gene activation, and also genotype (DNA base sequence) of each organism. For the environment variables should include macroscopic (i.e., thermodynamic) variables such as temperature, pressure, mole numbers, and mechanical forces, where these are specified for each local region of the environment. Thus the entire geometrical space of the ecosystem may be thought of as covered by a grid and the state of the physical environment as being specified by specifying the macroscopic state of each grid unit (excluding that portion occupied by living matter).

These variables may be incorporated into the formalism by writing the states of the community and environment as many-tuples of the variables which characterize these states (admitting, of course, the assumption that the variables indicated above form a complete set). A convenient way of doing this is in terms of ecosystem hierarchy, i.e., in terms of units at various levels of organization. Thus the community consists of populations, populations of organisms, and organisms consist of a genome and a phenome (which might be further decomposed into organs, cells, etc.). The top level (or partial state) of the community is described in terms of the species composition and the foodweb structure; the populations (or species, as restricted to the environmental space being considered) are described in terms of number and location of organisms, the organisms in terms of pattern of gene activation, the phenome in terms of physiological state, and the genome in terms of base sequence.

The hierarchical (or, more precisely, compartmental) organization of the ecosystem is undoubtedly quite subtle, since it is clear that the degree of isolation and flow of information between levels and across compartments is subtle. However, it can be thought of as a way of structuring descriptions of ecosystems. In practice it corresponds to the way biologists (and biology departments) actually structure their descriptions and later we will see that such structured descriptions are in a sense self-justifying from the standpoint of adaptability theory.

3. HIERARCHICAL ADAPTABILITY THEORY

3.1. The Decomposition Scheme

Following the above idea of a structured description, partial state f of the ith subcompartment at level j is denoted by α_{ij}^f, and the gth state of region h of the environment is denoted by β_{h0}^g. The state of the community can be specified by specifying its set of partial states (e.g., with level 3 for the community, level 2 for the population, level 1 for the organisms, and level 0 for the genome). Also, the state of the environment is given by the set of states of each region, all taken as being at level 0. The transition scheme for a subsystem p at level q is given by

$$\omega_{pq} = \left\{ p \left[\alpha_{pq}^u(t+\tau) \middle| \bigcap_{i,j} \alpha_{ij}^f(t), \bigcap_h \beta_{h0}^g(t) \right] \right\}, \tag{8}$$

where the intersections either run over all subcompartments of the community or over all environmental regions, and similar local transition schemes can be written for regions of the environment. However, it should be noted that, once the compartmental point of view is adopted, it is necessary to use a reference structure to construct the complete state of the community (since with time organisms and other units come into and pass out of existence).

The transition scheme of the entire community, expressed in terms of local schemes is given by

$$\omega = \left\{ p \left[\bigcap_{i,j} \alpha_{ij}^d(t+\tau) \middle| \bigcap_{i,j} \alpha_{ij}^f(t), \bigcap_h \beta_{h0}^g(t) \right] \right\} = \prod_{i,j} \omega_{ij}, \tag{9}$$

where the product symbol should be taken to express a joint probability. Thus, the entropy of the community (and also of the environment) transition scheme can be expressed as a sum of unconditional and conditional local entropies. However, since there are many possible expansions it is desirable to choose a single canonical form. The expansion of choice is the one which puts all subcompartments on an equal footing. To do this, define the effective entropy, $H_e(\omega_{ij})$, as the sum of the unconditional and all possible conditional entropies of the subsystem, but normalized by normalizing the linear combination of all the possible expansions. For example, for $H(\omega_{pq}\omega_{rs})$ the effective entropies are

$$H_e(\omega_{pq}) = \tfrac{1}{2}[H(\omega_{pq}) + H(\omega_{pq}|\omega_{rs})], \tag{10a}$$
$$H_e(\omega_{rs}) = \tfrac{1}{2}[H(\omega_{rs}) + H(\omega_{rs}|\omega_{pq})]. \tag{10b}$$

More generally

$$H\left(\prod_{i,j} \omega_{ij}\right) = \sum_{i,j} H_e(\omega_{ij}) \tag{11}$$

and therefore Eq. (7) can be written

$$\sum_{i,j} H_e(\hat{\omega}_{ij}) - \sum_{i,j} H_e\left(\hat{\omega}_{ij} \middle| \prod_h \hat{\omega}_{h0}'\right) + \sum_h H_e\left(\hat{\omega}_{h0}' \middle| \prod_{i,j} \hat{\omega}_{ij}\right) \to \sum_h H_e(\omega_{h0}'). \tag{12}$$

Equation (12) expresses a principle of compensation among adaptabilities, viz., that change in the adaptability of one subsystem tends to be compensated by opposite changes in the adaptability of other subsystems, at the same or different levels, or by opposite changes in the indifference to the environment.

3.2. Dependencies among Different Levels of Organization

The unconditional parts of the effective entropy (i.e., terms of the type $H(\hat{\omega}_{ij})$) represent the uncertain modifiability of the particular subsystem in question, whereas the conditional parts (terms of the type $H(\hat{\omega}_{ij}|\hat{\omega}_{pq})$, $H(\hat{\omega}_{ij}|\hat{\omega}_{pq}\hat{\omega}_{rs})$, and so on) represent the degree to which modifiability is independent of the behavior of other subsystems (which, depending on the conditioning terms, may be at the same or at different levels of organization). An immediate and fundamental conclusion is that the behavioral repertoire component of total adaptability is not in general equal to the algebraic sum of the uncertainty measures on the behavioral repertoires of the component systems. This could only be the case if the component systems are all independent of one another. The proper statement is thus that *total adaptability is highest when the behavioral repertoires of the component subsystems are most independent and decreases to its minimum value (for given repertoires) when these repertoires are completely correlated.*

On the face of it, it might seem that independence of component systems would be incompatible with compensation since there would be no possibility for rerouting the effects of disturbance. However, it must be remembered that independence is statistically defined, in terms of conditional behavioral uncertainty, and does not imply the nonexistence of coordination between the conditioned system and the system on which it is conditioned. For example, thermoregulative responses (e.g., feather fluffing) may prevent the effects of temperature change from ramifying to the population level (e.g., in terms of changes in population numbers). To the extent that this is the case, physiological modifiability is statistically independent of population modifiability. To the extent that it is not the case, functionally significant modifiability at the population level is not independent of physiological events at the organism level and therefore the additional contribution to total adaptability cannot be commensurate with this modifiability. (It is also possible for events at one of the levels to involve only the most subtle, barely observable changes, but in this case they would appear only in the finer states, either informationally distinct or macroscopically equivalent, of the information transfer picture.)

An analogous consideration extends to the anticipation term. This can also be expressed in two parts, with terms such as $H(\hat{\omega}_{ij}|\prod_{h,0}\hat{\omega}'_{h0})$ determining the extent to which the corresponding modifiability can in fact contribute to adaptability and terms such as $H(\hat{\omega}_{ij}|\hat{\omega}_{pq}\prod_{h,0}\hat{\omega}'_{h0})$ determining the extent to which one subsystem anticipates another. If the uncertainty in the modifiability of a subsystem is high, but not reduced when the environment is specified, then this modifiability makes no contribution to adaptability. Similarly, if the independence of a subsystem from others is high, but not reduced when the environment is specified, then this independence makes no contribution to adaptability. For example, modifiability or independence concomitant to lack of integration among the parts of a system (arising, e.g., in competitive game-playing interactions) certainly does not contribute to adaptability. Such processes would contribute equally to both the behavioral uncertainty and independence terms.

3.3. Self-Justifiability of Hierarchical and Compartmental Descriptions

The informal considerations of the previous section suggest that the assumption of levels of organization (and also of compartmental structure) is in a very definite sense self-justifying. Ecological systems are built (or appear to be built) hierarchically (and compartmentally), since this increases the effective adaptability concomitant to any given amount of total observable structural and functional alterability of the system. There are undoubtedly other reasons for the hierarchical structure of ecological systems (e.g., specialization of labor) but it is interesting to note that within the framework of adaptability theory the hierarchical and compartmental structurability of ecosystem descriptions has the status of a bootstrap principle.

The above consideration is of key importance for the construction and interpretation of ecological models. Thus, it is worthwhile to restate it in more formal terms, at least in terms of one simple, model situation (which could easily be generalized by induction on indices).

Consider a three-level description of a community, i.e., in terms of genome and phenome (level 0), organism (level 1), and population (level 2). For simplicity, assume that the community consists of only one population (thereby eliminating level 3) and also combine the description of the phenome and the organism (which corresponds more closely to the ordinary use of the word phenome than the definition which is most natural in adaptability theory). Just distinguishing levels (i.e., looking only at genomes in the population, all organisms and phenomes, and at the population level itself):

$$H(\hat{\omega}) = H\left(\bigcap_i \hat{\omega}_{i0}, \bigcap_j \hat{\omega}_{j0}, \bigcap_k \hat{\omega}_{k1}, \hat{\omega}_{12}\right), \tag{13}$$

where indices are defined on a reference structure, i runs over genomes (by convention odd numbered), j runs over phenomes (by convention an even number), k runs over all the organism indices, and $\hat{\omega}_{12}$ is the partial transition scheme of the population. Introducing the following definition for simplicity:

$$\bigcap_i \hat{\omega}_{i0} = \hat{\omega}_g, \quad \text{(for gene level)}, \tag{14a}$$

$$\left(\bigcap_j \hat{\omega}_{j0}, \bigcap_k \hat{\omega}_{k1}\right) = \hat{\omega}_0, \text{(for phenome-organism level)}, \tag{14b}$$

$$\hat{\omega}_{12} = \hat{\omega}_p, \quad \text{(for population level)}. \tag{14c}$$

Now, defining effective entropies for these levels

$$H(\hat{\omega}) = H_e(\hat{\omega}_g) + H_e(\hat{\omega}_0) + H_e(\hat{\omega}_p), \tag{15}$$

where

$$H_e(\hat{\omega}_g) = \tfrac{1}{3}\{H(\hat{\omega}_g) + \tfrac{1}{2}H(\hat{\omega}_g|\hat{\omega}_0) + \tfrac{1}{2}H(\hat{\omega}_g|\hat{\omega}_p) + H(\hat{\omega}_g|\hat{\omega}_0\hat{\omega}_p)\}, \tag{16a}$$

$$H_e(\hat{\omega}_0) = \tfrac{1}{3}\{H(\hat{\omega}_0) + \tfrac{1}{2}H(\hat{\omega}_0|\hat{\omega}_g) + \tfrac{1}{2}H(\hat{\omega}_0|\hat{\omega}_p) + H(\hat{\omega}_0|\hat{\omega}_g\hat{\omega}_p)\}, \tag{16b}$$

$$H_e(\hat{\omega}_p) = \tfrac{1}{3}\{H(\hat{\omega}_p) + \tfrac{1}{2}H(\hat{\omega}_p|\hat{\omega}_g) + \tfrac{1}{2}H(\hat{\omega}_p|\hat{\omega}_0) + H(\hat{\omega}_p|\hat{\omega}_g\hat{\omega}_0)\}. \tag{16c}$$

The conditions for a maximum behavioral uncertainty component of adaptability are

$$H(\hat{\omega}_g) = H(\hat{\omega}_g|\hat{\omega}_0) = H(\hat{\omega}_g|\hat{\omega}_p) = H(\hat{\omega}_g|\hat{\omega}_0\hat{\omega}_p) \tag{17}$$

and similarly for all the terms in Eqs. (16b) and (16c). This means (as stated in Section 3.2) that the adaptability is greatest when the uncertainty of behavior at the various levels is the same whether or not the contemporaneous behavior of other levels is specified, i.e., when no information about the level in question is given by the other levels. In general such complete informational isolation may only rarely obtain. However, to the extent that it does, the behavior of the various levels (or other compartments) will appear autonomous and it is just this appearance of autonomy which justifies their being treated as levels (or relatively isolated systems) in the first place. Thus level (and more generally compartmental) structure is a necessary condition for effective adaptability and in this sense the initial assumption of a structured description of ecological systems is self-justifying within the framework of adaptability theory.

3.4. Dynamical Independence of Levels

Suppose that a level (or more generally a compartment) is completely independent of other levels (or compartments), i.e., that the conditions specified in Eq. (17) are realized. In this case no information is provided by specifying the behavior at other levels. Thus the partial transition scheme of the level is as good a predictor of the behavior of that level as the complete biota transition scheme (e.g., in the case of the population level ω_p would be as good a predictor as ω). Under these circumstances there exists a law (or the possibility of formulating a law) governing the behavior of the level which appears autonomous in the sense that only those variables associated with the level (and possibly also environmental variables) appear in the law. To the extent that informational isolation does not hold (e.g., to the extent that Eq. (17) does not hold) this law will become increasingly less effective for prediction. In effect the law will be "broken" and the degree of breaking will depend on the degree of dependence. A broken law might of course be restored to good predictive status by amending it with the addition of variables from other levels, but in this case the integrity of the level (the internal autonomy of its behavior) is violated, with concomitant mathematical complication of the law.

From the above considerations emerges a fundamental but seemingly paradoxical consequence, viz., *that as a predictive law for behavior at any given level requires fewer assumptions about other levels in order to be accurate, the contribution of processes at these other levels becomes more important for understanding how this behavior is possible.* In effect, any compartment or level of the biota, insofar as it appears to be a free and autonomous system, is actually under the influence of other, unviewed, compartments or levels. The reason of course is that as independence increases adaptability increases. This allows either for unviewed components of the system to control the level or conversely. There are four possibilities:

(i) The modifiability term (e.g., $H(\omega_p)$) is small and the independence terms (e.g., $H(\omega_p|\omega_0)$) are comparable to the modifiability. In this case the law governing the level will in general be quite simple— indeed independent of the environment and in the extreme case even constant. As an example, suppose that organismic and genetic adaptabilities absorb or dissipate most environmental disturbance, preventing it from ramifying to the population level. Thus population size may follow a law which incorporates no variables from other levels (or even from the environment), but this is possible only because these other levels are playing an especially important role from the standpoint of adaptability.

(ii) The modifiability term is small and the independence terms are negligible. This is basically the same situation as case (i), except that a more accurate law can be obtained by incorporating variables from other levels. It may be noted that in this case the adaptability must be less effective, for insofar as processes at other levels give information about population size, the latter cannot be restricted to a single trajectory.

(iii) The modifiability term is large and the independence terms comparable to the modifiability. In this case the law governing the level will not require incorporation of variables from other levels, but will in general be better if environmental variables are specified (since in general $H(\omega_p) > H(\omega_p|\omega')$). The law may be either deterministic or stochastic, depending on the actual magnitude of the anticipation term. If the anticipation term is relatively small, processes at this level play an especially important role from the standpoint of other levels, which, however, may appear to have fairly autonomous dynamics. Insofar as the level allows other levels to absorb different classes of disturbances, these other levels also provide an invisible support for its dynamics. An example for this type of situation would be a population in which unexpected variations in energy input were absorbed in population size but other types of variations were absorbed at organismic or, over the longer run, genetic levels.

(iv) The modifiability term is large and the independence terms smaller than the modifiability. In this case an accurate law can only be obtained by incorporating variables from other levels and also (as above) environmental variables. For example, suppose that rapid growth (or culturing) of a microbial population in response to an increase in energy input dilutes out duplicate nuclei. This is a grossly observable physiological modification which is causally correlated to the dynamical behavior of the population and which would almost certainly have to be incorporated into any law accurately predictive of these dynamics, yet from the standpoint of adaptability organismic processes are not playing as efficient a compensating role as in the (imaginary) example of (iii). (Alternatively stated, population size alone is not in this example an adequate state variable for a law governing the dynamics of the population.)

It is important that a single compartment may be independent or dependent and controlled or controlling with respect to certain variables and not others. Also, independence implies informational isolation of levels as regards correlations among functionally distinct states, not informational isolation in the sense of the more fine-grained information transfer picture.

4. CROSS-CORRELATION WITH DYNAMICAL ECOLOGICAL MODELS

4.1. Dynamical Notions of Stability

Now it is possible to cross-correlate elements of adaptability analysis with elements of dynamical modeling, beginning with some preliminary remarks on dynamical concepts of stability and then turning to a simple, classical dynamical model as paradigm.

For a system to appear stable it must be able either to absorb or dissipate disturbances—which is basically the same as the criterion for adequate adaptability. However, from the dynamical standpoint the various notions of stability are usually defined relative to trajectorial behavior (cf. Rosen, 1970). Thus if a system is (asymptotically) orbitally stable it dissipates disturbance, i.e., asymptotically returns to the trajectorial behavior which it would have exhibited had it not been perturbed. If it is weakly stable it absorbs perturbation, i.e., it never returns to the unperturbed trajectorial behavior but never diverges to a greater than prespecified extent from it either. A system is structurally stable if its trajectorial behavior is qualitatively invariant to slight alterations of parameter or equation structure (cf. Thom, 1970; also Güttinger, 1974). The motive is that systems without this property are unlikely to persist and therefore will inevitably change until they assume a persistent form. Some caution is necessary, however, since this argument is of course only good for structural stability to the class of perturbations which actually affect the system. The argument for the importance of orbital stability is that a system will keep changing until it falls into some basin of attraction, in which it will remain until sufficiently perturbed. The assumption, within the framework of any given model, is that there are basins of attraction and that the time spent in them is relatively long in comparison to the time which it takes to reach them. The broader assumption is that the natural system will keep changing until it assumes a structure describable by a model with reachable basins of attraction.

An important point is that orbitally stable systems and in general structurally stable systems are only partial descriptions. The laws of physics are Hamiltonian, i.e., conservative and in detail reversible. It is only when details are ignored that irreversibility obtains. Thus for any real system described by dissipative (gradient-type) dynamics, the dissipated (forgotten) disturbances are disturbances that are exported to some other system (the heat bath) for absorption. This is just a way of saying that the disturbances are eventually dissipated into an ensemble of microscopic states (of the surroundings) equivalent from the macroscopic point of view. Systems which are structurally stable to perturbation of the dissipative terms must also be dissipative and therefore only partial descriptions in the above sense; but of course a nondissipative system could (in an admittedly restrictive sense) be structurally stable relative to a change in parameter corresponding to a sign-preserving change in a constant of the motion.

The various notions of stability indicated above could be made precise. However, this does not guarantee that they are by themselves adequate for ecological and other complex biological systems. For example, a system may be able to absorb disturbances in terms of an ensemble of accessible steady states or limit cycles, or in terms of a continuum of weakly stable trajectories. The disturbance is absorbed if it causes the system to undergo transitions from one acceptable state to another. The criterion for acceptability, however, is not itself definable in terms of strictly dynamical constructs, i.e., in terms of measures of convergence, divergence, or topology of trajectories. The reason is that acceptability entails the notion of biological function and this returns us immediately to the framework of adaptability theory and the problem of operationally defining maximum tolerable uncertainty (cf. Section 2). In particular, the ideas of a continuum of weakly stable trajectories, multiple attractors (e.g., steady states, limit cycles), and qualitative similarity of trajectories in structurally stable systems correspond to the absorption of disturbance in adaptability theory; the idea of asymptotic orbital stability corresponds to the absorption and dissipation of disturbance, with dissipation corresponding to the absorption of disturbance in a heat bath; and degree of independence of levels corresponds to the degree to which these absorption and dissipation processes are hidden from view. If levels (or compartments) are not independent, disturbances absorbed and dissipated at one level must also be absorbed or dissipated at other levels, or ramifications of these disturbances must be absorbed and dissipated.

4.2. Interpretation of Dynamical Models (Simplest Example)

The simplest example is the classical Lotka–Volterra equations. For a two-species system without crowding these are

$$dN_1/dt = a_1 N_1 - b_1 N_1 N_2, \quad (18a)$$
$$dN_2/dt = -a_2 N_2 + b_2 N_1 N_2, \quad (18b)$$

where N_1 is prey, N_2 is predator, a_1 is the growth rate of prey in absence of predation, a_2 is the death rate of predator in the absence of prey, and b_1 and b_2 express the encounter and utilization rates which lead to decrease of the prey population (in the case of b_1) and increase in the predator population (in the case of b_2). As is well known, separation of variables gives a functional relationship between number of predator and number of prey (see, e.g., Rescigno and Richardson, 1973). Thus the system is conservative, i.e., has a constant of the motion and forever pursues a periodic oscillation not describable in terms of elementary functions. It is important that this oscillation is not of the limit cycle type. Rather it is weakly stable in the sense that perturbation to either the number of predator or prey results in a shift of the trajectory by a distance which will forever be periodically reestablished.

The biological and mathematical unrealisms of the Lotka–Volterra equations have frequently been discussed. Here, however, our interest is restricted to illustrating the cross-correlation between adaptability theory and dynamical models and in its use for interpreting the latter. The main points are:

(i) Only population variables enter, no environmental variables enter, and all other properties (e.g., organismic properties) are reflected in constants. Thus the modifiability and anticipation terms are zero and the condition for the applicability of the equation is that all environmental uncertainty be absorbed by adaptability at other levels or by indifference (or that there be no environmental uncertainty at all). Thus the population dynamics are completely controlled and appear autonomous and independent of the environment.

(ii) Suppose that an extra condition is added to Eqs. (18a) and (18b) which expresses the effect of environmental variability, e.g., by imposing a probability distribution on the constant of motion. The behavioral uncertainty at the population level (i.e., modifiability plus independence terms) is then the measure of the uncertainty of the ensemble of constants. If the modifiability is high this means that environmental disturbances are being absorbed (but not dissipated) and that this absorption is preventing disturbances from ramifying to other levels of organization, and no correlated lower level changes occur which are not reflected in these constants. Thus adaptability is achieved with a minimum of total observable modifiability and the dynamics of the population level appear autonomous (but not independent of the environment). As before, modifiability at other levels might absorb or dissipate other types of disturbance.

The existence of a constant of the motion in the Lotka–Volterra equations makes it possible to construct a statistical mechanics of many-species systems (Kerner, 1957). From the standpoint of adaptability theory the implication of such models is that the ensemble of possible weakly stable trajectories is providing an isolated bath for the absorption of environmental disturbances.

(iii) Suppose that the constants (a_1, b_1, a_2, b_2) are treated as variables, e.g., that extra equations are added that contain these variables but for simplicity not population variables. In this case the modifiability of the population is not independent of modifiability at the organismic level, but not conversely. This breaks the appearance of autonomy at the population level. However, the dynamical behavior of the population is structurally stable in the limited sense that its qualitative character does not change as long as the "constants" do not change sign or become equal to zero. As before disturbances may be absorbed in modifiability of the population sizes, in modifiability at lower levels, or in indifference. If the absorption is in terms of organismic modifiability, however, it will entail modifiability at the population level as well and therefore in general reliance on the structural stability of the population dynamics. Such reliance implies that adaptability is achieved with a greater than minimum total observable modifiability and therefore that the adaptabilities at other levels are not playing as effective a support role (unless the increase in structural stability appears as an increase in independence).

Conservative models such as the Lotka–Volterra model are, of course, not structurally stable in the sense that the addition of even slight damping will radically alter their qualitative behavior. The addition of such damping terms (e.g., crowding terms of the form $c_1 N_1^2$ and $c_2 N_2^2$) may allow the system to be asymptotically orbitally stable or (with the addition of a source term) to have limit cycle behavior. The essential added element is that such systems are capable not only of absorbing but also of dissipating environmental disturbances. As previously discussed this ultimately means that the disturbance is absorbed into a physical heat bath (ensemble of microstates) from which the behavior of the system is independent. The heat bath thus is the hidden, controlling system in gradient-type dynamics and is the basis, along with support by adaptability at other levels, for the appearance of autonomy of the system (including eventual autonomy relative to environmental disturbances which directly affect the population level).

Note that with the addition of a dissipative term the possibility of using weak stability is lost (since the conservative character of the system is broken). However, absorption is still possible if the system has multiple steady states. Unless the population dynamics are really highly regulated, this is the more plausible basis for absorption (since dissipative systems are potentially structurally stable to a larger class of perturbations).

4.3. Relevance to Construction of Dynamical Models

The interpretation of dynamical models from the standpoint of adaptability theory suggests (by inversion of the argument) a consideration relevant to the construction of dynamical models: In writing an equation for a level of interest it is justified to ignore variables associated with all other levels which block any disturbances affecting them from ramifying to the level of interest. If such blockage fails for any such level, variables associated with and disturbances affecting this level must be incorporated into the equation (i.e., it becomes a level of interest). The predictive accuracy of any equation omitting such variables will decrease as the failure to block increases, but how serious this decrease is depends on the structural stability of the trajectorial behavior of the level of interest.

With suitable squeezing of the parameters many dissipative equation structures are in general capable of approximating any given trajectorial behavior. Some possibilities can be eliminated, however, if the interpretations differ from the standpoint of adaptability.

5. CONCLUSION

Hierarchical adaptability theory is a formal apparatus for representing and analyzing the entire spectrum of adaptability processes in ecological and other complex, multilevel, compartmentally organized biological systems. In this paper the emphasis has been on the relationship between adaptability and degree of independence (or apparent autonomy) of levels and on the significance of this for the interpretation of dynamical ecological models. Such models are often formulated, sometimes quite fruitfully, in terms of variables associated with one or a few levels of organization. Yet, in general, it is the adaptabilities associated with other levels of organization which make this possible. These adaptabilities provide the support for dynamical ecological models formulated in terms of a restricted set of variables, and it is only to the extent that they are adequate to the task that they can be ignored and nevertheless a predictive model obtained. On the other side, to the extent that they are inadequate they of necessity become relevant to the dynamical behavior of the level of interest and therefore the variables with which they are associated should be added to the list of obligatory variables.

One might initially think, to take a concrete example, that extremely effective genetic adaptation would certainly eventually be quite relevant to the population dynamics or that extremely effective populational adaptability would have strong effects on gene frequencies. However, in the former case phenotypic changes would take place which absorb environmental variation, but without this becoming manifest at the population level; in the latter case the absorbing and dissipating capabilities at the population level would prevent the effects of disturbance from ramifying to the structure of the gene pool. Such extreme effectiveness is, however, rare and, in general, adaptation at the genetic level does to a greater or lesser extent become manifest at the population level and conversely. Thus, peculiar as it may at first appear, the limitation on dynamical models formulated in terms of an incomplete set of biological variables is not the ubiquitous occurrence of adaptive processes in biology, but rather the failure of such processes to attain an ideal relative to one criterion of effectiveness. This failure may lead to evolution or other significant changes which break the dynamic model, or circumscribe its range of validity. On the other hand, the inherent advantages of level independence suggest that if consideration is taken of the interplay of adaptabilities in a system, predictive and tractable models can be developed; more important, from the structure of phenomenologically predictive models and a knowledge of the environment, conclusions can be drawn about the role of nonexplicitly represented adaptabilities.

Such hidden adaptabilities are essential for a correct interpretation of dynamic ecological models, even more essential than adaptabilities explicitly represented in these models. This is particularly true for the interpretation of the relationship between stability and complexity. Increasing the structural complexity of a model may decrease its stability to perturbation, but increasing hidden adaptabilities may decrease the range or variety of perturbations. To the issue of stability versus complexity there must therefore be added the dimension of hidden adaptability. To the issue of the propriety and applicability of specific dynamic notions of stability for the explicitly represented dynamics must be added the functional significance of these notions within the framework of adaptability theory.

REFERENCES

Conrad, M. (1972a). Statistical and hierarchical aspects of biological organization. In "Towards a Theoretical Biology" (C. H. Waddington, ed.), Vol. 4, pp. 189–221. Edinburgh Univ. Press, Edinburgh.
Conrad, M. (1972b). Can there be a theory of fitness? *Int. J. Neurosci.* **3**, 125–134.
Conrad, M. (1975). Analyzing ecosystem adaptability. *Math. Biosci.* **27**, 213–230.
Conrad, M. (1976a). Biological adaptability: The statistical state model. *BioScience* **27**, 319–324.
Conrad, M. (1976b). Patterns of biological control in ecosystems. In "Systems Analysis and Simulation in Ecology" (B. C. Patten, ed.), Vol. 4, pp. 431–456. Academic Press, New York.
Conrad, M. (1977a). Functional significance of biological variability. *Bull. Math. Biol.* **39**, 139–156.
Conrad, M. (1977b). Biological adaptability and human ecology. *Proc. Int. Congr. Hum. Ecol., 1st,* 1975, 467–473.
Güttinger, W. (1974). Catastrophe geometry in physics and biology. In "Physics and Mathematics of the Nervous System" (M. Conrad, W. Güttinger, and M. Dal Cin, eds.), pp. 2–30. Springer Verlag, Berlin and New York.
Kerner, E. (1957). A statistical mechanics of interacting biological species. *Bull. Math. Biophys.* **19**, 121–146.
Khinchin, A. I. (1957). "Mathematical Foundations of Information Theory." Dover, New York.

Rescigno, A., and Richardson, I. W. (1973). The deterministic theory of population dynamics. *In* "Foundations of Mathematical Biology" (R. Rosen, ed.), Vol. 3, pp. 283–359. Academic Press, New York.

Rosen, R. (1970). "Dynamical System Theory in Biology," Vol. 1. Wiley (Interscience), New York.

Thom, R. (1970). Topological models in biology. *In* "Towards a Theoretical Biology" (C. H. Waddington, ed.), Vol. 1, pp. 89–116. Edinburgh Univ. Press, Edinburgh.

Chapter 7

STRUCTURE AND STABILITY OF MODEL ECOSYSTEMS

D. D. Šiljak

1. Introduction 151
2. System Structure 154
   2.1 Digraphs and Interconnection Matrices 154
   2.2 Input and Output Reachability 155
3. Partitions and Condensations 160
4. Vulnerability of Structure 164
   4.1 Structural Perturbations: Invulnerability 165
   4.2 Condensations: Connective Reachability 166
5. Vulnerability of Stability 169
   5.1 Global Stability 171
   5.2 Stability Regions. 174
   5.3 Lotka–Volterra Model 175
   5.4 Hierarchical Models 177
6. Conclusions. 179
   References. 179

1. INTRODUCTION

The essential characteristic of ecological models which distinguishes them from dynamic models used extensively in other natural, physical, and social sciences is the importance of interaction structure. As pointed out by Patten and Witkamp (1967), "... to understand ecosystems, ultimately will be to understand networks." The fundamental processes in ecological communities such as matter or energy transfers, feeding relations or trophic characteristics, selective effects of pests or epidemics control, and so on, are intrinsically structural in nature. "Yet only the beginnings of a mathematical theory of ecosystem structure and function are emerging" (Levin, 1975). Our objective in this chapter is to associate directed graphs with dynamic systems and provide a suitable framework for analysis of basic structural properties of ecological processes.

In Section 2, we fix the relationship among dynamic systems, directed graphs, and interconnection matrices (Šiljak, 1977a). In this way, inputs, states, and outputs of an ecosystem are represented by input, state, and output points of the corresponding digraph. The lines of the digraph and entries of the related interconnection matrix are determined in the usual way by occurrence of inputs, states, and outputs in the system equations. Various ecological meanings can be given to the points and lines of the graph depending on the specific modeling objective, such as trophic web representations (Gallopin, 1972) or general model ecosystem descriptions (Caswell et al., 1972). Then, *input* and *output reachability* is defined in order to systematically study patterns of material or energy transport through ecosystems. Computer algorithms are available for calculating path matrices (Bowie, 1976) which appear in this context.

Section 3 is devoted to *partitions* and *condensations* of digraphs as related to dynamic systems, which provide an appropriate setting for analysis of larger models of communities of species (or organisms) and their hierarchical structures. In this context, we formulate a *canonical structure of nonlinear systems*, which is of conceptual importance in relating the internal ecosystem process with its external stimulus–response behavior.

In Section 4, we consider vulnerability of ecosystem–environment interaction due to changes in the internal coupling topology. Having modeled an ecosystem as a number of interconnected elements, one can introduce structural perturbation to study the effects of adding or deleting interconnections among elements, or even deleting or adding the elements from or to the existing model. A natural question to ask in this case is how connectivity of the corresponding directed graph is affected by the structural disturbances. That is: Are input and output reachability invariant under structural perturbations? This constitutes the notion of *structural vulnerability* of dynamic systems (Šiljak, 1977a). Various implications of structural vulnerability of input–output reachability may arise in control of pollutants (Levin, 1975) and pests (Wickwire, 1977).

Section 5 is concerned with *vulnerability of stability* in dynamic ecosystems. Due to a virtual continuum of all sorts of perturbations acting on ecological processes, a natural question to ask is: How much of a certain kind of perturbation can a given ecosystem absorb before collapsing? Since dynamic processes are involved, a natural characteristic that describes a breakdown (or a blowup) of an ecosystem is stability. A great deal of internal and environmental perturbations (May, 1973) can be appropriately embedded in the classical stability theory (Hahn, 1967). Then, vulnerability (or resilience, persistence, robustness, sensitivity) can simply be defined by a suitable measure of the perturbation that the system can take without going unstable (e.g., May, 1973; Innis, 1975; Goh, 1975, 1976). Since the essential aspect of ecosystems is the interconnection structure of its constituents, we argue in this section that vulnerability of stability should be defined with respect to structural perturbations (Šiljak, 1974, 1975). In this way, the introduced notion is consistent with the classical notion of vulnerability of directed graphs and networks (e.g., Harary *et al.*, 1965; Boesch and Thomas, 1970; Chan and Frank, 1970). This opens up considerable opportunities for formal structural analysis as well as computational simplifications in model ecosystem studies.

Section 5 is subdivided into four subsections. In the first subsection we derive conditions for connective global stability of discrete-time dynamic models which parallel those obtained for continuous-time systems (Šiljak, 1975). The conditions allow suitable interpretations of the effects of interaction complexity on ecosystem stability. Besides implying stability under a variety of structural perturbations, the conditions guarantee a considerable robustness of ecosystems with regard to a wide range of nonlinear and time-varying effects in the interconnections. Most importantly, they confirm the fact that while complexity may improve community stability by keeping the size of interactions over a certain level, it is when complexity is limited so that each element is stable by itself and is weakly coupled to the others, that stability of the community is invulnerable to structural perturbations.

Since common population models (May, 1973) are not globally stable due to multiple equilibrium states, in Section 5.2 we consider regions of connective stability. The Liapunov function used in the preceding subsection to prove global stability is utilized in computations of the stability region estimates for general model ecosystems. These results are then applied to a discrete-time version of the Lotka–Volterra model to delineate a stability region in the corner of the positive quadrant in the species space, which contains the nontrivial equilibrium. The geometric shape of the region is similar to the positive quadrant, and a better fit can be obtained than that available by quadratic forms (MacArthur, 1970) or Volterra's functions (Goh, 1977).

Finally, in Section 5.4, we consider vulnerability of stability in hierarchical models of ecosystems. By lumping a number of elements of ecosystems into subsystems, we can explore important structural and stability properties of multilevel ecological organizations. In particular, we confirm by rigorous analysis the conclusion of Simon (1962), on the basis of intuitive arguments, that vulnerability of stability of complex systems is low if they are formed as a hierarchy of interconnected subsystems. Besides the vulnerability aspect of hierarchical models, partitioning of ecosystem opens up real possibilities for lumping unstable species (or organisms) into stable *coalitions* (such as predator–prey interaction), and thus gaining further insights into structural properties of complex ecosystems.

2. SYSTEM STRUCTURE

2.1. Digraphs and Interconnection Matrices

Consider a system $\mathscr{S}$ which is described by the equations

$$\mathbf{z}(t) = \mathbf{f}[t, \mathbf{x}(t), \mathbf{u}(t)], \qquad \mathbf{y}(t) = \mathbf{g}[t, \mathbf{x}(t)], \tag{1}$$

where $\mathbf{x} \in \mathscr{R}^n$ is the state, $\mathbf{u} \in \mathscr{R}^m$ is the input, and $\mathbf{y} \in \mathscr{R}^l$ is the output of $\mathscr{S}$ at time $t \in \mathscr{T}$. In (1), $\mathbf{z} \in \mathscr{R}^n$ represents the change of the state $\mathbf{x}$ in time. When $\mathbf{z}(t) \equiv d\mathbf{x}(t)/dt$, $\mathscr{T}$ is the nonnegative real line $\mathscr{R}_+$; when $\mathbf{z}(t) \equiv \mathbf{x}(t+1)$, $\mathscr{T}$ is the set of nonnegative integers $\{0, 1, 2, \ldots\}$. The functions $\mathbf{f}: \mathscr{T} \times \mathscr{R}^n \times \mathscr{R}^m \to \mathscr{R}^n$, $\mathbf{g}: \mathscr{T} \times \mathscr{R}^n \to \mathscr{R}^l$ are sufficiently smooth, so that in the former case $\mathscr{S}$ is a *continuous-time dynamic system* (Kalman et al., 1969), and in the later case $\mathscr{S}$ is a *discrete-time dynamic system* (Kalman and Bertram, 1960).

With the dynamic system $\mathscr{S}$ we associate a *directed graph* (*digraph*) (Harary, 1969) as the ordered pair $\mathscr{D} = (V, R)$, where $V = \{U \cup X \cup Y\}$ and $U = \{u_1, u_2, \ldots, u_m\}$, $X = \{x_1, x_2, \ldots, x_n\}$, $Y = \{y_1, y_2, \ldots, y_l\}$ are disjoint sets of input, state, and output points, respectively. R is a relation in V, that is, R is a set of ordered pairs which represent the lines (u_j, x_i), (x_j, x_i), and (x_j, y_i) joining the points of $\mathscr{D}$. We make an important assumption about $\mathscr{S}$ by requiring that $\mathscr{D}$ does not contain lines of the type (u_j, u_i), (x_j, u_i), (u_j, y_i), (y_j, u_i), (y_j, x_i), and (y_j, y_i). This requirement may seem to be overly restrictive, but in fact it is not since it reflects the structure of what we ordinarily consider as a dynamic system $\mathscr{S}$. We merely assume that there are no lines joining the input points, no lines from the input points to the output points, etc.

A convenient way to represent a digraph $\mathscr{D}$ associated with $\mathscr{S}$ is to use *interconnection matrices*. We propose that the $p \times p$ interconnection matrix $M = (m_{ij})$ which we define as a composite matrix

$$M = \begin{bmatrix} E & L & 0 \\ 0 & 0 & 0 \\ F & 0 & 0 \end{bmatrix}, \tag{2}$$

such that $i, j = 1, 2, \ldots, p$ and $p = n+m+l$, be used to describe the *basic structure of* $\mathscr{S}$. In (2), the $n \times n$ state connection matrix $E = (e_{ij})$ is defined as a binary matrix with elements e_{ij} specified by

$$e_{ij} = \begin{cases} 1, & (x_j, x_i) \in R, \\ 0, & (x_j, x_i) \notin R, \end{cases} \quad (3)$$

where $i, j = 1, 2, \ldots, n$. That is, $e_{ij} = 1$ if x_j "occurs" in $f_i(t, \mathbf{x}, \mathbf{u})$, and $e_{ij} = 0$ if x_j "does not occur" in $f_i(t, \mathbf{x}, \mathbf{u})$. Similarly, we define the $n \times m$ input connection matrix $L = (l_{ij})$ as

$$l_{ij} = \begin{cases} 1, & (u_j, x_i) \in R, \\ 0, & (u_j, x_i) \notin R, \end{cases} \quad (4)$$

where $i = 1, 2, \ldots, n$ and $j = 1, 2, \ldots, m$. In other words, $l_{ij} = 1$ if u_j occurs in $f_i(t, \mathbf{x}, \mathbf{u})$, and $l_{ij} = 0$ if u_j does not occur in $f_i(t, \mathbf{x}, \mathbf{u})$. Finally, the $l \times n$ output connection matrix $F = (f_{ij})$ is defined by

$$f_{ij} = \begin{cases} 1, & (x_j, y_i) \in R, \\ 0, & (x_j, y_i) \notin R, \end{cases} \quad (5)$$

where $i = 1, 2, \ldots, l$ and $j = 1, 2, \ldots, n$. Again, $f_{ij} = 1$ if x_j occurs in $g_i(t, \mathbf{x})$, and $f_{ij} = 0$ if x_j does not occur in $g_i(t, \mathbf{x})$.

A description of the ecosystem structure by graphs without a reference to dynamics is outlined by Gallopin (1972). Various important topological properties of food or trophic webs are also discussed with regard to interpretations of vertices and lines of the relevant graphs, partitioning of the webs into subunits, as well as lumping of a number of species or species' subgroups into a single food web vertex.

By associating digraphs with dynamic systems we plan to go a step further and explore the interplay between the structure and the dynamic process in model ecosystems. In this way, we will be able to set up a suitable framework for analyzing the effects of various topological features and configurations on the behavioral properties of relevant ecosystems. A useful concept in this context is that of reachability which is considered next.

2.2. Input and Output Reachability

There are numerous ways the environment can influence the internal behavior of an ecosystem. They range from natural factors, such as weather, to a virtual continuum of pollutants which are man-made perturbations to ecosystems—pesticides, oil, carbon, sulfur, and biphenyls. To be able to estimate their impact on an ecosystem, it is of primary importance to find out the pattern of transport of these perturbations (Levin, 1975). This fact leads naturally to the reachability concept in digraphs.

To express the reachability properties of the system $\mathscr{S}$ in graph-theoretic terms, let us recall several well-known notions from the theory of directed graphs (Harary, 1969). We consider again a digraph $\mathscr{D} = (V, R)$, where the set $V = \{v_1, v_2, \ldots, v_p\}$ is specified as $V = \{U \cup X \cup Y\}$ and $p = n + m + l$. If a collection of distinct points $v_1, v_2, \ldots, v_k$, together with the lines $(v_1, v_2), (v_2, v_3), \ldots, (v_{k-1}, v_k)$ can be placed in a sequence, then the ordered set $\{(v_1, v_2), (v_2, v_3), \ldots, (v_{k-1}, v_k)\}$ is a (*directed*) *path* from v_1 to v_k. Then, v_i is *reachable* from v_j if there is a path in $\mathscr{D}$ from v_j to v_i. A *reachable set* $V_i(v_j)$ of a point v_j is a set of points v_i reachable from v_j. Carrying this a step further, we define a *reachable set* $V_i(V_j)$ *of a set* V_j as a set of points v_i reachable from any point $v_j \in V_j$. An *antecedent set* $V_j(V_i)$ *of a set* V_i is the set of points v_j from which some point v_i of V_i is reachable.

Now we need the following (Šiljak, 1977a):

Definition 1. A system $\mathscr{S}$ with a digraph $\mathscr{D} = (U \cup X \cup Y, R)$ is input reachable if and only if X is a reachable set of U.

The "directional dual" of Definition 1 is:

Definition 2. A system $\mathscr{S}$ with a digraph $\mathscr{D} = (U \cup X \cup Y, R)$ is output reachable if and only if X is an antecedent set of Y.

From Definition 1 we conclude that an ecosystem represented by $\mathscr{S}$ and $\mathscr{D}$ is input reachable if the external (environmental) stimuli or disturbances **u** affect all the internal (ecosystem) variables **x**. Similarly, by Definition 2, we consider the same ecosystem as output reachable if all the internal states **x** are reflected by the observation variables **y**. This may be more than we are interested in, and partial information about what variables reach other variables is more desirable. A complete solution to this partial problem is available by forming the $p \times p$ path matrix $P = (p_{ij})$ which corresponds to the digraph $\mathscr{D} = (V, R)$ and is defined as

$$p_{ij} = \begin{cases} 1, & \text{there is a path from } v_j \text{ to } v_i \\ 0, & \text{there is no path from } v_j \text{ to } v_i. \end{cases} \quad (6)$$

In definition (6) of P, the trivial paths of zero length are excluded; the length of a path being the number of lines in the path.

To determine the path matrix for a given digraph (Harary et al., 1965), we assume that the $p \times p$ matrix $M = (m_{ij})$ of (2) is given and we denote by $N = (n_{ij})$ a $p \times p$ matrix such that $N = M^d$ where $d \in \{1, 2, \ldots, p\}$. Then, n_{ij} is the total number of distinct sequences $(v_j, \ldots), \ldots, (\ldots, v_i)$ of length d in the corresponding digraph $\mathscr{D} = (V, R)$. Now, we calculate the $p \times p$ matrix $Q = (q_{ij})$ as

$$Q = M + M^2 + \cdots + M^p, \quad (7)$$

7. Structure and Stability of Model Ecosystems

and conclude that $p_{ij} = 1$ if and only if $q_{ij} \neq 0$. From (2), we get

$$M^d = \begin{bmatrix} E^d & E^{d-1}L & 0 \\ 0 & 0 & 0 \\ FE^{d-1} & FE^{d-2}L & 0 \end{bmatrix}, \qquad (8)$$

where again $d \in \{1, 2, \ldots, p\}$. The matrix Q of (7) has the form

$$Q = \begin{bmatrix} A & B & 0 \\ 0 & 0 & 0 \\ C & D & 0 \end{bmatrix}, \qquad (9)$$

where

$$A = E + E^2 + \cdots + E^p, \quad B = (I + E + \cdots + E^{p-1})L,$$
$$C = F(I + E + \cdots E^{p-1}), \quad D = F(I + E + \cdots + E^{p-2})L. \qquad (10)$$

By calculating P and Q in (9) we know what stimulus influences which state **x** and what state **x** influences which response **y**.

To illustrate the application of the path matrix P, consider the digraph on Fig. 1. The interconnection matrix M of (2) is given as

$$M = \begin{bmatrix} & x_1 & x_2 & x_3 & x_4 & u_1 & y_1 & \\ & 0 & 1 & 0 & 1 & 1 & 0 & x_1 \\ & 1 & 0 & 0 & 1 & 0 & 0 & x_2 \\ & 0 & 0 & 0 & 1 & 0 & 0 & x_3, \\ & 0 & 0 & 1 & 0 & 0 & 0 & x_4 \\ & 0 & 0 & 0 & 0 & 0 & 0 & u_1 \\ & 0 & 0 & 1 & 0 & 0 & 0 & y_1 \end{bmatrix} \qquad (11)$$

and the corresponding path matrix P is calculated via the matrix Q of (9) as

$$P = \begin{bmatrix} & x_1 & x_2 & x_3 & x_4 & u_1 & y_1 & \\ & 1 & 1 & 1 & 1 & 1 & 0 & x_1 \\ & 1 & 1 & 1 & 1 & 1 & 0 & x_2 \\ & 0 & 0 & 1 & 1 & 0 & 0 & x_3. \\ & 0 & 0 & 1 & 0 & 0 & 0 & x_4 \\ & 0 & 0 & 0 & 0 & 0 & 0 & u_1 \\ & 0 & 0 & 1 & 1 & 0 & 0 & y_1 \end{bmatrix} \qquad (12)$$

From (12), we conclude that the stimulus u_1 affects the states x_1 and

158   D. D. Šiljak

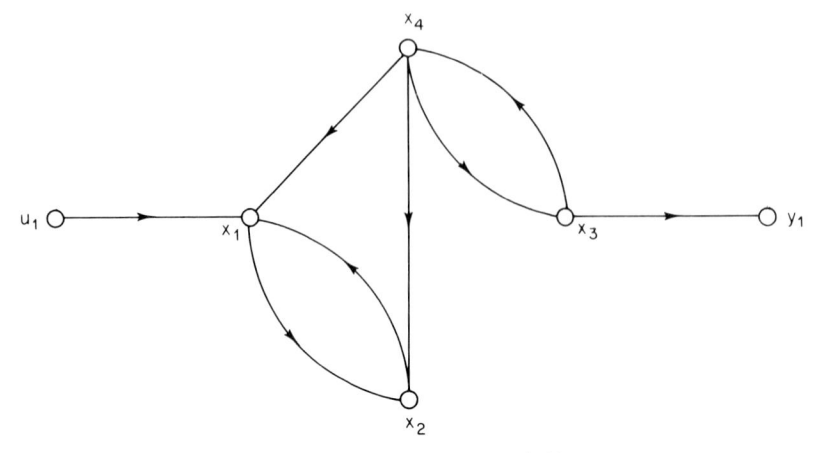

Figure 1. Input and output unreachable system.

x_2 but not x_3 and x_4. On the other hand, the changes of the states x_1 and x_2 cannot be observed by measuring the response y_1, while the changes in states x_3 and x_4 could. We also notice that the stimulus u_1 has no effect on the response y_1. This leads to:

Definition 3. A system $\mathscr{S}$ with a digraph $\mathscr{D} = (U \cup X \cup Y, R)$ is input–output reachable if and only if Y is a reachable set of U and U is an antecedent set of Y.

In Definition 3, we ignore the fact that $X \cup Y$ may also be a reachable set of U and $U \cup X$ may also be an antecedent set of Y. From Definitions 1 to 3, we conclude that if a system $\mathscr{S}$ is both input and output reachable, then it is also input–output reachable, but the converse is not true in general.

Now, we have the following (Šiljak, 1977a):

Theorem 1. A system $\mathscr{S}$ with an interconnection matrix M defined in (2), is input reachable if and only if the matrix B of (10) has no zero rows, it is output reachable if and only if the matrix C of (10) has no zero columns, and it is input–output reachable if and only if the matrix D of (10) has neither zero rows nor zero columns.

Proof. By constructing the path matrix P using Q of (9), the proof of Theorem 1 is automatic.

Applying Theorem 1 to the path matrix P of (12) we conclude that the corresponding digraph of Fig. 1 is not input, output, and input–output

7. Structure and Stability of Model Ecosystems

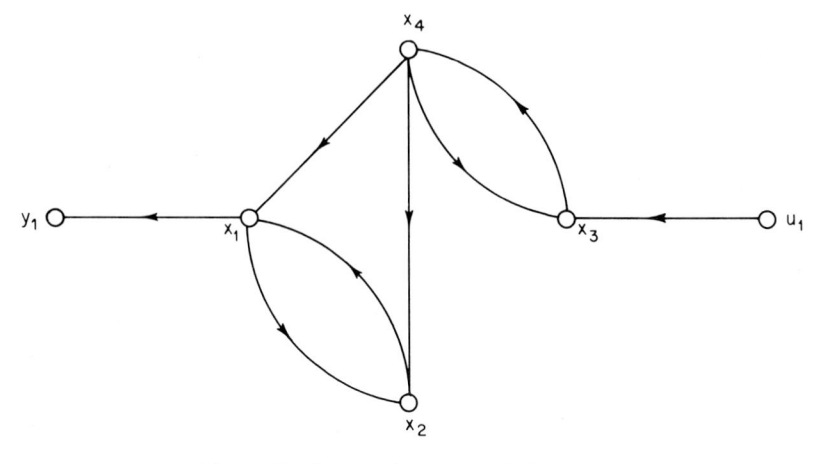

Figure 2. Input and output reachable system.

reachable. If we interchange input and output in the digraph of Fig. 1 to get the digraph shown in Fig. 2, then we obtain a system which is both input and output reachable. The matrix M corresponding to the digraph of Fig. 2 is

$$M = \begin{bmatrix} & x_1 & x_2 & x_3 & x_4 & u_1 & y_1 \\ & 0 & 1 & 0 & 1 & 0 & 0 \\ & 1 & 0 & 0 & 1 & 0 & 0 \\ & 0 & 0 & 0 & 1 & 1 & 0 \\ & 0 & 0 & 1 & 0 & 0 & 0 \\ & 0 & 0 & 0 & 0 & 0 & 0 \\ & 1 & 0 & 0 & 0 & 0 & 0 \end{bmatrix} \begin{matrix} x_1 \\ x_2 \\ x_3 \\ x_4 \\ u_1 \\ y_1 \end{matrix} \tag{13}$$

which produces the path matrix

$$P = \begin{bmatrix} & x_1 & x_2 & x_3 & x_4 & u_1 & y_1 \\ & 1 & 1 & 1 & 1 & 1 & 0 \\ & 1 & 1 & 1 & 1 & 1 & 0 \\ & 0 & 0 & 1 & 1 & 1 & 0 \\ & 0 & 0 & 1 & 0 & 1 & 0 \\ & 0 & 0 & 0 & 0 & 0 & 0 \\ & 1 & 1 & 1 & 1 & 1 & 0 \end{bmatrix} \begin{matrix} x_1 \\ x_2 \\ x_3 \\ x_4 \\ u_1 \\ y_1 \end{matrix} \tag{14}$$

160   D. D. Šiljak

confirming the statement about the digraph of Fig. 2. Since the corresponding system $\mathscr{S}$ is both input and output reachable, it is also input–output reachable, which is confirmed by P in (14) and verified by inspection of Fig. 2.

The computation of the path matrix by generating powers of the interconnection matrix as performed above, is not a numerically attractive procedure for varifying reachability properties of the corresponding dynamic system. There are numerous algorithms developed to avoid various numerical difficulties in identifying the paths of a digraph, which started by the well-known Boolean representation algorithm of Warshall (1962) and culminated by the depth-first search method of Tarjan (1972). A survey of these algorithms was given by Bowie (1976).

Similar concepts in the context of causality has been used by Patten and Finn (Chapter 8, this volume) as well as Orava (Chapter 9, this volume), wno provides a considerable insight into modeling of ecosystems by exploiting the tools of mathematical system theory.

3. PARTITIONS AND CONDENSATIONS

In model building in ecology, it has long been recognized that partitioning of the models into components helps to overcome their intrinsic complexity. Partitioning of the population models can be on the basis of sex and age (Keyfitz, 1968) or other biological characteristics (Gallopin, 1972). If food web is used as a topological representation of a population, it can be partitioned into subwebs as proposed by Paine (1966), which can be used to incorporate a detailed study of subcommunities interactions into the global food web of the whole ecosystem. Each subweb can be aggregated (condensed) into a single vertex of a digraph that represents the food web of the population on a higher hierarchical level. These aggregated digraphs are called condensations (Harary, 1969) of the original digraphs, that is, food webs.

To see how partitions and condensations can be used in the context of the proposed dynamic system-directed graph association, let us consider again an ecosystem $\mathscr{S}$ described by

$$\mathbf{z}(t) = \mathbf{f}[t, \mathbf{x}(t), \mathbf{u}(t)], \qquad \mathbf{y}(t) = \mathbf{g}[t, \mathbf{x}(t)], \tag{1}$$

where $\mathbf{x} \in \mathscr{R}^n$ is the state, $\mathbf{u} \in \mathscr{R}^m$ is the input, $\mathbf{y} \in \mathscr{R}^l$ is the output of $\mathscr{S}$, and $\mathbf{z} \in \mathscr{R}^n$ is either $d\mathbf{x}(t)/dt$ or $\mathbf{x}(t+1) - \mathbf{x}(t)$. We assume that $\mathscr{S}$ is decomposed into s interconnected subsystems $\mathscr{S}_i$ described by the equations

$$\mathbf{z}_i(t) = \mathbf{f}_i[t, \mathbf{x}(t), \mathbf{u}(t)], \qquad \mathbf{y}_i(t) = \mathbf{g}_i[t, \mathbf{x}(t)], \tag{15}$$

where $x_i \in \mathcal{R}^{n_i}$ is the state, $u_i \in \mathcal{R}^{m_i}$ is the input, $y_i \in \mathcal{R}^{l_i}$ is the output of $\mathcal{S}_i$, and $z_i \in \mathcal{R}^{n_i}$ is either $dx_i(t)/dt$ or $x_i(t+1) - x_i(t)$. We also assume that

$$\begin{aligned} \mathcal{R}^n &= \mathcal{R}^{n_1} \times \mathcal{R}^{n_2} \times \cdots \times \mathcal{R}^{n_s}, \\ \mathcal{R}^m &= \mathcal{R}^{m_1} \times \mathcal{R}^{m_2} \times \cdots \times \mathcal{R}^{m_r}, \\ \mathcal{R}^l &= \mathcal{R}^{l_1} \times \mathcal{R}^{l_2} \times \cdots \times \mathcal{R}^{l_q}, \end{aligned} \quad (16)$$

so that

$$\begin{aligned} \mathbf{x} &= (\mathbf{x}_1^T, \mathbf{x}_2^T, \ldots, \mathbf{x}_s^T)^T, \\ \mathbf{u} &= (\mathbf{u}_1^T, \mathbf{u}_2^T, \ldots, \mathbf{u}_r^T)^T, \\ \mathbf{y} &= (\mathbf{y}_1^T, \mathbf{y}_2^T, \ldots, \mathbf{y}_q^T)^T. \end{aligned} \quad (17)$$

are the state, input, and output vectors of $\mathcal{S}$, respectively.

We associate again a digraph $\mathcal{D} = (V, R)$ with the dynamic system $\mathcal{S}$ described by (15). Then, we partition each set U, X, Y of V into disjoint subsets $u_1, u_2, \ldots, u_r$; $x_1, x_2, \ldots, x_s$; $y_1, y_2, \ldots, y_q$ whose union is all of V. Thus, each element of V is in exactly one subset of V. The condensation $\mathcal{D}^* = (V^*, R^*)$ of $\mathcal{D} = (V, R)$ with respect to this partition is the digraph whose points V^* are those subsets of V, that is, $V^* = \{U^* \cup X^* \cup Y^*\}$ and

$$\begin{aligned} X^* &= \{x_1, x_2, \ldots, x_s\}, \\ U^* &= \{u_1, u_2, \ldots, u_r\}, \\ Y^* &= \{y_1, y_2, \ldots, y_q\}. \end{aligned} \quad (18)$$

The lines of the condensation $\mathcal{D}^*$ are determined by the following rule: There is a line (u_j, x_i) in R^* if and only if there is a line in R from a point of the subset u_j of U to a point of the subset x_i of X. Similarly, the rule holds for the lines (x_j, x_i) and (x_j, y_i) of $\mathcal{D}^*$. In this way, the condensation $\mathcal{D}^*$ represents (uniquely) the structure of the composite system $\mathcal{S}$ described by (15) with points of $\mathcal{D}^*$ standing for the subsystems and the lines of $\mathcal{D}^*$ standing for the interconnections among them.

Another way to represent a partition of $\mathcal{S}$ is to use interconnection matrices in the same way as they are used to describe the original system $\mathcal{S}$. Rewriting the matrix M of (2), but otherwise using different submatrices we can define the $p^* \times p^*$ matrix

$$M^* = \left[\begin{array}{c|c|c} E^* & L^* & 0 \\ \hline 0 & 0 & 0 \\ \hline F^* & 0 & 0 \end{array} \right] \quad (19)$$

which we associate with the condensation $\mathcal{D}^*$ in an obvious way. In (19),

E^*, L^*, F^* are the $s \times s$, $s \times r$, $q \times s$ matrices, respectively, and $p^* = s+r+q$. Now, the reachability properties of the condensation $\mathscr{D}^*$ can be determined by applying Theorem 1 to the matrix M^* of (19).

There are many ways in which a dynamic system $\mathscr{S}$ and the corresponding digraph $\mathscr{D}$ may be partitioned into subsystems and subgraphs. In the pure theory of structures (Harary, 1969), it is common to partition a digraph $\mathscr{D}$ into its *strong components* so that each point of the condensation $\mathscr{D}^*$ corresponds to one and only one strong component of $\mathscr{D}$; a strong component of a digraph $\mathscr{D}$ being a subgraph of $\mathscr{D}$ in which every two points are mutually reachable.

In order to conclude input and output reachability of such special condensations, we denote by $\mathscr{D}_x = (X, R_x)$ the *state truncation* (Šiljak, 1977b), which is the subgraph of $\mathscr{D}$ obtained from $\mathscr{D}$ by deleting the points of U and Y together with the lines incident to the points of U and Y. Now, we prove the following:

Theorem 2. Let the condensation $\mathscr{D}_x^*$ be constructed with respect to the strong components of $\mathscr{D}_x$. Then, the digraph $\mathscr{D}$ is input (output) reachable if and only if the condensation $\mathscr{D}^*$ is input (output) reachable.

Proof. We prove only the input reachability part of the theorem since the output reachability part is its directional dual. If $\mathscr{D}^*$ is input reachable then X^* is the reachable set of U^* and there is a path to each point X^* from a point U^*. Since $\mathscr{D}_x^*$ is a condensation with respect to strong components of $\mathscr{D}_x$, it follows from reachability of the components that there is a path to each point of X from a point U. That is, X is the reachable set of U and $\mathscr{D}$ is input reachable. Conversely, if $\mathscr{D}^*$ is not input reachable, then X^* is not a reachable set of U^*, and there are points of X^* that cannot be reached by any point of U^*. Obviously, by the definition of condensation, those points of X that correspond to the unreachable points of X^* cannot be reached by any point of U, and $\mathscr{D}$ is not input reachable. This proves Theorem 2.

Algorithms of Purdom (1970), Munro (1971), and Kevorkian (1975) can be used to compute the strongly connected components of the digraph $\mathscr{D}$. Then, input and output reachability can be determined by forming the condensation $\mathscr{D}_x^*$ and applying Theorem 2.

Partitions of ecosystems into strong components may be used to identify the subcommunities of organisms that are tightly interrelated to each other. This may require that young and adults of the same species belong to two distinct components. Nevertheless, some formal structural properties in this case may outweigh the importance of lumping naturally the organisms of the same species into one subsystem. However, in most

7. Structure and Stability of Model Ecosystems

cases, the partitions and condensations of ecosystem structure are guided by feeding relations, mode of nutrition, degree of mobility, etc., as surveyed by Gallopin (1972).

A condensation of a digraph, which is of special interest in the context of input and output reachability, is the *canonical structure* introduced by Šiljak (1977b). We consider again the dynamic system $\mathscr{S}$ of (1) together with its interconnection matrix M of (2). By permutation of rows and columns of M on the basis of the path matrix P, the matrix M can be transformed into a matrix $\hat{M}$ which has the following form

$$\hat{M} = \begin{bmatrix} \hat{x}_{io} & \hat{x}_{jo} & \hat{x}_{i\phi} & \hat{x}_{j\phi} & u & y \\ E_{11} & E_{12} & 0 & 0 & L_1 & 0 \\ 0 & E_{22} & 0 & 0 & 0 & 0 \\ E_{31} & E_{32} & E_{33} & E_{34} & L_3 & 0 \\ 0 & E_{42} & 0 & E_{44} & 0 & 0 \\ \hline 0 & 0 & 0 & 0 & 0 & 0 \\ F_1 & F_2 & 0 & 0 & 0 & 0 \end{bmatrix} \begin{matrix} \hat{x}_{io} \\ \hat{x}_{jo} \\ \hat{x}_{i\phi} \\ \hat{x}_{j\phi} \\ u \\ y \end{matrix} \qquad (20)$$

The partition of the state vector of the transformed system $\hat{\mathscr{S}}$,

$$\hat{x} = (\hat{x}_{io}^T, \hat{x}_{jo}^T, \hat{x}_{i\phi}^T, \hat{x}_{j\phi}^T)^T, \qquad (21)$$

into four components, which can be identified from $\hat{M}$ in (20), represent the four subsystems with the following properties:

($\hat{\mathscr{S}}_{io}$) input reachable and output reachable,
($\hat{\mathscr{S}}_{jo}$) input unreachable and output reachable,
($\hat{\mathscr{S}}_{i\phi}$) input reachable and output unreachable,
($\hat{\mathscr{S}}_{j\phi}$) input unreachable and output unreachable.

By using condensation, we can represent each component of $\hat{x}$ in (21) as a point of the condensation $\hat{\mathscr{D}}^*$ shown in Fig. 3. The digraph $\hat{\mathscr{D}}^*$ and the corresponding matrix $\hat{M}$ identify the states of a given ecosystem which can be observed by the response **y**; the states which can be influenced by a stimulus **u**, but cannot be identified by **y**; the states which can be measured using **y**, but cannot be excited by **u**; and finally the states which form a subsystem that is isolated from the environment and neither can be influenced by **u** nor observed by **y**.

Before concluding this section, a few comments are in order. With an abuse of Definitions 1 and 2, in the above proof we referred to input (output) reachability of the digraphs $\mathscr{D}$ and $\mathscr{D}^*$. This was done to avoid ambiguity arising from the fact that both the digraph $\mathscr{D}$ and its condensation $\mathscr{D}^*$ are related to the same system $\mathscr{S}$. The change in

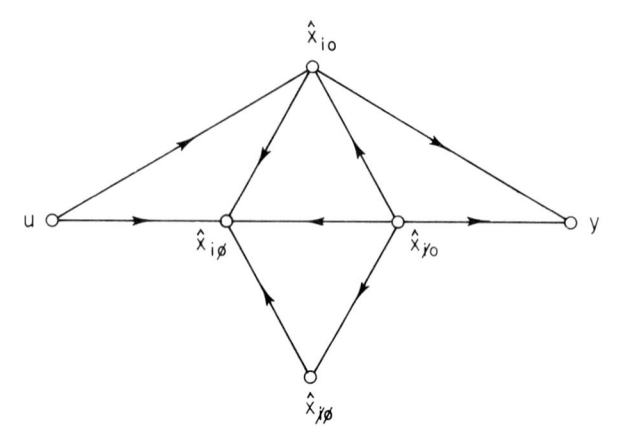

Figure 3. Canonical structure.

terminology should create no confusion since input (output) reachability is defined unambiguously in terms of digraphs. Still another abuse of notation is committed in (17) and (18), where it would be appropriate to use asterisks on all components of the vectors **x**, **u**, and **y** since they represent a partition and condensation rather than the individual states, inputs, and outputs of the system $\mathscr{S}$. This again should not pose any difficulty since the two entities are never mixed together in the same derivation and their proper identification is straightforward.

Finally, we should mention a possibility to use the condensation process as a preliminary analysis to stability of hierarchical models by aggregation, which is outlined in Section 5.4. Furthermore, it may be also possible to use the structural condensation in the general aggregation context proposed by Zeigler (Chapter 1, this volume).

4. VULNERABILITY OF STRUCTURE

In digraph theory, reachability is the basic concept and vulnerability of a digraph stands for liability of reachability caused by the removal of lines or points from a given digraph. As such, vulnerability found a considerable application in communication networks (Boesch and Thomas, 1970), where it plays a considerable role in their "survival" (Chan and Frank, 1970). These results, however, were obtained under no particular classification of points and lines of a communication net. For these reasons, the results cannot be directly applied to dynamic systems where a special structure is imposed by qualifying the points as inputs, states, and outputs, and limiting the line arrangement to suit the notion of a dynamic system.

Vulnerability of input and output reachability of dynamic systems was introduced (Šiljak, 1977a) and studied (Šiljak, 1977b) with the intent to derive conditions under which a system preserves its reachability properties despite structural perturbations whereby a number of lines among states of the system are disconnected. In the context of an ecosystem food web structure such a study can be useful because of "a permanent change in such structure in time" as pointed out by Gallopin (1972). Caswell *et al.* (1972) also emphasize this fact: "More importantly, having modeled a system as a set of interacting objects, the observer can now investigate the effect on system behavior of changing the coupling topology, the behavioral characteristics of an object or objects, or even the effects of adding or deleting objects." An example of such importance arises in the general ecological context when a certain pattern of transport of chemicals is established, and we are interested to find out the effects on the pattern of disconnecting parts of the basic ecosystem structure.

4.1. Structural Perturbations: Invulnerability

In terms of digraph theory (Harary *et al.*, 1965), disconnections in the system $\mathscr{S}$ can be identified by "line removals" from the corresponding digraph $\mathscr{D}$. We say that a system $\mathscr{S}$ is vulnerable if a removal of a line (x_j, x_i) connecting two points of the set X of $\mathscr{D}$ causes a loss of input or output reachability of $\mathscr{S}$. We introduce the following:

Definition 4. A line $(x_j, x_i) \in R$ of a digraph $\mathscr{D} = (U \cup X \cup Y, R)$ is an input (output) strengthening line if and only if its removal destroys input (output) reachability of $\mathscr{D}$. Otherwise, the line (x_j, x_i) is an input (output) neutral line.

It is intuitively clear that if in an input reachable digraph $\mathscr{D}$ there is a line (x_j, x_i) which is present in every path from U to x_i, then it is an input strengthening line. Therefore, to have an input invulnerable system $\mathscr{S}$, the corresponding digraph $\mathscr{D}$ should have no input strengthening lines, that is, all of its lines should be input neutral lines. To this extent, we recall that two paths of a digraph are distinct if they have no common points, and we prove the following:

Theorem 3. A digraph $\mathscr{D} = (U \cup X \cup Y, R)$ has no input (output) strengthening lines if there are more than one distinct path from any point of U to each point of X (from each point of X to any point of Y).

Proof. Obviously, if to each point of X there is more than one distinct path from some point of U, by the definition of a distinct path it follows that no line in $\mathscr{L}$ is on every path to any point of X from any point of U. Therefore, there are no input strengthening lines in $\mathscr{L}$. The dual of the conclusion proves the output part and, thus, Theorem 3.

To determine distinct paths in any directed graph, one can use the interconnection matrix and a method described by Harary et al. (1965). A simple adjustment is necessary to take into account the special structure of dynamic systems described by the matrix M of (2).

4.2. Condensations: Connective Reachability

The concept of input and output invulnerability can be extended to include a removal of a number of lines. Such structural perturbations can be conveniently described by interconnection matrices as shown by Šiljak (1975). Furthermore, the structural perturbations may involve condensations rather than original graphs as in the case of large-scale systems (Šiljak, 1978). For such purposes, we need the notion of the $s \times s$ fundamental interconnection matrix $\bar{E}^* = (\bar{e}_{ij}^*)$ associated with a condensation $\bar{\mathscr{D}}^* = (U^* \cup X^* \cup Y^*, \bar{R}^*)$ of $\mathscr{S}$ in (19) as follows:

$$\bar{e}_{ij}^* = \begin{cases} 1, & (x_j, x_i) \in \bar{R}^*, \\ 0, & (x_j, x_i) \notin \bar{R}^*. \end{cases} \quad (22)$$

That is, $\bar{e}_{ij}^* = 1$ if $\mathbf{x}_j$ occurs in $f_i(t, \mathbf{x}, \mathbf{u})$, and $\bar{e}_{ij}^* = 0$ if $\mathbf{x}_j$ does not occur in $f_i(t, \mathbf{x}, \mathbf{u})$. Now, a structural perturbation is represented by a removal of a line (or number of lines) of the condensation $\bar{\mathscr{D}}^*$ between points of X^*. That results in a spanning subgraph $\mathscr{D}^*$ of $\bar{\mathscr{D}}^*$, that is, a subgraph with the same set of points as $\bar{\mathscr{D}}^*$. All spanning subgraphs of $\bar{\mathscr{D}}^*$ obtained this way, can be represented uniquely by an interconnection matrix E^* which is obtained from $\bar{E}^*$ as follows: $\bar{e}_{ij}^* = 0$ in $\bar{E}^*$ implies $e_{ij}^* = 0$ in E^* for any $i, j = 1, 2, \ldots, s$; and removal of a line (x_j, x_i) of $\bar{\mathscr{D}}^*$ implies that $\bar{e}_{ij}^* = 1$ in $\bar{E}^*$ is replaced by $e_{ij}^* = 0$ in E^*. The fact that an interconnection matrix E^* is generated in this way by the fundamental interconnection matrix $\bar{E}^*$ is denoted by $E^* \in \bar{E}^*$. Finally, without loss of generality, a point removal can be treated in the same way as a special case of line removals. If a kth point of $\bar{\mathscr{D}}^*$ is removed, then $e_{ik}^* = e_{kj}^* = 0$ for all $i, j = 1, 2, \ldots, s$.

On the basis of the above considerations, we introduce the following:

Definition 5. A system $\mathscr{S}$ with a condensation $\bar{\mathscr{D}}^* = (U^* \cup X^* \cup Y^*, \bar{R}^*)$ is connectively input (output) reachable if and only if it is input (output) reachable for all interconnection matrices $E^* \in \bar{E}^*$.

We recall from the work of Šiljak (1977b) that the input truncation

$\mathcal{D}_u^* = (U^* \cup X^*, R_u^*)$ of a digraph $\mathcal{D}^* = (U^* \cup X^* \cup Y^*, R^*)$ is the maximal subgraph of $\mathcal{D}^*$ not containing point of Y^*, and prove the following:

Theorem 4. A system $\mathscr{S}$ with an input truncation condensation $\bar{\mathcal{D}}_u^* = (U^* \cup X^*, \bar{R}_u^*)$ is connectively input reachable if and only if the set U^* is a 1-basis of $\mathcal{D}_u^*$ for $E^* = 0$.

Proof. The "if" part for $E^* = 0$ follows directly from the definition of the 1-basis (Harary, 1969), which is the minimal collection of mutually nonadjacent points in $\bar{\mathcal{D}}_u^*$ such that every point of $\bar{\mathcal{D}}_u^*$ is either in the collection or adjacent to a point of the collection. For $E^* \neq 0$, we get the corresponding digraph by adding lines to the one that corresponds to $E^* = 0$. But, it is obvious that there are no lines in any digraph whose addition can destroy its reachability property. For if (x_j, x_i) is any line of $\bar{\mathcal{D}}_u^*$, any path in $\bar{\mathcal{D}}_u^* - (x_j, x_i)$ is also in $\bar{\mathcal{D}}_u^*$.

On the other hand, if U^* is not a 1-basis for $E^* = 0$, then there would be points of X^* that are not reachable from U^* for all $E^* \in \bar{E}^*$, and $\mathscr{S}$ is not connectively input reachable. This proves the "only if" part and, thus, Theorem 4.

By principle of duality of digraphs (Harary, 1969) from Theorem 4 we get:

Theorem 5. A system $\mathscr{S}$ with an output truncated condensation $\bar{\mathcal{D}}_y^* = (X^* \cup Y^*, \bar{R}_y^*)$ is connectively output reachable if and only if the set Y^* is a 1-contrabasis of $\mathcal{D}_y^*$ for $E^* = 0$.

Structural perturbations are illustrated in Fig. 4. The top digraph $\bar{\mathcal{D}}_y^*$ is an output truncation which represents the basic structure of a composite system $\mathscr{S}$ when all of its inputs are removed, and which corresponds to the fundamental interconnection matrix $\bar{E}^*$. All possible structural perturbations formed by line removals are represented by the digraphs below the digraph $\bar{\mathcal{D}}_y^*$. From Fig. 4, it is clear that the first three perturbations are output reachable while the remaining four are not. Thus, the system $\mathscr{S}$ is not connectively output reachable. Under the first three structural perturbations, it may be possible to identify the state x_1 by observing the response y, while in the remaining situations such a task would be impossible.

The result of Theorems 4 and 5 is intuitively clear: If we want reachability to be preserved under structural perturbations, we should check the "worst case," that of $E^* = 0$. This may seem to be an overly conservative result, if the freedom in partitioning and condensing the ecosystem is not taken into account. Furthermore, one can generalize the above result by defining a partial connective reachability with respect to a

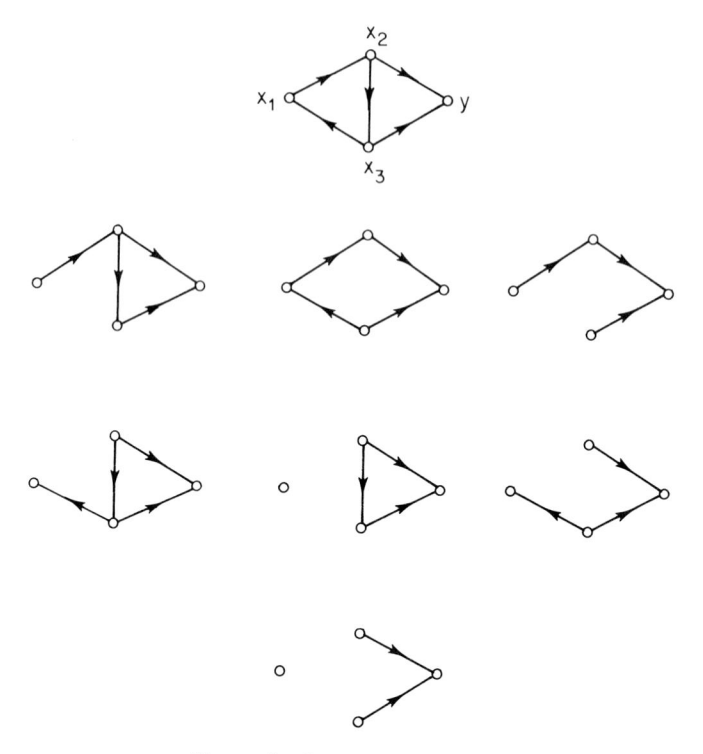

Figure 4. Structural perturbations.

pair of interconnection matrices $(\bar{E}^*, \hat{E}^*)$, where a fixed matrix $\hat{E}^* \in \bar{E}^*$ takes the role of $\hat{E}^* = 0$. Again, it is intuitively clear that if U^* is a 1-basis for $\hat{E}^*$, this being the worst case, then $\mathscr{S}$ is partially connectively input reachable for all $E^* \in \bar{E}^* - \hat{E}^*$, where by the difference $\bar{E}^* - \hat{E}^*$ we mean all interconnection matrices generated by $\bar{E}^*$ which have always all the unit elements of $\hat{E}^*$. It should also be noted here that there is no reason why the notion of structural perturbations should not be enlarged to include the removal of lines which connect the inputs with states and states with outputs of the relevant digraphs.

In the context of vulnerability of dynamic systems with regard to reachability, the following useful structural characteristics should be determined: (i) the lines (or maximum number of lines) of $\bar{\mathscr{D}}^*$ whose removal does not affect input (output) reachability of $\mathscr{S}$; (ii) the lines (the minimum number of lines) of $\bar{\mathscr{D}}^*$ which are required to preserve input (output) reachability of $\mathscr{S}$; and (iii) the lines (or the minimum number of lines) of $\bar{\mathscr{D}}^*$ whose removal destroys input (output) reachability of $\mathscr{S}$.

7. Structure and Stability of Model Ecosystems

After defining properly the structural perturbations, we can turn our attention to the effects that such perturbations can have on the stability of dynamic systems.

5. VULNERABILITY OF STABILITY

Due to highly detrimental and largely irreversible effects of ecosystem breakdowns on biotic resources of the earth, vulnerability of model ecosystems is one of the central issues in theoretical ecology. Being under a barrage of disturbances that vary considerably in both type and size, ecosystems are not safe if they are merely stable. This problem is further complicated by complexity of the internal structure of the ecosystems, and the question clearly is (Šiljak, 1975): "How vulnerable is stability of complex ecosystems?"

It is simple to show that increased complexity can stabilize an otherwise unstable system. Consider the one-prey–one-predator Lotka–Volterra model (Lotka, 1925; Volterra, 1926),

$$\dot{y}_1 = \alpha y_1 - \gamma y_1 y_2, \qquad \dot{y}_2 = \delta y_1 y_2 - \beta y_2 - \theta y_2^2, \tag{23}$$

where $y_1(t)$ is the prey and $y_2(t)$ is the predator population at time t.

As usual, the parameter α represents the birth rate of the prey; β represents the death rate of the predator; γ, δ represent the interaction between the two species; and θ represents the death rate of the predator due to direct competition within predator population. All of these parameters are assumed positive. Obviously, in the absense of the predator, the prey population would grow as a Malthusian exponential process at a rate determined by α. Similarly, if the prey population disappears, the predator would become extinct. It is a well-known fact that although the two separate populations do not have a stable constant population, their union can have a stable nontrivial equilibrium.

There are two equilibrium populations $\mathbf{y}^e = \mathbf{0}$ and $\mathbf{y}^e = \{(\alpha\theta\gamma^{-1} + \beta)\delta^{-1}, \alpha\gamma^{-1}\}$, and none of them can be globally stable. Therefore, we proceed to study small deviations $\mathbf{x}(t)$ about the nontrivial equilibrium using linearization. Substituting the perturbed populations

$$\mathbf{y}(t) = \mathbf{y}^e + \mathbf{x}(t), \tag{24}$$

into (23) and neglecting the nonlinear terms of $\mathbf{x}(t)$, we get the linear model

$$\dot{x}_1 = a_{11}x_1 + a_{12}x_2, \qquad \dot{x}_2 = a_{21}x_1 + a_{22}x_2, \tag{25}$$

where

$$a_{11} = 0, \quad a_{12} = -(\alpha\theta + \beta\gamma)^{-1},$$
$$a_{21} = \alpha\gamma^{-1}\delta, \quad a_{22} = -\alpha\gamma^{-1}\theta. \tag{26}$$

The equilibrium $\mathbf{x}^e = \mathbf{0}$ of the system (25), which corresponds to the equilibrium $\mathbf{y}^e \neq \mathbf{0}$ of (23), is stable if (and only if)

$$a_{22} < 0, \quad a_{12}a_{21} < 0. \tag{27}$$

Due to positivity of the system parameters of the system (23), conditions (27) are satisfied, and $\mathbf{x}^e = \mathbf{0}$ of (25) is locally stable. That is, the equilibrium $\mathbf{y}^e \neq \mathbf{0}$ of (23) can be established provided the deviations of population vector $\mathbf{y}(t)$ from $\mathbf{y}^e \neq \mathbf{0}$ are sufficiently small. Stability of $\mathbf{y}^e \neq \mathbf{0}$, however, hinges on the conditions in (27), which means that the predator–prey coalition has to be permanent, that is, both interaction coefficients a_{12} and a_{21} have to be different from zero. No structural perturbation of the community can be tolerated.

If each species is density dependent, that is, we have in (25),

$$a_{11} < 0, \quad a_{22} < 0, \tag{28}$$

then the necessary and sufficient condition for stability of $\mathbf{x}^e = \mathbf{0}$ is

$$a_{11}a_{22} - a_{12}a_{21} > 0. \tag{29}$$

Taking into account (28), we conclude from (29) that a_{12} and a_{21} can have any value including zero so long as the absolute value of their product is bounded by $a_{11}a_{22}$, that is,

$$|a_{12}a_{21}| < |a_{11}a_{22}|. \tag{30}$$

Therefore, under the condition (30), the equilibrium $x^e = 0$ is stable under structural perturbations whereby the two species are disconnected and again connected together, which includes extinction of any one of them, change of the nature of their interaction (competitive, symbiotic, predator–prey), and so on. While these conclusions are fairly straightforward, it takes a more refined analysis (Šiljak, 1975) to show that under the conditions (29), stability is a robust property of the equilibrium $\mathbf{x}^e = \mathbf{0}$ and can tolerate a wide range of nonlinear and time-varying phenomena in the species interactions as long as the interactions are properly limited, that is,

$$|a_{12}(t, x)| < |a_{11}|, \quad |a_{21}(t, x)| < |a_{22}| \tag{31}$$

for all t and $\mathbf{x}$.

The above example points to a conjecture that complexity can bring about stability to an otherwise unstable system, but it is autonomy of each constituent of the system and limited interactions, thus limited complexity, that eliminate vulnerability to structural perturbations and are fundamental to system robustness with regard to time-varying nonlinear interconnection effects. This conjecture was proved to be true in wide variety of model ecosystems, which included deterministic (Šiljak, 1975) as well as stochastic disturbances (Ladde and Šiljak, 1976) of the environment, in the context of boundedness and stability of the ecosystem internal dynamic processes. In the rest of this section, these properties will be established for discrete-time dynamic ecosystems.

5.1. Global Stability

Consider a discrete dynamic system $\mathscr{S}$ described by equation

$$\mathbf{x}(t+1) = \mathbf{f}[t, \mathbf{x}(t)], \tag{32}$$

which is obtained from Eqs. (1) by ignoring the input u and output y, and setting $\mathbf{z}(t) = \mathbf{x}(t+1)$. In (32), the function $\mathbf{f}: \mathscr{T} \times \mathscr{R}^n \to \mathscr{R}^n$ is continuous with respect to all of its arguments, and one to one for any fixed $t \in \mathscr{T}$, so that solutions $\mathbf{x}(t; t_0, \mathbf{x}_0)$, which start at $(t_0, \mathbf{x}_0) \in \mathscr{T} \times \mathscr{R}^n$, satisfy $\mathbf{x}(t_b; t_0, \mathbf{x}_0) = \mathbf{x}[t_b; t_a, \mathbf{x}(t_a; t_0, \mathbf{x}_0)]$ for all $t_a, t_b \in \mathscr{T}_0$, where $0 \leqslant t_0 \leqslant t_a \leqslant t_b$ and $\mathscr{T}_0 = [t_0, +\infty)$.

We also assume that $\mathbf{f}(t, \mathbf{0}) = \mathbf{0}$ for all $t \in \mathscr{T}$, and that $\mathbf{x}^e = \mathbf{0}$ is the unique equilibrium of the system $\mathscr{S}$. If we have $\mathbf{y}(t+1) = \mathbf{g}[t, \mathbf{y}(t)]$ and $\mathbf{g}(t, \mathbf{y}^e) = \mathbf{y}^e$ for all $t \in \mathscr{T}$, but $\mathbf{y}^e \neq \mathbf{0}$, then we use the transformation $\mathbf{y}(t) = \mathbf{y}^e + \mathbf{x}(t)$ as in the above example, and consider again Eq. (32) where $\mathbf{f}[t, \mathbf{x}(t)] \equiv \mathbf{g}[t, \mathbf{y}^e + \mathbf{x}(t)] - \mathbf{y}^e$.

In order to suitably describe the interconnective features of ecosystems, we consider Eq. (32) in the form

$$\mathbf{x}(t+1) = A[t, \mathbf{x}(t)]\mathbf{x}(t), \tag{33}$$

where $A = (a_{ij})$ is an $n \times n$ matrix function with the coefficients

$$a_{ij}(t, x) = e_{ij}(t)\phi_{ij}(t, x). \tag{34}$$

In (34), $e_{ij}: \mathscr{T} \to [0, 1]$ are elements of the $n \times n$ interconnection matrix $E = (e_{ij})$ defined in Section 1. To get (34) from (32) we can always choose

$$A(t, \mathbf{x}) = \text{diag}\{f_1(t, \mathbf{x})/x_1, f_2(t, \mathbf{x})/x_2, \ldots, f_n(t, \mathbf{x})/x_n\}, \tag{35}$$

but this choice is by no means unique nor best.

If the function $\mathbf{f}(t, \mathbf{x})$ is continuously differentiable with respect to x, then we recall from Apostol (1957) that for $0 \leqslant \mu \leqslant 1$ we can get

$$\mathbf{f}(t, \mathbf{x}+\mathbf{h}) - \mathbf{f}(t, \mathbf{x}) = \int_0^1 \frac{d}{d\mu}[\mathbf{f}(t, \mathbf{x}+\mu\mathbf{h})]\, d\mu = \int_0^1 J(t, \mathbf{x}+\mu\mathbf{h})\, d\mu, \tag{36}$$

where $J = (\partial f_i/\partial x_j)$ is the $n \times n$ Jacobian matrix of $f(t, \mathbf{x})$, and we choose

$$A(t, \mathbf{x}) = \int_0^1 J(t, \mu\mathbf{x}) \, d\mu. \tag{37}$$

The desired stability characteristics of ecosystems which can be represented by the system $\mathcal{S}$ of (33), is the following:

Definition 6. The equilibrium $\mathbf{x}^e = \mathbf{0}$ of the system $\mathcal{S}$ is connectively and exponentially stable in the large if and only if there exist numbers $\Pi \geq 1$, $\pi > 0$, which do not depend on initial conditions $(t_0, \mathbf{x}_0)$, and such that

$$\|\mathbf{x}(t; t_0, \mathbf{x}_0)\| \leq \Pi \|\mathbf{x}_0\| \exp[-\pi(t - t_0)], \quad \forall t \in \mathcal{T}_0 \tag{38}$$

for all $(t_0, \mathbf{x}_0) \in \mathcal{T} \times \mathcal{R}^n$ and all $E \in \bar{E}$.

The meaning of Definition 6 is fairly obvious: We want that stability of the system $\mathcal{S}$ to be invulnerable to structural perturbations. Furthermore, we want a certain degree of stability in such a way that after any impulse-type perturbation of the system state, the system returns to the equilibrium $\mathbf{x}^e = \mathbf{0}$ faster than an exponential. In linear systems (the matrix A is constant and independent of $\mathbf{x}$), this requirement is equivalent to asking that the largest eigenvalue of the matrix A be less or equal to $-\pi$. The constant π may be viewed as a measure of robustness (Harte, Chapter 18, this volume) since it indicates how much of environmental perturbations can be absorbed by the ecosystem as shown by May (1973), and Ladde and Šiljak (1976).

To establish the kind of stability specified in Definition 6, we introduce the following constraints on the interconnection functions $\phi(t, \mathbf{x})$ which appear in the coefficients $a_{ij}(t, \mathbf{x})$ in (34),

$$|\phi_{ij}(t, \mathbf{x})| < \alpha_{ij}, \quad \forall (t, \mathbf{x}) \in \mathcal{T} \times \mathcal{R}^n, \tag{39}$$

where α_{ij} are nonnegative numbers. Furthermore, we define the $n \times n$ constant matrix $\bar{A} = (\bar{a}_{ij})$ as

$$\bar{a}_{ij} = \bar{e}_{ij}\alpha_{ij}, \tag{40}$$

where $\bar{e}_{ij}$ are binary elements of the fundamental interconnection matrix $\bar{E}$.

Now, we again consider the system $\mathcal{S}$ of (33) and prove the following:

Theorem 6. The equilibrium $\mathbf{x}^e = \mathbf{0}$ of the system $\mathcal{S}$ is connectively and exponentially stable in the large if there exist numbers $d_i > 0$, $i = 1, 2, \ldots, n$, such that

$$d_i^{-1} \sum_{j=1}^n d_j \bar{a}_{ij} \leq \pi, \quad \forall i = 1, 2, \ldots, n, \tag{41}$$

where $0 < \pi < 1$.

Proof. Consider the function $v: \mathscr{R}^n \to \mathscr{R}_+$ defined by

$$v(\mathbf{x}) = \max_i \{d_i^{-1}|x_i|\}, \tag{42}$$

as a candidate for Liapunov's function of the system $\mathscr{S}$. We use (33) and (41) to compute

$$v[\mathbf{x}(t+1)] = \max_i \{d_i^{-1}|x_i(t+1)|\} = \max_i \left\{ d_i^{-1} \sum_{j=1}^n a_{ij} x_j(t) \right\}$$

$$\leq \max_i \left\{ d_i^{-1} \sum_{j=1}^n d_j \bar{a}_{ij} d_j^{-1} |x_j(t)| \right\} \tag{43}$$

$$\leq \max_i \left\{ d_i^{-1} \sum_{j=1}^n d_j \bar{a}_{ij} \right\} \max_j \{d_j^{-1}|x_j(t)|\}$$

$$\leq \pi v[\mathbf{x}(t)], \qquad \forall (t, \mathbf{x}) \in \mathscr{T} \times \mathscr{R}^n, \quad \forall E \in \bar{E}.$$

From (43), we get

$$v[\mathbf{x}(t+1)] \leq v(\mathbf{x}_0) \exp[-\pi(t - t_0)], \qquad \forall (t, \mathbf{x}) \in \mathscr{T} \times \mathscr{R}^n, \quad \forall E \in \bar{E}. \tag{44}$$

Finally, using (42) and (44), we get (38) where $\Pi = n d_M d_m^{-1}$, $d_M = \max_i d_i$, and $d_m = \min_i d_i$. This proves Theorem 6.

It is easy to show (Newman, 1961) that the condition (41) of Theorem 6 is equivalent to the negative quasidominant diagonal property of the matrix $\bar{B} = \bar{A} - I$, where I is the identity matrix, which was used by Šiljak (1975) to establish stability under structural perturbations of the continuous-time version of the system $\mathscr{S}$ in (33). From Newman (1961), it follows then that the condition (41) is equivalent to the condition

$$(-1)^k \begin{vmatrix} \bar{b}_{11} & \bar{b}_{12} & \cdots & \bar{b}_{1k} \\ \bar{b}_{21} & \bar{b}_{22} & \cdots & \bar{b}_{2k} \\ \bar{b}_{k1} & \bar{b}_{k2} & \cdots & \bar{b}_{kk} \end{vmatrix} > 0, \qquad \forall k = 1, 2, \ldots, n \tag{45}$$

which is easier to test than (41).

From (37), it follows that if the $n \times n$ Jacobian matrix $J = (\partial f_i/\partial x_j)$ satisfies (41) so does the corresponding matrix $A(t, \mathbf{x})$. Therefore, we consider the system $\mathscr{S}$ described by (32) and establish from Theorem 6 the following:

Corollary. *The equilibrium* $\mathbf{x}^e = \mathbf{0}$ *of the system* $\mathscr{S}$ *is exponentially stable in the large if there exist numbers* $d_i > 0$, $i = 1, 2, \ldots, n$ *such that*

$$d_i^{-1} \sum_{j=1}^{n} d_j |\partial f_i/\partial x_j| \leqslant \pi, \quad \forall\, (t, x) \in \mathcal{T} \times \mathcal{R}^n, \quad \forall\, i = 1, 2, \ldots, n \quad (46)$$

where $0 < \pi < 1$.

Using Theorem 6, it is simple to introduce the connectivity aspect of stability in the above corollary.

In their book on competitive economic analysis, Arrow and Hahn (1971) wrote: "The kind of result that we need here is one that would allow us to deduce global stability from the postulate that the Jacobian of excess supplies has everywhere DD (diagonal dominance)." The above corollary is a discrete-time version of the desired result. A continuous-time result can be easily obtained in terms of the negative quasi-dominance of the matrix $\bar{B} = \bar{A} - I$ (Šiljak, 1978).

5.2. Stability Regions

When a nonlinear model ecosystem has more than one equilibrium, it is not globally stable. Such is the case of the classical Lotka–Volterra used in the above example. If stability of a nontrivial equilibrium is established locally by linearization, then it is of interest to find out the extent of the stability region containing the nontrivial equilibrium. For continuous-time ecosystems, this problem was resolved by Šiljak (1975) and applied to the Lotka–Volterra model in the context of the arms race (Šiljak, 1977c).

Let us assume that the constraints on interactions (39) are not valid globally, but are restricted to a finite region $\mathcal{B} \subseteq \mathcal{R}^n$, which is a rectangle defined by

$$\mathcal{B} = \{\mathbf{x} \in \mathcal{R}^n : |x_i| < \mu_i, i = 1, 2, \ldots, n\}, \quad (47)$$

where all $\mu_i > 0$. For the Liapunov function we again choose $v(\mathbf{x})$ defined in (42), and compute a stability region $\mathcal{A} \subseteq \mathcal{B}$ of $\mathbf{x}^e = \mathbf{0}$ as

$$\mathcal{A} = \{\mathbf{x} \in \mathcal{R}^n : v(\mathbf{x}) < v^0\}, \quad (48)$$

where the number v^0 is computed by

$$v^0 = \min_{i} \{d_i^{-1} \mu_i\} \quad (49)$$

as shown by Weissenberger (1973). If the condition (41) is satisfied, then (43) holds in the entire region $\mathcal{A}$, and $v(\mathbf{x})$ is a decreasing function of t on $\mathcal{A}$. Since the boundary of $\mathcal{A}$ is a level surface of $v(\mathbf{x})$, the solutions that start in $\mathcal{A}$ remain in $\mathcal{A}$ and converge exponentially to the equilibrium $\mathbf{x}^e = \mathbf{0}$ as $t \to +\infty$. The set $\mathcal{A}$ is a region of exponential stability for the equilibrium $\mathbf{x}^e = \mathbf{0}$ of the system $\mathcal{S}$ described by (33). Since the function $v(\mathbf{x})$ has the same geometric shape as the region $\mathscr{B}$, it is possible to show that $\mathscr{A}$ is all of $\mathscr{B}$ (that is, $\mathscr{A} = \mathscr{B}$). Furthermore, on the basis of Theorem 6, it is easy to see that $\mathscr{A}$ is invariant under structural perturbations, that is, it is also a region of connective stability.

5.3. Lotka–Volterra Model

Let us consider a discrete version of the classical Lotka–Volterra model for n interacting populations,

$$y_i(t+1) = y_i(t)\left[1 + c_i + \sum_{j=1}^{n} b_{ij} y_j\right], \tag{50}$$

where $y_i(t)$ is the population of the ith species at time t, and all c_i's and b_{ij}'s are given numbers. We assume that all numbers b_{ii} are negative, but place no sign restrictions on the numbers b_{ij} ($i \neq j$). We suppose that the nontrivial equilibrium $\mathbf{y}^e \neq \mathbf{0}$ which is the solution of equations

$$c_i + \sum_{j=1}^{n} b_{ij} y_j = 0, \quad i = 1, 2, \ldots, n, \tag{51}$$

is such that $y_i^e > 0$, for all $i = 1, 2, \ldots, n$.

To investigate stability of $\mathbf{y}^e \neq \mathbf{0}$ using the model (33), we apply the transformation

$$\mathbf{y}(t) = \mathbf{y}^e + \mathbf{x}(t) \tag{52}$$

to Eqs. (50), and get

$$\mathbf{x}(t+1) = A[\mathbf{x}(t)]\mathbf{x}(t), \tag{53}$$

where the coefficients of the $n \times n$ functional matrix $A = (a_{ij})$ are defined by

$$a_{ij}(\mathbf{x}) = \begin{cases} 1 + (y_i^e + x_i) b_{ii}, & i = j \\ (y_i^e + x_i) b_{ij}, & i \neq j. \end{cases} \tag{54}$$

From (34) and (54), we conclude that the region $\mathscr{B}$ defined in (47) can be chosen as

$$\mathscr{B} = \{\mathbf{x} \in \mathscr{R}^n : |x_i| < y_i^e - \varepsilon, i = 1, 2, \ldots, n\}, \tag{55}$$

where $\varepsilon > 0$ can be selected as an arbitrarily small number.

Imitating the development in Šiljak (1977c), we can show that the conditions (45) on the numbers

$$\bar{b}_{ij} = \begin{cases} -\bar{e}_{ii} b_{ii}, & i = j, \\ \bar{e}_{ij} b_{ij}, & i \neq j, \end{cases} \tag{56}$$

imply connective and exponential stability throughout the entire region $\mathscr{B}$

and that, in fact, $\mathscr{A} = \mathscr{B}$. To see this we recall from Newman (1959) that if an $n \times n$ matrix $\bar{B} = (\bar{b}_{ij})$ satisfies conditions (45) so does the matrix $D\bar{B}$, where $D = \mathrm{diag}\{\varepsilon_1, \varepsilon_2, \ldots, \varepsilon_n\}$, which, in turn, implies (Newman, 1961) that the matrix $A = \bar{D} + I$ satisfies the conditions (41) of Theorem 6. Therefore, $\mathscr{A}$ is all of $\mathscr{B}$ defined in (55).

The obtained result can be interpreted in many different ways (Šiljak, 1975). It includes species extinction or addition of species considered by MacArthur (1970) but without prescribing the symmetry conditions on B. Furthermore, by observing that the result is valid even if the number b_{ij} are replaced by functions $b_{ij}(t, \mathbf{x})$, which are bounded by these numbers, we conclude considerable robustness of community stability established by conditions (45). Stability is invariant to changes in fundamental interconnection variables, which include predator switching, saturation in predator attack capacities, etc.

We should make a note that stability of the nontrivial equilibrium of the Lotka–Volterra model was established in a finite rectangular region containing the equilibrium, which is located in the corner of the positive quadrant of the species space. In an earlier work of MacArthur (1970), a claim is made that similar conditions imply *global* stability in the entire positive quadrant, since a positive definite quadratic form can be shown to have a negative definite derivative throughout the first quadrant. Due to the fact that the boundary of the quadrant is not a level surface of a quadratic form, it is not clear that the claim is valid. Goh (1977) improved MacArthur's conditions and, at the same time, repeated his claim by referring to a *feasible region* as the entire first quadrant of the species space. The stability region determined here is not the full first quadrant, but it is invariant (invulnerable) to structural perturbations and simultaneously robust with respect to nonlinearities in the interactions among species in the community (Šiljak, 1976). Furthermore, the stability conditions (45) can be easily tested.

It may be argued at this point that the dominant diagonal condition for the community matrix, which established vulnerability of stability and robustness of the community models, is an overly restrictive condition. It requires that each species (or organism) is density-dependent in (at least) finite region of the state space around the equilibrium. This restriction can be removed by showing (Šiljak, 1976, 1978) that the populations are ultimately bounded. That is, they converge to a region in the state space and if once there, they stay there for all future times. This region takes the role of the equilibrium, and it can be estimated by Liapunov functions proposed here and in Šiljak (1978).

The density-dependence restriction can also be removed by hierarchical analysis proposed by Šiljak (1975). This is considered next.

5.4. Hierarchical Models

For various structural or methodological reasons, it is desirable to lump groups of species (or organisma) together and form interconnected subcommunities (see Webster, Chapter 5, this volume). Analysis of composite communities can provide insights into population processes that may be otherwise hidden from the observer so long as the entire community is treated as a whole. For this reason, a hierarchical stability analysis was initiated (Šiljak, 1975), which combines the detailed analysis of the subcommunities with much coarser analysis of the aggregate community model formed on the higher level of the community interconnection structure. The approach has been extended (Ladde and Šiljak, 1976) to include stochastic stability of multispecies communities in randomly varying environment and under structural perturbations. In this section, we extend these results to discrete-time ecosystems relying on the work by Bitsoris and Burgat (1977).

Suppose that the system $\mathscr{S}$ of (32) is decomposed into s $(s \leqslant n)$ interconnected subsystems $\mathscr{S}_i$ described by the difference equations

$$\mathbf{x}_i(t+1) = \mathbf{g}_i[t, \mathbf{x}_i(t)] + \mathbf{h}_i[t, \mathbf{x}], \qquad i = 1, 2, \ldots, s, \qquad (57)$$

where $\mathbf{x}_i \in \mathscr{R}^{n_i}$ is the state of $\mathscr{S}_i$ at time t so that the state of $\mathscr{S}$ is $\mathbf{x} = (\mathbf{x}_1^T, \mathbf{x}_2^T, \ldots, \mathbf{x}_s^T)^T$ and $\mathscr{R}^n = \mathscr{R}^{n_1} \times \mathscr{R}^{n_2} \times \cdots \times \mathscr{R}^{n_s}$. The functions $\mathbf{g}_i : \mathscr{T} \times \mathscr{R}^{n_i} \to \mathscr{R}^{n_i}$ specify the free (isolated) subsystems $\mathscr{S}_i^0$ described by the difference equations

$$\mathbf{x}_i(t+1) = \mathbf{g}_i[t, \mathbf{x}_i(t)], \qquad i = 1, 2, \ldots, s, \qquad (58)$$

where $\mathbf{g}_i(t, \mathbf{0}) = \mathbf{0}$ for all $t \in \mathscr{T}$, and $\mathbf{x}_i = \mathbf{0}$ is the unique equilibrium of $\mathscr{S}_i^0$.

We assume that each free subsystem $\mathscr{S}_i^0$ is exponentially stable in the large and that we can associate with each $\mathscr{S}_i^0$ a function $v_i : \mathscr{T} \times \mathscr{R}^{n_i} \to \mathscr{R}_+$, which is continuous in both of its arguments and satisfies the inequalities

$$\eta_{i1} \|\mathbf{x}_i\| \leqslant v_i(t, \mathbf{x}_i) \leqslant \eta_{i2} \|\mathbf{x}_i\|$$
$$\Delta v_i(t, \mathbf{x}_i)_{(58)} \leqslant -\eta_{i3} \|\mathbf{x}_i\| \qquad (59)$$
$$\|v_i(t, \mathbf{x}_i') - v_i(t, \mathbf{x}_i'')\| \leqslant \eta_{i4} \|\mathbf{x}_i' - \mathbf{x}_i''\|, \qquad \forall\, (t, \mathbf{x}_i) \in \mathscr{T} \times \mathscr{R}^{n_i},$$

where $\eta_{i1} > 0$, $\eta_{i2} > 0$, $\eta_{i4} > 0$, $0 < \eta_{i3} < 1$ are all positive numbers. In (59), $\Delta v_i(t, \mathbf{x}_i)_{(58)} \equiv v_i[t+1, \mathbf{x}_i(t+1)] - v_i[t, \mathbf{x}_i(t)]$ is the forward difference of the function $v_i(t, \mathbf{x}_i)$ computed with respect to (58). That is,

$$\Delta v_i(t, \mathbf{x}_i)_{(58)} = v_i\{t+1, \mathbf{g}_i[t, \mathbf{x}_i(t)]\} - v_i[t, \mathbf{x}_i(t)]. \qquad (60)$$

As for the interconnections $\mathbf{h}_i(t, \mathbf{x})$ among the subsystems $\mathscr{S}_i^0$, we assume the following constraint:

$$\|\mathbf{h}_i(t, \mathbf{x}_i)\| \leq \sum_{j=1}^{s} \bar{e}_{ij}^{*} \xi_{ij} v_j(t, \mathbf{x}_j), \quad \forall (t, \mathbf{x}) \in \mathcal{T} \times \mathcal{R}^n, \tag{61}$$

where $\bar{e}_{ij}^{*}$ are the elements of the $s \times s$ fundamental interconnection matrix $\bar{E}^*$ corresponding to the condensation $\bar{\mathcal{D}}_x^*$ of (57).

Now, by using (59) and (61), we compute

$$\Delta v_i(t, \mathbf{x}_i)_{(57)} \leq -\eta_{i3} v_i(t, \mathbf{x}_i) + \eta_{i4} \sum_{j=1}^{s} \bar{e}_{ij}^{*} \xi_{ij} v_j(t, \mathbf{x}_j), \quad i = 1, 2, \ldots, s. \tag{62}$$

Introducing the vector Liapunov function $\mathbf{v}: \mathcal{T} \times \mathcal{R}^n \to \mathcal{R}_+^s$, as $\mathbf{v} = (v_1, v_2, \ldots, v_s)^T$, we can rewrite (62) as a vector difference inequality

$$\mathbf{v}(t+1) \leq \bar{A} \mathbf{v}(t), \tag{63}$$

which is valid for all $(t, \mathbf{x}) \in \mathcal{T} \times \mathcal{R}^n$ and all $E^* \in \bar{E}^*$. The elements $\bar{a}_{ij}$ of the $s \times s$ matrix $\bar{A}$ are defined as

$$\bar{a}_{ij} = \begin{cases} 1 - \eta_{i3} + \bar{e}_{ii}^{*} \xi_{ii}, & i = j, \\ \bar{e}_{ij}^{*} \xi_{ij} \eta_{i4}, & i \neq j. \end{cases} \tag{64}$$

The difference inequality (63) is an aggregate model which describes the processes in the ecosystem $\mathcal{S}$ on a higher hierarchical level. In the aggregate model, each subcommunity is represented by a scalar Liapunov function v_i which is a component of the vector Liapunov function $\mathbf{v}$. Stability of each subcommunity and stability of the aggregate model imply stability of the overall community. This fact is established for the system $\mathcal{S}$ of (57) by:

Theorem 7. The equilibrium $\mathbf{x}^e = \mathbf{0}$ of the system $\mathcal{S}$ is connectively and exponentially stable in the large if there exist numbers $d_i > 0$, $i = 1, 2, \ldots, s$, such that

$$d_i^{-1} \sum_{j=1}^{s} d_j \bar{a}_{ij} \leq \pi, \quad \forall i = 1, 2, \ldots, s, \tag{65}$$

where $0 < \pi < 1$.

Proof. Let us consider the function $v: \mathcal{R}_+^s \to \mathcal{R}_+$ defined by

$$v(\mathbf{v}) = \max_{i} \{d_i^{-1} v_i\}, \tag{66}$$

as a candidate for Liapunov's function of the system $\mathcal{S}$. We use (63) and (65) to compute

$$v[\mathbf{v}(t+1)] = \max_{i} \{d_i^{-1} v_i(t+1)\} = \max_{i} \left\{ d_i^{-1} \sum_{j=1}^{s} \bar{a}_{ij} v_j(t) \right\}$$

$$\leqslant \max_i \left\{ d_i^{-1} \sum_{j=1}^{s} d_j \bar{a}_{ij} d_j^{-1} v_j(t) \right\}$$

$$\leqslant \max_i \left\{ d_i^{-1} \sum_{j=1}^{s} d_j \bar{a}_{ij} \right\} \max_j \left\{ d_j^{-1} v_j(t) \right\}$$

$$\leqslant \pi v[\mathbf{v}(t)], \qquad \forall\, (t,x) \in \mathcal{T} \times \mathcal{R}^n, \quad \forall\, E^* \in \bar{E}^*. \tag{67}$$

From (67), we get

$$v[\mathbf{v}(t+1)] \leqslant v(\mathbf{v}_0) \exp[-\pi(t-t_0)] \qquad \forall\, (t,x) \in \mathcal{T} \times \mathcal{R}^n, \quad \forall\, E^* \in \bar{E}^*. \tag{68}$$

By using the well-known relations among norms (Šiljak, 1977c), we can rewrite (68) as (38) where $\Pi = n d_m^{-1} d_M \eta_{m1}^{-1} \eta_{M2}$. This proves Theorem 7.

Conceptually, Theorem 7 says that if each free subsystem is stable and the interactions among the subsystems are properly bounded, then the entire multilevel system is stable. Furthermore, stability if established in this structure is inherently reliable characteristic of the system. An inherent robustness of hierarchical processes in ecology was promoted and explained by intuitive arguments in Simon (1962). Theorem 7 gives a rigorous support to Simon's arguments.

6. CONCLUSION

Directed graphs have been associated with dynamic systems in an essential way to formulate, analyze, and resolve various problems in the context of stability, complexity, and vulnerability of model ecosystems. It has been shown that vulnerability of stability in ecosystems, which are composed of interconnected subsystems, can be eliminated by increasing autonomy of each subsystem. That is, an ecosystem can undergo topological changes and remain stable if each subsystem is stable and the size of interactions is properly limited. This fact provides a partial answer to the central problem of complexity versus stability in ecological studies, where the structural aspects are essential. A great deal of work remains to be done to exploit further the association of dynamic systems and directed graphs and to come up with new and important results concerning the dynamic processes in complex ecosystems.

REFERENCES

Apostol, T. M. (1957). "Mathematical Analysis." Addison-Wesley, Reading, Massachusetts.
Arrow, K. J., and Hahn, F. H. (1971). "General Competitive Analysis." Holden-Day, San Francisco, California.

Bitsoris, G., and Burgat, C. (1977). Stability analysis of complex discrete systems with locally and globally stable subsystems. *Int. J. Control* **25**, 413–424.

Boesch, F. T., and Thomas, R. E. (1970). On graphs of invulnerable communication nets. *IEEE Trans. Circuit Theory* **ct-17**, 183–192.

Bowie, W. S. (1976). Application of graph theory in computer systems. *Int. J. Comput. Inf. Sci.* **5**, 9–31.

Caswell, H., Koenig, H. E., Resh, J. A., and Ross, Q. E. (1972). *In* "Systems Analysis and Simulation in Ecology" (B. C. Patten, ed.), Vol. 2, pp. 3–78. Academic Press, New York.

Chan, W., and Frank, H. (1970). Survivable communication networks and the terminal capacity matrix. *IEEE Trans. Circuit Theory* **ct-17**, 192–197.

Gallopin, C. C. (1972). Structural properties of food webs. *In* "Systems Analysis and Simulation in Ecology" (B. C. Patten, ed.), Vol. 2, pp. 241–282. Academic Press, New York.

Goh, B. S. (1975). Stability, vulnerability and persistence of complex ecosystems. *Ecol. Modell.* **1**, 105–116.

Goh, B. S. (1976). Nonvulnerability of ecosystems in unpredictable environments. *Theor. Pop. Biol.* **10**, 83–95.

Goh, B. S. (1977). Global stability in many species systems. *Am. Nat.* **111**, 135–143.

Hahn, W. (1967). "Stability of Motion." Springer-Verlag, Berlin and New York.

Harary, F. (1969). "Graph Theory." Addison-Wesley, Reading, Massachusetts.

Harary, F., Norman, R. Z., and Cartwright, D. (1965). "Structural Models: An Introduction to the Theory of Directed Graphs." Wiley, New York.

Kalman, R. E., and Bertram, J. E. (1960). Control system analysis and design via the 'second method' of Lyapunov. *Trans. ASME* **82**, Parts I and II, 371–393 and 394–400.

Kalman, R. E., Falb, P. L., and Arbib, M. A. (1969). "Topics in Mathematical System Theory." McGraw-Hill, New York.

Kevorkian, A. K. (1975). Structural aspects of large dynamic systems. *Proc. 6th IFAC Congr.*, Boston, Massachusetts, Section 19.3; pp. 1–7.

Keyfitz, N. (1968). "Introduction to the Mathematics of Polulations." Addison-Wesley, Reading, Massachusetts.

Innis, G. (1975). Stability, sensitivity, resilience, persistence. What is of interest? *In* "Ecosystem Analysis and Prediction" (S. A. Levin, ed.), pp. 131–139. SIAM, Philadelphia, Pennsylvania.

Ladde, G. S., and Šiljak, D. D. (1976). Stability of multispecies communities in randomly varying environment. *J. Math. Biol.* **1**, 165–178.

Levin, S. A. (1975). Pollutants in ecosystems. *In* "Ecosystem Analysis and Prediction" (S. A. Levin, ed.), pp. 4–8. SIAM, Philadelphia, Pennsylvania.

Lotka, A. J. (1925). "Elements of Physical Biology." Williams & Wilkins, Baltimore, Maryland (reissued as "Elements of Mathematical Biology." Dover, New York, 1956).

MacArthur, R. H. (1970). Species packing and competitive equilibrium for many species. *Theor. Pop. Biol.* **1**, 1–11.

May, R. M. (1973). "Stability and Complexity of Model Ecosystems." Princeton Univ. Press, Princeton, New Jersey.

Munro, I. (1971). Efficient determination of the transitive closure of a directed graph. *Inf. Proc. Lett.* **1**, 56–58.

Newman, P. (1959). Some notes on stability conditions. *Rev. Econ. Stud.* **72**, 1–9.

Newman, P. (1961). Approaches to stability analysis. *Economica* **28**, 12–29.

Paine, R. T. (1966). Food web complexity and species diversity. *Am. Nat.* **100**, 65–75.

Patten, B. C., and Wikamp, M. (1967). System analysis of 134 cesium kinetics in terrestrial microcosm. *Ecology* **48**, 813–824.

Purdom, P. (1970). A transitive closure algorithm. *BIT* **10**, 76–94.

Šiljak, D. D. (1974). Connective stability of complex ecosystems. *Nature (London)* **249**, 280.
Šiljak, D. D. (1975). When is a complex ecosystem stable? *Math. Biosci.* **25**, 25–50.
Šiljak, D. D. (1976). Competitive economic systems: Stability, decomposition, and aggregation. *IEEE Trans. Autom. Control* **ac-21**, 149–160.
Šiljak, D. D. (1977a). On reachability of dynamic systems. *Int. J. Syst. Sci.* **8**, 321–338.
Šiljak, D. D. (1977b). On pure structure of dynamic systems. *Nonlinear Anal. Theory, Methods Appl.* **1**, 397–413.
Šiljak, D. D. (1977c). On the stability of the arms race. *In* "Mathematical Systems in Internation Relations Research" (J. V. Gillespie and D. A. Zinnes, eds.), pp. 264–304. Praeger, New York.
Šiljak, D. D. (1978). "Large-scale Dynamic Systems: Stability and Structure." North-Holland Publ., Amsterdam.
Simon, H. A. (1962). The architecture of complexity. *Proc. Am. Philos. Soc.* **106**, 467–482.
Tarjan, R. E. (1972). Depth-first search and linear graph algorithms. *SIAM J. Comput.* **1**, 146–160.
Volterra, V. (1926). Variazioni e fluttuazioni del numero d'individui in specie animali conviventi. *Atti R. Accad. Naz. Lincei, Mem. Cl. Sci. Fis., Mat. Nat.* **2**, 31–113 (translated in R. N. Chapman, "Animal Ecology." McGraw-Hill, New York, 1931).
Warshall, S. (1962). A theorem on Boolean matrices. *J. Assoc. Comput. Mach.* **9**, 11–12.
Weissenberger, S. (1973). Stability regions of large-scale systems. *Automatica* **9**, 653–663.
Wickwire, K. (1977). Mathematical models for the control of pests and infectious diseases: A survey. *Theor. Pop. Biol.* **11**, 182–238.

Chapter **8**

SYSTEMS APPROACH TO CONTINENTAL SHELF ECOSYSTEMS

Bernard C. Patten and John T. Finn

1. Introduction	184
2. Causal Theory of Environment	185
2.1 Causal Objects and Systems	185
2.2 Duality of Environment	190
2.3 Subject/Environment Unity	191
3. Causal Analysis of Ecosystems	192
3.1 Marine Coprophagy Model	192
3.2 Transitive Closure Property	193
3.3 Flow Analysis	194
3.4 Ecosystem Partition into Component Environments	194
3.5 Other Methods of Causal Analysis	198
4. Flow Analysis of the Ross Sea Pelagic Ecosystem	199
4.1 Model Description	199
4.2 Genon Analysis	202
4.3 Creaon Analysis	206
4.4 Carbon Cycling in the Ross Sea Model	207
4.5 Comments on the Ross Sea Model	208
5. Summary	209
References	210

1. INTRODUCTION

The proposition advanced in this paper is that continental shelves are ecosystems whose systematic use by man requires whole-system synthesis. All social, economic, legal, and political aspects of food, mineral, and energy exploitation of the world's shelves are circumscribed by ecological constraints. Human involvement with continental shelves is not and cannot be extraecological.

Whole-ecosystem synthesis is a science in prospect, not a reality. Figure 1 shows its elements displayed as a cybernetic feedback process. Objectives and qualitative and quantitative knowledge are converted into predictions and explanations of behavior, structure, function, and organization. *Modeling* may be defined as iterative, nonunique mapping from some physical domain to some logical domain, with reference to objectives, and usually requiring reduction in size or complexity. It is an abstraction process, common to all perceptual contact with and synthesis of the real world by any observer.

Ecosystem modeling involves the following steps:

(a) Choosing conceptual and physical boundaries of the system to be studied;

(b) Identifying system components, including those of direct interest (such as fisheries) and those of interest because they influence the former (e.g., detritus and dissolved organic carbon);

(c) Identifying interactions between system components and factors controlling these interactions; and

(d) Describing the above in a practical and realistic mathematical form.

The mathematical model may be designed to *simulate* behavior and make predictions, or to *analyze* for properties that give explanatory understanding (Caswell, 1976). The same model is not necessarily optimal for both purposes. In the recent history of systems ecology (e.g., IBP Analysis of Ecosystem "Biome" Projects; see Patten *et al.*, 1975; Vol. 3, Part I), step (d) above has received the greatest attention. This is because of the generally quantitative, data-oriented emphasis of ecosystem ecology. It is being learned, however, that qualitative, conceptual phases of modeling are of utmost significance; decisions that influence a model's behavior and realism most occur during the first three steps.

The above approach to modeling is mechanistic in that it constructs wholes from parts. This is an essential difference between a model and its real-world counterpart. In the latter, wholes exist before parts. Any given ecosystem (except, possibly, one in early stages of primary succession)

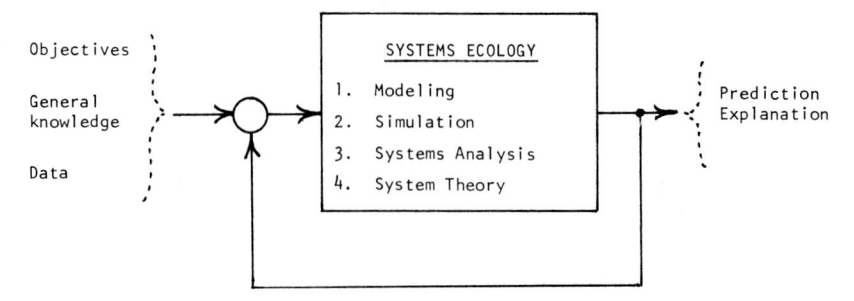

Figure 1. Elements of systems ecology.

predates its contemporary components, which serve only to carry past system organization into a future, perhaps slightly changed, organization. The organization is persistent over parts, and a methodology for constructing models holistically to reflect this is not apparent at the present time. What is apparent is that the ecologists' reflex appeal to mechanisms, in the IBP and elsewhere, has failed to produce on first attempt satisfactory ecosystem models. That ecology has overlooked something basic in its view of the ecosystem as a system is possible (Patten, 1975a) and indicates need for the fourth element of systems ecology listed in Fig. 1, system *theory*. To achieve whole-system synthesis, for continental shelves as well as for other ecosystems, ecology must develop a special ecosystem theory that is consistent with modern, formal General Systems Theory.

Our current version of a special ecosystem theory (Patten *et al.*, 1976) is based on the philosophical principle of causality. This theory begins by not taking the definition of environment for granted, with the consequence that the organism/environment relation comes under scrutiny in a proper domain of theoretical ecology. The purposes of this paper are twofold:

(1) To outline this causal theory of the organism/environment relation, and to show its relevance to origin of ecosystems; and

(2) To illustrate application of the theory by causal analysis of a continental shelf ecosystem model.

2. CAUSAL THEORY OF ENVIRONMENT

2.1. Causal Objects and Systems

The Institute of Ecology (TIE, 1973) defines *environment* as "the sum total or the resultant of all the external conditions which act upon an organism." Whatever virtue this definition has, it cannot be said to have rigor or formality. Official ecology (i.e., TIE) treats the definition of

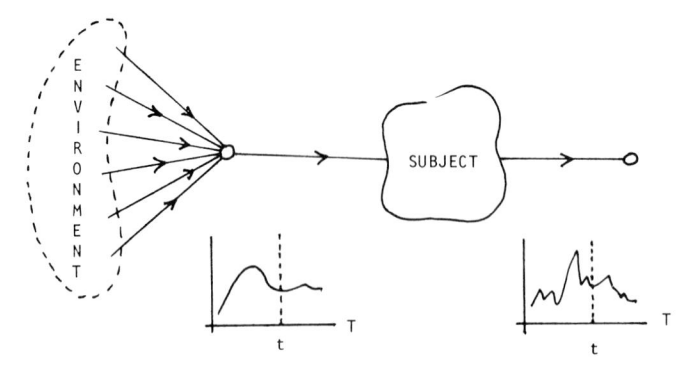

Figure 2. Causal concept of environment.

environment as a trivial matter. Two important features of an environmental concept are encompassed by the definition in question, however; environment is (a) external to a defined subject (organism) which it (b) influences. As environment changes the subject reacts; environmental change is causal in that subject change is produced by it (Fig. 2).

To model a causal link between two entities requires some kind of

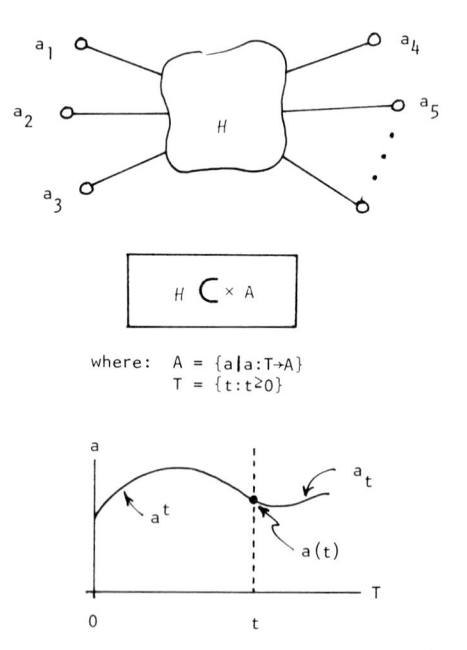

$$H \subset \times A$$

where: $A = \{a \mid a : T \to A\}$
$T = \{t : t \geq 0\}$

Figure 3. Concept of a holon (Koestler, 1967), abstract object (Zadeh and Desoer, 1963), or general time system (Mesarovic and Takahara, 1975).

process or object whose action converts cause to effect. This object, as the thing of most immediate interest, corresponds to subject in the subject/environment pair. Such objects are formally modeled in general system theory, two of which (Zadeh and Desoer, 1963; Mesarovic and Takahara, 1975) will be combined here to meet present purposes.

Figure 3 illustrates the canonical notion of a general systems object for which Koestler's (1967) term *holon* will be used. The object $\mathcal{H}$ is a relation on a set A of attributes that are time functions in a time domain T. For each $a \in A$, a is a behavior $\forall\, t \in T$, $a(t)$ is the value of a at time $t \in T$, a^t is the segment of a prior to t, and a_t is the behavior segment of a beginning at and following t. This object definition gives complete latitude in selecting the set A of behavioral attributes.

The holon becomes oriented when its attribute set is partitioned into inputs Z and outputs Y (Fig. 4). The relation $\mathcal{H}$ on A is then expressed as a

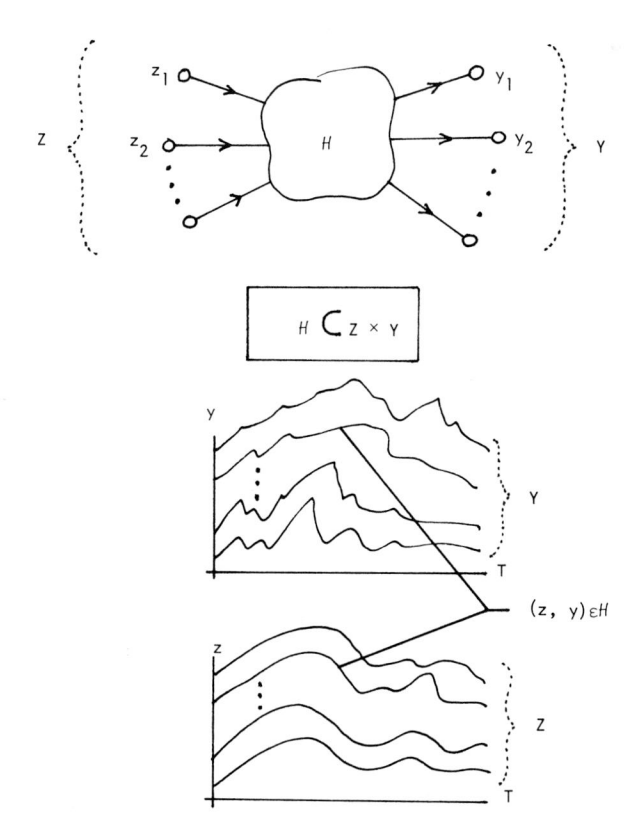

Figure 4. Oriented input–output holon, $\mathcal{H}$ a set of (z, y) pairs.

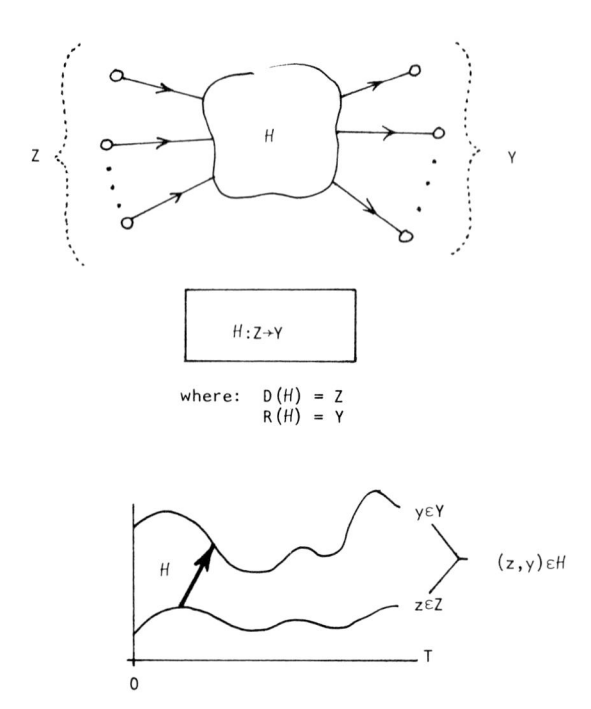

Figure 5. Functional holon, $\mathscr{H}$ a function (with domain D and range R) mapping inputs z into outputs y.

set of input–output time segments, $(z, y) \in \mathscr{H}$, $z \in Z$, and $y \in Y$. The oriented holon associates response (output) time sequences with stimulus (input) histories. In developing the holon as a causal object, given output sequences must be uniquely associated with given input segments. This property is embodied in the functional holon, where $\mathscr{H}$ is construed as a map (function) of inputs z into outputs y (Fig. 5). Such an object is said to be *determinate*, that is, a time series of inputs from its environment uniquely determines a corresponding time series of outputs.

Thus, dynamic behavior of a determinate object occurs in response to the object's environment's behavior which is received as input. This is modeled by introducing a third set X of object variables, states. Heuristically, inputs $z \in Z$ serve to map time $t \in T$ into states $x \in X$, and the states take inputs $z \in Z$ uniquely into outputs $y \in Y$. States are generated by a state transition function $f: Z \times X \to X$, and outputs by a response function $g: Z \times X \to Y$.

The only other requirement for a determinate holon to be causal is that it not respond at time t to inputs received after t. That is, the object

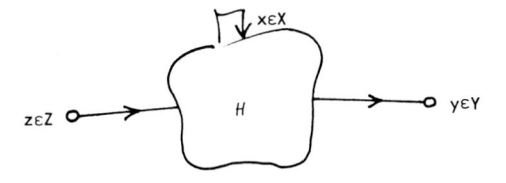

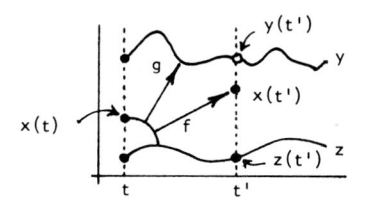

8. Systems Approach to Continental Shelf Ecosystems 189

Figure 6. The causal holon. Inputs z map time determinately into states x (function f), which in turn take inputs uniquely and without anticipation into outputs y (function g). Note in the case shown that input $z(t')$ (closed circle) arrives before output $y(t')$ (open circle) is generated.

cannot anticipate its future environment; it is *nonanticipatory*. If a determinate object were to generate more than one output sequence corresponding to a given input sequence, the only way it could do this (since it is determinate) would be based on information about the future. This possibility is precluded for the causal object, whose determinate and nonanticipatory characteristics are depicted in Fig. 6.

The causal holon is the building block of causal systems in a modeling paradigm in which systems (wholes) represent a synthesis of objects (parts). The ecosystem may be considered a special case of the general causal system whose essential property (Patten, 1973, 1975b; Patten *et al.*, 1976) is causal closure. That is, ecosystems are circular causal systems (Hutchinson, 1948) whose component objects are mutually causally related (Patten *et al.*, 1976) through nested feedback loops, the grandest of which are provided by biogeochemical cycles. Any cause introduced at the interface between an ecosystem and its environment propagates around the influence network defined by component interactions and ramifies throughout the system to return eventually in dissipated strength to the point of original introduction. Propagation of cause in complex networks is difficult to understand and interpret. In all cases, however, it is fundamental in the causal model that the elemental source of system behavior is exogenous input, that is, environment.

2.2. Duality of Environment

The causal model of subject/environment leads to not one, but to two equally plausible and useful concepts of environment. The first (Figs. 2 and 7) is that of an *input environment* $\mathcal{H}'$, defined by holon $\mathcal{H}$ in the act of a signal perception. Behavioral attributes of the real world that are not perceived as input by $\mathcal{H}$ cannot influence the state of the object. They go unrecorded by $\mathcal{H}$ and consequently are not part of its environment. Such attributes may as well not exist where $\mathcal{H}$ is concerned. So basic is this environment-defining function that this aspect of the holon is given (Patten, 1975c; Patten *et al.*, 1976) a special name, *creaon*, to signify an act of environment creation (Fig. 7). Mason and Langenheim (1957) restrict the

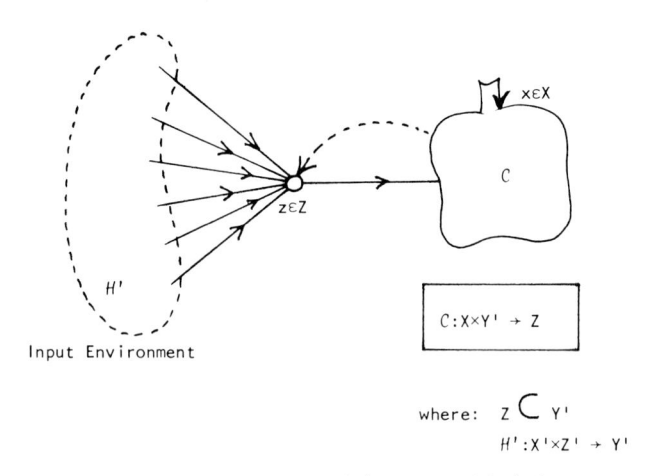

Figure 7. Input environment-defining aspect of the holon.

concept of (input) environment to phenomena that "directly impinge" upon the organism, but Patten *et al.* (1976) include both directly perceived inputs and the indirect nexus of causes from which they are generated. This is more consistent with a systems view and in context of finite ecosystem models does not produce an infinite causal regress, to which Mason and Langenheim objected. The regress is traceable only to the model boundary, becoming beyond this merely undifferentiated input. Metaphysically, the creaon is not passive in reception. Input environment definition is thus taken to be an active process (Fig. 7, broken arrow) of every influenced thing.

Reciprocally, the second concept of environment is that of an *output environment* $\mathcal{H}''$ (Fig. 8). This begins as a set of potential environments embodied in the states of holon $\mathcal{H}$. These states are converted to outputs

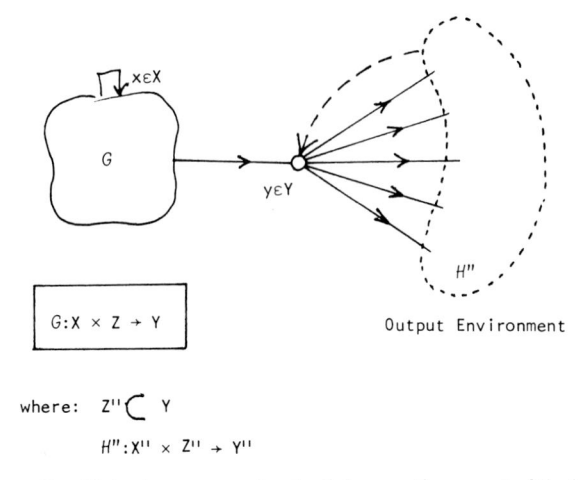

Figure 8. Output environment potential-generating aspect of the holon.

through interaction of $\mathcal{H}$ with other objects (creaons). That is, to produce an actual output environment from potential environments implicit in the state structure of $\mathcal{H}$ requires actualization as input to other holons. Output environment is the causality propagated from $\mathcal{H}$ as a network of direct and indirect effects. Mason and Langenheim (1957) exclude output environment completely from the general concept of environment, but von Uexküll (1926) includes it in his notion of the "function circles" of an organism. The environment-generating property of holons is equally basic to the creaon function and to distinguish it the name *genon* (Fig. 8) is given (Patten, 1975c; Patten *et al.*, 1976). As in the creaon case, an infinite progression of effects from $\mathcal{H}$ is implied, but in the context of finite models the progression terminates at the model boundary beyond which only undifferentiated output is recognized. Like input environment definition, output environment generation is also an active process. It requires genon production of potential attributes and then sequential creaon selections to achieve realization of these potentials.

Patten (1975c) and Patten *et al.* (1976) emphasize the distinctness of input and output environments $\mathcal{H}'$ and $\mathcal{H}''$, respectively, and demonstrate that both may serve as valid decompositions of an ecosystem. This is shown again here in Section 3 below.

2.3. Subject/Environment Unity

Figure 9 summarizes the foregoing theory of the subject/environment relationship. The subject model, holon $\mathcal{H}$, is depicted as a creaon/genon

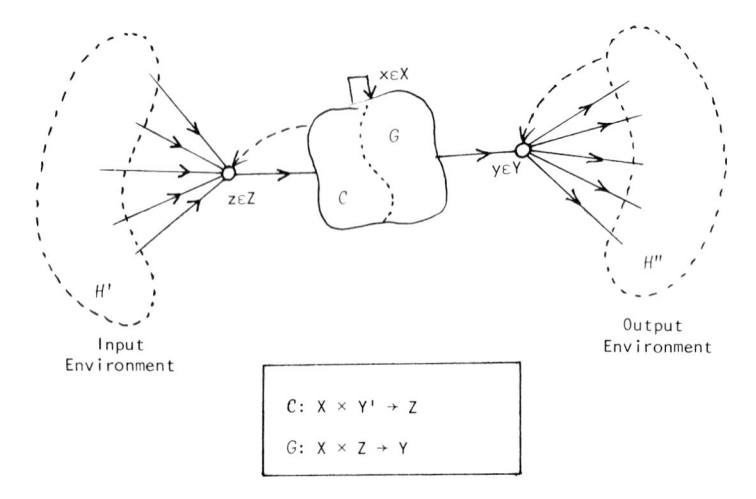

Figure 9. Holon as creaon/genon pair together with input and output environments.

pair. The creaon $\mathscr{C}$ maps holon states X and input environment potential outputs into holon inputs, $Y' = Z$. The genon $\mathscr{G}$ takes states X and inputs Z into potential outputs, subsequently to be realized as actual outputs, $Y = Z''$. The input and output environments resulting from these holon actions are so logically dependent upon the latter that in effect they are inseparable from it. On the same theme, Patten et al. (1976) provide a Goedel's theorem argument showing that the internal cause propagating structure of a system $\mathscr{H}$ cannot be determined to be complete (all pathways in the endogenous network accounted for) in absence of contact with an external input or output environment. Heuristically, it is meaningless to consider an entity, such as an organism, independently of its environments. Subject and environment are logically and ontologically inseparable, and they form a unit. This subject/environment unity provides a basis for composition or decomposition of ecosystems, as follows.

3. CAUSAL ANALYSIS OF ECOSYSTEMS

3.1. Marine Coprophagy Model

Figure 10 illustrates a simple steady-state model of marine coprophagy (Cale and Ramsey, 1970; description in Patten et al., 1976, Appendix). The model consists of four holons in series, with a feedback loop connecting $\mathscr{H}_3$ and $\mathscr{H}_4$. Causality is limited to expression as carbon flow

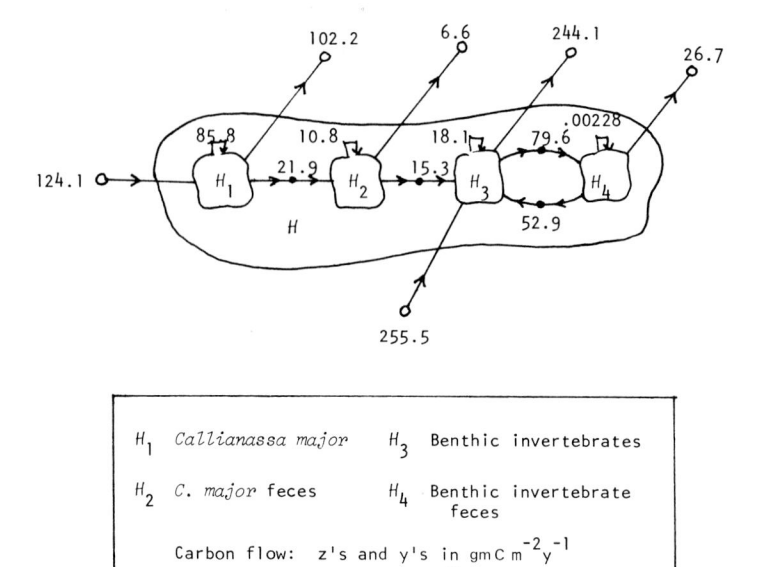

8. Systems Approach to Continental Shelf Ecosystems

Figure 10. Marine coprophagy model (Cale and Ramsay, 1970).

H_1 *Callianassa major* H_3 Benthic invertebrates

H_2 *C. major* feces H_4 Benthic invertebrate feces

Carbon flow: z's and y's in gm C m^{-2} y^{-1}

x's in gm C m^{-2}

(gm C m^{-2} yr^{-1}) and system state is represented by carbon storages (gm C m^{-2}). Inputs are received at $\mathcal{H}_1$ and $\mathcal{H}_3$, and outputs are generated (respiration) by all four holons. For illustrative purposes, this model may be taken as a reference "ecosystem." The carbon flows and contents indicated in Fig. 10 are actually measurable in nature. The point will be to decompose this example system into input and output environments associated with each of the four component holons.

3.2. Transitive Closure Property

To perform such an analysis requires that all causal pathways be accounted for. This so-called "transitive closure" property (Ore, 1962) is illustrated for the marine coprophagy model by the set of matrices shown in Table I. Let $B = (b_{ij})$ be a binary Boolean adjacency matrix denoting direct causal coupling (paths of length one) from $\mathcal{H}_j$ to $\mathcal{H}_i$, $i, j = 1, \ldots, 4$. Performing matrix multiplication, B^2 entries identify indirect couplings via paths of length two, B^3 via paths of length three, and in general B^k via paths of length k. The Table I matrices B, B^2, and B^3 may be readily verified by reference to Fig. 10.

Table I Boolean Matrices for the Marine Coprophagy Model

$$B = \begin{bmatrix} 0 & 0 & 0 & 0 \\ 1 & 0 & 0 & 0 \\ 0 & 1 & 0 & 1 \\ 0 & 0 & 1 & 0 \end{bmatrix} \qquad B^2 = \begin{bmatrix} 0 & 0 & 0 & 0 \\ 0 & 0 & 0 & 0 \\ 1 & 0 & 1 & 0 \\ 0 & 1 & 0 & 1 \end{bmatrix}$$

$$B^3 = \begin{bmatrix} 0 & 0 & 0 & 0 \\ 0 & 0 & 0 & 0 \\ 0 & 1 & 0 & 1 \\ 1 & 0 & 1 & 0 \end{bmatrix} \qquad B^* = \begin{bmatrix} 0 & 0 & 0 & 0 \\ 1 & 0 & 0 & 0 \\ 1 & 1 & 1 & 1 \\ 1 & 1 & 1 & 1 \end{bmatrix}$$

The matrix $B^* = \sum_{k=1}^{\infty} B^k$ denotes all causal paths of all lengths in a system (including diverging, converging, and feedback paths). This is the transitive closure property, meaning that all causality propagated within the system network is accounted for. B^* is a transitive closure matrix. This matrix for the marine coprophagy model is the last of the set that appears in Table I.

3.3. Flow Analysis

Leontief (1936, 1965) developed a method for steady-state analysis of economic systems that requires transitive closure. The procedure, as modified by Finn (1976), in effect defines within-system input and output environments of each component holon. A self-explanatory summary of creaon and genon derivations is provided in Fig. 11. The matrix $(I - Q')^{-1}$ which results from the creaon analysis contains elements that specify the causal influence propagated over all paths from $\mathcal{H}_j$ to $\mathcal{H}_i$ in the system network per unit of output observed at $\mathcal{H}_i$, $i,j = 1,\ldots,n$. Reciprocally, elements of matrix $(I - Q'')^{-1}$ which result from the genon analysis denote the causality propagated over all paths from $\mathcal{H}_j$ to $\mathcal{H}_i$ per unit of input received by $\mathcal{H}_j$, $i,j = 1,\ldots,n$. Both inverse matrices are transitive closure matrices. Each row i of $(I - Q')^{-1}$ represents the normalized input environment $\mathcal{H}_i'$ of holon $\mathcal{H}_i$, $i = 1,\ldots,n$, and each column j of $(I - Q'')^{-1}$ represents the normalized output environment $\mathcal{H}_j''$ of holon $\mathcal{H}_j$, $j = 1,\ldots,n$.

3.4. Ecosystem Partition into Component Environments

Consistent with the concept of environment as subject-defined, the set of input environments actualized by a system's collection of creaons is nonintersecting,

$$\mathcal{H}_i' \cap \mathcal{H}_j' = \phi, \qquad i,j = 1,\ldots,n,$$

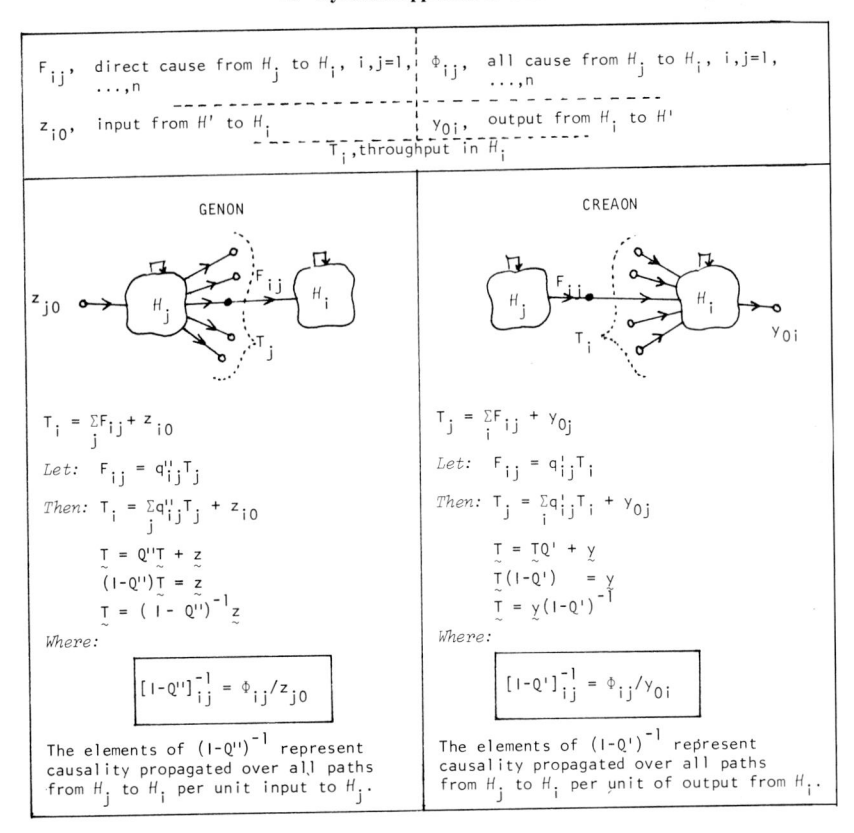

8. Systems Approach to Continental Shelf Ecosystems

Figure 11. Steady-state input–output analysis for creaon and genon cases. z, y, and **T** are input, output, and throughput vectors, respectively; $Q' = (q'_{ij})$ and $Q'' = (q''_{ij})$ are matrices of fractional inputs and outputs, respectively, from $\mathcal{H}_j$ to $\mathcal{H}_i$; $(I-Q')^{-1}$ is a matrix of causes propagated over all possible paths from $\mathcal{H}_j$ to $\mathcal{H}_i$ per unit of output from $\mathcal{H}_i$; and $(I-Q'')$ is a matrix of effects propagated over all possible paths from $\mathcal{H}_j$ to $\mathcal{H}_i$ per unit of input to $\mathcal{H}_j$.

ϕ the empty set. Similarly, the output environments generated by a system's set of genons is also nonoverlapping,

$$\mathcal{H}''_i \cap \mathcal{H}''_j = \phi, \quad i,j = 1,\ldots,n.$$

All the causality propagated within the structure of system $\mathcal{H}$ is accounted for by the union of either the input environments or output environments:

$$\mathcal{H} = \bigcup_{i=1}^{n} \mathcal{H}'_i = \bigcup_{j=1}^{n} \mathcal{H}''_j.$$

That is, both sets of holon environments form a partition of the whole into nonintersecting parts. These parts—holon/environment unities—constitute a basic system decomposition. Extrapolation to nature is straightforward. *Every ecosystem is neither more nor less than the sum of all the disjoint worlds of all acting or reacting things present.* The usefulness of this new way to compose reality remains to be seen.

The marine coprophagy model can serve as illustration. Let Fig. 10 be taken as analogous to a real ecosystem. Its decomposition into input and output environments is demonstrated in Figs. 12 and 13, respectively. Carbon flows and throughputs are normalized to unit outputs from and inputs to each holon of the model.

In Fig. 12 the holons are viewed as creaons. Consider the bottom diagram of the set. Observation (measurement) of one unit of carbon output from $\mathcal{H}_4$ specifies the indicated network of causes as input environment

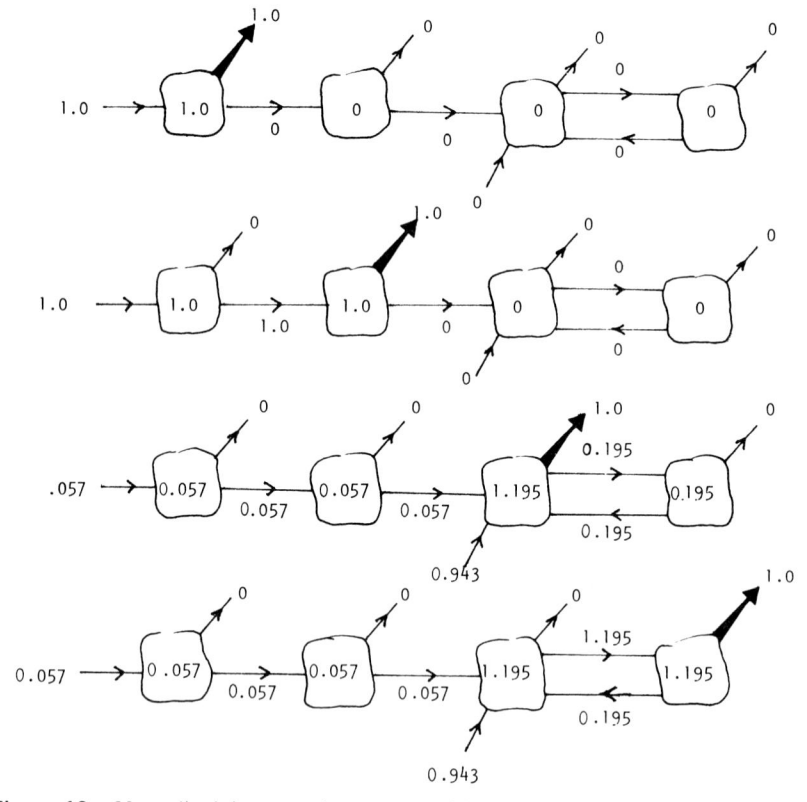

Figure 12. Normalized input environment partition of steady-state marine coprophagy model. Carbon flows (arrows) and throughputs (boxes) required to cause unit outputs (heavy arrows) from each holon are indicated.

8. Systems Approach to Continental Shelf Ecosystems

$\mathcal{H}_4'$. The actual dimensionalized input environment (flows and throughputs in gm C m^{-2} yr^{-1}) is obtained by multiplying all numbers by carbon output from $\mathcal{H}_4$ (26.7 gm C m^{-2} yr^{-1}; Fig. 10). Causality is traced back through the causal net to the system inputs at $\mathcal{H}_1$ and $\mathcal{H}_3$. Most of the output from $\mathcal{H}_4$ derives from $\mathcal{H}_3$ (94.3%) and only a small amount from $\mathcal{H}_1$ (5.7%). The relations depicted for the remaining three holons are self-evident. If the four normalized input environments $\mathcal{H}_i'$, $i = 1,\ldots,4$, are dimensionalized to actual carbon flows and summed together, the original system $\mathcal{H}$ of Fig. 10 is reconstructed:

$$\mathcal{H} = \sum_{i=1}^{4} \mathcal{H}_i'.$$

Figure 13 shows the holons as genons. Each diagram depicts

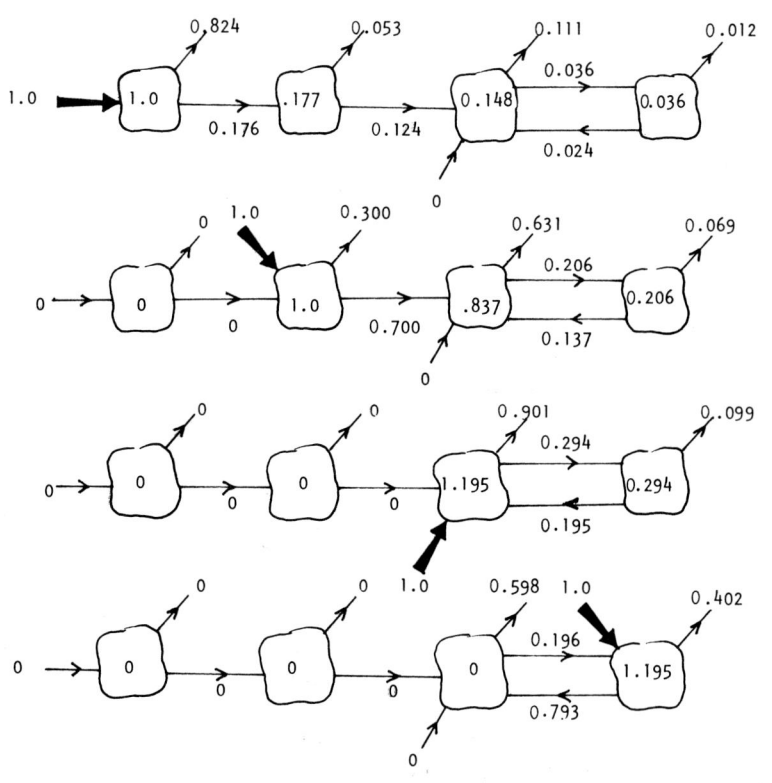

Figure 13. Normalized output environment partition of steady-state marine coprophagy model. Carbon flows (arrows) and throughputs (boxes) caused by unit inputs (heavy arrows) to each holon are indicated.

normalized output environments $\mathcal{H}_j''$ generated by a unit input (heavy arrows) to $\mathcal{H}_j$, $j = 1,\ldots,4$. Actual flows and throughputs in gm C m^{-2} yr^{-1} are calculated by multiplying all numbers in each figure by corresponding input values appearing in Fig. 10. Effects of each $\mathcal{H}_j$ are propagated to the system outputs, beyond which they can no longer be specified. In the upper diagram of the group, for example, 82.4% of input to $\mathcal{H}_1$ exits the system at $\mathcal{H}_1$, 5.3% at $\mathcal{H}_2$, 11.1% at $\mathcal{H}_3$, and 1.2% at $\mathcal{H}_4$. If the four illustrated output environments $\mathcal{H}_j''$, $j = 1,\ldots,4$, are dimensionalized to flows in gm C m^{-2} yr^{-1} and summed, the original system $\mathcal{H}$ of Fig. 10 is recomposed:

$$\mathcal{H} = \sum_{j=1}^{4} \mathcal{H}_j''.$$

In Section 4, flow analysis of a more realistic continental shelf ecosystem is described.

3.5. Other Methods of Causal Analysis

Sewall Wright (1921, 1960, 1968) developed a method for evaluating bloodlines for inbreeding called path analysis. Uses of path analysis extend well beyond calculating inbreeding coefficients. It is basically a statistical technique for testing plausibility of a hypothesized causal structure by finding the correlation coefficient between two variables with effects of all other variables removed. Path analysis becomes complex when causal loops are involved (Wright, 1960, 1968). The method was practically unknown outside of genetics and animal husbandry until the 1960s (Li, 1975) and has not yet been applied to ecosystems.

Richard Levins (1974, 1975) has recently developed a qualitative procedure for analyzing ecological structures called loop analysis. Loop analysis deals only with presence or absence of a causal link and the sign of that link (positive, negative, or zero). Loop analysis predicts how a system will react to changes in a component (as it evolves, for example) and how can it generate an expected correlation matrix whose elements denote signs of each correlation between components. These hypotheses, generated during model construction, can be readily tested. Loop analysis has only been applied to small models. Transitive closure is not assured for large models.

A third technique, influence analysis, is being developed by James Hill (1975). This method uses only presence or absence of a causal link to determine the influence one variable has upon another. Influence analysis allows the number of paths of any lengths from one variable to another (or to itself) to be counted. Transitive closure for any arbitrarily complex system is guaranteed.

Path, loop, and influence analyses are all promising for elucidating causal webs in ecosystems, particularly during the first three model-building steps mentioned in the Introduction. Application of these methods to ecological systems is just beginning, however, and we can do no more than mention them here. Input–output flow analysis is further illustrated below.

4. FLOW ANALYSIS OF THE ROSS SEA PELAGIC ECOSYSTEM

4.1. Model Description

The model discussed herein is the work of Dr. Katherine A. Green, Texas A & M University (Green, 1975). An annual budget of carbon flows (metric ton C yr^{-1}) for the Ross Sea, Antarctica, was constructed with data from USNS Eltanin cruise 51 (1971–1972) and supplementary information from the literature. The Ross Sea (Fig. 14) is a triangular body of water between 70 and 85°S latitude and 165°E and 160°W longitude. It is bounded on the north by the Antarctic continental slope, and on the south by the Ross Ice Shelf, the largest in the world (5.4×10^5 km^2, 250–700 m thick). The average depth is 550 m and the volume is 2.75×10^{14} m^3. The temperature ranges from -2 to 0°C, with little vertical or horizontal variation. Salinity is 33.5–34.7‰, and the density is 1.0267–1.0291 gm cm^{-3}. Depth of the surface mixed layer is approximately 1 km. Principal water masses are the Antarctic surface water, circumpolar deep water, and Ross Sea Shelf water. Day length varies from 24-hr solar days in austral summer to a 14-week period of darkness in winter. Between April and January there is sea ice cover from several centimeters to 5 m thick. The Ross Sea is ice-free during February and March. As pack ice forms and melts each year, its edge forms a meandering coastline in an otherwise oceanic environment.

The model consists of 12 pelagic compartments. Benthos are excluded. While benthic biomass is high, growth is slow and nutrient recycling through the water column is negligible. The 12 component holons are as follows:

$\mathcal{H}_1$, *Ice community algae.* All phytoplankton cells in sea ice or attached to the brash ice layer (whose lower surface is brown with Aufwuchs). These algae are highly shade-adapted and therefore productive. They become light-saturated at 7 langley day^{-1} and photoinhibited at 58 langley day^{-1}. Heterotrophy and reduced respiration account for survival during the 14-week dark period. When ice melts, accumulated

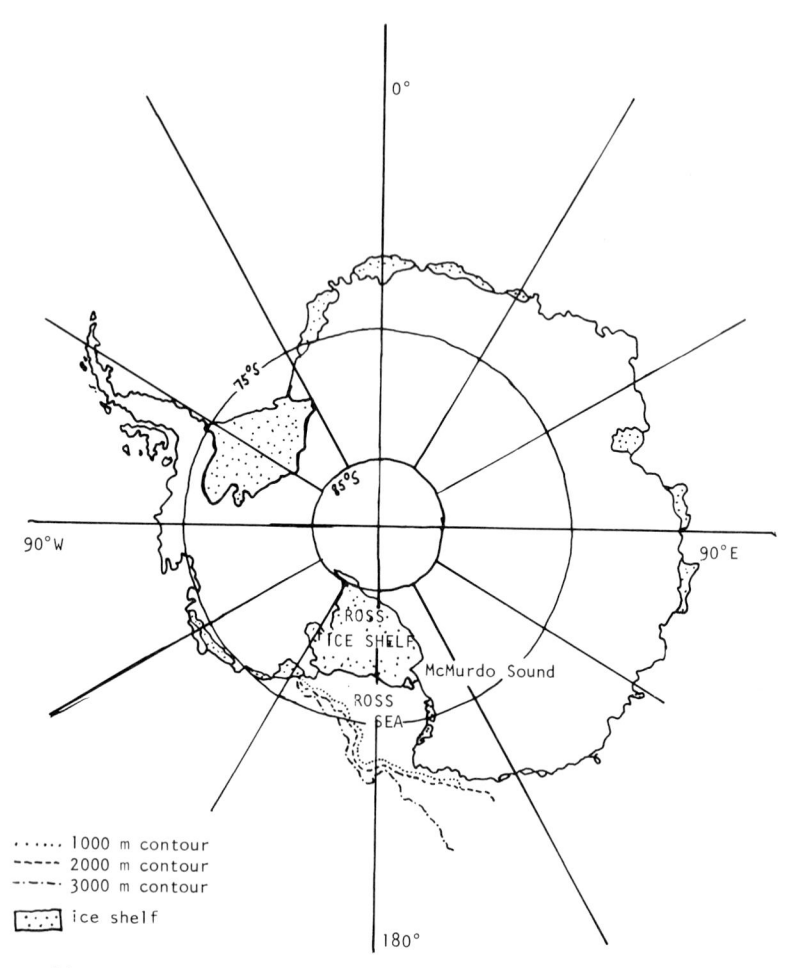

Figure 14. The Ross Sea, Antarctica (redrawn from Green, 1975, after American Geographical Society of New York, 1970).

production is released into the water causing swarming of zooplankton and their predators at the moving edge of pack ice.

$\mathcal{H}_2$, *Ice invertebrates.* These consist of predominantly polychaetes, copepods, and amphipods, up to 86 gm m^{-2}, associated with ice algae.

$\mathcal{H}_3$, *Phytoplankton.* These are water column forms, predominantly diatoms.

$\mathcal{H}_4$, *Zooplankton.* This group includes copepods, chaetognaths, and euphausids, with both herbivorous (90%) and carnivorous (10%) forms. Krill comprise 50% of the total standing crop. The principal krill species is

Euphausia superba, the Antarctic krill, but *Euphasia crystallorophias* predominates in shallow waters near the ice. Krill are central in the Antarctic food web.

$\mathcal{H}_5$, *Fishes*. Fish stocks are low compared to other seas. The dominant pelagic species is the notothenid *Pleurogamma antarcticum*. Demersal forms are excluded.

$\mathcal{H}_6$, *Cephalopods*. Squid are abundant, although no concentration estimates have been made.

$\mathcal{H}_7$, *Penguins*. Emperor and Adélie penguins breed in the Ross Sea. The Emperor is present year-round while the Adélie migrates from the coast to the northern limit of pack ice, a distance of up to 1000 km.

$\mathcal{H}_8$, *Seals*. The composition of seal populations is 67.5% crabeater seals which eat krill, 15.5% Weddell seals which feed on fishes and cephalapods, 15.5% Ross seals which eat cephalopods, and 1.5% leopard seals which consume other seals and penguins.

$\mathcal{H}_9$, *Whales*. Both baleen and toothed whales feed in the Antarctic but migrate to temperate or tropical waters for breeding and calving. Pack ice controls acess to krill. At peak baleen whale populations, 15×10^7 tons of krill were annually consumed in Antarctic waters.

$\mathcal{H}_{10}$, *Detritus*. This is dead particulate organic matter (POC) plus associated decomposer organisms (bacteria, yeasts, and molds).

$\mathcal{H}_{11}$, *Dissolved organic carbon*. DOC consists of any free organic molecules in the water column. A large, steady-state, old, stable sink which together with POC comprises 98% of oceanic organic carbon. The Ross Sea concentration is 0.81 mg C liter^{-1} versus 0.45 mg C liter^{-1} in other oceans.

$\mathcal{H}_{12}$, *Nutrients*. The nutrients are phosphate (range 0.5–2.0 µg-atm liter^{-1}) and nitrate (range 10–30 µg-atm liter^{-1}). Nutrients are not limiting to primary production, even during peaks of phytoplankton growth. Light is the limiting factor in the Ross Sea.

Figure 15 illustrates interactions between these holons of the Ross Sea model. The diagram represents causality propagated as carbon (or equivalent) flow through the ecosystem food web. This flow diagram differs from Green's original model in that no nutrients are fed back to the ice algae and phytoplankton (as carbon equivalents). This allows flows to be interpreted strictly as carbon flow instead of as combined carbon and nutrient information flow.

Three holons dominate flux through the system: $\mathcal{H}_4$ (zooplankton), $\mathcal{H}_{10}$ (detritus), and $\mathcal{H}_{11}$ (DOC). These three holons directly receive 49% of input and generate 52% of output and indirectly process most of the remaining carbon passing through the Ross Sea. Compared to 30% connectivity for the model as a whole, this three-component subsystem is 100% connected.

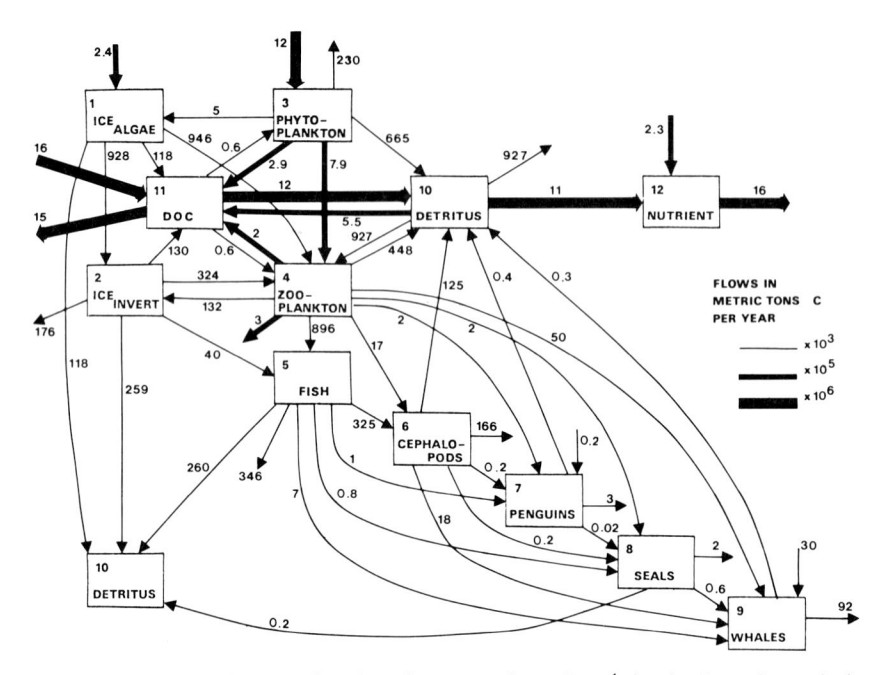

Figure 15. Annual budget of carbon flows, metric ton C yr^{-1} in the Ross Sea pelagic ecosystem model (modified from Green, 1975). Note that $\mathcal{H}_{10}$ is depicted twice to reduce crossing arrows.

Flow analysis, as described in Section 3.3, was performed on the Fig. 15 model. The specifics of matrix manipulation will not be described. Instead, some genon and creaon flow diagrams will be presented to illustrate results.

4.2. Genon Analysis

In this section, several holons will be examined as genons, generators of output environments. This analysis allows tracing a unit of input to a holon through the entire system until it leaves.

$\mathcal{H}_3$, *Phytoplankton.* Figure 16 shows the fate of one unit of carbon input to phytoplankton. Numbers associated with arrows represent flows normalized to one unit of input to $\mathcal{H}_3$ (including both z_{30} from the exogenous system environment plus intrasystem input to $\mathcal{H}_3$). Numbers inside boxes are throughputs. Flows less than 0.001 are omitted, and throughputs less than 0.001 are indicated by "tr," meaning trace.

There is often a large difference between direct flow and total flow from one holon to another. Total flow caused by $\mathcal{H}_3$ is the throughput of

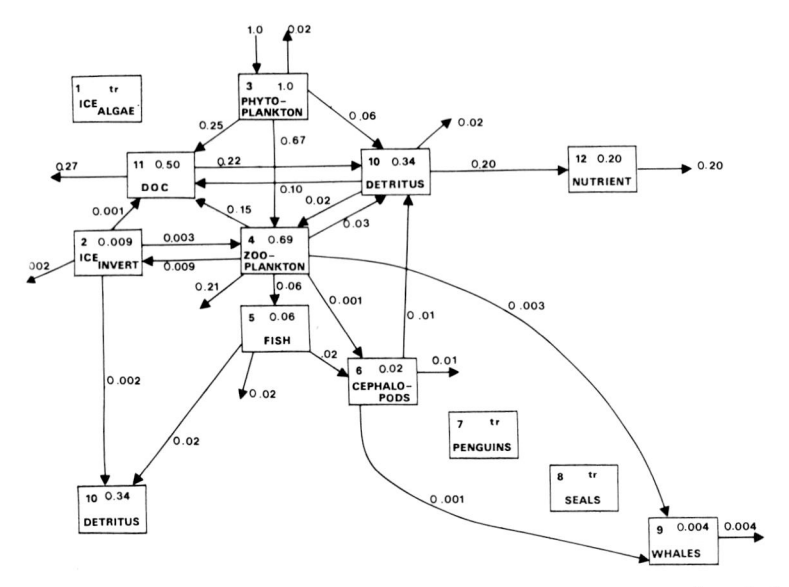

8. Systems Approach to Continental Shelf Ecosystems   203

Figure 16. Output environment $\mathcal{H}_3''$, showing fate of one unit of input to phytoplankton. Carbon flows (arrows) and throughputs (boxes) caused by one unit of input to $\mathcal{H}_3$ are indicated.

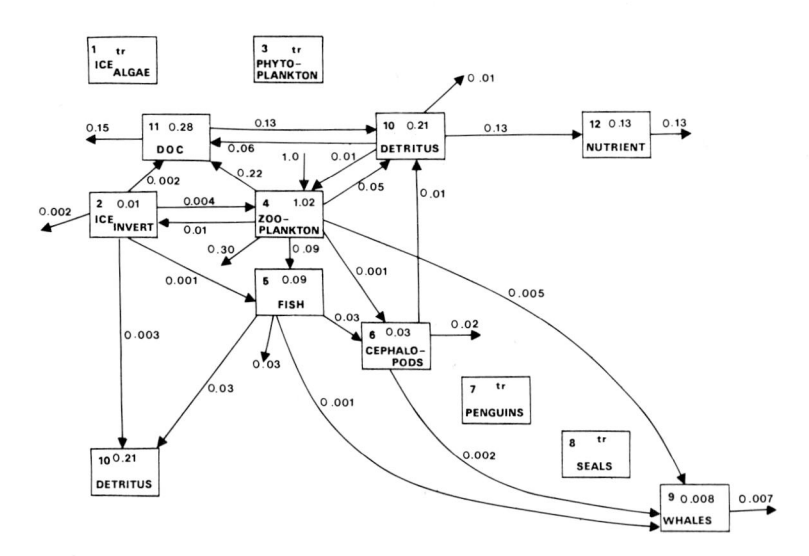

Figure 17. Output environment $\mathcal{H}_4''$, showing fate of one unit of input to zooplankton. Carbon flows (arrows) and throughputs (boxes) caused by one unit of input to $\mathcal{H}_4$ are indicated.

Table II Direct and Total Flow from $\mathcal{H}_3$ (Phytoplankton) to Other Holons, Normalized to Unit Input (z_{30}) to $\mathcal{H}_3$

Recipient holon	Direct flow	Total flow
$\mathcal{H}_4$ (zooplankton)	0.67	0.69
$\mathcal{H}_5$ (fish)	—	0.06
$\mathcal{H}_{10}$ (detritus)	0.06	0.34
$\mathcal{H}_{11}$ (DOC)	0.25	0.50
$\mathcal{H}_{12}$ (nutrients)	—	0.20

each holon shown in Fig. 16. $\mathcal{H}_3$ directly affects $\mathcal{H}_4$, $\mathcal{H}_{10}$, and $\mathcal{H}_{11}$ and indirectly affects these and all other holons. Table II lists direct and total flows from $\mathcal{H}_3$ to five other holons. The results demonstrate convincingly that considering direct causes alone and ignoring indirect causes can lead to a gross misunderstanding of a system.

$\mathcal{H}_4$, *Zooplankton.* Zooplankton ($\mathcal{H}_4$), detritus ($\mathcal{H}_{10}$), and DOC ($\mathcal{H}_{11}$) account for most of the carbon flow through the Ross Sea system as depicted in Fig. 15. The remaining food chain receives very little of phytoplankton production (e.g., only 6% reaches fish; Fig. 16). Zooplankton output environment $\mathcal{H}_4''$ is shown in Fig. 17. All numbers with arrows are flows normalized to one unit of input (intrasystem) to $\mathcal{H}_4$. The normalized throughput of $\mathcal{H}_4$ in Fig. 17 is 1.02. A "return cycling efficiency" (Finn, 1978) for $\mathcal{H}_4$ may be calculated according to the equation

$$RE = [(I-Q'')_{ii}^{-1} - 1]/(I-Q'')_{ii}^{-1}.$$

For $\mathcal{H}_4$,

$$RE = 0.02/1.02 = 0.0196,$$

meaning that 1.96% of what enters zooplankton eventually returns. Table III shows the difference between direct and total contribution of $\mathcal{H}_4$ to five other holons. Direct flow is a major factor in causality from $\mathcal{H}_4$ to $\mathcal{H}_5$, $\mathcal{H}_9$ and $\mathcal{H}_{11}$, but a minor factor in total cause from $\mathcal{H}_4$ to $\mathcal{H}_{10}$ and $\mathcal{H}_{12}$. The reader may make similar comparisons for remaining examples.

In the output environment of zooplankton, the remainder of the food chain is a relatively minor element (Fig. 17). Less than 10% of what enters $\mathcal{H}_4$ reaches higher trophic levels. Respiration and other losses account for 30%, and flows to detritus and DOC together account for about 27% of what enters $\mathcal{H}_4$. Remaining carbon is stored by zooplankton. In this model, zooplankton seem to act less as a food source for upper trophic levels than as a source of DOC and detritus.

8. Systems Approach to Continental Shelf Ecosystems

Table III Direct and Total Flow from $\mathcal{H}_4$ (Zooplankton) to Other holons, Normalized to Unit Input to $\mathcal{H}_4$

Recipient holon	Direct flow	Total flow
$\mathcal{H}_5$ (fish)	0.09	0.09
$\mathcal{H}_9$ (whales)	0.005	0.008
$\mathcal{H}_{10}$ (detritus)	0.05	0.21
$\mathcal{H}_{11}$ (DOC)	0.22	0.28
$\mathcal{H}_{12}$ (nutrient)	—	0.13

$\mathcal{H}_{11}$, *Dissolved organic carbon.* The output environment of $\mathcal{H}_{11}$ (DOC) is shown in Fig. 18. Again, all flows are normalized to one unit of input to $\mathcal{H}_{11}$ ($z_{11,0}$ exogenously plus intrasystem inputs). This output environment $\mathcal{H}''_{11}$ traces the fate of the largest carbon input, $z_{11,0}$, to the Ross Sea. The most striking feature of Fig. 18 is its simplicity. It contains flows between DOC, detritus, and zooplankton, but almost nothing else. The conclusion suggested is that the Ross Sea pelagic ecosystem does not much utilize carbon entering it as DOC (63% of what enters leaves as DOC), although 31% of DOC input is mineralized to nutrients. The pelagic

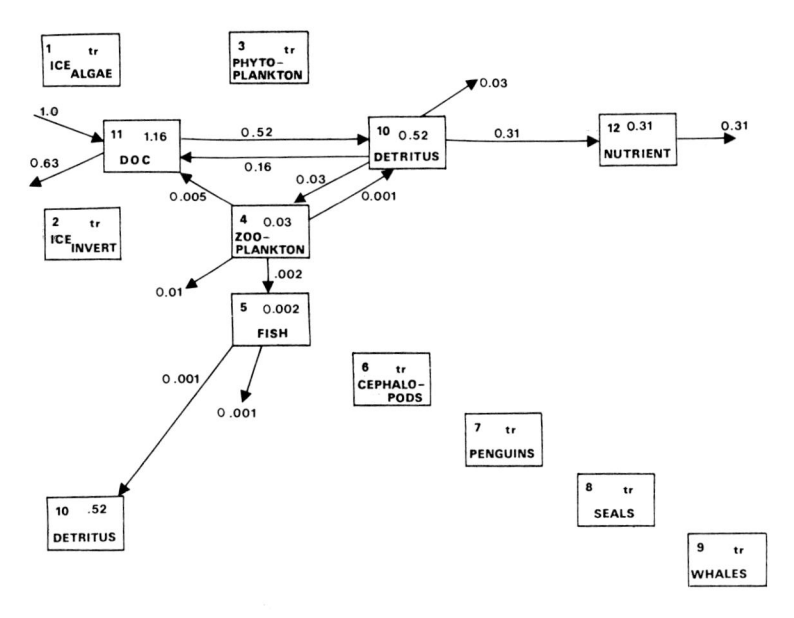

Figure 18. Output environment $\mathcal{H}''_{11}$, showing fate of one unit of input to DOC. Carbon flows (arrows) and throughputs (boxes) caused by one unit of input to $\mathcal{H}_{11}$ are indicated.

ecosystem may thus use dissolved organic matter as a nutrient rather than as a carbon source. $\mathscr{H}_{11}$ cycles a relatively large amount of carbon. The return cycling efficiency is

$$RE_{11} = 0.16/1.16 = 0.14.$$

That is, 14% of what enters $\mathscr{H}_{11}$ eventually returns. The principal loop involved is $\mathscr{H}_{11} \to \mathscr{H}_{10} \to \mathscr{H}_{11}$.

4.3. Creaon Analysis

The object of analyzing holons as creaons is to determine the origin of inputs. The holon is thus examined as a selector of input environments. Discussion here will be restricted to the input environment of dissolved organic carbon.

$\mathscr{H}_{11}$, *Dissolved organic carbon*. The input environment $\mathscr{H}'_{11}$ of DOC is shown in Fig. 19. All flows are normalized to one unit of output from $\mathscr{H}_{11}$ ($y_{0,11}$ exogenously plus intrasystem outputs). About 25% of unit

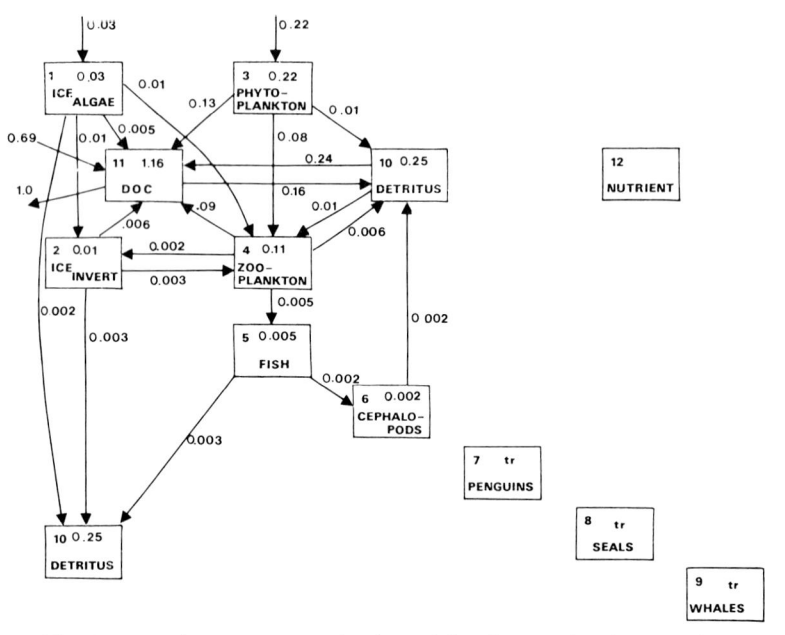

Figure 19. Input environment $\mathscr{H}'_{11}$, showing origin of one unit of output from DOC. Carbon flows (arrows) and throughputs (boxes) necessary to cause one unit of output from $\mathscr{H}_{11}$ are indicated.

output from $\mathcal{H}_{11}$ originates with primary production, and almost 70% comes from exogenous input to $\mathcal{H}_{11}$. The three components $\mathcal{H}_4$, $\mathcal{H}_{10}$, and $\mathcal{H}_{11}$ along with $\mathcal{H}_3$ (phytoplankton) account for most of the flow.

Cycling is as important in creaon analysis of $\mathcal{H}_{11}$ as in genon analysis. This is always the case since RE_i is calculated from diagonal elements of $(I-Q')^{-1}$ and $(I-Q'')^{-1}$, which are always identical. The interpretation differs, however. In genon analysis $RE_{11} = 0.14$ means that 14% of what enters $\mathcal{H}_{11}$ will eventually return. In creaon analysis it means that 14% of what leaves $\mathcal{H}_{11}$ has been there before.

Compare the output environment $\mathcal{H}''_{11}$ of $\mathcal{H}_{11}$ (Fig. 18) with its input environment $\mathcal{H}'_{11}$ (Fig. 19). The output environment is much simpler, even though all 12 holons are involved and $\mathcal{H}_{12}$ was excluded from the input environment of $\mathcal{H}_{11}$. The number of major flows and the number of holons involved in major flows are fewer in the output environment. This example illustrates the dual nature of input and output environments; each is a distinct and different entity.

4.4. Carbon Cycling in the Ross Sea Model

Return cycling efficiency, as calculated above for $\mathcal{H}_4$ and $\mathcal{H}_{11}$, can be computed for all holons in the Ross Sea model. A measure of the importance of cycling in a system can be derived from this information (Finn, 1978).

The amount of throughput T_i in $\mathcal{H}_i$ that cycles (T_{ci}) is

$$T_{ci} = RE_i T_i.$$

Summing for all holons, total-system throughput cycled (TST_c) is

$$TST_c = \sum_{i=1}^{n} T_{ci}.$$

Total-system throughput (TST) is the sum of all n throughputs,

$$TST = \sum_{i=1}^{n} T_i.$$

Cycling index (CI) is then the fraction of total-system throughflow that is cycled,

$$CI = TST_c/TST.$$

Table IV Calculation of Cycling Index for the Ross Sea Pelagic Ecosystem Model

Return cycling efficiencies

$RE_1 = 0.0$ $RE_2 = 4.2 \times 10^{-3}$ $RE_3 = 1.1 \times 10^{-5}$
$RE_4 = 1.4 \times 10^{-2}$ $RE_5 = 2.1 \times 10^{-3}$ $RE_6 = 6.9 \times 10^{-4}$
$RE_7 = 1.9 \times 10^{-6}$ $RE_8 = 9.4 \times 10^{-6}$ $RE_9 = 9.3 \times 10^{-7}$
$RE_{10} = 0.14$ $RE_{11} = 0.14$ $RE_{12} = 0.0$

$$TST_c = \sum_{i=1}^{12} RE_i T_i = 6,567,113 \text{ metric ton C yr}^{-1}$$

$$TST = \sum_{i=1}^{12} T_i = 88,914,910 \text{ metric ton C yr}^{-1}$$

$$CI = TST_c/TST = 0.074$$

Table IV shows calculations for determining the cycling index of the Ross Sea model ($CI = 0.074$). Thus, 7.4% of total-system throughflow is cycled, which is relatively high for a carbon model (Finn, 1978).

4.5. Comments on the Ross Sea Model

What does flow analysis of the Ross Sea pelagic ecosystem model indicate about the Ross Sea and about continental shelves generally?

First, indirect causation can be more important than direct causation. That is, one holon can affect another more through other holons than directly. The three components $\mathscr{H}_4$ (zooplankton), $\mathscr{H}_{10}$ (detritus), and $\mathscr{H}_{11}$ (DOC) are important direct or indirect propagators of cause in the Ross Sea model, where they receive about half the total system input and generate half the output. In typically more shallow continental shelves these three components should be even more important because (a) stronger currents carry more DOC through the shelf ecosystem, and (b) inputs of detritus and DOC from continental sources are major factors. Indirect effects of the zooplankton–detritus–DOC complex are thus expected to figure significantly in the causal ecology of continental shelves.

Second, the present Ross Sea model represents only the pelagic subsystem. In reality, benthos are important recipients of carbon from above and serve to propagate indirect effects to the pelagic system. The Ross Sea is so deep that benthos may be reasonably omitted. In most continental shelves, however, interplay between bottom and open water communities is potentially greater. Many organisms have both benthic and pelagic life stages. Benthos live almost exclusively off the detritus rain from above. To evaluate causal influences in a general shelf ecosystem realistically, a model containing both pelagic and benthic holons should be used.

Third, genon analysis of $\mathcal{H}_4$ indicated that zooplankton do not serve in the ecosystem context primarily as food for higher trophic levels. Direct flow from zooplankton to detritus ($\mathcal{H}_{10}$) and DOC ($\mathcal{H}_{11}$) is three times as great as that to fish ($\mathcal{H}_5$) (Table III, Fig. 17; note that Fig. 15 shows this). The effects of opening nutrient ($\mathcal{H}_{12}$)-to-primary producers ($\mathcal{H}_1, \mathcal{H}_2$) loops require evaluation before such a result can be extended to other shelf ecosystems, however.

Finally, not all causal events are reflected in carbon flow. The present model represents just one aspect of ecosystem causation. Green's (1975) original model combined nutrient and carbon flows more realistically, but flow analysis interpretations for such a combined model are difficult. Other aspects of causation, such as flow of genetic material, predator–prey interactions, and catastrophes like storms and epizootics, should also be included in a comprehensive model of continental shelf ecosystems. Only such more complete models can aid realistic assessment of ecological effects of various shelf development schemes.

5. SUMMARY

The mechanistic approach to systems ecology, that of building wholes from parts, has so far failed to develop realistic and practical ecosystem models. Too much attention has been lavished on processes at the component level and on mathematical finery, and too little on phenomenological understanding at the whole-system level. A theoretical base is needed to assist modelers in (a) choosing conceptual and physical boundaries, (b) identifying system components, and (c) identifying interactions between components. Causality theory is an attempt to meet this need.

The object of interest in causality theory is a *holon*, composed of an input half (*creaon*) and an output half (*genon*). A creaon defines (creates) its own input environment to which it will respond. A genon generates its own potential output environment. That is, it generates attributes which creaons in the output environment may select as inputs. Input and output environments are distinct entities and may be qualitatively as well as quantitatively different. Even if ecosystems are circular causal systems, input and output environments of each component will contain the same elements but still remain quantitatively and conceptually different.

Causal analysis of a system must account for all causal paths of all lengths between system components. This is the transitive closure property. Transitive closure requires that a system be connected to an environment, through input, output, or both. Four techniques of causal analysis are path, loop, influence, and input–output flow analysis. The first three should prove helpful in early stages of ecological model building. Flow analysis quantifies causality propagation in system networks; analyses of a simple marine coprophagy model and a more complex continental shelf ecosystem model are presented for illustration.

Flow analysis of the latter model representing the pelagic subsystem of the Ross Sea suggested that DOC, detritus, and zooplankton are the holons most involved in processing carbon. A relatively high proportion (7.4%) of carbon fluxing through the pelagic ecosystem represents cycled flow. Indirect causation may be more important than direct influences in many instances of holon interactions. Zooplankton contribute more to DOC and detritus than to the rest of the food chain, suggesting that krill as a food source for whales may not have as important an ecological role as generating DOC and detritus.

Continental shelves are ecosystems whose systematic use by man requires whole-system synthesis to discriminate direct and indirect causation. Because the whole/part relationship is not straightforward, such synthesis will require a special ecosystem theory consistent with modern General Systems Theory. This paper has attempted to stress the need for theory as a basis for ecosystem modeling, simulation, and analysis (Fig. 1). Causality theory, as one approach that emphasizes interconnectedness of systems, accommodates both direct and indirect influences. Incorporation of complex causation in scientific and management models is advocated as the surest way to gain ecosystem understanding required for sound utilization of the world's continental shelves. Noncausal treatments of quantitative data, and loose coupling of such results to intuition, are not acceptable as sole methodology for a future technology of whole-ecosystem design and management.

ACKNOWLEDGMENT

We gratefully acknowledge Dr. Katherine Green's cooperation in preparing this paper. She discussed her Ross Sea pelagic ecosystem model with us at length, and provided computer output essential for carrying out flow analysis of the model. This research is from University of Georgia, *Contributions in Systems Ecology*, No. 36.

REFERENCES

American Geographical Society of New York (1970). "Antarctica and Adjacent Seas." Hoen Co., Baltimore, Maryland.

Cale, W. G., and Ramsey, P. R. (1970). Trophic significance of coprophagy by benthic organisms. Unpublished Systems Ecology course report, University of Georgia, Athens.

Caswell, H. (1976). The validation problem. *In* "Systems Analysis and Simulation in Ecology" (B. C. Patten, ed.), Vol. 4, pp. 313–325. Academic Press, New York.

Finn, J. T. (1976). Measures of ecosystem structure and function derived from analysis of flows. *J. Theor. Biol.* **56**, 363–380.

Finn, J. T. (1978). Cycling index: A general definition for cycling in compartment models. *In* "Environmental Chemistry and Cycling Processes Symposium" (D. C. Adriano and I. L. Brisbin, eds.). ERDA (in press).

Green, K. A. (1975). Simulation of the pelagic ecosystem of the Ross Sea, Antarctica: A time varying compartmental model. Ph.D. Dissertation, Texas A & M University, College Station.

Hill, J. (1975). Influence: A topological hypothesis in systems ecology and general systems theory. Unpublished Ph.D. Dissertation proposal, University of Georgia, Athens.

Hutchinson, G. E. (1948). Circular causal systems in ecology. *Ann. N.Y. Acad. Sci.* **50**, 221–246.

Koestler, A. (1967). "The Ghost in the Machine." Macmillan, New York.

Leontief, W. W. (1936). Quantitative input–output relations in the economic system of the United States. *Rev. Econ. Statist.* **18**, 105–125.

Leontief, W. W. (1965). The structure of the U.S. economy. *Sci. Am.* **212**, 25–35.

Levins, R. (1974). The qualitative analysis of partially specified systems. *Ann. N.Y. Acad. Sci.* **231**, 123–138.

Levins, R. (1975). Evolution in communities near equilibrium. *In* "Ecology and Evolution of Communities" (M. Cody, ed.), pp. 16–50. Belknap Press, Harvard University, Cambridge, Massachusetts.

Li, C. C. (1975). "Path Analysis—A Primer." Boxwood Press, Pacific Grove, California.

Mason, H. L., and Langenheim, J. H. (1957). Language and the concept of environment. *Ecology* **38**, 325–340.

Mesarovic, M. D., and Takahara, Y. (1975). "General Systems Theory: Mathematical Foundations." Academic Press, New York.

Ore, O. (1962). "Theory of Graphs," Colloq. Publ. No. 38. Am. Math. Soc., Providence, Rhode Island.

Patten, B. C. (1973). Need for an ecosystem perspective in eutrophication modeling. *In* "Modeling the Eutrophication Process" (E. J. Middlebrooks, ed.), pp. 83–87. Utah State University, Logan.

Patten, B. C. (1975a). Ecosystem linerization: An evolutionary design problem. *Am. Nat.* **109**, 529–539.

Patten, B. C. (1975b). A reservoir cove ecosystem model. *Trans. Am. Fish. Soc.* [N.S.] 594–617.

Patten, B. C. (1975c). Ecosystem as a coevolutionary unit: A theme for teaching systems ecology. *In* "New Directions in the Analysis of Ecological Systems" (G. S. Innis, ed.), Part 1, Simul. Counc. Proc. Ser., Vol. 5, pp. 1–8. Soc. Comput. Simul., La Jolla, California.

Patten, B. C., Egloff, D. A., Richardson, T. H., and 38 co-authors (1975). Total ecosystem model for a cove in Lake Taxoma. *In* "Systems Analysis and Simulation in Ecology" (B. C. Patten, ed.), Vol. 3, pp. 205–421. Academic Press, New York.

Patten, B. C., Bosserman, R. W., Finn, J. T., and Cale, W. G. (1976). Propagation of cause in ecosystems. *In* "Systems Analysis and Simulation in Ecology" (B. C. Patten, ed.), Vol. 4, pp. 457–579. Academic Press, New York.

The Institute of Ecology (1973). "An Ecological Glossary for Engineers and Resource Managers." TIE, Madison, Wisconsin.

von Uexküll, J. (1926). "Theoretical Biology." Kegan Paul, Trench, Tubner & Co., Ltd., London.

Wright, S. (1921). Correlation and causation. *J. Agric. Res.* **20**, 557–585.

Wright, S. (1960). The treatment of reciprocal interaction, with or without lag, by path analysis. *Biometrics* **16**, 423–445.
Wright, S. (1968). "Evolution and the Genetics of Populations," Vol. 1. Univ. of Chicago Press, Chicago, Illinois.
Zadeh, L. A., and Desoer, C. A. (1963). "Linear System Theory, the State Space Approach." McGraw-Hill, New York.

Chapter 9

A FRAMEWORK FOR DYNAMICAL SYSTEM MODELS: CAUSE–EFFECT RELATIONSHIPS AND STATE REPRESENTATIONS

P. Jussi Orava

1. Introduction	214
2. Input–Output System	215
2.1 Basic Formalism	215
2.2 Interconnection of Systems	220
2.3 Time System	222
3. Causality	224
3.1 Cause–Effect and Functionality	224
3.2 Nonanticipation	225
3.3 Discussion	227
4. State	227
4.1 Dynamical State	227
4.2 State Representation	228
4.3 State Transition	229
5. Discussion of Ecological Examples	231
References	232

1. INTRODUCTION

The subject of this chapter is included in the so-called mathematical systems theory. The formalisms and mathematical methods of this theory are increasingly used in many fields of empirical and applied sciences. Systems ecology is one of these fields. Engineering and economics are other typical and traditional examples. Systems theory, in general, is being established as an independent scientific discipline. Some authors even consider it as a completely new approach to the phenomena of nature (see, e.g., Klir, 1972; von Bertalanffy, 1972; Patten *et al.*, 1976). Similarly, it can be regarded as a collection of several particular theories with detailed structures in each of which the basic subject under consideration possesses the general traits of the common notion of system.

Generally speaking, a *system* is a collection of objects together with some relevant relations between them. The term is extensively used in many contexts, beginning with the everyday language up to sophisticated scientific theories. In this chapter the modeling of *input–output systems* is considered and formalized in a specific way. Other central concepts here are *time*, *causality*, and *state*. All the concepts here are adaptations of the corresponding notions in traditional sciences and natural language. The concepts have specific meanings, defined in terms of the present theory, but the basic ideas of the common notions have been preserved. The formalism is essentially aimed at modeling cause–effect relationships and dynamical phenomena in empirical and applied sciences, involving relations with respect to time. For brevity, the term *system* is frequently used in place of *system model* in this mathematical formalism, although elsewhere in this book the former term actually means the real world or real system.

Klir (1972), among others, considers two main features in the development of general systems theory: the *inductive* and the *deductive* approach. The inductive development begins with classifying and analyzing various "traits" which are associated with notions of the system in various fields. The formalization of the concepts will be done afterward in various phases. On the other hand, the deductive development begins with stating a well-defined logico-mathematical concept of system. Then one defines various specifications and properties in the formalism. The results of the theory then consists of theorems which will be deduced by using purely logical and mathematical procedures.

According to the above classification, the formalism of this article belongs to the deductive development. The formalism follows from two previous works by Orava (1973, 1974). It includes many conventional notions presented in some established works on the subject; among others there are those by Windeknecht (1967), Zadeh (1969), Kalman *et al.* (1969),

Mesarovic (1972), Blomberg (1973), and Mesarovic and Takahara (1975). As to the main structure, the formalism resembles, in the first place, the theories of Zadeh and of Mesarovic, although the different concept of system gives rise to considerable dissimilarities in details. The logical terminology and apparatus used follows the textbook presentation by Rogers (1971). The greatest part of the formalism is presented by the aid of standard set-theoretic expressions (consult, e.g., Suppes, 1960).

This chapter is only an introduction and discussion of the formalism in question. Deductive results, theorems, and proofs are not included in the text. Because of the generality also, the structure of the formalism is quite barren in regard to calculus (not even an algebraic addition operation appears). Therefore, the theory here cannot give any direct aid for solving particular problems in applications. However, this kind of general formalism on the whole may be valuable in developing common language for various specialists in particular fields of applications.

2. INPUT–OUTPUT SYSTEM

Some authors, typically Zadeh and Mesarovic, define the *general system* as follows. A system S is a set of k-tuples such that $S \subset \mathscr{V}_1 \times \cdots \times \mathscr{V}_k$ where $\mathscr{V}_1, \ldots, \mathscr{V}_k$ are some abstract sets ("system objects" by Mesarovic; $\subset$ denotes inclusion and $\times$ the Cartesian product; for simplicity, only finite products are considered here). An input–output system S in the above formalism is then simply a *directed binary relation*, a set of input–output pairs with $S \subset \mathscr{U} \times \mathscr{Y}$, where $\mathscr{U}$ is an input set and $\mathscr{Y}$ an output set; $\mathscr{U}$ and $\mathscr{Y}$ themselves can be some Cartesian products (e.g., consisting of some $\mathscr{V}_1, \ldots, \mathscr{V}_k$ above).

The above definition is quite clear and certainly general. However, in commonplace systems theory and applications the system models are most often given by more detailed representations, with formulas or equations, hardly ever directly as sets. For that reason many modelers may not be interested in the general theory. In the formal (metatheoretic) sense every theory based on the extensional first-order predicate logic can be identified with a certain collection of set-theoretic expressions; in practical sense, however, there is a considerable dissimilarity between the conceptions. This question is the starting point in the development of the present definition below.

2.1. Basic Formalism

The system concept of the present formalism includes the various logical and mathematical representations discussed above. Only input–output systems are considered here, but the definition could be easily extended for general systems, too. Some basic terminology of formalized languages and theories is needed for obtaining the adequate precision and generality (consult, e.g., Rogers, 1971). The examples presented in the sequel will desirably illustrate the sound relation of the concept to the everyday system models.

2.1.1. Formal Definition

An *input–output system* (model) $\mathscr{S}$ is (given by, represented by) a triplet

$$(u, F, y), \tag{1}$$

where u is the *input*, y is the *output*, and F is the *formula*, or *relationship*, of the system; the input is an m-tuple and the output is an n-tuple of component variables,

$$u = (u_1, \ldots, u_m), \qquad y = (y_1, \ldots, y_n). \tag{2}$$

The above objects are formally interpreted (in a suitable formalized theory and first-order predicate logic) as follows:

(a) The u_i's and y_j's are distinct *individual variables* (letters of the language).

(b) The F is a logical *formula* (sentence, clause, condition, etc., in the theory) the satisfaction (truth) of which represents the relationship between u and y in $\mathscr{S}$.

(c) All of the occurrences of u_i's and y_j's in F are *free* in the sense of the predicate logic (i.e., not bounded by either the existential quantifier $\exists$ or the universal quantifier $\forall$; $\exists$ should be read *there exists* and $\forall$ *for all*).

The definition is a verbose version of the one introduced by Orava (1974). It still needs some comments and additional terminology, as follows. First, the triplet (1) is only a condensed notation where the three main objects of $\mathscr{S}$ are named; their detailed expressions can then be given elsewhere. Second, the symbols u_i and y_j may stand for very abstract and complicated objects; they should not be interpreted only as scalar components of vectors. Condition (c) means that the basic form of a systems is an *open-loop* representation, without any interconnections yet; in detailed cases the fulfillment of this condition is probably guaranteed by common sense. There may also occur other individual variables and constants in the formula F in (1); they are called *parameter variables* and *constant parameters*, respectively, or jointly *parameters*, of $\mathscr{S}$. Then, define a "set-formation" *operation* $(\ )^\circ$ by

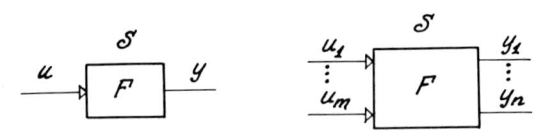

9. Dynamical System Models: Cause–Effect Relationships and State Representations

Figure 1. Graphical representation of system $\mathscr{S}$ in (1)–(2).

$$(\mathscr{S})^\circ = \{(u, y) | F^\circ\}, \tag{3}$$

where F° is a formula obtained from F by prefixing F with the existential quantifier ($\exists$) *on each of its still free parameter variable* (or $F^\circ = F$, if the prefixing already exists); the order of the prefixings is immaterial, because they commute. The set $(\mathscr{S})^\circ$, i.e., the set of all input–output pairs satisfying the condition F°, clearly corresponds to the set-theoretic system discussed before. The operation ( )$^\circ$ is a very convenient aid in the sequel when the set-theoretic language is extensively used in definitions and discussions. The system $\mathscr{S}$ is graphically depicted by either form of the diagrams in Fig. 1, that is, by the conventional block diagram, with $\mathscr{S}$ denoting now the whole triplet of the named objects.

2.1.2. Discussion

The definition above is clearly very general. Conventional mathematical expressions are, however, most frequent in applications. Also the set-theoretic system concept discussed before can be clearly interpreted in the present definition in the following natural way: A set S of input–output pairs is a system (u, F, y) where the formula F reads

$$(u, y) \in S. \tag{4}$$

The present definition necessitates the use of fixed "names" for the input and output; however, the system can be loosely denoted by the same symbol S if the interpretation (4) is agreed upon (the set S is the only constant parameter of the system). Conversely, the set $(\mathscr{S})^\circ$ in (3) can be considered as a system S above, "derived" from $\mathscr{S}$. As pointed out earlier, the set-theoretic system concept is as general as the present one in the metatheoretic sense, because the latter one is based on the extensional first-order predicate logic.

The generality of the definitions discussed above is worth emphasizing once more. The variables in (1)–(2) may stand for very abstract and highly structured objects; thereby, the components of the pairs of $(\mathscr{S})^\circ$ in (3) may be elements with complicated inner structures (sets, sets of sets, etc.). For example, the *stochastic systems* are also embedded in the definition. In such a system the formula contains the relevant probabilistic machinery represented by parameters, and the input and output stand for stochastic variables or processes together with their relevant probability distributions. In this case the elements of the set $(\mathscr{S})°$ may be really complicated when expressed ultimately in set-theoretic language. As a second example, a system formulation can be given in an implicit form containing an *optimization* condition in the inner structure. The minimization of a functional, for example, can be easily expressed exactly in relevant theory by the aid of suitable parameters and boundings. The *goal-seeking* systems of Mesarovic can be included in this class of systems. The actual optimal control and game-theoretic problems, on the other hand, would rather be interpreted as "outer" problems in which the systems models belong to the premises.

In many cases in systems theory and applications the representations of systems are expressed rather incompletely in the formal sense; many details are to be interpreted from the context in question or even by intuition. The exact formulation would often be impractical, and verbose lingual expressions in some places can make the presentation more readable; however, every writer should make sure that the readers can perceive the correct interpretation of the presentation without unreasonable effort. This is especially important in connection with large or unconventional system models. For example, the models of economic and ecological macrosystems often contain large numbers of various kinds of variables and constants; see Patten *et al.* (1976) and Patten and Finn (Chapter 8, this volume). The interpretation of systems representation is illustrated by a simple example below; a more qualified example will be presented in Section 2.3.1.

The input–output system thus far is a logico-mathematical concept, without any empirical interpretation of the inputs and outputs. For example, its use for modeling cause–effect phenomena in empirical sciences presupposes knowledge or hypothesis of causal relations.

2.1.3. Example 1

According to the familiar Pythagorean theorem for a right-angled triangle, the lengths x, y, and z of its sides (z for the longest side) satisfy the equation

$$x^2 + y^2 = z^2. \tag{5}$$

Imagine this triangular concept as a system $\mathscr{S}$ with x and z, say, as the input variables and y as the output; i.e., $\mathscr{S}$ is a triplet $((x, z), F, y)$. This abstraction could be understood as a "cause–effect" model for the drawing

9. Dynamical System Models: Cause–Effect Relationships and State Representations   219

construction where x and z are given, and y is generated. The core part of F is given by Eq. (5). However, according to the origin of $\mathscr{S}$, the formula F must also contain somehow the condition: "x, y, and z are positive real numbers." The system does not contain parameter variables; the $(\ )^2$ and $+$ are certain operations in the theory in question. The set $(\mathscr{S})^\circ$ is obviously a function; i.e., for given positive values of x and z, if there exists a positive value of y satisfying (5), then it is unique.

Consider then modifications of $\mathscr{S}$ in the following respects: (a) x is considered as the input in place of (x, z), or/and (b) $x, y,$ and z can be *any* real or complex numbers in (5), deleting the geometric origin. The resulting system, say $\mathscr{S}'$, is now necessarily different from $\mathscr{S}$, because in any case $(\mathscr{S}')^\circ \neq (\mathscr{S})^\circ$. In the case (a), z is now a parameter, variable or constant. If z is considered as a variable, then its bounding by $\exists$ is immaterial; on the contrary, its bounding by $\forall$ would lead to a trivial system in which $(\mathscr{S}')^\circ$ is empty. In the case (b) the set $(\mathscr{S}')^\circ$ is certainly not a function, because the algebraic solutions of Eq. (5) are not unique.

This example, even though very elementary, illustrates the importance of the adequate definition of systems; agreement of inputs and outputs, decision of constraints, and bounding of variables, etc. These questions are especially important in connection with large economic and ecological systems, as mentioned before.

2.1.4. Input–Output Mappings

A systems as a set of input–output pairs is a primitive form of models in the sense that the mere set does not give any arrangement or classification of elements. It can be considered as a *black-box model*, because any inner structure is not given. This is also in accord with the empirical study of a system by plain measurements of input–output pairs; the measurement table or graph represents a black-box model.

A system S as a set of pairs [and $(\mathscr{S})^\circ$ in (3)] need not to be a function, or mapping, in general. However, representation of systems by the aid of some functions is quite attractive both conceptually and notationally, partly due to our scientific and mathematical tradition. For example, every system S can be parametrized, or covered, by the aid of some family $\mathscr{F}$ of mappings f_i, $i \in I$ (an index set) so that

$$S = \cup \mathscr{F} = \bigcup_{i \in I} f_i, \tag{6}$$

where $\cup$ denotes the union operation. The elements f_i of $\mathscr{F}$ can be called here *input–output mappings*. It is not necessary, or even judicious, to presuppose that f_i's have the same domain. Some authors (e.g., Blomberg, 1973) consider the family $\mathscr{F}$ as a basic form of input–output systems in general. A similar representation is obtained by a so-called *parametric mapping* (Orava, 1974), by some mapping $r(\cdot,\cdot)$ with a set Q so that

$$(u, y) \in S \Leftrightarrow (\exists q \in Q)(y = r(q, u)). \tag{7}$$

It is easily seen that there is a natural correspondence between the representations (6) and (7); the set $\{f_i | i \in I\}$ serves as a Q, and conversely, the partial mappings $r(q,\cdot)$ serve as elements f_i of an $\mathscr{F}$. The representation type (7) has been extensively used as an auxiliary concept by many authors, among others Windeknecht, Mesarovic, and Orava, especially in connection with the concept of state. In fact, the elements $q \in Q$ (like $f_i \in \mathscr{F}$) can be interpreted as abstract *initial states* of S.

2.2. Interconnection of Systems

The basic concept of the system before is presented in the open-loop form where the variables of the inputs and outputs are free in the sense of predicate logic. To be useful in modeling systems in applications, the formalism should also admit the interconnection (and decomposition) of systems. The interconnection is here illustrated briefly by a special example in the following, without any general definition.

Let $\mathscr{S}_1$ and $\mathscr{S}_2$ be systems given by the triplets

$$((u_1, u_2), F_1, (v_1, v_2)), \qquad ((x_1, x_2, x_3), F_2, (y_1, y_2)), \tag{8}$$

respectively. An agreement upon the graphical representation of systems was made in the basic definition before; thereby, interconnections of systems can be clearly depicted in the conventional way. Now let the system $\mathscr{S}_1$ and $\mathscr{S}_2$ be interconnected according to the broken lines in the diagram in Fig. 2. It is tacitly assumed that F_1 and F_2 do not contain the same free parameter variables, because such variables would bring forth hidden connections; this restrictive condition is guaranteed in detailed cases by common sense.

The interconnection by Fig. 2 contains three single loops which exemplify typical cases appearing in applications. It can be interpreted as a system in many ways, depending on the choice of input and output variables. An alternative is as follows: The interconnection is a system

$$((u_1, x_1), F, (v_1, v_2, y_1, y_2)), \tag{9}$$

where the original u_2, x_2, and x_3 are now parameter variables, and the

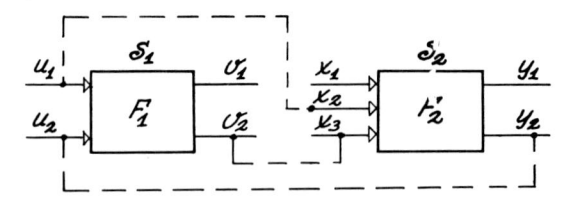

9. Dynamical System Models: Cause–Effect Relationships and State Representations

Figure 2. An interconnection of systems $\mathscr{S}_1$ and $\mathscr{S}_2$ in (8).

formula F reads

$$F_1 \,\&\, F_2 \,\&\, (u_1 = x_2) \,\&\, (v_2 = x_3) \,\&\, (u_2 = y_2); \tag{10}$$

the mark & denotes the logical conjunction. The representation (9)–(10) is in accord with the basic definition. The expression (10) is a correct formula in the language in question, and the input and output variables in (9) are really free in F. Other alternative representations for the interconnection can be obtained by interchanging the roles of interconnected variables (e.g., u_1 and x_2), or by altering the set of output variables (e.g., changing v_1 and v_2 into parameters), or by eliminating variables by substitution (e.g., v_2 for x_3 everywhere).

The single interconnection loops represented by the conjuncts in parentheses in Eq. (10) are examples only, and any general formalization is not studied in the present article. The conjunct $u_1 = x_2$ is clearly a *parallel* connection; two input components of the original systems are connected together. On the contrary, the other two conjuncts can be considered either as *series* or *feedback* modes depending on the succession between $\mathscr{S}_1$ and $\mathscr{S}_2$. Without this interpretation, both the connections are feedback loops; some original input and output components are connected together. A very general formalism of interconnections is presented by Blomberg (1973); the connections there are expressed with the aid of a Boolean matrix. Thereby, the analysis and transformations of interconnections can be studied in a theoretically elegant algebraic way.

A reverse procedure of the interconnection is obviously the *decomposition* of systems into subsystems. The above example shows that some input and output variables may change into parameters or disappear in an interconnection. Conversely, in a decomposition some parameters may obviously change into input or output variables, and new variables appear. The decomposition technique is a very central tool, for example, in hierarchical solution methods for control and optimization problems in large-scale systems (see, e.g., Singh, Chapter 17, this volume).

2.3. Time System

The concept of time is conventionally formalized as a set T, a *time domain*, together with a suitable order relation $<$ (in $T \times T$). In some presentations in literature, $(T, <)$ is quite axiomatically and generally characterized. However, T as a *subset of real numbers* with the natural order $<$ of the reals is completely adequate for the present purposes. Note that this definition includes both the continuous- and discrete-time cases as well as mixed cases. Any subset of T is clearly a time domain too. For the subsequent use, the following additional notation is defined; for each $t, t' \in T$:

$$T_{t]} = \{t'' | t'' \in T \,\&\, t'' \leq t\},$$
$$T_{[t} = \{t'' | t'' \in T \,\&\, t'' \geq t\}, \qquad (11)$$
$$T_{[t,t']} = \{t'' | t'' \in T \,\&\, t \leq t'' \leq t'\}.$$

This notation is easily memorizable in analogy with the interval notation of reals; also $T_{(t}$, $T_{t)}$, etc., would then be self-evident. Make an agreement that a time domain T is given and fixed throughout the remaining part of this article.

The concept and notation of time functions and their restrictions are quite standard as follows. Let V be any set, and let T' and T'' be subsets of T. The expression $V^{T'}$ denotes the set of all functions $v: T' \to V$, i.e., functions defined on T' with values in V. The elements of $V^{T'}$ are now called *time functions*. For $v \in V^{T'}$, the expression $v|T''$ denotes the *restriction* of v on T'', and for $t, t' \in T'$ the restriction $v|T'_{[t,t']}$ is called a *segment* of v on $T'_{[t,t']}$; in symbols,

$$v|T'' = \{(t,w) | t \in T' \cap T'' \,\&\, w = v(t)\}. \qquad (12)$$

Note that V is an arbitrary set; therefore, the time function is still a very general concept. For example, elements of V may consist of stochastic variables and/or probability distributions.

The concept of time system can now be defined by the aid of the above tools. At this point there is some variety among the formalisms presented in literature with regard to generality. For facilitating the presentation of ideas in the sequel, a very simple form of time systems is considered in this article as follows. A system $\mathscr{S}$ is called a *(uniform) time system* if

$$(\mathscr{S})^\circ \subset U^T \times Y^T, \qquad (13)$$

where U and Y are some sets. By (13), $\mathscr{S}$ is said to be *in* $U^T \times Y^T$; this will be a very handy expression in the sequel. The specification is given here in backward direction by "digging out" the set $(\mathscr{S})^\circ$ from the system. Therefore, it is tacitly assumed that the set T and its elements are completely distinguishable from all the other sets and elements appearing in this context.

The essential restriction for a system by (13) is that all the time functions associated with it have a common time domain. This simple form is extensively used by Mesarovic (1972; Mesarovic and Takahara, 1975), and it appears frequently in descriptive and tutorial texts. On the contrary, Zadeh (1969) and Patten *et al.* (1976), for example, even *presuppose* explicitly that a general time system is *closed under segmentation*: In the present notation, for every $(u', y') \in (\mathscr{S})^\circ$ and for every $t, t' \in T$, the pair $(u'|T_{[t,t']}, y'|T_{[t,t']})$ of segments of u' and y' also belongs to $(\mathscr{S})^\circ$. This kind of strong condition in the very basic definition seems questionable. A very wide and general form of time system is introduced by Orava (1974). The question about the time domain is not only scholastic; for example, in nonlinear differential equations there may exist solutions with the so-called finite escape-times which vary with the initial conditions. Then the requirement about the common time domain may unduly restrict the set of solutions. Second, in the automata theory the varying length of the input and output lists is an inherent property of the formalism.

2.3.1. Example 2

The (state) *differential system* is a model which is frequently used in many branches of science, probably because the classical laws of physics involve the concept of continuous time. Such a system $\mathscr{S}$ is customarily given by the expressions

$$\dot{x}(t) = f(x(t), u(t), t), \qquad y(t) = h(x(t), u(t), t), \qquad t \in T, \tag{14}$$

where u, y, and x are the input, output, and state of $\mathscr{S}$, respectively; f and h are the generator mapping and the output mapping of $\mathscr{S}$, respectively; the dot $\cdot$ denotes the derivative operation with respect to time t on a given real interval T.

The formal interpretation of $\mathscr{S}$ as a system follows the same lines as in Example 1 before. First, u, y, x, and t are variables, whereas f, h, and T are constant parameters. The dot $\cdot$, the parentheses ( ), and $\in$ are constant operations or predicates of the theory. The variables u, y, and x stand for time functions on T. The system $\mathscr{S}$ is now a triplet (u, F, y). The core part of the formula F is given by the conjunction of the *first two* formulas in (14), together with some relevant analytical restrictions. There may also exist some initial conditions for the motion x. Altogether, denote this core formula with its completions by the symbol A. There are now two parameter variables in this system: x and t. Customarily, F is now obtained by bounding t by the universal quantifier as follows: $(\forall t)(t \in T \Rightarrow A)$. By convention, the antecedent $t \in T$ can also be removed into the prefixing as follows: $(\forall t \in T)(A)$, corresponding to the usual prefixing phrase "for every $t \in T$." On the other hand, any prefixing on the parameter variable x is not necessary in the sense of the general definition. The set of input–output pairs of $\mathscr{S}$ is thus given by

$$(\mathscr{S})^\circ = \{(u, y) | (\exists x)(\forall t \in T)(A)\}. \qquad (15)$$

It is now very instructive to observe the role of the bounding of t; if t were free in the definition (14), then a very silly and unconventional interpretation of differential system would be obtained:

$$(\mathscr{S})^\circ = \{(u, y) | (\exists x)(\exists t)(A)\};$$

i.e., Eqs. (14) need to be satisfied only at one point $t \in T$! The definition of $\mathscr{S}$ was purposely written down in the ambiguous form (14) for exemplifying the careless language in the field. The danger of misunderstanding is more severe in connection with large or untraditional system models.

3. CAUSALITY

The input–output system defined in the preceding section is still an abstract concept without any empirical interpretation. When it is used for modeling cause–effect relations in empirical sciences, the input is naturally identified with the cause, or stimulus, and the output with the effect or response. The corresponding variables are then to be interpreted as *measures* or *information* of the respective phenomena. These inputs and outputs may not correspond directly to material or energetic input and output *flows* in a real system. In this respect there is a great danger of confusion with the *input–output flow analysis* in economics (see, e.g., Patten *et al.*, 1976; Patten and Finn, Chapter 8, this volume). There are numerous examples in many fields where a material output flow of a real system corresponds to an input variable of a system model, and conversely, a material input flow may correspond to an output variable.

3.1. Cause–Effect and Functionality

Some authors, like Patten *et al.* (1976), require that the input–output system $\mathscr{S}$ as a model of cause–effect relationship is functional, that is, the set $(\mathscr{S})^{\circ}$ is a mapping. They interpret the nonfunctional systems as the *multiple causation*. The functionality is, however, a very strong requirement in general. The requirement may be understandable in natural sciences, if the very initial states of systems cannot be manipulated. The input–output mapping family and the parametric mapping defined in Section 2.1.4 serve as more realistic models of the causal determination; the mappings of the family, or equivalently, the parameters of the parametric mapping, correspond to the *initial causes* of the system. Accordingly, each mapping can be interpreted as a *determinate holon* of Patten et al. (1976).

3.2. Nonanticipation

A common (physical) notion of the dynamical causality involves the time order of the cause and effect: The effect cannot precede its cause. The concept of nonanticipation in dynamical systems and control theory is a special form of the idea. The concept corresponds to the *analytical causality* mode by Domotor (1972). In the present formalism it can be defined quite generally as follows.

Let $\mathscr{S}(p)$ be a system which depends on a parameter p, the value of which belongs to some index set. In the sense of the formal definition (Section 2.1), $\mathscr{S}(p)$ may be obtained from a system $\mathscr{S}$ by interpreting a logically free parameter variable of $\mathscr{S}$, or a tuple of such variables, as a constant parameter p, different values of which are, however, considered. Now, let $\mathscr{S}(p)$ be a time system in a set $U^T \times Y^T$ (with a fixed-time domain T as before) for each value of p. The system $\mathscr{S}(p)$, with the parameter p, is said to be *nonanticipatory*, if for each fixed value of p, for every (u', y'), $(u'', y'') \in (\mathscr{S}(p))^{\circ}$, and for every $t \in T$:

$$(u'|T_{t]} = u''|T_{t]}) \Rightarrow (y'|T_{t]} = y''|T_{t]}). \tag{16}$$

In other words, the restriction $y'|T_{t]}$ is uniquely determined by $u'|T_{t]}$. It is also said that the system does not *anticipate* its future behavior (i.e., $u'(t'), y'(t')$ for $t' > t$). In terms of causality, the "effect" $y'(t)$ does not precede its dynamical "cause" $u'|T_{t]}$.

It can be easily shown that (16) implies the functionality of $\mathscr{S}(p)$ for each value of p. Therefore, the parametrization by p is necessary for preserving the generality. The different values of p can be interpreted as different initial states, or causes. The expression of the nonanticipation is just clearer in connection with those detailed system representations which include explicitly some natural quantities for the role of the initial state. The concepts of the mapping family and parametric mapping (Section 2.1.4) are customarily used in the definition in standard literature.

The nonanticipation in stochastic systems is somewhat problematic; there are some alternatives for defining it. If the input and output of such a system stand for stochastic processes only, then condition (16) is too weak to yield a (physically) meaningful nonanticipation. Proper forms of the concept are obtained by associating the relevant probability distributions with the inputs and outputs. A strong and conceptually clear form of the nonanticipation is in turn obtained by an *outcomewise* definition; this form turns out to be reduced back to the deterministic interpretation of systems involving realizations of stochastic processes. These questions will not, however, be considered more in this article.

3.2.1. Example 3

Consider the familiar linear differential system given by the equation

$$\dot{x}(t) = A(t)x(t) + B(t)u(t) \tag{17}$$

for every $t \in [0, t_1]$, where t_1 is a positive number, x and u are n- and m-vector-valued functions on the interval $[0, t_1]$, and A and B are appropriate matrix-valued functions (with suitable regularity conditions). The system is a simple special case of the differential system in Section 2.3.1. By the standard theory, solutions of Eq. (17) can be given by

$$x(t') = \Phi(t', t)x(t) + \int_t^{t'} \Phi(t', t'')B(t'')u(t'')\,dt'', \tag{18}$$

where Φ is the conventional transition matrix function of the system determined by the coefficient function A.

Relation (18) directly induces input–output mappings for system (17). For example, a parametric mapping $r(\cdot, \cdot)$ can be defined by

$$r(x_0, u)(t) = \Phi(t, 0)x_0 + \int_0^t \Phi(t, t'')B(t'')u(t'')\,dt''. \tag{19}$$

The mapping r, with x_0 as the parameter, is obviously nonanticipatory because the upper limit of the integral in (19) is t. The quadrature (18) is also reversible in the sense that t may be greater than t'. Thereby, also anticipatory mappings r may be defined by parametrizing the system with initial states at some $t > 0$. This illustrates the fact that the *pure mathematical* solution of the system equations is neutral with respect to the nonanticipation; the interpretation of the positive direction of the time t is optional. The system (17)–(19) will also serve as an explicit example for the state and state transition concepts later on.

3.3. Discussion

The nonanticipation is clearly a form of the dynamical causality, if the input segments of a system are interpreted as causes. However, some authors make a distinction between the causality and nonanticipation, whereas some authors do not include the concepts in formalisms explicitly. The causality of Windeknecht (1967) is essentially involved with dynamical state concepts. Zadeh (1969), in his theory of *aggregates*, does not mention the nonanticipation, but it is implicitly included in the structure. Also, the theories which use less general basic definitions for systems may not need these concepts at all, for example, the *dynamical system* of Kalman et al. (1969) and the automata theory as a whole.

4. STATE

Some primitive notions of state appeared already in the concepts of input–output mapping family and the parametric mapping given before. In general, the state is defined as an object which represents the relevant status of the system. As it is well known, the state is a central concept in thermodynamics. For example, in a homogenous material system like a gas without electrical properties the pressure, volume, and temperature are quantities, any two of which may serve as a state of the system. In addition, some derived quantities, like the entropy according to the second law, may also serve as proper components of the state. This example also illustrates the fact that the state is a relative concept depending on the purpose of the description.

4.1. Dynamical State

Zadeh (1969) has presented a clear lingual definition of the dynamical state which sharply describes the essence of the common notion: "The state (at a time moment) contains all the information about its (the system) past history that is relevant to the prediction of its future behavior." The definition of the state to be presented in the following has a close relation to the above idea. The concept is first given in a very primitive form which, however, turns out to be almost equivalent to concepts of other authors.

Let $\mathscr{S}$ be a system, with an input u and output y, in a set $U^T \times Y^T$ (where T is a time domain as before). Then let V be a set and let $\mathscr{S}_0$ be a system in $U^T \times (V^T \times Y^T)$ with the same input u and an output (v, y). The object v is now called a (*dynamical*) *state* of $\mathscr{S}$ if

(a) For each $(u', y') \in (\mathscr{S})^\circ$ there exists a $v' \in V^T$ such that $(u', (v', y')) \in (\mathscr{S}_0)^\circ$.

(b) For every $(u',(v',y')) \in (\mathscr{S}_0)^\circ$ and for every $t \in T$, the value $v'(t)$ together with the restriction $u'|T_{[t}$ uniquely determines the restrictions $v'|T_{[t}$ and $y'|T_{[t}$.

The set V and the system $\mathscr{S}_0$ are called a *state space* and an *input–state–output* system of $\mathscr{S}$.

The concept $\mathscr{S}_0$ only introduces an object v for $\mathscr{S}$ without giving any inner structure for $\mathscr{S}_0$. The conditions express the basic property given in the above-mentioned lingual definition of Zadeh. Naturally, the $\mathscr{S}_0$ and v are not unique for a given system $\mathscr{S}$ in this formalism. It can be easily shown that for every system an $\mathscr{S}_0$ can be constructed, for example, as a set of input–output pairs. However, in detailed cases and applications a v with physical or similar interpretations is given in advance.

4.2. State Representation

The state is often introduced directly by a more structured representation than that given before. A familiar example was already discussed in Section 2.3.1. This can be formalized in general terms of the present formalism as follows.

A pair $(\mathscr{S}_1, \mathscr{S}_2)$ of systems is called a *state representation* of a system $\mathscr{S}$ in a set $U^T \times Y^T$ if:

(a) V is a set and $\mathscr{S}_1$ is a system in $U^T \times V^T$ with an input u and an output v.

(b) For every $(u',v') \in (\mathscr{S}_1)^\circ$ and for every $t \in T$ the value $v'(t)$ together with the input restriction $u'|T_{[t}$ uniquely determines the restriction $v'|T_{[t}$.

(c) $\mathscr{S}_2$ is a system in $(V^T \times U^T) \times Y^T$ with the input (v,u) and an output y.

(d) For every $((v',u'),y') \in (\mathscr{S}_2)^\circ$ and for every $t \in T$ the restrictions $v'|T_{[t}$ and $u'|T_{[t}$ together uniquely determine the restriction $y'|T_{[t}$ of the output.

The objects $\mathscr{S}_1$, v, V, and $\mathscr{S}_2$ are called a *state system, state, state space*, and *output system*, respectively. Condition (b) is called a *state property* of $\mathscr{S}_1$ (also independently for any separate $\mathscr{S}_1$). The pair $(\mathscr{S}_1, \mathscr{S}_2)$ represents the system $\mathscr{S}$ so that for every $(u',y') \in (\mathscr{S})^\circ$ there exists a v':

$$(u',v') \in (\mathscr{S}_1)^\circ, \qquad ((v',u'),y') \in (\mathscr{S}_2)^\circ. \tag{20}$$

The state representation above can be graphically illustrated by the diagram of Fig. 3, according to the rules agreed upon in Section 2.1 (the broken lines and the symbol $\mathscr{S}_0$ should be neglected for the moment). The

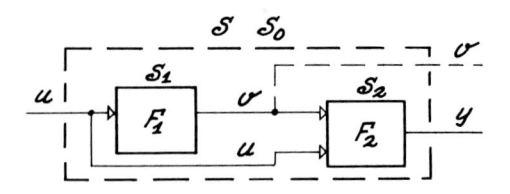

9. Dynamical System Models: Cause–Effect Relationships and State Representations

Figure 3. State representation ($\mathscr{S}_1$, $\mathscr{S}_2$) of system $\mathscr{S}$.

representation is interpreted as an interconnection of the systems $\mathscr{S}_1$ and $\mathscr{S}_2$; for simplicity, the same symbols u and v are used in the interconnection loops. Condition (d) is very wide because in many cases in applications $\mathscr{S}_2$ is a static, or memoryless system. It is easily observed that $\mathscr{S}_2$ is in every case a functional system.

Consider now the relation of the state representation in Fig. 3 to the input–state–output system $\mathscr{S}_0$ before. Obviously, the block enveloped in the broken lines in Fig. 3 represents an $\mathscr{S}_0$ of $\mathscr{S}$; the conditions (a)–(d) obviously imply all the properties for $\mathscr{S}_0$. Conversely, from a given $\mathscr{S}_0$ a state representation can be constructed via obvious rearrangements of variables. This shows that the definitions are in accord with each other.

The systems in Examples 2 and 3 in Sections 2 and 3 can be easily interpreted as state representations. In Example 2 the differential equation part of (14) represents a state system $\mathscr{S}_1$ in the definition, with x in the role of v. The solutions of this differential equation obviously satisfy the state property. The mapping h there also represents a *static output system* $\mathscr{S}_2$; for such a system condition (d) has the stronger form: For every $((v', u'), y') \in (\mathscr{S}_2)^\circ$ and for every $t \in T$, the value $v'(t)$ together with $u'(t)$ uniquely determines the value $y'(t)$. The interpretation of Example 3 is more straightforward, because an explicit quadrature (18) is known, and the state x also serves as the output.

4.3. State Transition

The state concepts include implicitly also the notion of transition of state, point by point in time. The transition is a familiar concept in many detailed system models. In Section 3.2.1, the quadrature (18) gives an explicit representation for the transition of x from a time point t to t'. This example already illustrates general principles for formalization, even though it also has certain strong properties which should not be included in the basic definition.

The formalization of the notion *from a time point t to t'* necessitates the use of a set of systems, because the uniform time systems is applied in the present formalism. Let $\{\mathscr{S}_t | t \in T\}$ be a family of time systems such that for each $t \in T$:

(a) $\mathscr{S}_t$ is a system in a set $U^{T_{[t}} \times V^{T_{[t}}$.

(b) For every $(u^*, v^*) \in (\mathscr{S}_t)^\circ$ and for every $t' \in T_{[t}$ the value $v^*(t')$ together with the restriction $u^*|T_{[t'}$ uniquely determines the $v^*|T_{[t'}$.

(c) For every $(u^*, v^*) \in (\mathscr{S}_t)^\circ$ and for every $t' \in T_{[t}$,

$$(u^*|T_{[t'}, v^*|T_{[t'}) \in (\mathscr{S}_{t'})^\circ. \tag{21}$$

The set $\{\mathscr{S}_t | t \in T\}$ is now called a *state transition family*. Condition (b) expresses anew that $\mathscr{S}_t$'s are state systems, because (b) is simply the state property for $\mathscr{S}_t$. Condition (c) expresses that the input–output motions of $\mathscr{S}_t$ "go through" every subsequent system $\mathscr{S}_{t'}$ with $t' > t$.

The above definition is quite general, because the inner structure of the time systems $\mathscr{S}_t$ is not specified. Condition (c), which expresses the essential property of transition, may seem very abstract, because it is given at a simple set-theoretic level. However, together with the state property of the $\mathscr{S}_t$'s it implies stronger properties and representations. Many authors, among others Windeknecht, Mesarovic, and Kalman *et al.* (also Orava, 1973, 1974), present the state systems and transitions explicitly by the aid of mappings corresponding to the concept of parametric mapping of Section 2.1.4. The principle of these representations can be roughly illustrated in symbols of (21) as follows. Each $\mathscr{S}_{t'}$ is represented by a parametric mapping $s_{t'}(\cdot, \cdot)$, where values of state motions v^* at t' serve as parameters so that

$$v^*|T_{[t'} = s_{t'}(v^*(t'), u^*|T_{[t'}). \tag{22}$$

This property and its close relatives appear in literature under various names: the "continuation", "consistency", "transition", "semi-group" properties, etc.

Condition (22) clearly implies property (21), if the representations $s_{t'}(\cdot, \cdot)$ are identified with $\mathscr{S}_{t'}$'s. On the other hand, the *existence* of the family $\{s_t | t \in T\}$ satisfying (21)–(22) can be easily concluded from conditions (a)–(c) of the definition. Moreover, the mappings s_t have the state property (the interpretation is probably self-evident). In fact, the present definition in its abstract set-theoretical form corresponds to Zadeh's theory of *bundles and aggregates*; the condition (c) corresponds to the "continuation" and the (b) to the "uniqueness" in the bundles. Zadeh's axioms are, however, stronger, and the nonanticipation of the state transition is also implied there. On the whole, the arguments of this paragraph show that the present definition, although seemingly superficial, has an intimate relation to other theories and more detailed representations.

5. DISCUSSION OF ECOLOGICAL EXAMPLES

The mathematical methods for modeling and analysis of ecosystems can be included in systems theory in general, as discussed in the Introduction. However, the special features of ecology may emphasize the use of a certain class of models and methods. In many cases the ecosystems are distributed variable systems involving elementary phenomena which are not yet known satisfactorily. Therefore, the aggregation and simplification of phenomena is necessary. In addition, a steady-state model is often used, when the ecosystem under consideration is nearly in a (quasi)static equilibrium, or the identification of an accurate dynamical model would require much too difficult experimentation and numerical procedure. In this section two ecological examples are briefly discussed in the light of the preceding formalism.

The *marine coprophagy model* described by Patten and Finn (Chapter 8, this volume) and by Patten et al. (1976) is a simple representation for the carbon flow in marine benthic ecosystems. The model consists of four compartments: *Callianassa major*, *C. major* feces, benthic invertebrates, and benthic invertebrate feces. The carbon storages in the compartments are state variables of the system, denoted by x_1, x_2, x_3, x_4, respectively. The *mass balance* of the system is now given by the equations (see Patten et al., 1976)

$$\dot{x}_1 = z_{10} - \Phi_{21} - y_{01}, \quad \dot{x}_2 = \Phi_{21} - \Phi_{32} - y_{02},$$
$$\dot{x}_3 = z_{30} + \Phi_{32} + \Phi_{34} - \Phi_{43} - y_{03}, \quad \dot{x}_4 = \Phi_{43} - \Phi_{34} - y_{04}, \quad (23)$$

where z_{10} and z_{30} are inflows, $y_{01}, \ldots, y_{04}$ are outflows, and each Φ_{ij} is the flow from compartment j to i; for simplicity, the time arguments are deleted. Equations (23) express, among other things, that the compartments 1–4 are in "series," and there is a "feedback loop" from 4 to 3.

Equations (23) resemble the differential systems (14) and (17) (in Sections 2.3.1 and 3.2.1). However, (23) *is not yet* a state model or an input–output system in the sense of the present formalism. First, the internal variables Φ_{ij} obviously depend on some of the variables $x_1, \ldots, x_4$, $z_{10}, z_{30}, y_{01}, \ldots, y_{04}$ and possibly some external parameters and time. These dependences should be modeled by the aid of some functional relations which are then substituted for Φ_{ij}'s in (23). In the treatment of the steady state by the aid of the *input–output flow analysis* these relations are given by suitable linear equations with proportional constants. Second, the definition of the external inputs and outputs is still an open question, at least in regard to the true *causal* explanation of the systems behavior. This question was mentioned at the beginning of Section 3. For example, it is very hard to imagine that the inflows z_{10} and z_{30} were the actual input quantities of this system in the classical causal interpretation. The total behavior of the system is obviously determined, among other things, by the food and carbon concentrations in its environment (and by some external parameters such as temperature, sunlight, and currents). Thereby, the inflows z_{10} and z_{30} may be actually *output* quantities of the causal system. On the whole, the above system is a part of the decomposition of the entire marine system.

The model of ^{60}Co *kinetics in an aquatic microcosm* presented by Bargmann and Halfon (1977) is concerned with the tracer identification of transfer parameters of an ecosystem in aquarium. The system is modeled by 12 compartments: five species of rooted plants, two of floating plants, plankton, snails, newts, water, and detritus, excluding the sand and glass slides. The radioactive material was injected into the aquarium and observed for 114 days in the compartments.

The above ecosystem is obviously assumed to be in static equilibrium with constant and homogeneous food flows between the compartments (possibly except the detritus, sand, and glass). By some simplifying assumptions, the vector y of ^{60}Co amounts in the compartments satisfies the autonomous differential equation

$$\dot{y}(t) = Ay(t), \tag{24}$$

where A is a constant matrix with the column sums equal to zero. The elements of A are now estimated from the experimental data by using some type of numerical method. The food transfer flows can then be calculated if the total masses of the compartments are known.

The ecological model in the latter example is static. It does not involve any input–output or causality interpretation at all. The dynamical model (24) is only secondary; it describes the migration of the tracer material which has no influence on the life of the ecosystem. This example shows that the (control theoretic) input–output model or causality interpretation is sometimes irrelevant in natural sciences.

REFERENCES

Bargmann, R. E., and Halfon, E. (1977). Efficient algorithms for statistical estimation in compartmental analysis: Modelling ^{60}Co kinetics in an aquatic microcosm. *Ecol. Modell.* **3**, 211–226.

Blomberg, H. (1973). On set theoretical and algebraic systems theory. Part 1. *Proc. Eur. Meet. Cybernet. Syst. Res.*, 1972, Vol. 1, pp. 17–39.

Domotor, Z. (1972). Causal models and space–time geometries. *Synthese* **24**, 5–57.

Kalman, R. E., Falb, P. L., and Arbib, M. A. (1969). "Topics in Mathematical System Theory." McGraw-Hill, New York.

Klir, G. J. (1972). The polyphonic general systems theory. *In* "Trends in General Systems Theory" (G. J. Klir, ed.), pp. 1–18. Wiley (Interscience), New York.

Mesarovic, M. D. (1972). A mathematical theory of general systems. *In* "Trends in General Systems Theory" (G. J. Klir, ed.), pp. 251–269. Wiley (Interscience), New York.

Mesarovic, M. D., and Takahara, Y. (1975). "General Systems Theory: Mathematical Foundations." Academic Press, New York.

Orava, P. J. (1973). Causality and state concepts in dynamical systems theory. *Int. J. Syst. Sci.* **4**, 679–691.

Orava, P. J. (1974). Notion of dynamical input–output systems: Causality and state concepts. *Int. J. Syst. Sci.* **5**, 793–806.

Patten, B. C. and Finn, J. T. (1978). Systems Approach to Continental Shelf Ecosystems, this *volume, pp.* 183–212.

Patten, B. C., Bosserman, R. W., Finn, J. T., and Cale, W. G. (1976). Propagation of cause in ecosystems. *In* "Systems Analysis and Simulation in Ecology" (B. C. Patten, ed.), Vol. 4, pp. 457–579. Academic Press, New York.

Rogers, R. (1971). "Mathematical Logic and Formalized Theories." North-Holland Publ., Amsterdam.

Suppes, P. (1960). "Axiomatic Set Theory." Van Nostrand-Reinhold, Princeton, New Jersey.

von Bertalanffy, L. (1972). The history and status of general systems theory. *In* "Trends in General Systems Theory" (G. J. Klir, ed.), pp. 21–41. Wiley (Interscience), New York.

Windeknecht, T. G. (1967). Mathematical systems theory: Causality. *Math. Syst. Theory* **1**, 279–288.

Zadeh, L. A. (1969). The concepts of system, aggregate, and state in system theory. *In* "System Theory" (L. A. Zadeh and E. Polak, eds.), pp. 3–42. McGraw-Hill, New York.

Part **III**

SYSTEM IDENTIFICATION

Chapter **10**

STRUCTURAL IDENTIFIABILITY OF LINEAR COMPARTMENTAL MODELS

C. Cobelli, A. Lepschy, and G. Romanin-Jacur

1. Introduction 237
2. Linear Time-Invariant Compartmental Models 239
3. The Problem of Structural Identifiability 241
   3.1 Introduction 241
   3.2 Definition. 242
4. Structural Properties Related to Identifiability. 243
5. The Analysis of Structural Identifiability 246
   5.1 Introduction 246
   5.2 Evaluation of v. 248
6. Examples and Conclusions 251
   6.1 General. 251
   6.2 Calcium Cycle 251
   6.3 Phosphorus Dynamics 254
   References. 257

1. INTRODUCTION

Compartmental analysis is a phenomenological and macroscopic approach for modeling physicochemical processes. The choice of the compartments and the level of aggregation needed to model an ecosystem depends on the goal of the modeling exercise and the information available from the system. Zeigler and Cale and Odell (Chapters 1 and 2, this volume) analyze this problem from a theoretical point of view. The next step in model development is the identification problem which concerns the determination of the structure and the parameter values of a mathematical model in such a way that it describes the system behavior in accordance with some predetermined criteria. Some papers in this volume (e.g., Beck, Klir, and Ivakhnenko, Chapters 11–13) describe some identification methods available for models of ecological systems.

In this chapter an *a priori* problem related to the identification of compartmental models from input–output experiments is discussed, i.e., the so-called identifiability problem. This may heuristically be defined as the *a priori* evaluation of the possibility of identifying a given model from the planned experiment. In this chapter only linear time-invariant compartmental models are considered and we assume to have enough information about the system to choose the compartmental structure of the model and those compartments which are accessible for the identification experiment. The choice of the compartmental structure refers to the number of the compartments and to the existence of fluxes among them. The identification experiment usually consists of applying to the system suitable perturbations and measuring the consequent time variations of the relevant variables. Generally, different structures may be assumed to model the same system and several experiments may be designed to identify the chosen model, that is, to estimate its parameters. After having assumed a compartmental model for the ecosystem, it is necessary to test whether the experiment designed for its identification allows the estimation of all unknown parameters of the model. This structural (*a priori*) identifiability problem has to be solved before performing the experiment and only on the basis of the knowledge of model structure.

The problem of structural identifiability has been formalized by Bellman and Åström (1970). Cobelli and Romanin-Jacur (1975, 1976a, b) and Cobelli *et al.* (1976, 1977a) studied this problem for the case of linear time-invariant compartmental systems in a system theory context and proposed conditions for local identifiability for all parameter values, i.e., the existence of a finite number of solutions at most in the parameter space. In the above papers compartmental configurations of increasing complexity were considered and a testing procedure to check the suggested conditions directly from the chosen compartmental structure and the planned input–output experiment was given.

The problem of parameter identifiability of dynamical systems (linear and nonlinear), not only for the compartmental ones, has recently received great attention and several definitions and results are available (Glover and

10. Structural Identifiability of Linear Compartmental Models

Willems, 1974; Grewal and Glover, 1976). Grewal and Glover (1976) give a precise general definition of parameter identifiability in terms of output distinguishability and propose two criteria to check local identifiability of linear time-invariant dynamical systems.

These definitions and criteria are used in this chapter for the purpose of creating a more rigorous framework for the results previously obtained on identifiability of compartmental models. The recently defined concepts of input and output connectability (Davidson, 1977) are employed to give necessary conditions for structural identifiability instead of the formerly used concepts of structural controllability and observability.

Section 2 will be devoted to review some fundamentals of linear time-invariant compartmental models and some comments are made about their applicability to the modeling of ecosystems. In Section 3 the role of structural identifiability in an identification procedure is evidenced, and formal definitions of local and global identifiability are given. Section 4 deals with some structural properties related to identifiability of compartmental models; in particular, reference is made to the recently introduced notions of input and output connectability and to their relations with the well-known properties of controllability and observability. Section 5 summarizes the results previously obtained by the authors, relating them to the notions of output indistinguishability and of input and output connectability. In Section 6, the structural identifiability test procedure for compartmental models is applied to some models of ecosystems.

2. LINEAR TIME-INVARIANT COMPARTMENTAL MODELS

The dynamical equation of the generic ith compartment of a linear time-invariant compartmental model is written as follows (Fig. 1):

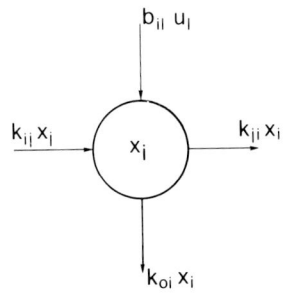

Figure 1. A graphical representation of the generic ith compartment. Material flows from other compartments ($k_{ij}x_j, j = 1 \ldots n, j \neq i$), to other compartments ($k_{ji}x_i, j = 1 \ldots n, j \neq i$), to the external environment ($k_{oi}x_i$) and from an external input ($b_{il}u_l$).

$$\dot{x}_i(t) = -\sum_{\substack{j=0 \\ j \neq i}}^{n} k_{ji} x_i(t) + \sum_{\substack{j=1 \\ j \neq i}}^{n} k_{ij} x_j(t) + \sum_{l=1}^{r_b} b_{il} u_l, \qquad (1)$$

where x_i is the state variable associated to the ith compartment, $\dot{x}_i = dx_i(t)/dt$, k_{ij} is the constant nonnegative transport rate parameter from compartment j to compartment i, k_{0i} is the constant nonnegative transport rate parameter from compartment i to the external environment, u_l is the lth input, b_{il} is the fraction of the lth input entering the ith compartment, n is the number of compartments, and r_b is the number of inputs. The set of Eq. (1) can be usefully put in the input-state form:

$$\dot{\mathbf{x}} = A\mathbf{x} + B\mathbf{u}, \qquad (2)$$

where

$\mathbf{x} = [x_1, x_2, \ldots, x_n]^T$,

$\mathbf{u} = [u_1, u_2, \ldots, u_{r_b}]^T$,

$A = \{a_{ij} : i, j = 1, 2, \ldots, n\}$, with $a_{ij} = k_{ij}, \quad i \neq j$,

$$a_{ii} = -\sum_{j=0}^{n} k_{ji},$$

$B = \{b_{il} : i = 1, 2, \ldots, n; l = 1, 2, \ldots, r_b\}$.

A general state-output (measurement) equation can be written assuming that r_c measurements are made on the system which are linear combinations of the state variables of compartments

$$\mathbf{y} = C\mathbf{x}, \qquad (3)$$

where

$\mathbf{y} = [y_1, y_2, \ldots, y_{r_c}]^T$,

$C = \{c_{mi} : m = 1, 2, \ldots, r_c; i = 1, 2, \ldots, n\}$.

A noise-free measurement is assumed in (3) as we will deal only with structural (a priori) identifiability problems. Vectors $\mathbf{x}$ and $\mathbf{y}$ are usually related to the amount of material or tracer in the compartments. The fundamental properties of this class of compartmental models have been proved by Hearon (1963) and Thron (1972).

The set of equations (2) and (3) can represent, at least in a first approximation, the dynamics of a compartmentalizable ecosystem with the input–output experiment designed for its identification. In fact several justifications of using linear modeling techniques in ecology have been given (e.g., Waide et al., 1974; Patten, 1975; Patten et al., 1975; Waide and Webster, 1976). Let us emphasize that the class of models described by (2) and (3) represents the small signal model that arises naturally from a nonlinear model of an ecosystem in steady state and a nonlinear observation equation if an input–output experiment of the first-order perturbation type (for instance, a tracer experiment) is employed for its identification (e.g., DiStefano, 1976).

3. THE PROBLEM OF STRUCTURAL IDENTIFIABILITY

3.1. Introduction

For clearly stating the problem of structural identifiability it is helpful to review briefly the steps usually performed in identification and parameter estimation of compartmental models from input–output data.

The following steps are to be performed: (i) *compartmentalization*: to identify ecologically meaningful compartments and flows among them; (ii) *input–output experiment design*: to select the compartments for perturbation inputs and for measurements; (iii) *modeling*: on the basis of (i) and (ii) to write the model equations (2) and (3); (iv) *structural identifiability check*: to check whether the planned experiment allows us the estimation of the nonzero transport rate parameters (matrix A) and, if necessary, the unknown elements of matrices B and C. Only if the check gives a positive answer are the following steps valid from a theoretical point of view and are they capable of being successful; otherwise steps (i) and (ii) are to be reconsidered; (v) *performance of the experiment*; (vi) *parameter estimation*: to estimate the transport rate parameters from input–output data by using suitable estimation procedures; (vii) *a posteriori identifiability check*: to evaluate the reliability of the estimates (taking into account noises and measurement errors) and the consistency of the model response to experimental data; if the check gives no positive answer, the whole procedure is to be suitably reconsidered; and (viii) *model validation.*

This chapter discusses step (iv). Structural identifiability analysis is important as it is strictly linked to the problem of the design of a meaningful experiment. It should also be noticed that the structural identifiability check gives an *a priori* quantitative evaluation of the degree of identifiability of the model. As a consequence of a negative test answer, it is necessary either to modify the compartmentalization (e.g., by aggregating or neglecting from a dynamical point of view some compartments) and/or to modify the experiment or to look for some external constraints (e.g., nondynamical relations among the parameters); if the test response is positive, the possibility of performing simpler experiments, adopting more complex compartmental structures, and obtaining uniqueness of some parameter values may be considered.

3.2. Definition

Consider a linear time-invariant model. According to Lin (1974) the model is said to be a *fixed-structure model* if the identically null elements of matrices *A*, *B*, and *C* are *a priori* stated, while other elements are free (i.e., they can assume any value and are mutually independent). The free elements of system matrices can be ordered in a *parametrization vector*, $\mathbf{p} \in P$; its value characterizes a particular model by distinguishing it from all the others having the same structure.

The above definition may be generalized to any suitable analytical description of the system. In particular a compartmental model associated to a physical system and to an input–output experiment can be considered as a fixed-structure model if the space *P* of parameters k_{ij}, b_{il}, and c_{im} is considered; the parametrization vector **p** is formed by all nonzero transport rate parameters k_{ij} and by nonzero parameters b_{il} and c_{mi}.

In order to rigorously define structural identifiability of compartmental models it is helpful to use some definitions from Grewal and Glover (1976).

Definition 1. Given a fixed-structure model two parametrization vectors $\mathbf{p}'$ and $\mathbf{p}''$ are said to be (output) *indistinguishable* if the outputs of the related models $\mathbf{y}(\mathbf{p}')$ and $\mathbf{y}(\mathbf{p}'')$ are identical for every input **u** and every initial state $\mathbf{x}_0 = \mathbf{x}(0)$. Otherwise, the two parametrization vectors are said to be (output) *distinguishable*.

Definition 2. A fixed-structure model is said to be *locally identifiable* in $\mathbf{p}'$ if there exists a neighborhood $\varepsilon(\mathbf{p}')$ such that for every $\mathbf{p}''$ in it the pair $(\mathbf{p}', \mathbf{p}'')$ is distinguishable.

Definition 3. A fixed-structure model is said to be *globally identifiable* in $\mathbf{p}'$ if for every $\mathbf{p}'' \neq \mathbf{p}'$ in the parameter space *P* the pair $(\mathbf{p}', \mathbf{p}'')$ is distinguishable.

Definition 4. A fixed-structure model is said to be *almost everywhere locally identifiable* if it is locally identifiable for every $\mathbf{p}'$ in the whole space *P* except for at most a subset of zero measure.

Definition 5. A fixed-structure model is said to be *almost everywhere globally identifiable* if it is globally identifiable for every $\mathbf{p}'$ in the whole space *P* except for at most a subset of zero measure.

With reference to the problem of (*a priori*) structural identifiability of compartmental models, we will now state the following definitions:

Definition 6. A compartmental model is said to be *structurally identifiable* if it is almost everywhere locally identifiable.

Definition 7. A compartmental model is said to be *uniquely structurally identifiable* if it is almost everywhere globally identifiable.

In this chapter techniques are given to test structural identifiability of compartmental models. In order to avoid possible misunderstanding (Delforge, 1977) let us emphasize that structural properties hold almost everywhere in the parameter space, and also the difference between structural identifiability and unique structural identifiability. The very nature of the problem of uniqueness and the difficulty connected with its solution will be shown in Section 6.

4. STRUCTURAL PROPERTIES RELATED TO IDENTIFIABILITY

In this section some structural properties which are strictly related to identifiability will be presented; in particular, controllability, observability, and connectability will be defined, and a necessary condition for identifiability will be given.

Controllability may be defined heuristically as the possibility of transferring the model from any state to any other state in a finite time by means of a suitable control. Observability may be defined as the possibility of reconstructing all the state variables of the model from its outputs (e.g., Chen, 1970).

Correspondingly, a compartmental model is *completely controllable* (c.c.) if all variables associated to the compartments can be varied independently by the inputs and *completely observable* (c.o.) if all variables can be reconstructed from the measured outputs (Cobelli and Romanin-Jacur, 1976a).

As it is necessary to state whether a compartmental model is c.c. and/or c.o. only on the basis of its structure, the notion of structural controllability and observability (e.g., Lin, 1974; Davidson, 1977) is employed.

A linear time-invariant fixed-structure model, particularly a compartmental one, is said to be *structurally controllable* (*structurally observable*) if and only if there exists a completely controllable pair (A, B) (a completely observable pair (A, C)) which has the same structure. Consequently, the model is c.c. (c.o.) for every parametrization vector $\mathbf{p} \in P$ except for a subset of zero measure (almost everywhere controllable (observable)).

244   C. Cobelli, A. Lepschy, and G. Romanin-Jacur

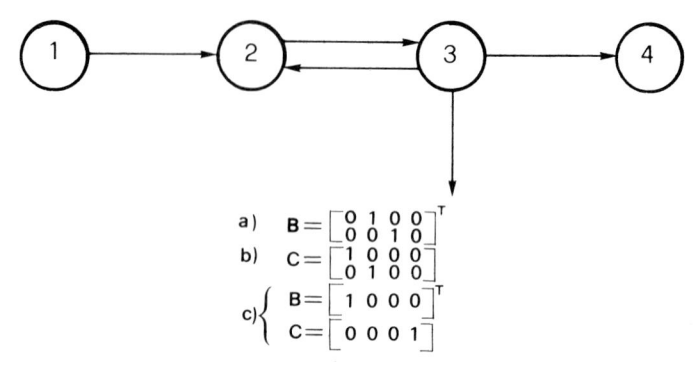

```
          1  ──▶  2  ⇄  3  ──▶  4
                         │
                         ▼
```

a) $B = \begin{bmatrix} 0 & 1 & 0 & 0 \\ 0 & 0 & 1 & 0 \end{bmatrix}^T$

b) $C = \begin{bmatrix} 1 & 0 & 0 & 0 \\ 0 & 1 & 0 & 0 \end{bmatrix}$

c) $\begin{cases} B = \begin{bmatrix} 1 & 0 & 0 & 0 \end{bmatrix}^T \\ C = \begin{bmatrix} 0 & 0 & 0 & 1 \end{bmatrix} \end{cases}$

Figure 2. Input and output connectability of a four-compartment model. (a) The model is not input connectable (no path to compartment 1). (b) The model is not output connectable (no path from compartment 4). (c) The model is input and output connectable.

Two other important structural properties have been recently defined by Davidson (1977): input and output connectability. We will present these properties for the case of compartmental models.

With reference to the graph of a compartmental model let us define a *path* as a succession of directed arcs, such that the compartment entered by an arc is the same from which the next arc is started; the length of the path is defined as the number of its intercompartmental arcs. The model is then said to be *input connectable* if there exists a path to every compartment not directly entered by an input from at least one compartment directly entered by an input (Figs. 2a, c). The model is said to be *output connectable* if there exists a path from every compartment not directly influencing any measurement output to at least one compartment directly influencing an output (Figs. 2b, c).

Necessary conditions for structural identifiability have been presented (Cobelli and Romanin-Jacur, 1976a, b) with reference to the notions of structural controllability and observability; here they will be presented with reference to the notions of input and output connectability, and mutual implications between the two approaches will be examined.

It is easy to prove that input and output connectability are necessary conditions for structural identifiability (Cobelli et al., 1978a). In fact, the experiment allows us to determine only the input–output relations that are obviously not influenced by transport rate parameters entering and/or starting from those compartments which are not connected to any input or output.

Input and output connectability are also related to structural controllability and structural observability by the following theorem (Davidson, 1977).

10. Structural Identifiability of Linear Compartmental Models

Theorem. A fixed-structure model is structurally controllable (observable) if and only if the following two conditions both hold

(i) $\quad \text{rank}\,[A \mid B] = n, \quad \left(\text{rank}\begin{bmatrix} A \\ \hline C \end{bmatrix} = n\right)$

almost everywhere in the parameter space

(ii) the model is input (output) connectable.

In the previous papers by the authors, the class of the input and output connectable models (i.e., the class of models for which the above-mentioned necessary identifiability condition holds) was obtained by grouping structurally controllable and observable models and a set of "pathological" models. It is now easy to realize that these pathological models are the ones for which condition (ii) but not condition (i) of the above theorem holds. The above considerations (see, also, Cobelli *et al.*, 1978a, b) supply the answer to some questions about the relationships between controllability, observability, and structural identifiability (Di Stefano, 1977; DiStefano and Mori, 1977; Zazworsky and Knudsen, 1977).

As a simple example for these cases let us consider Figs. 3 and 4. The model of Fig. 3 is input connectable but not structurally controllable as three state variables x_2, x_3, x_4 cannot be varied independently by the input; correspondingly

$$\text{rank}\,[A \mid B] = 3 < 4.$$

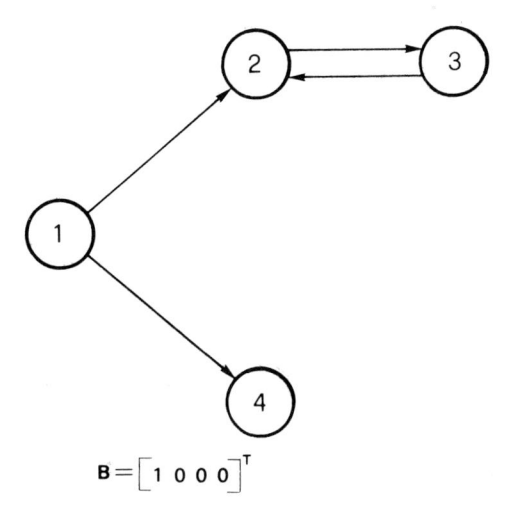

Figure 3. An example of an input connectable, but not structurally controllable model.

$$C = \begin{bmatrix} 0 & c_{12} & c_{13} \end{bmatrix}$$

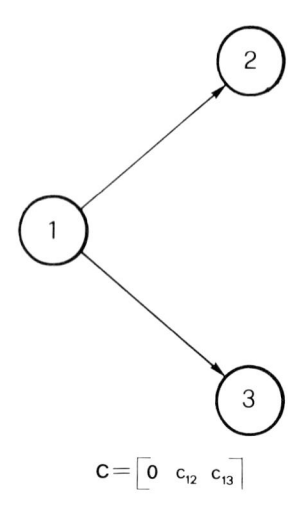

Figure 4. An example of an output connectable, but not structurally observable model.

Similarly the model of Fig. 4 is output connectable but not structurally observable:

$$\text{rank} \begin{bmatrix} A \\ \hline C \end{bmatrix} < 3.$$

A class of models largely employed in compartmental analysis is the one of strongly connected models. A compartmental model is said to be strongly connected if there exists a path from every compartment to every other compartment of the model. It has been proved that strong connection is a sufficient condition for structural controllability and observability (Cobelli and Romanin-Jacur, 1975).

In concluding this section it is helpful to note that if the designed experiment corresponds to a non-input-connectable and/or non-output-connectable model, the input–output relations depend only on the input- and output-connectable submodel. If the transport rate parameters of the nonconnectable submodel are of interest, a different experiment is to be planned.

5. THE ANALYSIS OF STRUCTURAL IDENTIFIABILITY

5.1. Introduction

Consider a linear time-invariant model, made up by n compartments, with r_b inputs and r_c outputs. Relationships between the input vector $\mathbf{u}(t)$

and the output vector $\mathbf{y}(t)$ may be represented by referring to their Laplace transforms $\mathbf{U}(s)$ and $\mathbf{Y}(s)$, namely, by the transfer function matrix:

$$G(s) = \frac{\mathbf{Y}(s)}{\mathbf{U}(s)} = C(sI - A)^{-1}B = \frac{C \operatorname{adj}(sI - A)B}{\det(sI - A)}. \tag{4}$$

$G(s)$ is an $r_c \times r_b$ matrix; its generic element G_{ij} is the transfer function between input j and output i (j and i are the order numbers of the input and output) and therefore it refers to the submodel made up by the n_{ij} compartments connectable to input j (u_j connectable) and to output i (y_i connectable). G_{ij} is expressed by the ratio of two polynomials. In its reduced form, that is, after simplification of common factors between numerator and denominator, the degree of the denominator is n_{ij}. If the nonreduced form derived from Eq. (4) is adopted, the denominator of all G_{ij}, $\det(sI - A)$, has degree n; in this case, if $n_{ij} < n$, a numerator factor corresponds to the submodel made up by the non-u_j-y_i-connectable compartments.

The problem is now to be faced of evaluating *a priori* whether the numerator and denominator coefficients of all G_{ij} (that represent the total information about the model behavior with the given experiment) allow the determination of the unknown entries of the triple (A, B, C). This result can be achieved by considering the set of equations relating numerator and denominator coefficients of all G_{ij} to their analytical expressions which are functions of the elements of (A, B, C).

The equations which can be written are generally not mutually independent; the number v of independent coefficients is to be *a priori* determined and compared with the number μ of the unknown elements of (A, B, C).

If $v < \mu$, the equations set exhibits infinitely many solutions and the model is therefore *structurally not identifiable*. The solutions form a manifold in the parameter space, that is, for $\mu = 2$, they form a curve in the parameter plane; for $\mu = 3$, a surface in the three-dimensional parameter space; and generally for $\mu > 3$, a hypersurface (manifold) in the parameter space. Therefore in any neighborhood of a solution there are always other indistinguishable parametrization.

If $v = \mu$, the number of solutions is surely finite and the model is therefore *structurally identifiable* (the solutions form a discrete set in the parameter space, and therefore, for every solution there exists a neighborhood which does not contain any other indistinguishable parametrization). In general the equations of the above considered set are nonlinear and therefore the solution is not unique; external constraints (as for instance, positivity and boundedness of each k_{ij}, b_{jl}, or c_{mi}) may reduce the number of acceptable solutions (even if it does not necessarily bring them to uniqueness).

5.2. Evaluation of v

In order to evaluate the number v of independent coefficients of $G(s)$ the following considerations hold.

The transfer function matrix $G(s)$ given by Eq. (4) can be presented in a suitable form by resorting to the expressions (Chen, 1970):

$$C \operatorname{adj}(sI - A)B = CB(s^{n-1} + \alpha_1 s^{n-2} + \cdots + \alpha_{n-1})$$
$$+ CAB(s^{n-2} + \alpha_1 s^{n-3} + \cdots + \alpha_{n-2}) + \cdots + CA^{n-1}B, \quad (5)$$

$$\det(sI - A) = s^n + \alpha_1 s^{n-1} + \cdots + \alpha_{n-1}s + \alpha_n. \quad (6)$$

Consequently,

$$G_{ij} = ([CB]_{ij}s^{n-1} + \{[CB]_{ij}\alpha_1 + [CAB]_{ij}\}s^{n-2}$$
$$+ \cdots + \{[CB]_{ij}\alpha_{n-1} + [CAB]_{ij}\alpha_{n-2}$$
$$+ \cdots + [CA^{n-1}B]_{ij}\})/(s^n + \alpha_1 s^{n-1} + \cdots + \alpha_{n-1}s + \alpha_n). \quad (7)$$

For the evaluation of v we must consider the common denominator and the $r_b r_c$ numerators of $G(s)$.

The common denominator is a polynomial of degree n, where n is the number of compartments of the model. It has at most n independent coefficients because the coefficient of s^n is always equal to 1: if m closed submodels are present, the coefficients $\alpha_1, \alpha_2, \ldots, \alpha_{n-m}$ are greater than zero while the coefficients $\alpha_{n-m+1}, \alpha_{n-m+2}, \ldots, \alpha_n$ are equal to zero (Hearon, 1963). In this case only $n-m$ independent equations may be written.

Each numerator is a polynomial the degree of which is at most $n-1$ and in this case it exhibits n coefficients; in order to obtain the number of independent equations, first of all the coefficients equal to zero are to be excluded and the factors, if any, common to the denominator or common to previously considered numerators are also to be excluded.

The coefficients equal to zero may be the ones of highest and lowest orders.

If $[CB]_{ij} \neq 0$, then the coefficients of s^{n-1}, s^{n-2}, etc., in the numerator of G_{ij} are different from zero; on the contrary if $[CB]_{ij} = 0$, then the coefficient of s^{n-1} is null; if $[CB]_{ij} = 0$ and $[CAB]_{ij} = 0$, the coefficients of s^{n-1} and s^{n-2} are null and so on. Let p_{ij} be the number of the coefficients of s^{n-1}, s^{n-2}, etc., in the numerator of G_{ij} that are equal to zero; that corresponds to $[CA^tB]_{ij} = 0$ for $0 \leq t \leq p_{ij} - 1$ and in particular, as seen above, if $[CB]_{ij} \neq 0$, p_{ij} is assumed to be null.

On the basis of these considerations, it may be proved that p_{ij} is the length of the shortest path between one of the compartments directly entered by u_j and one of the compartments directly measured via y_i; if u_j and y_i take place in the same compartment, the length of the path is zero

10. Structural Identifiability of Linear Compartmental Models

and therefore $p_{ij} = 0$. The interpretation of p_{ij} in terms of length of a path of the compartmental graph gives a simple rule for its evaluation.

For what concerns the coefficients of s^0, s^1, etc., in the numerator and denominator polynomials, the following considerations may be made.

From system theory (e.g., Chen, 1970) it is well known that each transfer function in its reduced form cannot present a factor s^γ with $\gamma > 0$, that is, the submodel cannot behave as a differentiator. Moreover, Hearon (1963) has proved that, for a compartmental model, a null eigenvalue of matrix A, if any, has multiplicity one; therefore, each transfer function G_{ij}, in its reduced form, cannot present a factor $(1/s)^\delta$ with $\delta > 1$, that is, any $i-j$ submodel cannot behave as a multiple integrator. As a conclusion, in the reduced form of each G_{ij}, the coefficient of s^0 in the numerator is always different from zero, and the coefficient of s^0 in the denominator may be null or different from zero; the first case occurs either if the $i-j$ submodel is closed or if it contains at least one closed submodel; the coefficient of s^1 in the denominator is always different from zero even if the $i-j$ submodel contains more than one closed submodel. Coming back to $G(s)$ in its extended form given by Eq. (4) if $\det(sI - A)$ contains the factor s^m, the numerator also presents either the factor s^m or the factor s^{m-1} and accordingly the coefficients of the powers s^t in the numerator are equal to zero for $0 \leqslant t \leqslant q_{ij} - 1$ with $q_{ij} = m$ or respectively $q_{ij} = m - 1$. As seen above, the first situation occurs if the $u_j - y_i$ submodel does not contain closed submodels, and the second situation occurs if the submodel contains one or more closed submodels.

It may be easily seen from Eq. (7) that the coefficients of the intermediate powers between $s^{n-p_{ij}-1}$ and $s^{q_{ij}}$ cannot be equal to zero. Therefore the number of nonzero coefficients is $n - p_{ij} - q_{ij}$.

Now, for each G_{ij} the factors common to the denominator (except for $s^{q_{ij}}$) and to previously considered numerators are to be taken into account.

The degree r_{ij} of the factors common to the $i-j$ numerator and the common denominator is given by the number of compartments which are not $u_j - y_i$ connectable (in fact these compartments do not affect the minimal expression of G_{ij}, but they affect the denominator if it is written in the form $\det(sI - A)$). As r_{ij} includes q_{ij}, the number $r_{ij} - q_{ij}$ represents the degree of the factor common to the denominator, except for $s^{q_{ij}}$.

Finally, for evaluating the factors common to $i-j$ and $l-m$ numerators, let us define a *common cascade submodel* (c.c.s.) between $u_j - y_i$ and $u_l - y_m$ submodels. A common cascade submodel exists if there are two compartments f and g such that each compartment of the c.c.s. is influenced by compartments outside the c.c.s. only through f and influences compartments outside the c.c.s. only through g. Each compartment of the c.c.s. is u_j and u_l connectable only through a path entering f and it is y_i and

y_m connectable only through a path outgoing from g. Moreover a c.c.s. may exist if the two submodels have the same input (respectively output) and a compartment g (respectively f) defined as above exists. The transfer function relative to the c.c.s. (i.e., between f and g, input and g, f and output, respectively) is obviously a factor of both G_{ij} and G_{ml}. The degree of its numerator may be computed as previously stated for a generic submodel; as the common factor always has gain equal to unity, the number of independent relations is reduced by one.

Taking into account the considered causes of dependency among the equations obtainable from the transfer matrix $G(s)$, the number of equations becomes:

$$n_e = (n-m) + \sum_{\substack{i=1,r_c \\ j=1,r_b}} w_{ij} - \sum_{\substack{i=1,r_c \\ j=1,r_b}} z_{ij} \qquad (8)$$

where:

$$w_{ij} = n - p_{ij} - r_{ij}$$

is the number of nonzero coefficients of the $i-j$ numerator (if $p_{ij} = 0$ and if the output coefficient is known, w_{ij} is reduced by one) and where z_{ij} is the number of coefficients related to the c.c.s. between current numerator $i-j$ and previously considered ones.

The number v of independent equations is $\leq n_e$ as other causes of dependency may exist beside the considered ones (Cobelli et al., 1978a). Unfortunately no rules are presently available for detecting directly on the compartmental diagram the causes of dependency different from the ones taken into account by (8).

However, the knowledge of n_e is useful as $n_e \geq \mu$ is a necessary condition (beside input and output connectability) both for structural identifiability and unique structural identifiability, which is easy to be tested.

These necessary conditions refer to a check of almost everywhere identifiability in the parameter space, i.e., if no other dependency causes are present then $n_e = v$ and the condition $n_e = v = \mu$ assures identifiability almost everywhere (a zero measure set in the parameter space where the system is not identifiable can exist). In some cases it is possible to detect *a priori* this type of nonidentifiability situation on the basis of some known constraints. A typical case is the one where an equally weighted sum of all the state variables associated to the compartments of a closed submodel is measured.

In order to evaluate the number v of actually independent coefficients it is necessary to consider the rank of the Jacobian matrix of the Markov parameters (Grewal and Glover, 1976); equivalently, on the basis of the above considerations, it is possible to refer to the Jacobian matrix of the n_e equations of (8).

6. EXAMPLES AND CONCLUSIONS

6.1. General

The identifiability test of a compartmental model may be performed by evaluating the number v of independent coefficients of the transfer function matrix and by comparing it with the number μ of unknown parameters of the model.

As the evaluation of v may be very cumbersome, it is useful to test directly on the compartmental diagram the set of necessary conditions presented in Sections 4 and 5, i.e., input–output connectability and $n_e \geqslant \mu$ (for very complex situations n_e may be evaluated via a computer procedure presented by Cobelli et al., 1977a). If this set of conditions is not satisfied, the model is not structurally identifiable and either the model structure or the experiment configuration are to be modified.

With the purpose of illustrating how the structural identifiability test works, we consider some compartmental models of ecosystems taken from the literature. In the reported examples, we will assume that input coefficients are known, which corresponds to the usual experimental situation.

6.2. Calcium Cycle

In Fig. 5 a four compartment model of calcium cycle in a forest watershed ecosystem (Waide et al., 1974) is reported, where: 1, vegetation; 2, litter; 3, available nutrients; and 4, soil and rock minerals. In order to identify the model a perturbation experiment is to be performed; a perturbation input can be applied to compartment 3 while all compartments are accessible for measurement.

For the choice of the outputs to be actually measured the different complexity of the measurements is to be taken into account: the measurement of 3 is less complex than that of 2, which is less complex than that of 1. Measurement of 4 is the least complex, but compartment 4 is not input connectable without an input in 4. Consider now the minimal experiment that insures input and output connectability, that is the one with input and output in compartment 3; the related model is shown in Fig.

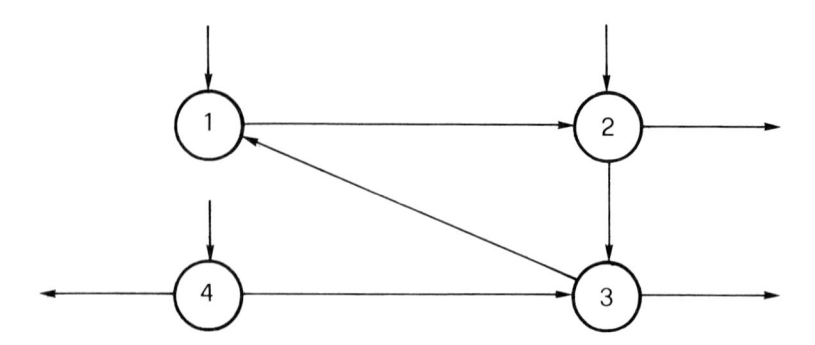

Figure 5. A four-compartment model of calcium cycle. 1, vegetation; 2, litter; 3, available nutrients; 4, soil and rock minerals.

6a where the constant inputs to compartments 1 and 2 and the constant flow from 4 to 3 are neglected as the system is in steady state.

The number μ of unknown parameters is 6. To test the second necessary condition, the number n_e has to be computed according to (8): n = number of compartments = 3; m = number of closed submodels = 0; $w_{11} = 3$ as p_{11} and r_{11} are equal to zero; z_{11} is obviously equal to zero as $G(s)$ is scalar. Therefore $n_e = 6 = \mu$.

As the necessary conditions are satisfied, the number of independent equations has to be computed. The following procedure may be adopted.

The transfer function $G_{11}(s)$ is given by:

$$G_{11}(s) = \frac{c_{13}(s^2 + \beta_1 s + \beta_0)}{s^3 + \alpha_2 s^2 + \alpha_1 s + \alpha_0} \tag{9}$$

where

$$\begin{aligned}
\beta_1 &= k_{21} + k_{02} + k_{32} \\
\beta_0 &= k_{21}(k_{02} + k_{32}) \\
\alpha_2 &= k_{21} + k_{02} + k_{32} + k_{03} + k_{13} \\
\alpha_1 &= k_{21}(k_{02} + k_{32}) + (k_{21} + k_{02} + k_{32})(k_{03} + k_{13}) \\
\alpha_0 &= k_{21}[k_{03}(k_{02} + k_{32}) + k_{02}k_{13}]
\end{aligned} \tag{10}$$

As c_{13} is given directly by the coefficient of s^2, it is sufficient to consider a 5×5 Jacobian matrix obtained by differentiating the coefficients, α_i and β_i, with respect to the unknowns k_{ij}. The evaluation of the Jacobian is cumbersome; in this particular case it is better to look directly at the set of equations. The model results in not being identifiable as the fourth equation is not independent from the first three, as it may be written in the

10. Structural Identifiability of Linear Compartmental Models

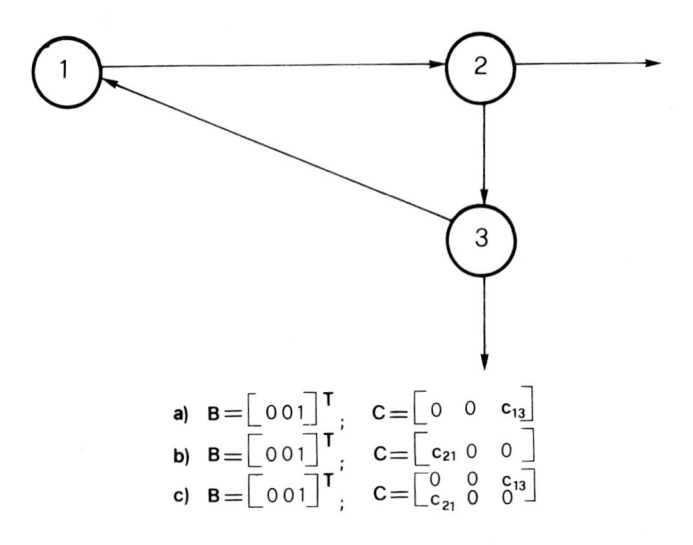

Figure 6. The model derived from Fig. 5 and the three considered experiments.

a) $B = \begin{bmatrix} 0 & 0 & 1 \end{bmatrix}^T$; $C = \begin{bmatrix} 0 & 0 & c_{13} \end{bmatrix}$

b) $B = \begin{bmatrix} 0 & 0 & 1 \end{bmatrix}^T$; $C = \begin{bmatrix} c_{21} & 0 & 0 \end{bmatrix}$

c) $B = \begin{bmatrix} 0 & 0 & 1 \end{bmatrix}^T$; $C = \begin{bmatrix} 0 & 0 & c_{13} \\ c_{21} & 0 & 0 \end{bmatrix}$

form:

$$\alpha_1 = \beta_0 + \beta_1(\alpha_2 - \beta_1).$$

Let us now consider another one input–one output experiment (Fig. 6b). The transfer function is:

$$G_{21}(s) = \frac{c_{21}(\beta_1's + \beta_0')}{s^3 + \alpha_2 s^2 + \alpha_1 s + \alpha_0} \tag{11}$$

where

$$\begin{aligned} \beta_1' &= k_{13} \\ \beta_0' &= k_{13}(k_{02} + k_{32}) \end{aligned} \tag{12}$$

and $\alpha_0, \alpha_1, \alpha_2$ are the same as before.

Assuming c_{21} as known, the determinant of the Jacobian matrix is $-k_{21}k_{13}^2(k_{21} - k_{03} - k_{13})$, which is different from zero almost everywhere in the parameter space. Therefore the model is structurally identifiable, i.e., almost everywhere locally identifiable. Only k_{13} can be determined uniquely; therefore unique structural identifiability is not achieved, as the equations do not have a unique solution for all the unknowns in the parameter space.

If both considered experiments are performed (Fig. 6c), uniqueness is achieved as from the sets of equations (10) and (12) and a set of five linear independent equations may be obtained.

6.3. Phosphorus Dynamics

A very complex model for the dynamics of phosphorus in water has been suggested by E. Halfon (Fig. 7). The model is based on two previously proposed models (Lean, 1973; Halfon *et al.*, 1978) which are represented in Figs. 8 and 9.

In Figs. 7, 8, 9 compartment 1 corresponds to soluble phosphorus, compartment 2 to colloidal phosphorus, compartment 3 to organic phosphorus compound, and compartment 4 to particulate phosphorus; in Fig. 9 a new compartment is added, i.e., 5, which represents the storage of particulate phosphorus and in Fig. 7 compartments 6–14 represent herbivores zooplankton, while compartments 15, 16 and 17 represent carnivores zooplankton.

The only feasible tracer input concerns compartment 1. Measurements may be performed in all compartments, except 5, but with different complexity (in order of increasing complexity: compartment 1; 4; 6=7=8 =9=10=11=12=13=14=15=16=17; 2=3).

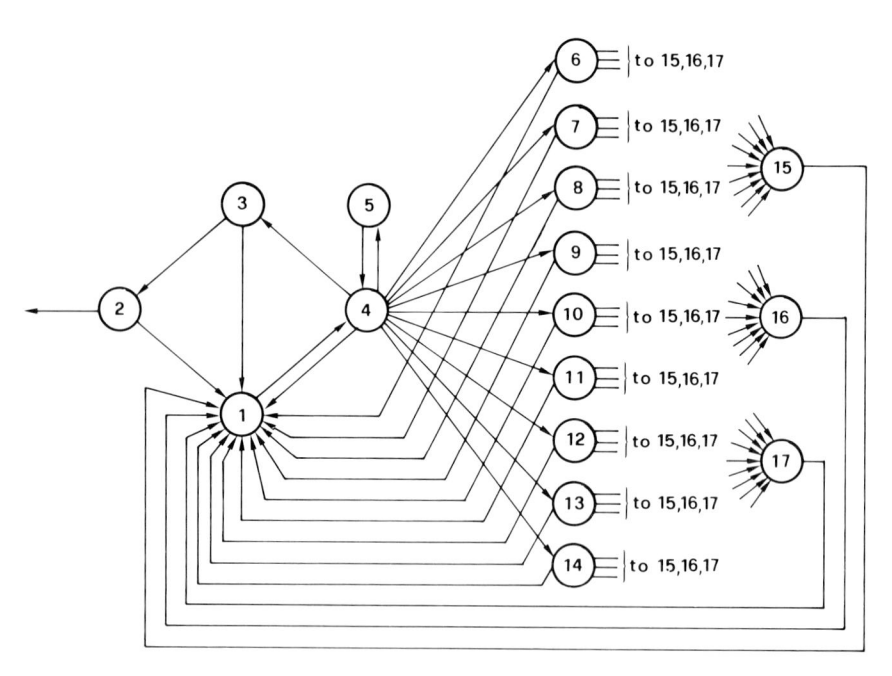

Figure 7. A seventeen-compartment model of phosphorus dynamics. 1, Soluble phosphorus; 2, colloidal phosphorus; 3, organic phosphorus compound; 4, particulate phosphorus; 5, storage of particulate phosphorus; 6–14, herbivores zooplankton; 15–17, carnivores zooplankton.

10. Structural Identifiability of Linear Compartmental Models 255

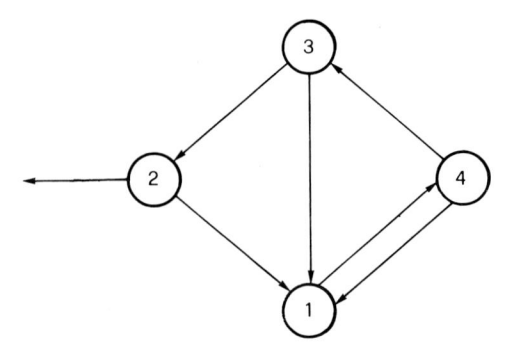

Figure 8. A four-compartment model of phosphorus dynamics. 1, Soluble phosphorus; 2, colloidal phosphorus; 3, organic phosphorus compound; 4, particulate phosphorus.

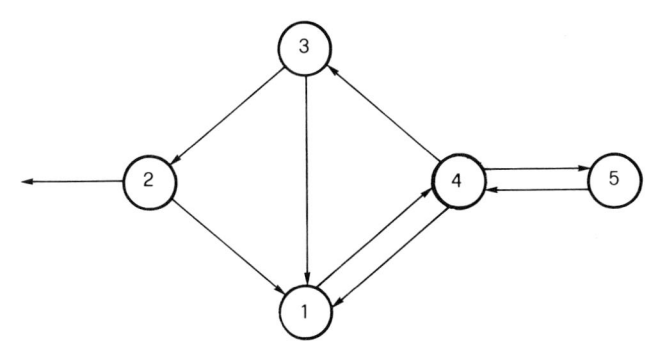

Figure 9. A five-compartment model of phosphorus dynamics. 1, Soluble phosphorus; 2, colloidal phosphorus; 3, organic phosphorus compound; 4, particulate phosphorus; 5, storage of particulate phosphorus.

Let us consider the case of Fig. 7. The model is always input and output connectable and the values of n_e and μ are reported in Table I for five feasible experiments. The necessary condition $n_e \geq \mu$ is not satisfied for the two simplest experiments and the least expensive possibly successful experiment (subsequent evaluation of v is needed for achieving structural identifiability) requires the measurement of compartments 1, 4, and of another compartment chosen among 6; 7; ... ; 17. The example illustrates the usefulness of the proposed set of necessary conditions in ruling out, directly from the analysis of the compartmental diagram, *a priori* unsuccessful experiments.

A complete identifiability analysis is presented with reference to the model of Fig. 8. In Table II the values of n_e and μ are reported for some

Table I Evaluation of n_e (Fig. 7)

Input	Output	n	m	w_{ij}	z_{ij}	n_e	μ
1	1	17	0	17	—	34	58
1	1, 4	17	0	17+16	0	50	59
1	1, 4, 6	17	0	17+16+15	0	65	60
1	1, 4, 6, 7, 8, 9, 10, 11, 12, 13, 14, 15, 16, 17	17	0	17+16 +9.15+3.14	0	227	71
1	1, 2, 3, 4, 6, 7, 8, 9, 10, 11, 12, 13, 14, 15, 16, 17	17	0	17+14 +15+16 +9.15+3.14	0	256	73

experiments among the feasible ones. The necessary conditions are satisfied for all considered experiment configurations; therefore v has to be evaluated. With reference to the first experiment (input and output in compartment 1) the transfer function is:

$$G_{11}(s) = \frac{\beta_3^{11} s^3 + \beta_2^{11} s^2 + \beta_1^{11} s + \beta_0^{11}}{s^4 + \alpha_3 s^3 + \alpha_2 s^2 + \alpha_1 s + \alpha_0} \quad (13)$$

where

$$\beta_3^{11} = c_{11}$$
$$\beta_2^{11} = c_{11}(k_{02} + k_{12} + k_{13} + k_{14} + k_{23} + k_{34})$$
$$\beta_1^{11} = c_{11}(k_{02}k_{13} + k_{02}k_{14} + k_{02}k_{23} + k_{02}k_{34} + k_{12}k_{13} + k_{12}k_{14}$$
$$+ k_{12}k_{13} + k_{12}k_{34} + k_{14}k_{23} + k_{23}k_{34} + k_{13}k_{14} + k_{13}k_{34})$$
$$\beta_0^{11} = c_{11}(k_{02}k_{14}k_{23} + k_{02}k_{23}k_{34} + k_{02}k_{13}k_{14} + k_{02}k_{13}k_{34}$$
$$+ k_{12}k_{14}k_{23} + k_{12}k_{23}k_{34} + k_{12}k_{13}k_{14} + k_{12}k_{13}k_{34})$$
$$\alpha_3 = k_{02} + k_{12} + k_{13} + k_{14} + k_{23} + k_{34} + k_{41} \quad (14)$$
$$\alpha_2 = k_{02}k_{41} + k_{12}k_{41} + k_{14}k_{23} + k_{23}k_{34} + k_{13}k_{14} + k_{13}k_{34}$$
$$+ k_{23}k_{41} + k_{13}k_{41} + k_{34}k_{41} + k_{02}k_{23} + k_{02}k_{13} + k_{02}k_{14}$$
$$+ k_{02}k_{34} + k_{12}k_{23} + k_{12}k_{13} + k_{12}k_{14} + k_{12}k_{34}$$
$$\alpha_1 = k_{02}k_{23}k_{41} + k_{02}k_{13}k_{41} + k_{02}k_{34}k_{41} + k_{12}k_{23}k_{41} + k_{12}k_{13}k_{41}$$
$$+ k_{12}k_{34}k_{41} + k_{02}k_{14}k_{23} + k_{12}k_{14}k_{23} + k_{23}k_{34}k_{41} + k_{02}k_{23}k_{34}$$
$$+ k_{12}k_{23}k_{34} + k_{02}k_{13}k_{14} + k_{12}k_{13}k_{14} + k_{02}k_{13}k_{34} + k_{12}k_{13}k_{34}$$
$$\alpha_0 = k_{02}k_{23}k_{34}k_{41}.$$

The unknown parameters are $c_{11}, k_{02}, k_{12}, k_{13}, k_{14}, k_{23}, k_{34}, k_{41}$. The rank of the Jacobian matrix is 8; therefore $v = 8$ and the model is

Table II Evaluation of n_e (Fig. 8)

Input	Output	n	m	w_{ij}	z_{ij}	n_e	μ
1	1	4	0	4	—	8	8
1	1, 4	4	0	$4+3=7$	0	11	9
1	1, 4, 2, 3	4	0	$4+3+1+2=10$	0	14	11

structurally identifiable. It may be noted, however, that c_{11}, k_{41}, and k_{14} are uniquely determined. For the evaluation of the rank it is convenient to manipulate previously the available equations trying to reduce the order of the set. In the considered case, for instance, c_{11}, k_{41}, and k_{14} are obtained by means of simple manipulation, thus reducing the Jacobian matrix to be considered to a 5×5 matrix.

If the second experiment configuration of Table II is considered, i.e., a measurement in 4 is added ($y_2 = c_{24}x_2$), a second transfer function $G_{21}(s)$ is available, where the numerator is:

$$\beta_2^{21} s^2 + \beta_1^{21} s + \beta_0^{21} \tag{15}$$

where:

$$\begin{aligned}
\beta_2^{21} &= c_{24}k_{41} \\
\beta_1^{21} &= c_{24}k_{41}(k_{02}+k_{12}+k_{23}+k_{13}) \\
\beta_0^{21} &= c_{24}k_{41}(k_{02}k_{23}+k_{02}k_{13}+k_{12}k_{23}+k_{12}k_{13}).
\end{aligned} \tag{16}$$

From both G_{11} and G_{21} structural identifiability is still assured; moreover three more parameters, c_{24}, k_{34}, and k_{13}, are uniquely determined. Also in this case, unique structural identifiability is not assured.

REFERENCES

Bellman, R., and Åström, K. J. (1970). On structural identifiability. *Math. Biosci.* **7**, 329–339.
Chen, C. T. (1970). "Introduction to Linear System Theory." Holt, New York.
Cobelli, C., and Romanin-Jacur, G. (1975). Structural identifiability of strongly connected biological compartmental systems. *Med. Biol. Eng.* **13**, 831–838.
Cobelli, C., and Romanin-Jacur, G. (1976a). Controllability, observability and structural identifiability of multi input and multi output biological compartmental systems. *IEEE Trans. Biomed. Eng.* **23**, 93–100.
Cobelli, C., and Romanin-Jacur, G. (1976b). On the structural identifiability of biological compartmental systems in a general input–output configuration. *Math. Biosci.* **30**, 139–151.
Cobelli, C., Lepschy, A., and Romanin-Jacur, G. (1976). On identifiability problems in biological systems. *In* "Preprints IV IFAC Symp. on Identification and System Parameter Estimation," N. S. Raybman, ed., Part 1, pp. 390–400.

Cobelli, C., Polo, A., and Romanin-Jacur, G. (1977a). A computer program for the analysis of controllability, observability and structural identifiability of biological compartmental systems. *Comput. Prog. Biomed.* **7**, 21–36.

Cobelli, C., Lepschy, A., and Romanin-Jacur, G. (1978a). Identifiability of compartmental systems and related structural properties. *Math. Biosci.* (to appear).

Cobelli, C., Lepschy, A., and Romanin-Jacur, G. (1978b). Comments on "On the relationships between structural identifiability and the controllability, observability properties." *IEEE Trans. on Autom. Control* (to appear).

Davidson, E. J. (1977). Connectability and structural controllability of composite systems. *Automatica* **13**, 109–123.

Delforge, J. (1977). The problem of structural identifiability of a linear compartment system: Solved or not? *Math. Biosci.* **36**, 119–125.

DiStefano, J. J. (1976). Tracer experiment design for unique identification of nonlinear physiological systems. *Am. J. Physiol.* **230**, 476–485.

DiStefano, J. J. (1977). On the relationship between structural identifiability and controllability, observability properties. *IEEE Trans. Autom. Control.* **22**, 652.

DiStefano, J. J., and Mori, F. (1977). Parameter identifiability and experiment design: Thyroid hormone metabolism parameters. *Am. J. Physiol.* **233**, R134–R144.

Glover, K., and Willems, J. C. (1974). Parametrization of linear dynamical systems: Canonical forms and identifiability. *IEEE Trans. Autom. Control.* **19**, 640–646.

Grewal, M. S., and Glover, K. (1976). Identifiability of linear and nonlinear dynamical systems. *IEEE Trans. Autom. Control.* **21**, 833–837.

Halfon, E., Unbehaven, H., and Schmid, C. (1978). Model order estimation and system identification theory and application to the modelling of ^{32}P kinetics within the trophogenic zone of a small lake. *Ecol. Model.* (in press).

Hearon, J. Z. (1963). Theorems on linear systems. *Ann. N.Y. Acad. Sci.* **108**, 36–68.

Lean, D. R. S. (1973). Phosphorus dynamics in lake waters. *Science* **179**, 678–680.

Lin, C. T. (1974). Structural controllability. *IEEE Trans. Autom. Control.* **19**, 201–208.

Patten, B. C. (1975). Ecosystem linearization: An evolutionary design problem. *Am. Nat.* **109**, 529–539.

Patten, B. C., Egloff, D. A., and Richardson, T. H., *et al.* (1975). Total ecosystem model for a cave in Lake Texoma. *In* "Systems Analysis and Simulation in Ecology" (B. C. Patten, ed.), Vol. 3, pp. 205–421. Academic Press, New York.

Thron, C. D. (1972). Structure and kinetic behavior of linear multi-compartment systems. *Bull. Math. Biophys.* **34**, 277–291.

Waide, J. B., and Webster, J. R. (1976). Engineering systems analysis: Applicability to ecosystems. *In* "Systems Analysis and Simulation in Ecology" (B. C. Patten, ed.), Vol. 4, pp. 329–371. Academic Press, New York.

Waide, J. B., Krebs, J. E., Clarkson, S. P., and Setzler, E. M. (1974). *Prog. Theor. Biol.* **3**, 261–345.

Zazworsky, R. M., and Knudsen, H. K. (1977). Comments on "Controllability, observability and structural identifiability of multi input and multi output biological compartmental systems." *IEEE Trans. Biomed. Eng.* **24**, 495–496.

Chapter **11**

MODEL STRUCTURE IDENTIFICATION FROM EXPERIMENTAL DATA

M. B. Beck

1. Introduction	260
2. System Identification: A Brief Review	261
2.1 Experimental Design	262
2.2 Choice of Model	262
2.3 Model Structure Identification: Problem Definition	266
2.4 Parameter Estimation	267
2.5 Verification and Validation	269
3. Model Structure Identification: Black Box Models	270
3.1 An Example: Anaerobic Digestion of Waste Organic/Biological Sludges	271
4. Model Structure Identification: Internally Descriptive Models	273
4.1 Formulation of the Combined State Parameter Estimation Problem	274
4.2 The Extended Kalman Filtering Algorithms	275
4.3 Operation of the EKF Algorithms	278
4.4 Intuitive Criteria for Model Structure Identification with the EKF	279
4.5 Parameter Dynamics and Model Structure Identification	280
4.6 Problems of Stability and Partially Observed State Vectors	281
4.7 Further Considerations of the Measurement Process	282
4.8 An Example: The Interaction between BOD and an Algal Population in a Freshwater River	284
5. Conclusions	287
References	287

1. INTRODUCTION

Dynamic model structure definition (identification) is arguably one of the major unresolved technical problems in the field of system identification and parameter estimation. It is certainly true in practice that model structure identification from experimental field data gives rise to all manner of difficulties. This chapter has the objective of presenting some theoretical techniques which can be applied to the solution of the identification problem.

Throughout the chapter we shall assume a pragmatic approach to modeling: Namely, the act of modeling implies the collection of experimental field data. In order to avoid confusion we may state that the term *system identification* is interpreted herein as the complete process of deriving mathematical models from, and by reference to experimental data; the term *identification* means the specific process of model structure identification. If the system under investigation can be represented by the (dynamic) model of Fig. 1, all variables thus being functions of time t, a broad definition of model structure identification can be given as the establishment of how the measured system inputs **u** are related to the system's state variables **x** and how these latter are in turn related both to themselves and to the measured system outputs **y**. The *dynamic* modeling context arises for the following reason: An experimenter studying a system under laboratory conditions wishes to keep that system as close to steady state as possible while he tests the relationship between, say, two particular variables. Such steady-state conditions, and especially so for ecological systems, rarely prevail in the field. Hence in order to establish any significant theory of the system's behavior it is necessary to set up the problem within a framework which recognizes the dynamic and stochastic (random) nature of experimental data. We are, however, concerned exclusively here with identifying a structure for the deterministic component

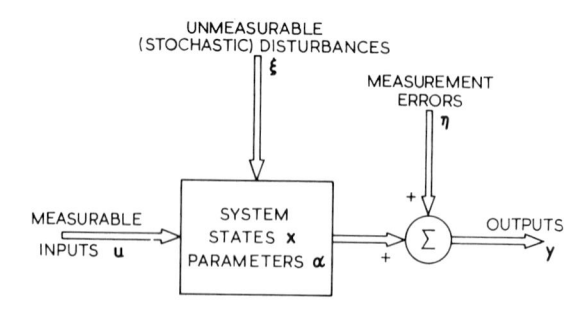

Figure 1. System and variables definition.

of the model. The techniques employed in the modeling analysis should therefore operate so as to discriminate effectively against the ever-present random noise component of measured signals. The motivation for solving the model structure identification problem stems from the experience of studying river water quality modeling and control (Beck, 1977); the following illustrative examples are drawn from this subject area.

2. SYSTEM IDENTIFICATION: A BRIEF REVIEW

The field of system identification has developed rapidly over the past decade, and like any other discipline which has emerged and matured so quickly, its accompanying literature is vast but not well coordinated. For the reader previously unacquainted with system identification a carefully guided introduction to the literature is appropriate.

The book by Eykhoff (1974) is to be recommended as giving the broadest and most comprehensive treatment of system identification; for a more brief survey of the subject and its earlier literature there is the review by Åström and Eykhoff (1971). Box and Jenkins' (1970) detailed account of discrete-time, input/output, black box modeling must also receive due

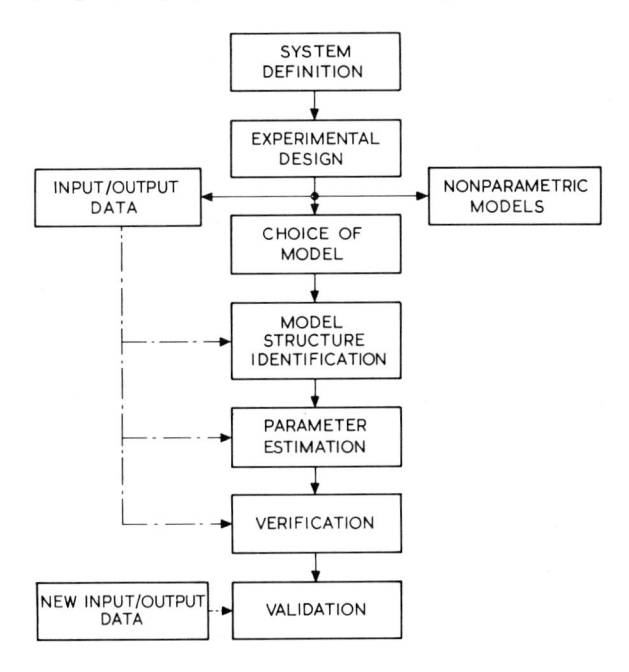

Figure 2. Individual steps in the procedure of system identification.

reference: This text, probably more so than others, has had a very significant impact on the application of time-series analysis in many diverse technical fields. Among the multitude of publications on methods of parameter estimation the easily readable article by Young (1974) provides an excellent introduction to recursive estimation techniques, or alternatively, these same techniques are given a rigorous treatment in Söderström et al. (1974).

Each of the above publications offers a suitable point of departure into the subject of system identification. The purpose of this section is to outline a scheme of individual steps in the procedure of system identification, thereby describing the context of the model structure identification problem (see Fig. 2).

2.1. Experimental Design

Besides the definition of the system and its variables, which we have assumed to be according to Fig. 1, a prerequisite of system identification is an appropriate record of the observed process dynamics. Any *a priori* knowledge of the system's dynamic behavior is an advantage, since this knowledge can be used in assessing the following important aspects of experimental design (Gustavsson, 1975): (a) major process time constants; (b) sampling (measurement) frequency; (c) duration of the experiment; (d) choice of input test signals **u**; (e) noise levels; (f) process nonlinearities. When confronted with a modeling problem it is thus not particularly encouraging to reflect upon the fact that a good experimental design, and hence the likelihood of useful results, is strongly dependent upon a good *a priori* knowledge (model) of the system! A particularly thorny problem with respect to ecological systems is the inability to probe the process dynamics with artificially manipulated signals such as step, impulse, or pseudorandom binary sequence (PRBS) inputs **u**. In other words, our experiments reduce simply to the observation of behavior without any intervention on behalf of the experimenter, that is, "normal operating conditions" (Eykhoff, 1974). Later sections of the chapter will illustrate how difficult it can be to undertake modeling exercises under the very limiting constraints of data derived from normal operating conditions.

2.2. Choice of Model

A distinction should be drawn between parametric and nonparametric classes of models since Fig. 2 assumes implicitly that only the former are to be dealt with here. Nonparametric models, such as Volterra series, impulse, and step response representations, are *intrinsically of infinite order*; they are characterized, in principle, by an *infinite number of parameters*. For instance, if a system with simple first-order dynamics (an exponential lag) were to be represented by its discrete-time impulse response, an infinite number of response coefficients (parameters) would be required to characterize those dynamics completely. Parametric models, in contrast, are characterized by a *finite* (and usually small) *number of parameters*. Indeed, we may remark that the translation of a nonparametric model into a parametric model representation constitutes the basis of the black box model identification problem (see Section 2.3).

Broadly speaking, a choice can be made between two parametric model forms: (a) a black box (or input/output) model and (b) an internally descriptive (or mechanistic) model. These two model representations reflect two opposite, yet complementary, approaches to modeling. Either one takes existing theory (that is, physical, chemical, biological, ecological theory) and develops this model so that it may be tested against experimental data—a deductive reasoning approach associated with a model of type b, or, assuming no *a priori* knowledge (theory) of process behavior, one attempts to develop the specific information acquired from the data into a more general model—an inductive reasoning approach closely related to black box model representations.

2.2.1. Black Box Model

For simplicity and brevity a *linear* form of the black box model is given by the discrete-time, difference equation

$$A(q^{-1})y(t_k) = \sum_{i=1}^{v} q^{-\delta_i} B_i(q^{-1}) u_i(t_k) + E(q^{-1}) e(t_k) \qquad (1)$$

in which $u_i(t_k)$, $i = 1, 2, \ldots, v$, and $y(t_k)$ are, respectively, observations of the multiple (v) system inputs and the system output at the kth sampling instant; $e(k)$ is a sequence of independent, Gaussian, random variables and q^{-1} is the backward shift operator

$$q^{-1}\{y(t_k)\} = y(t_{k-1}), \qquad \text{etc.}$$

$A(q^{-1})$ and $B_i(q^{-1})$ are polynomials in q^{-1}, of orders n and m_i, respectively, with parameters a_j and b_{ij} to be estimated

$$\begin{aligned} A(q^{-1}) &= 1 + a_1 q^{-1} + \cdots + a_n q^{-n}, \\ B_i(q^{-1}) &= b_{i0} + b_{i1} q^{-1} + \cdots + b_{im_i} q^{-m_i}; \qquad i = 1, 2, \ldots, v, \end{aligned} \qquad (2)$$

and δ_i, $i = 1, 2, \ldots, v$, represents a pure time delay in the response between output and input u_i. The form of $E(q^{-1})$ is left unspecified, except to state that it is in general a rational function. The precise description of the *lumped*

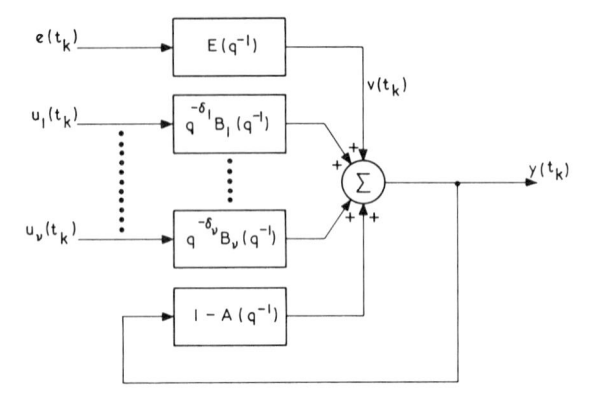

Figure 3. Schematic representation of the black box model [Eq. (1)].

stochastic process $v(t_k)$ in Fig. 3, which accounts for the combined effects of system noise ξ and measurement error η (in Fig. 1) as white noise $e(t_k)$ passed through this "shaping filter" $E(q^{-1})$, depends partly on the type of parameter estimation method to be applied (see Section 2.4). Details of this stochastic process description will not concern us greatly here since we are trying to establish the nature of deterministic relationships between **u** and y.

Equation (1) states essentially that the current value of the output $y(t_k)$ is a (scalar) function, f, of current and past measurements of the inputs u_i, of past measurements of the output, and of current and past realizations of the stochastic process v, as in Fig. 3,

$$y(t_k) = f\{y(t_{k-1}),\ldots,y(t_{k-n}), u_1(t_{k-\delta_1}),\ldots, u_1(t_{k-\delta_1-m_1}),\ldots,$$
$$u_\nu(t_{k-\delta_\nu}),\ldots, u_\nu(t_{k-\delta_\nu-m_\nu}), v(t_k),\ldots, v(t_{k-\tau})\}. \quad (3)$$

Here τ denotes that $y(t_k)$ depends upon a finite number of realizations of v. Such a black box model, being specific to the sample data set from which it is derived, is unlikely to be a universal description of a system's dynamics. Nor is this model necessarily amenable to interpretations on the perceived physical nature of process behavior. The *black box* is literally a fair reflection of our insight into the internal mechanisms of the system. As a model it is a first attempt at elucidating any observed basic cause/effect relationships, such as which inputs affect which output, by how much, and how quickly. Yet these are just the advantages that a black box modeling approach can offer: It is simple, and there are many situations where an internally descriptive model, although available, has a form which is too unwieldy or complex to be properly verified against field data.

We have, however, imposed a restriction on the model of Eq. (1) in that it refers to a *single*-output process. It is worth noting that the majority of applications of black box models have been similarly so restricted, although this is not a justification for imposing the constraint. Multivariate forms of the model are discussed in greater detail in Rowe (1970) and Young and Whitehead (1977).

2.2.2. Internally Descriptive Model

An internally descriptive model exploits much more, if not all, of the available *a priori* information on the physical, chemical, biological, and ecological phenomena governing process dynamics. As with Eq. (1) we confine the discussion to linear forms of the model for ease of illustration. The internally descriptive model may then be represented by the following *linear*, continuous-time, state vector differential equation (see also Fig. 4)

$$\dot{\mathbf{x}}(t) = \mathbf{F}\mathbf{x}(t) + \mathbf{G}\mathbf{u}(t) + \boldsymbol{\xi}(t) \tag{4a}$$

with sampled, noise-corrupted observations

$$\mathbf{y}(t_k) = \mathbf{H}\mathbf{x}(t_k) + \boldsymbol{\eta}(t_k) \tag{4b}$$

in which the dot notation refers to differentiation with respect to time t. The variables are defined as: $\mathbf{x}$, the l-dimensional state vector; $\mathbf{u}$, the v-dimensional input vector; $\mathbf{y}$, the p-dimensional vector of outputs; $\boldsymbol{\xi}$, l-dimensional vector of zero-mean, white, Gaussian disturbances; $\boldsymbol{\eta}$, p-dimensional vector of zero-mean, white, Gaussian measurement errors; $\mathbf{F}, \mathbf{G}, \mathbf{H}$, are accordingly $l \times l$, $l \times v$, and $p \times l$ matrices whose elements are the parameters that characterize the system.

The attractions of working with this type of model are its potentially universal applicability and its apparent grounding in theory or *the laws of nature*. But in a sense this latter feature is the source of many model structure identification problems because theory, at least in the ecological

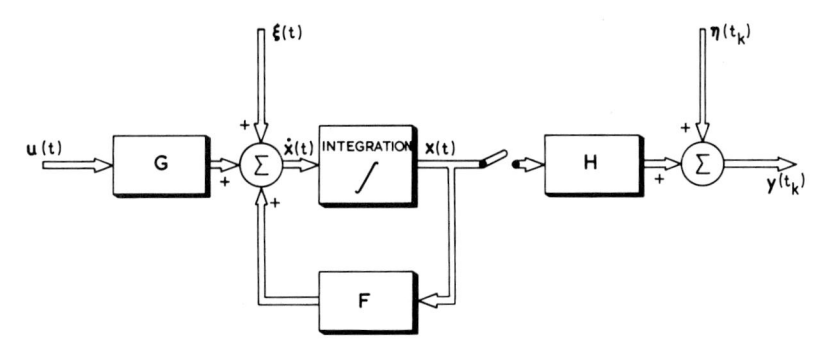

Figure 4. Schematic representation of the (linear) internally descriptive model [Eq. (4)].

microcosms of wastewater treatment processes (Curds, 1973; Olsson, 1975) and rivers (Thomann *et al.*, 1974), may diverge considerably from what is observed to happen in practice.

2.3. Model Structure Identification: Problem Definition

2.3.1. Black Box Model

Recalling the introductory definition of model structure identification we note that the problem for a black box model is considerably simplified since only input → output relationships are being sought. There are two specific identification problems to be solved with respect to Eqs. (1) and (2). The first sounds deceptively easy: For the multiple-input case it concerns the determination of which of these several input variables are in any way significantly related to the output y. Let us call this problem the identification of *cause/effect* relationships, that is, examination of the existence of a deterministic connection between inputs and output. The second identification problem is associated with defining the *time dependence* of the relationships between inputs and outputs. Of interest is the determination of factors such as the speed and nature of the output response to changes in a given input variable.

In formal terms we require a definition of the values of n, m_i, and δ_i in Eqs. (1) and (2); or, rather more precisely, we need to know further which of the b_{ij} parameters of the $B_i(q^{-1})$ polynomials are significantly nonzero. And last, although Eq. (1) is restricted to a linear form, it is also necessary to investigate possible nonlinearities in the terms of the model. Postulation of the correct structure for the nonlinearity, as an identification problem, is not at all trivial. However, providing the model remains "linear-in-the-parameters"—a term defined by Eykhoff (1974) and illustrated below—such nonlinearities present no additional difficulties in the subsequent parameter estimation phase of modeling.

The black box model structure identification problem may be loosely summarized as a problem of transferring from a nonparametric to a parametric representation with minimal loss of accuracy.

2.3.2. Internally Descriptive Model

If the principles of mass, momentum, and energy conservation are applied for the description of our system's behavior, we should be in a position to test the identifiability of the resulting internally descriptive model, as Eq. (4), with a view to *subsequent* planned experimentation. This kind of *a priori* identifiability analysis is presented rigorously elsewhere in this volume by Cobelli *et al.* (Chapter 10); it is not the model structure identification problem to be tackled here.

Our *a posteriori* identification problem is defined as follows: Given a set of measurements of **u** and **y**, determine an appropriate state vector **x**, the number of elements *l* in that vector, and which of the elements of the matrices **F, G, H** are significantly nonzero. The essence of the internally descriptive model structure identification problem is the testing of hypotheses and the evolution of a theory. So identification can be viewed as a procedure of repeated *hypothesis testing* and *decision making*—an intuitive interpretation which has been illustrated earlier in Beck (1978). There are two points about this view which are of some considerable importance: First, it reinforces the notion that modeling is subjective—it depends on the analyst's choice of criteria and his decision to accept or reject a hypothesis (model) on the basis of those criteria; second, it emphasizes the fact that the ultimate problem of modeling is the generation of a subsequent hypothesis given that the current hypothesis is inadequate.

The earlier assumption that an internally descriptive model derived from the application of basic theoretical principles has a linear structure is not a restriction on the following discussion. In fact, a nonlinear model structure arises frequently in the analysis of ecological (Di Cola *et al.*, 1976) and microbiological systems (Beck, 1977), although no explicit examples thereof are presented here. The assumption that the model form is lumped, thus enabling us to use ordinary differential equation representations in preference to partial differential equations, is much more restrictive. However, it is our intention to uphold this latter simplifying assumption in order to avoid consideration of the problem of identifying a correct lumping of the parameters from a distributed-parameter system. The problem already posed is quite sufficiently difficult, and not the least of these difficulties is that, unlike the black box model which may take a rather arbitrary structural form, the abstractions **x, F, G, H** of the internally descriptive model must bear some resemblance to the real world.

2.4. Parameter Estimation

The estimation of parameters is required in two different contexts: It is often implicit in the solution of the identification problem, as will be seen later; and parameter estimation is, of course, the means whereby the coefficients appearing in the finally identified differential/difference equations are accurately evaluated.

A basic principle of parameter estimation is that the estimates $\hat{\boldsymbol{\beta}}$, say, of the model parameters $\boldsymbol{\beta}$ are obtained by minimizing some function of the error

$$\varepsilon(t_k) = y(t_k) - \hat{y}\{\hat{\boldsymbol{\beta}}, t_k\} \tag{5}$$

between the output observation y and a (model) prediction $\hat{y}$ of that output variable. One of the simplest and most well known of parameter estimation schemes is that of *least-squares estimation* where the loss (error) function

$$J(t_N) = \sum_{j=1}^{N} \varepsilon^2(t_j) \qquad (6)$$

is minimized; N is the number of data samples. A more complete discussion of least-squares estimation and its fundamental role in time-series analysis is given in Young (1974).

2.4.1. Black Box Models

In most cases of practical interest least-squares estimation gives parameter estimates $\hat{\beta}$ that are biased, that is,

$$\mathscr{E}\{\hat{\beta}\} \neq \beta$$

in which $\mathscr{E}\{\cdot\}$ is the expectation operator, because the statistical properties of $v(t_k)$ (see Fig. 3) do not satisfy the conditions,

$$v(t_k) = e(t_k). \qquad (7)$$

But this is not to deny the importance of least-squares estimation; it is a ubiquitous technique and can be employed to good advantage as evidenced elsewhere in this book (see Chapter 13 by Ivakhnenko *et al.*). Indeed, the variety of parameter estimation methods stems from the many diverse attempts to overcome the problem of bias. Noting that Eq. (7) implies $E(q^{-1}) = 1$ in Eq. (1), the principal alternative methods of estimation are each associated with different noise process characterizations:

Generalized least-squares (Clarke, 1967; Hastings-James and Sage, 1969)

$$E(q^{-1}) = 1/C(q^{-1}).$$

Maximum likelihood (Åström and Bohlin, 1966)

$$E(q^{-1}) = D(q^{-1}).$$

Instrumental variable–approximate maximum likelihood as in Young (1976)

$$E(q^{-1}) = A(q^{-1})D(q^{-1})/C(q^{-1})$$

with the additional polynomials $C(q^{-1}), D(q^{-1})$ being defined in a fashion similar to $A(q^{-1})$ in Eq. (2).

2.4.2. Internally Descriptive Models

The number of techniques available for estimating parameters in Eq. (4) is remarkable for its smallness. To the best of our knowledge only a maximum likelihood (Källström et al., 1976) method and variants on the extended Kalman filtering (EKF) (Jazwinski, 1970) theme have been applied to the analysis of field data. This latter method, however, we shall consider in detail as a method for solving the model structure identification problem.

2.5. Verification and Validation

In deriving the models of Eqs. (1) and (4) some important assumptions (see Section 2.2) have been made about the statistics of e, ξ, and η. Model *verification*, in our terminology, sets out accordingly to check that the sample statistics of, say, the one-step-ahead prediction errors (residual errors, innovations process errors),

$$\varepsilon(t_k|t_{k-1}) = y(t_k) - \hat{y}(t_k|t_{k-1}) \tag{8}$$

approximate the conditions

$$\mathscr{E}\{\varepsilon(t_k|t_{k-1})\} = 0, \tag{9a}$$

$$\mathscr{E}\{\varepsilon(t_k|t_{k-1})\varepsilon(t_j|t_{j-1})\} = \sigma^2 \delta_{kj}, \tag{9b}$$

$$\mathscr{E}\{\varepsilon(t_k|t_{k-1})u_i(t_j)\} = 0; \quad \text{for all } k,j; \quad i = 1, 2, \ldots, \nu, \tag{9c}$$

where δ_{kj} is the Kronecker delta function such that

$$\delta_{kj} = \begin{cases} 0 & \text{for } k \neq j \\ 1 & \text{for } k = j \end{cases}$$

and $\hat{y}(t_k|t_{k-1})$ is the one-step-ahead prediction of $y(t_k)$ given all past sampled observations of the input and output time-series. Conditions (9a) and (9b) specify that the residuals are a zero-mean, white noise sequence, that is, not correlated with themselves in time, with variance σ^2; condition (9c) requires the residuals to be independent of the inputs u_i. If these conditions hold then our statistical assumptions are valid and it is reasonable to conclude that the model is an adequate characterization of the process behavior observed in the sampled data set from which the model is derived.

There is, however, no guarantee that the model's validity extends beyond this specific set of data. *Validation* is, then, the testing of the model's adequacy against a new set of field data and this will almost certainly entail the design and implementation of new experiments. So finally it can be seen how model building is properly accommodated within the easily recognizable scientific tradition of repeated experiment/analysis/and synthesis.

3. MODEL STRUCTURE IDENTIFICATION: BLACK BOX MODELS

Solutions to the problem of identifying a black box model structure are dealt with first since a black box modeling approach may sometimes be employed as a prelude to working with internally descriptive models.

The identification problems outlined in Section 2.3 (cause–effect; time-dependence) can both be partially solved by computing sample cross-correlation functions from the data

$$\rho_{uy}(\theta) = \frac{1}{N\sigma_u\sigma_y} \sum_{j=1}^{N-\theta} (u(t_j) - \mu_u)(y(t_{j+\theta}) - \mu_y); \qquad \theta = 0, 1, \ldots, \theta_{max};$$
$$\rho_{uy}(-\theta) = \rho_{yu}(\theta); \qquad \theta = 1, 2, \ldots, \theta_{max}. \tag{10}$$

Here μ_u, μ_y, and σ_u^2, σ_y^2, are, respectively, the sample means and variances of the chosen input and output observation sequences. If $\rho_{uy}(\theta)$ is not significantly nonzero for $\theta = -\theta_{max}, \ldots, 0, \ldots, \theta_{max}$, then it can be concluded that no dynamic relationship exists between u and y (cause/effect identification).

In an ideal situation it is desirable to have $u(t_k)$ approximating a white noise sequence, for it can then be shown (for example, Box and Jenkins, 1970) that $\rho_{uy}(\theta)$ approximates the *impulse response* $h(\theta)$ between input and output. Hence it is possible to see how the solution of the time-dependence identification problem is to be constructed as a matter of transferring from a nonparametric to a parametric model representation, as has already been mentioned in Section 2.3. The statistical properties of $u(t_k)$ do not generally, however, approximate those of white noise, although it may be justified to assume that $u(t_k)$ can be characterized by

$$u(t_k) = W(q^{-1})u^*(t_k) \tag{11}$$

in which $W(q^{-1})$, a rational function, is termed a "shaping filter," and $u^*(t_k)$ is a white noise sequence. If Eq. (11) is a valid assumption, and provided we can find $W(q^{-1})$, then (Box and Jenkins, 1970),

$$\rho_{u^*y^*}(\theta) \simeq h(\theta), \tag{12}$$

where $\rho_{u^*y^*}(\theta)$ is the cross-correlation function between the prefiltered (or prewhitened) time series

$$u^*(t_k) = W^{-1}(q^{-1})u(t_k); \qquad y^*(t_k) = W^{-1}(q^{-1})y(t_k).$$

In theory it is then possible to determine *by inspection* of the computed impulse response the pure time delay δ_i and appropriate orders n and m_i for the $A(q^{-1})$ and $B_i(q^{-1})$ polynomials of Eqs. (1) and (2). For the interested reader an exhaustive treatment of solving the identification problem in this manner is given in Box and Jenkins (1970). Note that the impulse response determined through Eqs. (12) and (10) is a truncated approximation of the true impulse response function since it has only a finite number of coefficients for $\theta = 0, 1, \ldots, \theta_{\max}$ (compare with Section 2.2).

The values so derived for n, m_i, and δ_i should at best be regarded as initial intelligent guesses. The cross-correlation function, while it is an indispensable component of any data analysis, has its limitations (see below). Moreover, it is important to note also that the use of input prewhitening involves a second subproblem of identification, namely, the specification of the orders of the numerator and denominator polynomials of $W(q^{-1})$. And even with δ_i, n, m_i specified we have yet to examine whether each b_{ij} parameter is significantly nonzero.

Not all methods of black box model structure identification require the use of input/output cross-correlation functions as described above. A distinguishing feature of this first approach to the identification problem is that it attempts to solve model order determination (that is, obtain values for δ_i, n, m_i) without recourse to any subsequent estimation of parameters. A transposed version of this approach, as it were, takes trial values (hypotheses) of n, δ_i, and m_i and analyzes the variance and statistical properties of the error sequences $\varepsilon(t_k)$ from the resulting fully estimated model (see, for instance, Åström and Eykhoff, 1971; Chan et al., 1974). Other methods which rely on the estimation of parameters as an index of a properly identified model structure include the novel *auxiliary system* method of Wellstead (1976) and the notion of "time-invariance of recursive parameter estimates" illustrated by Whitehead and Young (1975) and Whitehead (1976) (see also Section 4.4). Rather more unorthodox approaches to model structure identification include Ivakhnenko's (1968) group method of data handling (GMDH) algorithm (see also Chapter 13 by Ivakhnenko et al. in this volume) and the application of methods of pattern recognition (Kittler and Whitehead, 1976). For other reviews of specific details of this problem the reader is referred to the papers by Van den Boom and Van den Enden (1974) and Unbehauen and Göhring (1974).

3.1. An Example: Anaerobic Digestion of Waste Organic/Biological Sludges

The results presented in this example are taken from an analysis of gas production dynamics in the anaerobic digestion of waste municipal/domestic sludges (Beck, 1976). The principal biochemical feature of the process is the multistage breakdown of complex (insoluble) organic substrates to simple end products, primarily methane and carbon dioxide. The last stage of the overall reaction, in which methanogenic bacteria metabolize the volatile acid intermediates with the release of methane, is generally believed

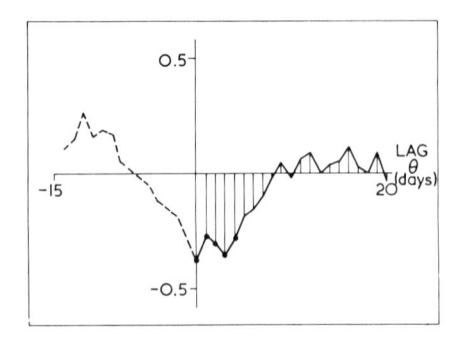

Figure 5. Cross-correlation function ρ_{uy} between volatile acids concentration (u) and gas production rate (y); correlation coefficients marked with ● denote (assumed) significant correlation between input and output.

to be rate limiting (Graef and Andrews, 1974) and thus crucial to an investigation of digester dynamics. Volatile acid concentration and gas production rate are frequently used to monitor process stability.

In the original study a multiple-input/single-output model representation is identified; for the purposes of illustration, however, merely the identification of a single-input/single-output model for volatile acid concentration, input u, and volumetric gas flow rate, output y, is selected. Figure 5 shows, thus, the cross-correlation function ρ_{uy}; the experimental data represent normal operating conditions at the Norwich Sewage Works in England. Initial conclusions from Fig. 5 are that, according to Eqs. (1) and (2), $\delta = 0$ and $m = 4$, approximately. So together with the assumption of $n = 1$ (by inspection of the autocorrelation function of the output time series) we can broadly state that there exists the following deterministic time-dependence relationship between volatile acids and gas production [compare with Eq. (3)]

$$y(t_k) = f\{y(t_{k-1}), u(t_k), u(t_{k-1}), u(t_{k-2}), u(t_{k-3}), u(t_{k-4})\}. \tag{13}$$

The inclusion of the term $u(t_k)$ in Eq. (13) probably occurs as a consequence of the relatively slow sampling frequency of the data which obscures some of the faster dynamic aspects of the relationship between volatile acids and gas production. This observation, apart from the several other attendant difficulties, is a cautionary message on the use of data from badly designed experiments, that is, normal operating conditions.

The interpretation of values for δ and m from Fig. 5 is clearly somewhat speculative. But any attempt at circumventing such imprecision by designing a prewhitening filter results in a cross-correlation function $\rho_{u^*y^*}$ which is equally inconclusive (Fig. 6). It is necessary, ultimately, to incorporate repeated parameter estimation of trial model structures, within

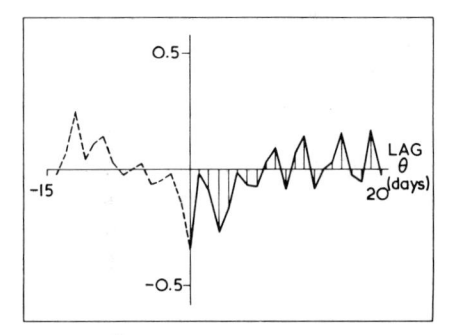

Figure 6. Cross-correlation function ρ_{u*y*} for the prewhitened volatile acid concentration series (u^*) and prewhitened gas production rate series (y^*).

the range of combinations allowed by Eq. (13), as a method of identification. At this stage we can exploit an intuitively useful criterion which states simply that if

$$(\hat{\sigma}_\beta / \hat{\beta}) \geq 1 \qquad (14)$$

for any model parameter estimate $\hat{\beta}$, where

$$\hat{\sigma}_\beta^2 \simeq \mathscr{E}\{(\beta - \hat{\beta})^2\}$$

is an estimate of the parameter estimation error, then the parameter β is not significantly nonzero and its associated term can be dropped from the model structure.

A final structure of the model obtained in such a fashion is given by

$$y(t_k) = a_1 y(t_{k-1}) + b_0 u'(t_k) + b_2 u'(t_{k-2})$$

with $u'(t_k) = (\mu_u / u(t_k))$ where μ_u is a sample mean value for $u(t_k)$. The interesting point here is the fact that the model remains linear-in-the-parameters (Section 2.3) but is quite nonlinear in terms of $u(t_k)$.

What can be concluded from the example presented? Primarily it is observed that solving the identification problem is subjective, clearly so in the inspection of cross-correlation functions and rather less obviously so in the use of Eq. (14). Second, it should be evident that there are good reasons for avoiding the analysis of normal operating data where at all possible.

4. MODEL STRUCTURE IDENTIFICATION: INTERNALLY DESCRIPTIVE MODELS

The technique to be applied exclusively to internally descriptive model structure identification is the extended Kalman filter (EKF). In order to see how the need for the EKF develops from the linear Kalman filter (Kalman, 1960; Kalman and Bucy, 1961) it is first necessary to pose the problem of combined state parameter estimation. Hence model structure identification in this case can be constructed as a problem of assessing diagnostic information on recursive parameter estimates and residual error sequences—a notion already introduced in Section 3.

A formal derivation of the EKF is given in the source reference of Jazwinski (1970). Alternatively, Young (1974) provides an outline of how the EKF algorithms can be obtained from an extension of linear regression analysis.

4.1. Formulation of the Combined State Parameter Estimation Problem

For the linear system of Eq. (4) the linear Kalman filter would provide recursive estimates $\hat{\mathbf{x}}(t_k|t_k)$ of the state vector $\mathbf{x}(t_k)$ conditioned upon all sampled process measurements up to and including those at time t_k.

Suppose now that some of the unknown, or imprecisely known elements of the matrices $\mathbf{F}, \mathbf{G}, \mathbf{H}$, that is, a vector of parameters $\boldsymbol{\alpha}$, say, are required to be estimated simultaneously with the estimation of the state vector. One approach to realizing a simultaneous state parameter estimator is to augment the state vector $\mathbf{x}$ with the parameter vector $\boldsymbol{\alpha}$ and accordingly to postulate a set of additional differential equations representing the parameter dynamics. If the augmented state vector $\mathbf{x}^*$ is defined by

$$\mathbf{x}^* \triangleq \begin{bmatrix} \mathbf{x} \\ \boldsymbol{\alpha} \end{bmatrix}$$

the state parameter dynamics and observation equation are given in the following general nonlinear form

$$\dot{\mathbf{x}}^*(t) = \mathbf{f}\{\mathbf{x}^*(t), \mathbf{u}(t)\} + \boldsymbol{\xi}^*(t), \tag{15a}$$

$$\mathbf{y}(t_k) = \mathbf{g}\{\mathbf{x}^*(t_k)\} + \boldsymbol{\eta}(t_k). \tag{15b}$$

The functions $\mathbf{f}\{\cdot\}$ and $\mathbf{g}\{\cdot\}$ are vector functions; they are nonlinear because of the product terms involving elements of $\boldsymbol{\alpha}$ with elements of $\mathbf{x}$ and $\mathbf{u}$. $\boldsymbol{\xi}^*(t)$ denotes that the vector of stochastic disturbances in Eq. (15a) is now of a different order to that defined for $\boldsymbol{\xi}(t)$ in Eq. (4a).

Let us consider the problem of specifying the dynamics of the parameters $\boldsymbol{\alpha}$. Of particular importance to the subsequent discussion are two such specifications: (a) we might assume that the parameters are constant, that is, time invariant

$$\dot{\boldsymbol{\alpha}}(t) = \mathbf{0}, \tag{16}$$

or (b) it might be proposed that they vary in an unknown "random walk" fashion,

$$\dot{\alpha}(t) = \xi(t). \tag{17}$$

Were there to be more *a priori* information on the parameter variations, then it would be appropriate, for instance, to define the dynamics as oscillatory in accordance with some diurnal or seasonal fluctuation.

4.2. The Extended Kalman Filtering Algorithms

The EKF is a linear approximation of the nonlinear filter which would ideally be needed to provide estimates of $\mathbf{x}^*$ in Eq. (15). The principal steps in its derivation are listed as follows.

(a) *Linearization of the nonlinear augmented state equations.* For small perturbations $\delta \mathbf{x}^*(t)$ of the state $\mathbf{x}^*(t)$ about some nominal reference trajectory $\bar{\mathbf{x}}^*(t)$, a set of linear dynamic equations in $\delta \mathbf{x}^*(t)$ are obtained by taking a first-order Taylor series expansion of $\mathbf{f}$ in Eq. (15). Here $\delta \mathbf{x}^*(t)$ is defined by

$$\delta \mathbf{x}^*(t) = \mathbf{x}^*(t) - \bar{\mathbf{x}}^*(t) \tag{18}$$

and

$$d\bar{\mathbf{x}}^*(t)/dt = \mathbf{f}\{\bar{\mathbf{x}}^*(t), \mathbf{u}(t)\}. \tag{19}$$

(b) *Linearization of the nonlinear observation equation.* By defining a nominal measurement trajectory in terms of $\bar{\mathbf{x}}^*(t)$ we can similarly derive the linear small perturbation observation equation for $\delta \mathbf{y}(t_k)$.

(c) *Application of a linear Kalman filter to the perturbational equations.* From step a we have

$$d\{\delta \mathbf{x}^*(t)\}/dt = \mathbf{F}^*\{\bar{\mathbf{x}}^*(t_0), \mathbf{u}(t)\} \delta \mathbf{x}^*(t) + \xi^*(t), \tag{20}$$

where

$$\mathbf{F}^*\{\bar{\mathbf{x}}^*(t_0), \mathbf{u}(t)\} \triangleq \left[\frac{\partial f_i\{\bar{\mathbf{x}}^*(t), \mathbf{u}(t)\}}{\partial x_j^*} \right]. \tag{21}$$

Integration of Eq. (20) over the interval $t_k \to t_{k+1}$ gives

$$\delta \mathbf{x}^*(t_{k+1}) = \mathbf{\Phi}\{t_{k+1}, t_k; \bar{\mathbf{x}}^*(t_k), \mathbf{u}(t_k)\} \delta \mathbf{x}^*(t_k) + \omega(t_{k+1}), \tag{22a}$$

and from step b we have

$$\delta \mathbf{y}(t_k) = \mathbf{H}^*\{\bar{\mathbf{x}}^*(t_k)\} \delta \mathbf{x}^*(t_k) + \eta(t_k) \tag{22b}$$

with the definitions

$$\Phi\{t_{k+1}, t_k; \bar{\mathbf{x}}^*(t_k), \mathbf{u}(t_k)\} \triangleq \exp(\mathbf{F}^*\{\bar{\mathbf{x}}^*(t_k), \mathbf{u}(t_k)\}(t_{k+1} - t_k)) \qquad (23)$$

and

$$\mathbf{H}^*\{\bar{\mathbf{x}}^*(t_k)\} \triangleq \left[\frac{\partial g_i\{\bar{\mathbf{x}}^*(t_k)\}}{\partial x_j^*}\right]. \qquad (24)$$

Note that Eqs. (20) and (21) imply that $\mathbf{F}^*\{\cdot\}$ and hence $\Phi\{\cdot\}$ are determined for all t by the choice of the initial conditions $\bar{\mathbf{x}}^*(t_0)$ of the reference trajectory; note also that $\omega(t_{k+1})$ is the discrete-time equivalent of $\xi^*(t)$.

By applying a linear (discrete-time) Kalman filter to the linear system of Eq. (22) estimates $\hat{\delta}\mathbf{x}^*$ of the small perturbations can be derived and hence through Eq. (18) we see a means of "reconstructing" estimates of the state $\mathbf{x}^*$, that is,

$$\hat{\mathbf{x}}^*(t_k|t_k) = \bar{\mathbf{x}}^*(t_k) + \hat{\delta}\mathbf{x}^*(t_k|t_k). \qquad (25)$$

(d) *A suitable choice of reference trajectory.* Clearly the choice of reference trajectory is crucial to the operation of the filter. If the choice of $\bar{\mathbf{x}}^*(t_0)$ were inaccurate then there is no guarantee that the perturbations about the reference trajectory are small, and thus the linearization is no longer a valid approximation. For the EKF the particular substitution of *the current state estimate as the reference trajectory* is made; in step e below we shall discuss how the term "current" is interpreted.

(e) *The algorithms.* The EKF algorithms provide for prediction of the estimates and estimation error covariances between sampling instants,

Prediction:

$$\hat{\mathbf{x}}^*(t_{k+1}|t_k) = \hat{\mathbf{x}}^*(t_k|t_k) + \int_{t_k}^{t_{k+1}} \mathbf{f}\{\hat{\mathbf{x}}^*(t|t_k), \mathbf{u}(t)\}\, dt, \qquad (26a)$$

$$\mathbf{P}(t_{k+1}|t_k) = \Phi\{t_{k+1}, t_k; \hat{\mathbf{x}}^*(t_k|t_k), \mathbf{u}(t_k)\}\mathbf{P}(t_k|t_k)$$
$$\times \Phi^T\{t_{k+1}, t_k; \hat{\mathbf{x}}^*(t_k|t_k), \mathbf{u}(t_k)\} + \mathbf{Q}(t_{k+1}), \qquad (26b)$$

and for corrections to be applied to those predictions at the sampling instant,

Correction:

$$\hat{\mathbf{x}}^*(t_{k+1}|t_{k+1}) = \hat{\mathbf{x}}^*(t_{k+1}|t_k) + \mathbf{K}(t_{k+1})[\mathbf{y}(t_{k+1}) - \mathbf{g}\{\hat{\mathbf{x}}^*(t_{k+1}|t_k)\}], \qquad (26c)$$

$$\mathbf{P}(t_{k+1}|t_{k+1}) = [\mathbf{I} - \mathbf{K}(t_{k+1})\mathbf{H}^*(t_{k+1})]\mathbf{P}(t_{k+1}|t_k)[\mathbf{I} - \mathbf{K}(t_{k+1})\mathbf{H}^*(t_{k+1})]^T$$
$$+ \mathbf{K}(t_{k+1})\mathbf{R}(t_{k+1})\mathbf{K}^T(t_{k+1}), \qquad (26d)$$

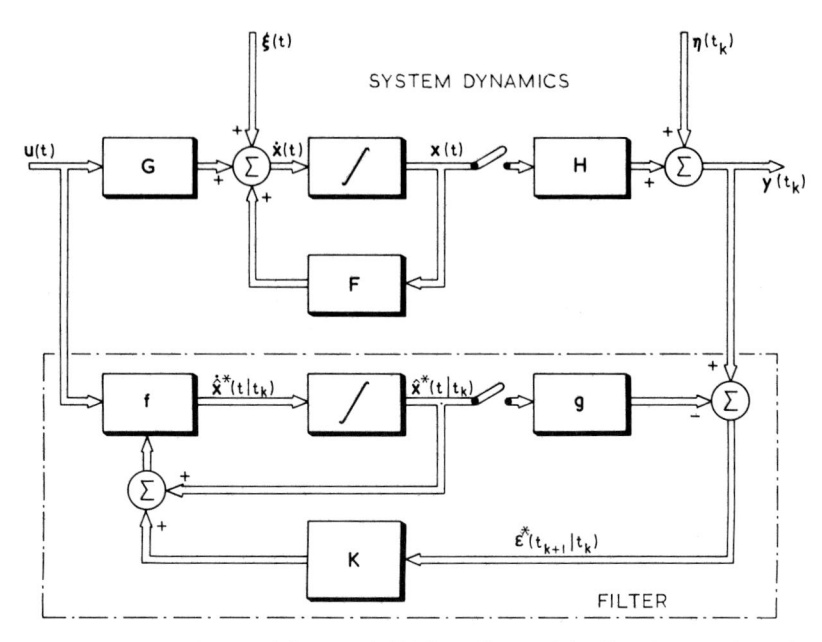

11. Model Structure Identification from Experimental Data

Figure 7. Block diagram of the extended Kalman filter and the (linear) system dynamics.

where **K**, the Kalman gain matrix, is given by

$$\mathbf{K}(t_{k+1}) = \mathbf{P}(t_{k+1}|t_k)\mathbf{H}^{*T}(t_{k+1})[\mathbf{H}^*(t_{k+1})\mathbf{P}(t_{k+1}|t_k) \\ \times \mathbf{H}^{*T}(t_{k+1}) + \mathbf{R}(t_{k+1})]^{-1}. \quad (26e)$$

I denotes the identity matrix and superscript T denotes the transpose of a vector or matrix.

In Eq. (26), $\mathbf{P}(t|t_k)$ is the estimation error covariance matrix defined as

$$\mathbf{P}(t|t_k) \triangleq \mathscr{E}\{(\mathbf{x}^*(t) - \hat{\mathbf{x}}^*(t|t_k))(\mathbf{x}^*(t) - \hat{\mathbf{x}}^*(t|t_k))^T\}$$

and $\mathbf{Q}(t_k)$ and $\mathbf{R}(t_k)$ are, respectively, the system noise covariance and measurement noise covariance matrices

$$\mathscr{E}\{\boldsymbol{\omega}(t_k)\boldsymbol{\omega}^T(t_j)\} = \mathbf{Q}(t_k)\delta_{kj} \quad \text{and} \quad \mathscr{E}\{\boldsymbol{\eta}(t_k)\boldsymbol{\eta}^T(t_j)\} = \mathbf{R}(t_k)\delta_{kj}$$

with $\mathscr{E}\{\boldsymbol{\omega}(t_k)\} = \mathscr{E}\{\boldsymbol{\eta}(t_k)\} = \mathbf{0}$.

To conclude the arguments leading to the EKF algorithms of Eq. (26) we note that the matrices $\boldsymbol{\Phi}\{\cdot\}$ and $\mathbf{H}^*\{\cdot\}$ of the perturbational system (Eq. 22) are required only in the computation of the covariances, Eqs. (26b) and (26d), and for the gain matrix, Eq. (26e). This is so since the substitution

$$\bar{\mathbf{x}}^*(t_{k+1}) = \hat{\mathbf{x}}^*(t_{k+1}|t_k)$$

for the evaluation of **H*** and the substitution

$$\bar{\mathbf{x}}^*(t_k) = \hat{\mathbf{x}}^*(t_k|t_k)$$

for the evaluation of **Φ** enable us to employ the original nonlinear system functions **f**{·} and **g**{·} in Eqs. (26a) and (26c). Hence also by these substitutions the nonlinear equations are effectively relinearized at each sampling instant t_k. Because of the prudent choice of reference trajectory the filtering algorithms can be formulated directly in terms of the augmented state vector **x*** instead of, as suggested at step c, a linear filter applied to the perturbation vector **δx*** together with the solution of Eq. (19) for $\bar{\mathbf{x}}^*$. A block diagram of the EKF is given in Fig. 7.

4.3. Operation of the EKF Algorithms

In order to implement the algorithms of Eq. (26) there are three matrices and one vector which must be quantified. These comprise the *initial conditions* of the filter, in other words, the *a priori* state parameter estimates $\hat{\mathbf{x}}^*(t_0|t_0)$; the *a priori* estimation error covariances $\mathbf{P}(t_0|t_0)$; and the *noise covariances*, system noise covariances $\mathbf{Q}(t_k)$; and measurement noise covariances $\mathbf{R}(t_k)$. Any identification (and parameter estimation) results obtained with the EKF are open to debate because the specification of these "unknowns," and especially that of $\mathbf{Q}(t_k)$, may depend strongly on the subjective judgement of the analyst. There are also analytical problems in that global convergence of the estimates is not guaranteed and thus the choice of $\hat{\boldsymbol{\alpha}}(t_0|t_0)$ in $\hat{\mathbf{x}}^*(t_0|t_0)$ should reflect a vector of *a priori* parameter estimates which are within the locality of the true parameter values.

Only a few guidelines can be offered on the mechanics of implementing the filter for any given system. First, it is probably common sense to evaluate $\hat{\mathbf{x}}^*(t_0|t_0)$, in particular $\hat{\boldsymbol{\alpha}}(t_0|t_0)$, by prior trial and error deterministic simulation comparisons with the experimental field data. Second, if

$$\mathbf{y}(t_k) = \mathbf{x}(t_k) + \boldsymbol{\eta}(t_k) \tag{27}$$

as is often the case, then $\mathbf{R}(t_k)$ and that submatrix of $\mathbf{P}(t_0|t_0)$ which refers to estimates of the state vector **x** can be quantified on the basis of standard instrumentation and laboratory analysis measurement errors. Third, it is customary to assume that **Q** and **R** are time invariant and further that $\mathbf{P}(t_0|t_0)$, **Q** and **R** are diagonal, unless there is evidence supporting an alternative choice.

For the quantification of **Q** there is indeed little that can be stated categorically. Loosely speaking, one might suggest that the **Q** matrix diagonal elements for **x** be evaluated from the *relative* accuracy (uncertainty) of the model dynamics (Eq. 4) with respect to the accuracy (uncertainty) of the measurements, that is, the corresponding elements of the **R** matrix. Quantification of those portions of $\mathbf{P}(t_0|t_0)$ and **Q** which refer to the parameter vector $\boldsymbol{\alpha}$ are discussed later in Section 4.5. Otherwise the reader is referred to Bowles and Grenney (1978) for further discussion of covariance matrices specification for the EKF.

4.4. Intuitive Criteria for Model Structure Identification with the EKF

With some understanding of the EKF and its limitations we are now in a position to consider how the filter can be used to solve the model structure identification problem. In the following our heuristic approach hinges primarily upon interpretation of the recursive parameter estimates $\hat{\boldsymbol{\alpha}}(t_k|t_k)$ as indices of an adequate/inadequate model structure.

An internally descriptive model will in general have an inadequate model structure if it does not contain explicit representations of all the significant physical, chemical, biological, or ecological processes associated with the system. Significance in this context implies that the effects of such relationships between inputs and states can be measured in the output observations **y**. The filter has a tendency to provide estimates $\hat{\mathbf{x}}$ of the state vector that track the observations **y** unless the system model is very accurate. If the model is inaccurate, which is more probable, then the filter attempts to adapt this model to the dynamic characteristics observed between **u** and **y**. Clearly the filter cannot adapt the model structure and thus *significant parameter adaptation* results.

On the basis of this argument it is possible to define a first intuitive criterion for model structure identification:

Criterion 1. A model structure is adequate if the recursive estimates $\hat{\boldsymbol{\alpha}}(t_k|t_k)$ of all parameters defined to be time invariant according to Eq. (16) display trajectories which are sensibly stationary once any initial transients have decayed away.

Now suppose we have estimated $\boldsymbol{\alpha}$ such that the matrices **F, G, H** in the original *linear* system dynamics of Eq. (4) are completely specified. In this event it would be possible to pass through the experimental data with the linear Kalman filter applied to Eq. (4) (note that for the functions **f** and **g** being linear the EKF algorithms of Eq. (26) reduce to those of the linear filter). The sequence of *innovations process residual errors*

$$\varepsilon(t_k|t_{k-1}) = \mathbf{y}(t_k) - \mathbf{H}\hat{\mathbf{x}}(t_k|t_{k-1}) \tag{28}$$

thereby generated should have certain statistical properties—providing our initial assumptions about ξ and η are valid—and a second intuitive criterion can be introduced:

Criterion 2. A model structure is adequate if the residual errors $\varepsilon(t_k|t_{k-1})$ of the linear Kalman filter for the original linear system model, Eq. (4), approximate zero–mean, white, Gaussian sequences.

Clearly the use of criterion 2 is somewhat restricted since the original system dynamics are required to be linear. Notice, however, that Eq. (28) is equivalent to

$$\varepsilon(t_k|t_{k-1}) = \mathbf{y}(t_k) - \hat{\mathbf{y}}(t_k|t_{k-1})$$

which has obvious similarities with Eq. (8)—compare also with the analogous situation for the EKF in Fig. 7.

It should be apparent that Criterion 1 is the more readily applicable criterion of model structure identification: Its use naturally precedes the application of Criterion 2. The reader should also note that, in principle, nonlinear system dynamics present no additional analytical problems for the implementation of the EKF. Nevertheless, having defined two criteria for model structure identification, it must be admitted that there will be few modeling exercises in which these criteria can be applied in any systematic manner!

4.5. Parameter Dynamics and Model Structure Identification

A familiar means of formulating a dynamic model from the available biological and ecological theory is the application of component mass balances across the system boundaries. For our specific purposes these components are usually the concentrations of dissolved substances, for example nutrients, or the magnitudes of microorganism populations. Thus Eq. (4a) might be rearranged to give

$$\dot{\mathbf{x}}(t) = \mathscr{S}\{\mathbf{x}(t), \mathbf{u}(t)\} + \mathscr{T}\{\mathbf{x}(t), \mathbf{u}(t), \boldsymbol{\alpha}_1\} + \mathscr{U}\{\boldsymbol{\alpha}_2(t)\} + \boldsymbol{\xi}(t), \qquad (29)$$

where $\mathscr{S}\{\cdot\}$ represents the bulk transport (flux) of components into and out of the system, and $\mathscr{T}\{\cdot\}$ includes *a priori well-known* theoretical relationships for population growth, death, nutrient uptake, respiration, and so on. $\mathscr{U}\{\cdot\}$ accounts for all physical, chemical, biological, and ecological phenomena whose presence in the observed data is a matter of speculation and for which no well-established formal mathematical relationships are available *a priori*. The distinction drawn between $\mathscr{T}$ and $\mathscr{U}$ is, of course, rather arbitrary. There tends to be a complete spectrum of shades of confidence in the theories incorporated in the model; the distinction serves primarily to illustrate both how to characterize the respective parameter dynamics and how to quantify the associated submatrices of $\mathbf{P}(t_0|t_0)$ and $\mathbf{Q}$.

Let us denote by $\mathbf{P}_1(t_0|t_0)$, $\mathbf{P}_2(t_0|t_0)$, and $\mathbf{Q}_1, \mathbf{Q}_2$ the submatrices of $\mathbf{P}(t_0|t_0)$ and $\mathbf{Q}$ corresponding to $\boldsymbol{\alpha}_1$ and $\boldsymbol{\alpha}_2$. In principle the parameter vector $\boldsymbol{\alpha}_1$ in Eq. (29) is defined as time invariant according to Eq. (16), whereas the dynamics of $\boldsymbol{\alpha}_2$ may be said to conform to Eq. (17), that is, random walk idealizations. The *a priori* estimates $\hat{\boldsymbol{\alpha}}_1(t_0|t_0)$ ought to be quantifiable from previous empirical evidence or prior simulation results. Suitable initial guesses for $\boldsymbol{\alpha}_2$ might be that

$$\hat{\boldsymbol{\alpha}}_2(t_0|t_0) = \mathbf{0}. \tag{30}$$

If $\mathbf{P}_1(t_0|t_0)$ is evaluated in terms of the (albeit subjective) confidence bounds on $\hat{\boldsymbol{\alpha}}_1(t_0|t_0)$, then $\mathbf{P}_2(t_0|t_0)$ should express the intuitively reasonable assumption that *relatively* less initial confidence is placed in the estimates $\hat{\boldsymbol{\alpha}}_2(t_0|t_0)$ than in $\hat{\boldsymbol{\alpha}}_1(t_0|t_0)$. Our objectives in so specifying $\mathbf{P}_2(t_0|t_0)$ are to permit the rapid adaptation of subsequent recursive estimates, $\hat{\boldsymbol{\alpha}}_2(t_k|t_k)$, to values more "realistic" than those of Eq. (30). It is important to note, however, that for the EKF the matrix $\mathbf{P}(t_k|t_k)$ cannot be interpreted as an *a posteriori* measure of the true estimation error covariances. Finally, by virtue of Eq. (16) it is easy to see that

$$\mathbf{Q}_1 = \mathbf{0},$$

and on a very approximate ad hoc basis $\mathbf{Q}_2$ could be chosen such that

$$\mathbf{Q}_2 \leqslant 0.1 \mathbf{P}_2(t_0|t_0).$$

The procedure for model structure testing according to Eq. (29) is formulated with the hope that various hypotheses about the form and combination of $\mathscr{T}$ and $\mathscr{U}$ can be assessed. The ideal objective would be to eliminate $\mathscr{U}$ from Eq. (29) by modification and/or expansion of the structure of $\mathscr{T}$. In this last respect it is particularly useful either to seek relationships, that is, correlated variations, between $\hat{\boldsymbol{\alpha}}_2(t_k|t_k)$ and $\mathbf{x}(t_k), \mathbf{u}(t_k)$; or, with $\mathscr{U}\{\cdot\} = \mathbf{0}$, to check any dependence of the innovations errors $\boldsymbol{\varepsilon}(t_k|t_{k-1})$ of Eq. (28) on variations in $\mathbf{u}(t_k)$.

4.6. Problems of Stability and Partially Observed State Vectors

The majority of the discussion so far has been centered implicitly upon the assumption that Eq. (27) is valid, namely,

$$\mathbf{y}(t_k) = \mathbf{x}(t_k) + \boldsymbol{\eta}(t_k).$$

If this assumption is not valid, for instance,

$$y(t_k) = [\mathbf{I} \vdots \mathbf{0}]\mathbf{x}(t_k) + \boldsymbol{\eta}(t_k),$$

then Criterion 1 cannot be applied with any confidence since there is the likelihood that filter estimates of those states not measured directly are adapted in preference to $\hat{\boldsymbol{\alpha}}(t_k|t_k)$. Model structure identification under these constraints becomes an almost impossible task of checking the statistical properties of the residual errors (Criterion 2) for each set of trial (known) values for the vector $\boldsymbol{\alpha}$. It is unfortunate, therefore, that such models arise naturally and readily in water quality and wastewater treatment systems where substrate and metabolic end product concentrations are measurable but the magnitudes of mediating enzyme and microorganism populations cannot be measured (see, for example, Beck, 1977).

We note, furthermore, as did Di Cola et al. (1976), that for certain microbiological and ecological constituents the state vector dynamics of Eq. (4) reduce to the (deterministic) form

$$\dot{x}(t) = \alpha(t)x(t). \tag{31}$$

For any $x(t_1) > x(t_2)$, $t_2 > t_1$, that is, population growth, it is implied that Eq. (31) exhibits temporary, marginal instability. Thus a small inaccuracy in the filter estimate $\hat{\alpha}(t_k|t_k)$ can lead to severe computational problems—in physical terms, estimates $\hat{x}(t_k|t_k)$ of organism populations, for example, assume erroneously large proportions. Yet even here such adversity can be turned to our advantage, for if the filter estimates remain reasonably bounded this is evidence of a kind that the chosen model structure is adequate.

4.7. Further Considerations of the Measurement Process

The primary objective of identification is to determine the structure of the state vector dynamics (Eq. 4a), rather than the nature of the measurement process (Eq. 4b). Nevertheless, a correct characterization of Eq. (4b) is clearly of fundamental importance to identification of the correct model structure. Two commonly occurring variations on the theme of Eq. (4b) which can be easily accommodated within the analytical framework of the EKF are: (a) the situation of systematic measurement error bias

$$y(t_k) = x(t_k) + \alpha(t_k) + \eta(t_k), \tag{32}$$

and (b) the case of a single observation of multiple state variables

$$y(t_k) = x_1(t_k) + x_2(t_k) + x_3(t_k) + \eta(t_k). \tag{33}$$

In Eq. (32) the bias is estimated as an additional parameter α, which may or may not be time varying. A typical example of Eq. (33) is the measurement of suspended solids in an aquatic environment which may embrace a number of bacterial populations at different states in their life cycle (see, for instance, Busby and Andrews, 1975).

A rather more improbable version of the measurement process concerns the analytical determination of biochemical oxygen demand (BOD), a macromeasure best defined as the amount of oxygen consumed in supporting the breakdown of organic matter by aerobic bacteria. (In fact, BOD is frequently interpreted even more loosely as a measure of the polluting strength of typical municipal/domestic effluents.) A BOD measurement is carried out over a period of 5 days under laboratory conditions in a sealed vessel, at constant temperature, and in the absence of light. The BOD of the sample, for instance river water, is defined as the change in dissolved oxygen (DO) concentration of that sample between the beginning and end of the 5-day period. This measurement, therefore, is itself a dynamic process since it resembles a batch reaction in a closed system. In the example of the following section we shall deal with a modeling exercise which leads ultimately to a description of DO–BOD–algae interaction in a reach of river. The correct characterization of the BOD measurement process turns out to be crucial in interpreting the results of model structure identification. Suppose the BOD measurement is represented as

$$y(t_k) = x_1(t_k) + \int_0^5 g_1\{x_2'(\zeta), x_3'(\zeta)\} \, d\zeta + \eta(t_k), \qquad (34)$$

where

$$x_2'(\zeta) = x_2(t_k) \quad \text{for} \quad \zeta = 0$$

and where y is the measured value of BOD, x_1 is the *in situ* river BOD concentration, and g_1 is some function of x_2', the concentration of live algae in the bottle, and of x_3', the bottled sample DO concentration. x_2 is the concentration of live algae in the river, and ζ is a dummy variable of time (in days). It is thus not at all clear which of the following two mechanisms is responsible for an apparent increase in river BOD concentration, x_1: death and decay of algal matter in the river—a phenomenon which would be described by an appropriate term in $\mathcal{T}$ (Eq. 29); or respiration (in the absence of light), and subsequent death, of a live algal population caught in the river water sample of the BOD measurement, as described by Eq. (34).

Although we shall not discuss these latter stages of the identification analysis below, it may be noted that it is impossible to resolve the above issue for this particular example.

4.8. An Example: The Interaction between BOD and an Algal Population in a Freshwater River

This example, as has been mentioned already, forms a part of a larger model for river water quality, that is, DO–BOD–algae interaction (Beck, 1975); the experimental data are taken from a field study of the River Cam in eastern England (Beck and Young, 1975). The DO balance in a river is generally believed, among other factors, to be determined by: (i) the withdrawal of oxygen by BOD decay; and (ii) the photosynthetic/respiratory activity of plants and algae. Our treatment here is necessarily brief and sets out to examine how a dead and decaying algal population places an additional BOD load on the river's oxygen resources. We assume, therefore, that this is the dominant of the two alternative mechanisms suggested in Section 4.7. A comprehensive report on the use of the EKF in this specific model structure identification context is given in Beck and Young (1976). The primary aim in presenting the example is to illustrate some of the basic principles of internally descriptive model structure identification summarized in Sections 4.4 and 4.5.

An *a priori* model for the BOD dynamics can be derived from the classical studies of Streeter and Phelps (1925)

$$\dot{x}(t) = \mathscr{S}(t) - \alpha_1 x(t) + \xi(t) \tag{35}$$

in which, according to the scheme of Eq. (29),

$$\mathscr{T}\{x(t), u(t), \alpha_1\} = -\alpha_1 x(t); \qquad \mathscr{U}\{\alpha_2(t)\} = 0.$$

x is the concentration of BOD in the river, and α_1 is a parameter representing the BOD decay rate constant under the assumption of simple,

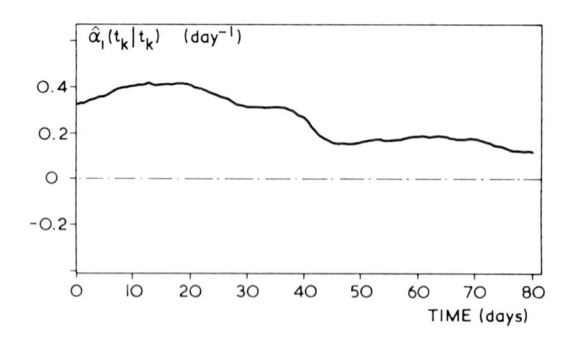

Figure 8. Recursive EKF estimates of the BOD decay rate constant, α_1, in the model of Eq. (35).

Table I *A Priori* Estimates and Covariance Specifications for the Example of Section 4.8

State parameter	$\hat{x}_i^*(t_0\|t_0)$	$p_{ii}(t_0\|t_0)$	q_{ii}	r_{ii}
x (gm^{-3})	1.4	1.0	0.4	0.4
α_1 (day^{-1})	0.32	0.005	0	—
α_2 (gm^{-3} day^{-1})	0	2.0	0.05	—

first-order kinetics. The augmented state vector of the EKF is

$$\mathbf{x}^*(t) = [x(t), \alpha_1]^T$$

with the observation equation $y(t_k) = x(t_k) + \eta(t_k)$ and the *a priori* estimates and covariance specifications are given in Table I. The recursive estimation trajectory for $\hat{\alpha}_1(t_k|t_k)$ shown in Fig. 8 is clearly nonstationary and considerable adaptation of the parameter occurs, in particular, at about day t_{40}. It is known that over this period of the experiment the weather was warm and sunny with very low flows in the river, in other words, conditions likely to stimulate algal growth. We may conclude that the *a priori* model does not satisfy Criterion 1 of Section 4.4.

By means of a modification of the Streeter–Phelps theory due to Dobbins (1964) a term

$$\mathcal{U}\{\alpha_2(t)\} = \alpha_2(t)$$

can be incorporated into Eq. (35) such that

$$\dot{x}(t) = \mathcal{S}(t) - \alpha_1 x(t) + \alpha_2(t) + \xi(t), \tag{36}$$

where $\alpha_2(t)$ is, in effect, a time-variable parameter which accounts for the rate of addition (removal) of BOD in the river by unknown physical, chemical, or biological mechanisms. The estimates $\hat{\alpha}_1(t_k|t_k)$ in Fig. 9 are thereby much improved, that is, they are more stationary. The variations in $\hat{\alpha}_2(t_k|t_k)$, on the other hand, can be shown to be strongly correlated with the day-to-day variations in the hours of sunlight, $u(t_k)$, incident on the river system during each 24-hr period. However, for the purpose of illustrating the use of Criterion 2 from Section 4.5 this same result can be presented in an alternative fashion. Assuming $\alpha_1 = 0.32$ a linear Kalman filter applied to Eq. (35) generates the residual errors $\varepsilon(t_k|t_{k-1})$ from Eq. (28) shown in Fig. 10. The cross-correlation function between these $\varepsilon(t_k|t_{k-1})$ and $u(t_k)$ is given in Fig. 11.

Beyond this point the analysis becomes less straightforward. It is sufficient to say that we now have to establish how and why u is related through α_2 to x. In fact a final version of the DO–BOD–algae interaction model assumes that algal population growth obeys Monod (1942) kinetics

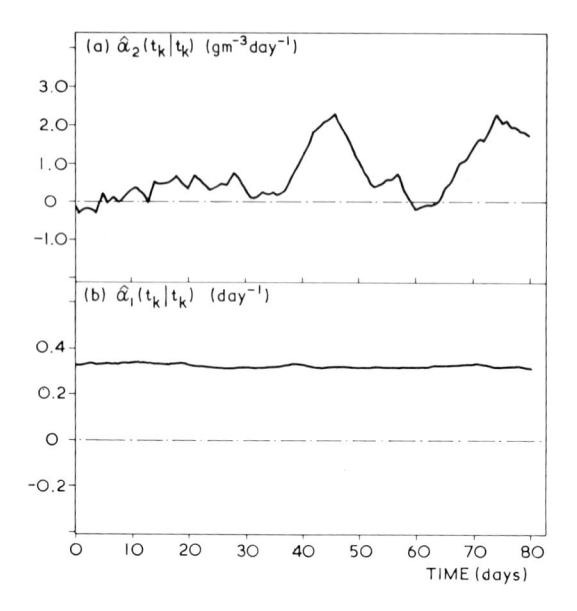

Figure 9. Recursive EKF estimates of (a) α_2 and (b) α_1 in the model of Eq. (36).

with sunlight, u, being growth-rate limiting; after death some of the particulate, algal cell material redissolves, thus creating an apparent additional BOD load in the river (Beck, 1975). During the course of the complete analysis (Beck, 1978) the order of the overall model state vector l (see Section 2.2) is expanded from two states [DO, BOD] to four states [DO, BOD, live algae, dead algae].

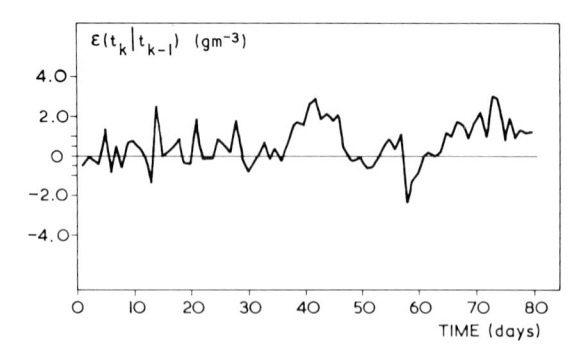

Figure 10. Residual errors $\varepsilon(t_k|t_{k-1})$ for the linear Kalman filter applied to Eq. (35); $\alpha_1 = 0.32\,\text{day}^{-1}$.

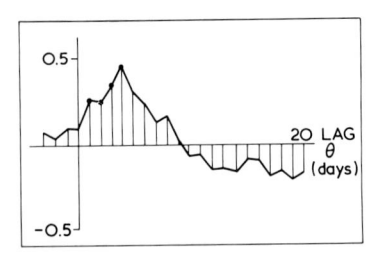

Figure 11. Cross-correlation function between $u(t_k)$, the hours of sunlight incident on the river during each day, and $\varepsilon(t_k|t_{k-1})$; correlation coefficients marked with ● denote (assumed) significant correlation between $u(t_k)$ and $\varepsilon(t_k|t_{k-1})$.

5. CONCLUSIONS

There are three points to be stressed about the model structure identification problem. The first, and most fundamental, is that its solution is in many ways unavoidably subjective. There is no "best" model of a system and, indeed, different types of models are required to fulfill different roles and objectives. Our specific objectives in this instance have been to increase the degree of understanding of a system's dynamic behavior. The second point concerns the distinction made between black box and internally descriptive models. Such a distinction is necessary only in terms of presenting the various theoretical techniques for model structure identification. In an actual problem-solving context solutions evolve from the interplay between both approaches to modeling (see, for example, Beck, 1978). Third, emphasis has been placed upon the extended Kalman filter as a method of identification. It is not an easy technique with which to work and perhaps this is one reason why it appears to have received less-than-fair treatment in the system identification literature. Another reason may be that the EKF lacks certain theoretical guarantees (convergence, efficiency of estimation) on its performance. But then, in a more general sense, the analysis of field data rarely yields elegant solutions, especially when these data are derived from the predominant "normal operating conditions" of microbiological and ecological systems.

REFERENCES

Åström, K. J., and Bohlin, T. (1966). Numerical identification of linear dynamic systems from normal operating records. *In* "Theory of Self-adaptive Control Systems" (P. H. Hammond, ed.), pp. 96–111. Plenum, New York.

Åström, K. J., and Eykhoff, P. (1971). System identification—a survey. *Automatica* **7**, 123–162.

Beck, M. B. (1975). The identification of algal population dynamics in a freshwater stream. *In* "Computer Simulation of Water Resources Systems" (G. C. Vansteenkiste, ed.), pp. 483–494. North-Holland Publ., Amsterdam.

Beck, M. B. (1976). "An Analysis of Gas Production Dynamics in the Anaerobic Digestion Process," Tech. Rep. CUED/F-CAMS/TR135. Eng. Dep., University of Cambridge, Cambridge, England.

Beck, M. B. (1977). Problems of river water quality modelling and control: A review. *Prepr., IFAC Symp. Environ. Syst. Plann., Des., Control, 1977*, pp. 341–349.

Beck, M. B. (1978). Random signal analysis in an environmental sciences problem. *Appl. Math. Modell.* **2**, 23–29.

Beck, M. B., and Young, P. C. (1975). A dynamic model for DO-BOD relationships in a non-tidal stream. *Water Res.* **9**, 769–776.

Beck, M. B., and Young, P. C. (1976). Systematic identification of DO-BOD model structure. *J. Environ. Eng. Div., Am. Soc. Civ. Eng.* **102**, 909–927.

Bowles, D. S., and Grenney, W. J. (1978). Steady-state river quality modelling by sequential extended Kalman filters. *Water Resour. Res.* **14**, 84–95.

Box, G. E. P., and Jenkins, G. M. (1970). "Time-series Analysis, Forecasting and Control." Holden-Day, San Francisco, California.

Busby, J. B., and Andrews, J. F. (1975). Dynamic modelling and control strategies for the activated sludge process. *J. Water Pollut. Control Fed.* **47**, 1055–1080.

Chan, C. W., Harris, C. J., and Wellstead, P. E. (1974). An order-testing criterion for mixed autoregressive moving average processes. *Int. J. Control* **20**, 817–834.

Clarke, D. W. (1967). Generalised least squares estimation of the parameters of a dynamic model. *Prepr., IFAC Symp. Ident. Syst. Parameter Estimation, 1967*, Paper 3.17.

Curds, C. R. (1973). A theoretical study of factors influencing the microbial population dynamics of the activated sludge process—I. *Water Res.* **7**, 1269–1284.

Di Cola, G., Guerri, L., and Verheyden, H. (1976). Parametric estimation in a compartmental aquatic ecosystem. *Prepr., IFAC Symp. Ident. Syst. Parameter Estimation, 4th, 1976* Part 2, pp. 157–165.

Dobbins, W. E. (1964). BOD and oxygen relationships in streams. *J. Sanit. Eng. Div., Am. Soc. Civ. Eng.* **90**, 53–78.

Eykhoff, P. (1974). "System Identification—Parameter and State Estimation." Wiley, New York.

Graef, S. P., and Andrews, J. F. (1974). Stability and control of anaerobic digestion. *J. Water Pollut. Control Fed.* **46**, 666–683.

Gustavsson, I. (1975). Survey of applications of identification in chemical and physical processes. *Automatica* **11**, 3–24.

Hastings-James, R., and Sage, M. W. (1969). Recursive generalised-least-squares procedure for online identification of process parameters. *Proc. Inst. Electr. Eng.* **116**, 2057–2062.

Ivakhnenko, A. G. (1968). The Group Method of Data Handling—a rival of the method of stochastic approximation. *Sov. Autom. Control (Engl. Transl.)* **1**(3), 43–55.

Jazwinski, A. H. (1970). "Stochastic Processes and Filtering Theory." Academic Press, New York.

Källström, G. C., Essebo, T., and Åström, K. J. (1976). A computer program for maximum likelihood identification of linear multivariable stochastic systems. *Prepr., IFAC Symp. Ident. Syst. Parameter Estimation, 4th, 1976* Part 2, pp. 508–521.

Kalman, R. E. (1960). A new approach to linear filtering and prediction problems. *J. Basic Eng., Am. Soc. Mech. Eng.* **82**, 35–45.

Kalman, R. E., and Bucy, R. S. (1961). New results in linear filtering and prediction theory. *J. Basic Eng.* **83**, 95–108.

Kittler, J., and Whitehead, P. G. (1976). Determination of the model structure using pattern recognition techniques. *Prepr. IFAC Symp. Ident. Syst. Parameter Estimation, 4th, 1976* Part 3, pp. 22–30.

Monod, J. (1942). "Recherches sur la croissance des cultures bacteriennes." Hermann, Paris.

Olsson, G. (1975). "Activated Sludge Dynamics—I. Biological Models," Tech. Rep. 7511(C). Dep. Autom. Control, Lund Inst. Technol., Sweden.

Rowe, I. H. (1970). A bootstrap method for statistical estimation of model parameters. *Int. J. Control* **12**, 721–738.

Söderström, T., Ljung, L., and Gustavsson, I. (1974). "A Comparative Study of Recursive Identification Methods," Tech. Rep. 7427. Dep. Autom. Control, Lund Inst. Technology, Sweden.

Streeter, H. W., and Phelps, E. B. (1925). "A Study of the Pollution and Natural Purification of the Ohio River," Bull. No. 146. U.S. Public Health Serv., Washington, D.C.

Thomann, R. V., Di Toro, D. M., and O'Connor, D. J. (1974). Preliminary model of Potomac estuary phytoplankton. *J. Environ. Eng. Div., Am. Soc. Civ. Eng.* **100**, 699–715.

Unbehauen, H., and Göhring, B. (1974). Tests for determining model order in parameter estimation. *Automatica* **10**, 233–244.

Van den Boom, A. J. W., and Van den Enden, A. W. M. (1974). The determination of the orders of process and noise dynamics. *Automatica* **10**, 245–256.

Wellstead, P. E. (1976). Model order testing using an auxiliary system. *Proc. Inst. Elect. Eng.* **123**, 1373–1379.

Whitehead, P. G. (1976). Application of recursive estimation techniques to time variable water resource systems. *Prepr., IFAC Symp. Ident. Syst. Parameter Estimation, 4th, 1976.*

Whitehead, P. G., and Young, P. C. (1975). A dynamic stochastic model for water quality in part of Bedford-Ouse river system. *In* "Computer Simulation of Water Resources" (G. C. Vansteenkiste, ed.), pp. 417–438. North-Holland Publ., Amsterdam.

Young, P. C. (1974). A recursive approach to time-series analysis. *Bull. Inst. Math. Appl.* **10**, 209–224.

Young, P. C. (1976). Some observations on instrumental variable methods of time-series analysis. *Int. J. Control* **23**, 593–612.

Young, P. C., and Whitehead, P. G. (1977). A recursive approach to time-series analysis for multivariable systems. *Int. J. Control* **25**, 457–482.

Chapter 12

COMPUTER-AIDED SYSTEMS MODELING

George J. Klir

1. Introduction 291
   1.1 Conceptual Framework 291
   1.2 Structure Modeling 294
2. Relevant Concepts 295
   2.1 Dataless Systems 295
   2.2 Data Systems 297
   2.3 Generative Systems 297
   2.4 Structure Systems 301
3. Systems Modeling 302
   3.1 General Discussion 302
   3.2 Experimental Investigation 304
   3.3 Generative System Identification 304
   3.4 Structure System Identification 307
4. Examples of Systems Modeling in Ecology 312
   4.1 Lake Ontario 312
   4.2 Andrews Experimental Forest 317
5. Conclusions 320
   References 322

1. INTRODUCTION

1.1. Conceptual Framework

A *conceptual framework for systems problem solving* has been presented by Cavallo and Klir (1978) as a natural and operational culmination of the development of a hierarchy of epistemological levels which the author has been working on since the late 1960s (Klir, 1969). The motivation for developing the framework has been fourfold: (a) to identify and operationally describe as large a class of genuine systems problems as possible; (b) to establish a useful taxonomy of systems, systems problems, and associated methodological tools; (c) to identify areas of systems problems which are methodologically underdeveloped and to initiate research in these areas; and (d) to provide a basis through which strategies could be developed for assisting the human problem solver (scientist, engineer, manager) in identifying and formulating his problem.

The *hierarchy of epistemological levels of systems*, which forms the kernel of the conceptual framework, seems vital, in one form or another, to the development of any organized package of methodological tools for systems problem solving (Cavallo and Klir, 1978; Klir, 1978; Zeigler, 1974, 1976b; Gaines, 1977).

At the lowest level of the hierarchy, denoted as *level 0*, a system is defined by a set of variables, a set of potential states declared for each variable, and some way of describing the meaning of both the variables and their states in terms of some associated real-world attributes and their manifestations.

The set of variables and the set of corresponding attributes are usually partitioned into two subsets, referred to as *basic and supporting variables* or attributes. Aggregate states of all supporting variables form a support set, usually referred to as a *parameter set*, within which changes in states of the individual basic variables occur.

Although supporting variables most frequently represent *time* and/or *space*, these are not the only possibilities. For instance, states of a supporting variable may represent individuals of a particular *population*, for example, persons in a social group or copies of a manufactured product. Basic variables represent various characteristics of the individuals (for example, sex, income, occupation, education, etc., in the case of the social group).

The set of all attributes under consideration together with a family of sets of their potential appearances is called an *object system*. The term "object system" was chosen to indicate that this kind of system is defined directly on an object of interest by selecting some of its characteristics relevant to the problem of concern.

The set of all variables under consideration together with a family of sets of their potential states is called an *image system*. The term "image system" was chosen to indicate that the system is an abstract and usually simplified representation of some object system.

An object system, image system, and some correspondence between entities of the two systems is referred to as a *source system*. The term "source system" was chosen to indicate that it is appropriate to view such a system as a source of observable (experimental) data.

The object system involves primarily problems associated with observation or measurement procedures. The image system, on the other hand, involves primarily problems associated with data processing. The source system provides then a comprehensive frame for data gathering, data processing, as well as interpretation of both the data and results of processing. These three systems are all referred to as *dataless systems*; this term indicates that no data are available regarding the actual states of basic variables within the parameter set.

Any useful relation which can be recognized in the sets of appearances of the individual attributes (e.g., the appearances are ordered) should be included in the definition of the object system. Moreover, if there exists dependencies among the basic attributes which are solely due to observation or measurement procedures (e.g., one attribute represents an arithmetic average of others), then such dependencies should also be included in the definition. All of these relations and dependencies must be preserved, possibly in a simplified form, in the image system representing the object system. The image system is thus required to be a *homomorphic image* of the corresponding object system (see, also, Zeigler, Chapter 1, this volume).

Systems defined at different higher epistemological levels are distinguished from each other by the level of knowledge regarding the variables of the image system. A higher level system entails all knowledge of the corresponding systems at any lower level and contains some additional knowledge which is not available at the lower levels.

When an image system is supplemented by data, that is, by actual states of the basic variables within the parameter set, either observed (as in the problem of empirical investigation) or desirable (as in the problem of system design), we consider the new system defined at epistemological *level 1*; systems defined at this level are called *data systems*.

Higher epistemological levels involve knowledge of some *parameter-invariant properties* through which the data can be generated (in a deterministic or stochastic fashion) for appropriate initial conditions. At *level 2*, the parameter-invariant properties are represented by a single generative relation among the variables of the image system and, possibly, some other variables; each of the latter is defined in terms of a specific translation rule in the parameter set, applied either to a variable of the image system or to some additional, artificially introduced variable (unobserved), referred to as *internal variable*. As the parameter-invariant relation (time-invariant, space-invariant, etc.) provides a direct basis for generating data, systems defined at *level 2* are called *generative systems*.

At epistemological *level 3*, each individual system consists of a set of generative systems, referred to as elements of the larger system, and some kind of relation among them. For the purpose of this exposition, the relation is represented by direct couplings among the elements—variables shared by several elements. Such variables indicate a form of data exchange among the elements as well as the way in which generative relations associated with the individual elements are composed. Systems defined at this level are called *structure systems*.

Systems at higher epistemological levels are characterized by considering the parameter set partitioned into subsets associated with different lower level systems (Klir, 1978). However, modeling, as discussed in this exposition, does not involve systems defined at levels higher than the structure system. Therefore, no attempt is made here to describe systems associated with these higher levels. An approach to systems modeling at level 4 was suggested by Uyttenhove (1978).

Regardless of the level at which the system is defined, the variables involved in the image system may be classified into *input and output variables*. This classification means that states of the input variables are viewed as conditions under which states of output variables change. That is to say, statements regarding the variables are conditional: "If the input variables are in state x, then...." Input variables are thus not a subject of inquiry but are viewed as being determined by some agent which is not part of the system under consideration. Such an agent is referred to as the *environment*. Observe that the notion of input variables is different from the notion of independent variables.

Systems whose variables are classified into input and output variables are called *directed systems*; those for which no such classification is given are called *neutral systems*. This dichotomy of systems holds for each of the epistemological levels. The same distinction is used in systems ecology (Patten and Finn, Chapter 8, this volume).

1.2. Structure Modeling

The term "*structure modeling*" is used in a previous paper (Klir and Uyttenhove, 1976b) to refer to the problem of inferring some knowledge regarding structure systems through which a given data system can be best approximated at each level of structure refinement (see Section 3.4), within some specified limits of complexity and, possibly, subject to other constraints. In terms of the hierarchy of epistemological levels, the problem is characterized by climbing up the hierarchy from level 1 to level 3.

Two partial problems involved in structure modeling are suggested by the epistemological hierarchy: (a) *identification of generative systems*, subject to given constraints, by which the given data systems can be best approximated; and (b) *identification of structure systems*, subject to given constraints, through which the generative systems obtained in (a) can be best approximated at each individual level of structure refinement.

Procedures described in a previous paper (Klir, 1975) are directly applicable for solving problem (a). If desirable, they can be augmented by allowing the modeler to introduce hypothetical (unobserved) internal variables with the aim of improving the approximation provided by the generative system within the given constraints.

Problem (b), referred to as the structure identification problem, can more precisely be formulated as follows: Given a generative system (neutral or directed) which approximates a specific data system, determine structure systems (neutral or directed) whose elements are associated with subsets of variables of the data system and which at each level of structure refinement best approximate the generative system.

A method for solving the structure identification problem, which was previously developed by the author (Klir, 1976; Klir and Uyttenhove, 1976a, b) is characterized by a postulational approach to the problem: Hypothetical structure systems which are potential candidates for the desirable structure systems are first postulated at level 3, then analyzed (relations of their elements are composed according to the configuration of couplings) and, finally, the resulting hypothetical relations are compared with the corresponding empirical relation of the given generative system which was derived directly from the data. A specific measure of conformation between the empirical relation and each individual hypothetical relation is then used as a basis for comparing different hypothetical structure systems.

The method is applicable to variables of any scale and to both well-defined and fuzzy variables. It is also applicable to relations of any kind such as time-invariant, space-invariant, deterministic, probabilistic, memoryless and those involving memory, etc. It is assumed, in conformity with the nature of empirical data, that the variables involved are discrete.

2. RELEVANT CONCEPTS

2.1. Dataless Systems

Let $A = \{a_i | i \in I_n\}$ be a set of basic attributes chosen by the investigator to represent an object of interest for some specific purpose; $I_n = \{1, 2, \ldots, n\}$ is an index set determined by the number n of basic attributes. Let A_i denote the set of potential appearances of basic attribute a_i.

Let $B = \{b_j | j \in I_m\}$ be a set of supporting attributes (time, space, etc.) chosen by the investigator and let B_j stand for the set of potential appearances of supporting attribute b_j. Then, the *neutral object system* in the simplest form is defined by the pair

$$\Omega_N = (\{(a_i, A_i) | i \in I_n\}; \{(b_j, B_j) | j \in I_m\}).$$

Any useful relation (e.g., ordering) which can be recognized in the sets of appearances of the attributes should be added to the definition of the object system. Moreover, if there exist dependencies among the basic attributes which are solely due to observation or measurement procedures (e.g., one attribute represents an arithmetic average of others), then these dependencies should also be included in the definition.

Let v_i and V_i ($i \in I_n$) denote basic variables and sets of states of the variables, respectively, and let supporting variables and their sets of states be denoted by w_j and W_j ($j \in I_m$), respectively. A *neutral image system* Γ_N compatible with the neutral object system Ω_N is then defined as the pair

$$\Gamma_N = (\{(v_i, V_i) | i \in I_n\}; \{(w_j, W_j) | j \in I_m\}).$$

This definition may be supplemented by (a) some recognized relations in the sets of states, and/or (b) dependencies among the basic variables due solely to the measurement procedures. The set $\times_{j \in I_m} W_j$ is referred to as the *parameter set*.

Assume now that a resolution level is introduced for each attribute of the object system which characterizes the meaning of data to be collected for the attribute. This can be accomplished by defining a partition $\pi_i(A_i)$ on each set A_i ($i \in I_n$) and a partition $\pi_j(B_j)$ on each set B_j ($j \in I_m$). The form each of these partitions takes depends primarily on the knowledge of the object of investigation, the purpose of the investigation, and the measuring instruments, as well as computing facilities which are available.

In order to get a meaningful basis for data gathering and data interpretation, a correspondence between the entities involved in the object and image systems must be introduced. This is accomplished by: (a) a one-to-one correspondence $f_a: A \to V$; (b) a one-to-one correspondence $f_b: B \to W$; (c) a family of one-to-one correspondences $G = \{g_i : \pi_i(A_i) \to V_k | i, k \in I_n; v_k = f_a(a_i)\}$; and (d) a family of one-to-one correspondences $H = \{h_j : \pi_j(B_j) \to W_l | j, l \in I_m; w_l = f_b(b_j)\}$. The collection of an object system, an image system, and a correspondence between them expressed in terms of the one-to-one correspondences (a)–(d) form a complete frame for data gathering and interpretation referred to as the *neutral source system*.

Directed dataless systems can be obtained from the corresponding neutral dataless systems by distinguishing the input attributes/variables from the output attributes/variables. An easy way to formally distinguish them is to partition the set I_n, which is used in all dataless systems for the identification of the basic attributes/variables, into two subsets I_{n_1} and $I_n - I_{n_1}$ ($n_1 < n$) which identify the input and output variables, respectively.

2.2. Data Systems

Each image or source system implicitly contains all possible trajectories of states of basic variables in the parameter set, that is, all possible functions from $\times_{j \in I_m} W_j$ to $\times_{i \in I_n} V_i$. A meaningful restriction to one of the functions, say function δ, which in the modeling problem is determined by data gathering, constitutes *data* regarding the variables. When the neutral dataless system, say $\mathscr{S}_{0N}$, is augmented with δ, we obtain a *neutral data system*. Let $\mathscr{S}_{1N}$ stand for the neutral data system. Then,

$$\mathscr{S}_{1N} = (\mathscr{S}_{0N}, \delta).$$

A modification to the directed data system is obvious.

2.3. Generative Systems

Let a set of variables s_k ($k \in I_q$), referred to as *sampling variables*, be introduced by the equation

$$s_{k,\mathbf{w}} = v_{i, \lambda_r(\mathbf{w})},$$

where $s_{k,\mathbf{w}}$ stands for states which sampling variable s_k assumes for state $\mathbf{w}$ of the parameter set, and λ_r denotes a parameter-invariant *translation rule* which for any given state $\mathbf{w}$ of the parameter set determines one or several other states in the parameter set.

For instance, when the parameter set is totally ordered (as in the case of time parameter) and represented by the set of positive integers, each translation rule can be described by a simple equation

$$\lambda_r(\mathbf{w}) = \mathbf{w} + a,$$

where a is an integer; sampling variables are then defined by

$$s_{k,\mathbf{w}} = v_{i, \mathbf{w}+a}.$$

When the parameter set is partially ordered, $\lambda_r(\mathbf{w})$ may stand, for example, for states in the parameter set which are predecessors (or successors) or $\mathbf{w}$ with a particular distance from $\mathbf{w}$.

Let Λ denote the set of all translation rules under consideration and let the relation

$$M \subset V \times \Lambda$$

specify which translation rules are applied to which variables (including possible internal variables). Set M is called a *mask*; sets $M_i \subset M$ such that $(x, y) \in M_i$ iff $x = v_i$ ($i \in I_n$) are called *submasks* of the overall mask M, each associated with one basic variables.

Let $s_{k,w} \in S_k$; then clearly $S_k = V_i$ if sampling variable s_k is defined in terms of basic variable v_i and a translation rule λ_r.

Given an image or source system and a mask, a set of sampling variables is uniquely defined. A relation

$$R_1 \subset S$$

defined on

$$S = \underset{k \in I_q}{\times} S_k$$

can be then introduced. Elements of R_1 are q-tuples of states of the sampling variables defined by the mask. They are called *data samples*. When probabilities $p(\mathbf{s})$ are given with which samples $\mathbf{s} \in S$ appear, we obtain set

$$B_N = \{(\mathbf{s}, p(\mathbf{s})) | \mathbf{s} \in R_1, 0 < p(\mathbf{s}) \leq 1, \sum_\mathbf{s} p(\mathbf{s}) = 1\}$$

which is referred to as *basic behavior* of the generative system under consideration.

To employ the basic behavior for generating data for the given primitive system, some order of states $\mathbf{w} \in W$ must be chosen in which the data are generated. The order must be compatible with the natural order of the parameter set W. If the parameter set has no natural order, it may be artificially ordered in some suitable way. Given an order of states in W (linear or partial), let $\mathbf{w} \leq \mathbf{w}'$ denote that either state $\mathbf{w} \in W$ precedes state $\mathbf{w}' \in W$ or $\mathbf{w} = \mathbf{w}'$.

Once a generative order in set W is decided, a relation

$$R_2 \subset R_1 \times R_1$$

whose elements are pairs $(\mathbf{s}, \mathbf{s}')$ of successive samples with respect to the generative order can meaningfully be defined. When probabilities $p(\mathbf{s}, \mathbf{s}')$ are given with which pairs $(\mathbf{s}, \mathbf{s}')$ appear, we obtain a set

$$A_N = \{((\mathbf{s}, \mathbf{s}'), p(\mathbf{s}, \mathbf{s}')) | (\mathbf{s}, \mathbf{s}') \in R_2, 0 < p(\mathbf{s}, \mathbf{s}') \leq 1, \sum_{\mathbf{s},\mathbf{s}'} p(\mathbf{s}, \mathbf{s}') = 1\}.$$

This set is called the basic *state transition relation* (or *basic ST relation*, in abbreviation).

Neither A_N nor B_N can directly be used for generating data for the given primitive system although A_N can readily be modified into a generative relation solely by changing the basic probabilities $p(\mathbf{s},\mathbf{s}')$ into the appropriate conditional probabilities $p(\mathbf{s}'|\mathbf{s})$. We obtain a set

$$\mathscr{A}_N = \{((\mathbf{s},\mathbf{s}'), p(\mathbf{s}'|\mathbf{s}))|(\mathbf{s},\mathbf{s}') \in R_2, 0 < p(\mathbf{s}'|\mathbf{s}) \leq 1, \sum_{\mathbf{s}'} p(\mathbf{s}'|\mathbf{s}) = 1$$

for each particular $\mathbf{s}\}$

which is called the *generative ST relation*.

To modify the basic behavior B_N into a generative relation, the involved sampling variables must be partitioned into those which are generated and those which generate. When W is ordered, the partition is unique. The *generated sampling variables* are defined in terms of translation rules λ_{r_i} for each $i \in I_n$ such that $\lambda_{r_i}(\mathbf{w}) \geq \lambda_r(\mathbf{w})$ for all λ_r such that $(v_i, \lambda_r) \in M_i$. Indeed, data are generated by selecting an appropriate initial sample and then moving the chosen mask in W (applying the translation rules for different states $\mathbf{w}$) according to the given order. When the mask moves from $\mathbf{w}$ to an immediate successor of $\mathbf{w}$, say $\mathbf{w}'$, states of all sampling variables except the generated variables are given in terms of the sample associated with $\mathbf{w}$; states of the generated variables are not given and must be determined through some appropriate relation.

All sampling variables defined for a given mask which are not generated variables will be referred to as *generating variables*.

Let $\mathbf{s}_g$ denote that portion of sample $\mathbf{s}$ which is associated with the generated variables and let $\bar{\mathbf{s}}_g$ denote the rest of $\mathbf{s}$ ($\mathbf{s}$ without $\mathbf{s}_g$). Then a generative relation $\mathscr{B}_N$ can be defined as

$$\mathscr{B}_N = \{((\mathbf{s}_g, \bar{\mathbf{s}}_g), p(\mathbf{s}_g|\bar{\mathbf{s}}_g))|\mathbf{s}_g\bar{\mathbf{s}}_g = \mathbf{s}, \mathbf{s} \in R_1, 0 \leq p(\mathbf{s}_g|\bar{\mathbf{s}}_g) \leq 1, \sum_{\mathbf{s}_g} p(\mathbf{s}_g|\bar{\mathbf{s}}_g) = 1$$

for each particular $\bar{\mathbf{s}}_g\}$,

where $\mathbf{s}_g\bar{\mathbf{s}}_g$ denotes the concatenation of $\mathbf{s}_g$ and $\bar{\mathbf{s}}_g$ (assume for convenience that generated variables are identified by $k \in I_n$). This relation is called the *generative behavior*.

Although only sets $\mathscr{A}_N$ and $\mathscr{B}_N$ express explicitly rules for generating data, sets A_N and B_N contain such rules implicitly. Hence, the *neutral generative system* $\mathscr{S}_{2N}$ is defined as a quadruple.

$$\mathscr{S}_{2N} = (\mathscr{S}_{0N}; \Lambda; M; G_N),$$

where $\mathscr{S}_{0N}$ is a neutral dataless system with ordered set W, M denotes a mask, and G_N stands for an element taken from the set $\{A_N, B_N, \mathscr{A}_N, \mathscr{B}_N\}$.

Although the generative system does not explicitly contain any data, it contains a relation through which data can be generated. Hence, it consists of a dataless system, a generative relation invariant with respect to the state set W of supporting variables and, indirectly, a set of data systems, which can be generated through the relation.

One convenient representation of the various forms of G_N are probabilistic matrices. The meaning of rows, columns, and entries of these matrices for the individual forms of neutral generative systems is summarized in the following tabulation.

	Rows	Columns	Entries
A_N	s	s'	$p(s, s')$
$\mathscr{A}_N$	s	s'	$p(s'\|s)$
B_N	$\bar{s}_g$	s_g	$p(s_g \bar{s}_g)$
$\mathscr{B}_N$	$\bar{s}_g$	s_g	$p(s_g\|\bar{s}_g)$

When the dataless system included in a generative system is directed, then generated variables defined in terms of the input variables are generated by the environment. Let sampling variables s_{k_1} for $k_1 \in I_{n_1}$ be input variables (generated by the environment) and let sampling variables s_{k_2} for $k_2 \in I_n - I_{n_1}$ be the generated variables. Then $\mathbf{s}_g$ is naturally split into two disjoint portions, $\mathbf{s}_{g_1}$ and $\mathbf{s}_{g_2}$, based on variables s_{k_1} and s_{k_2}, respectively. Clearly, $\mathbf{s}_g = \mathbf{s}_{g_1} \mathbf{s}_{g_2}$ and $\mathbf{s} = \mathbf{s}_g \bar{\mathbf{s}}_g = \mathbf{s}_{g_1} \mathbf{s}_{g_2} \bar{\mathbf{s}}_g$. $\mathbf{s}_{g_1}$ is referred to as the *stimulus*, $\mathbf{s}_{g_2}$ is the generated portion, and $\bar{\mathbf{s}}_g$ is the generating portion of each sample $\mathbf{s}$.

When the various forms of G_N are modified to corresponding forms based on the split of $\mathbf{s}_g$ into $\mathbf{s}_{g_1}$ and $\mathbf{s}_{g_2}$, we obtain forms of behavior and an ST relation applicable to the directed generative system. They can conveniently be represented by three-dimensional probabilistic arrays with the following characteristics, where $\bar{\mathbf{s}}$ stands for $\mathbf{s}_{g_2} \bar{\mathbf{s}}_g$ and $\bar{\mathbf{s}}'$ stands for a next state of $\mathbf{s}_{g_2} \bar{\mathbf{s}}_g$, as shown in the following tabulation.

	Pages	Rows	Columns	Entries
A_D	$\mathbf{s}_{g_1}$	$\bar{\mathbf{s}}$	$\bar{\mathbf{s}}'$	$p(\bar{\mathbf{s}}, \bar{\mathbf{s}}'\|\mathbf{s}_{g_1})$
$\mathscr{A}_D$	$\mathbf{s}_{g_1}$	$\bar{\mathbf{s}}$	$\bar{\mathbf{s}}'$	$p(\bar{\mathbf{s}}'\|\mathbf{s}_{g_1}, \bar{\mathbf{s}})$
B_D	$\mathbf{s}_{g_1}$	$\bar{\mathbf{s}}_g$	$\mathbf{s}_{g_2}$	$p(\mathbf{s}_{g_2} \bar{\mathbf{s}}_g\|\mathbf{s}_{g_1})$
$\mathscr{B}_D$	$\mathbf{s}_{g_1}$	$\bar{\mathbf{s}}_g$	$\mathbf{s}_{g_2}$	$p(\mathbf{s}_{g_2}\|\mathbf{s}_{g_1}, \bar{\mathbf{s}}_g)$

The *directed generative system* $\mathscr{S}_{2D}$ is then defined as a quadruple

$$\mathscr{S}_{2D} = (\mathscr{S}_{0D}; \Lambda; M; G_D),$$

where $\mathscr{S}_{0D}$ is a directed dataless system, M is a sampling mask, and G_D is an element taken from the set $\{A_D, \mathscr{A}_D, B_D, \mathscr{B}_D\}$.

2.4. Structure Systems

Let $E = \{e_1, e_2, \ldots, e_p\}$ be a nonempty set of identifiers of *elements* involved in a structure system and let e_0 stand for the environment of the system. Let Z_{2N} denote a nonempty set of neutral generative systems.

Each element of the structure system is one of these generative systems; the assignment of a particular system to each element is given in terms of function

$$f_N : E \rightarrow Z_{2N}.$$

Let X_i denote the set of external (observed) variables involved in the generative system associated with element e_i and let X_0 denote the set of variables of the environment. Then a *neutral coupling* $C_{i,j}$ between elements e_i and e_j is defined as the set of variables the two elements share, i.e.,

$$C_{i,j} = X_i \cap X_j \quad \text{for } i \neq j \quad (i,j = 0, 1, \ldots, p),$$

and by convention,

$$C_{i,i} = \varnothing.$$

Let $\mathbf{C}$ denote the matrix $[C_{i,j}]$ of couplings between all pairs of elements as well as between the individual elements and the environment. Variables in couplings of structure systems will be called *coupling variables*. Matrix $\mathbf{C}$, which will be called a *coupling matrix*, must satisfy the following conditions:

$$X_i = \bigcup_j C_{i,j} \quad \text{for each } i,$$

$$X_j = \bigcup_i C_{i,j} \quad \text{for each } j,$$

$$C_{i,i} = \varnothing \quad \text{for each } i.$$

On the basis of the introduced concepts, the *neutral structure* system $\mathscr{S}_{3N}$ can be now defined as a quintuple

$$\mathscr{S}_{3N} = (E; e_0; Z_{2N}; f_N; \mathbf{C}).$$

Directed structure systems differ from their neutral counterparts in two respects: (i) Each element is a directed generative system; (ii) couplings between the elements are directed.

A directed coupling $D_{i,j}$ between elements e_i, e_j is defined by

$$D_{i,j} = Y_i \cap X_j \quad (i,j = 0, 1, \ldots, p),$$

where Y_i, X_j stand for the set of output variables of element e_i and the set of input variables of element e_j, respectively. Matrix $\mathbf{D} = [D_{i,j}]$ of directed couplings must satisfy the following conditions

$$X_j = \bigcup_i D_{i,j} \quad \text{for each } j,$$

$$Y_i = \bigcup_j D_{i,j} \quad \text{for each } i.$$

The directed structure system $\mathscr{S}_{3D}$ is defined as quintuple

$$\mathscr{S}_{3D} = (E; e_0; Z_{2D}; f_D; \mathbf{D}),$$

where Z_{2D} is a set of directed systems defined at level 2, f_D is a mapping $E \rightarrow Z_{2D}$, and $\mathbf{D}$ is a matrix of directed couplings.

A convenient form of representing elements and couplings among them is a *block diagram*. Elements (including the environment) are represented in the block diagram by blocks (rectangulars); couplings are indicated by connections between the blocks, each labeled by a symbol of one variable. If the entries of each individual block include arrows pointing toward or away from the block, the block diagram represents a directed structure system; the arrows indicate the source of control of each variable. If no arrows are attached to the entries, the block diagram represents a neutral system. There are a number of alternative ways of representing structure systems; some of them will be described later.

3. SYSTEMS MODELING

3.1. General Discussion

At each epistemological level, systems are defined by certain specific types of systems traits which uniquely identify a category of systems; they are referred to as *primary traits* of the individual categories of systems. It is a general property of the epistemological hierarchy that the set of primary traits characterizing a particular level is a subset of the set of primary traits associated with any higher level. At each level, the primary traits are exactly those through which systems are defined at that level.

Besides the primary traits, a system may be supplemented by some additional traits which are not identificatory for that system. For instance, a generative system may be supplemented by a structure system which conforms to it. Since there is a set of different structure systems which all conform to the same generative system, the latter is invariant with respect to any change in the structure system within this set. The additional traits which are associated with a system but are not a part of its definition are called *secondary traits* of the system.

The primary traits of a system are required to be completely known and invariant with respect to the chosen parameter set in order to identify the system. No such requirements are imposed upon the secondary traits. They may be completely unknown or known only partially, and are not required to be parameter invariant.

If a primary trait of a system changes, then, by definition, the system changes too. On the other hand, if a secondary trait changes, the identity of the system is not affected. For example, while a source system is not affected by the set of available data regarding its variables, a particular data system changes when some data are excluded from it or added to it.

In the process of solving problems associated with transitions from one epistemological level to higher levels, it is often useful to redefine the system in the sense that some secondary traits are accepted as primary traits. In case of systems modeling, such changes in the definitions of systems correspond to inductive inferences.

In *pure empirical investigation,* the system is defined at level 0 as a source system; all other traits such as data regarding variables of the source system, behaviors representing the data, and structures representing the behaviors are viewed as secondary traits. When the investigator becomes confident that the collected set of data is sufficient to fully characterize the variables of the source system, he may accept the data set as a primary trait. This means that he redefines the source system, making it now a data system. Such change represents an *inductive step* because it is based on the assumption that the available set of data contains all information regarding the investigated variables. The investigator may decide to make this inductive step on the basis of both the data and some other aspects such as a comparison with similar investigations performed previously, relation of the data to the existing body of knowledge, investigator's intuition, and the like; these latter aspects are usually referred to as *extra evidential considerations.*

When the source system is redefined to the data system, the investigation becomes theoretical. Its objective is to find a suitable behavior to represent the data system. The behavior is a secondary trait of the data system at this level of investigation. There are different behaviors conforming to the data and each of them generates not only the data array involved but other data arrays as well. Whether the best behavior (for whatever adopted criterion of "goodness") with respect to the data system is also the best one for describing all data the source system may produce depends on how well justified the inductive step was. The transition from the data system to a justifiable generative system is called the *generative system identification.*

When the investigator becomes confident that he determined the right behavior, he may redefine the data system by including the behavior as a primary trait. This again involves an inductive step because the given data set is extended to all sets of data which conform to the behavior. Whether these are sets of data producible by the source system depends on the justification of both the previous inductive steps.

The system is now defined at level 2 (generative system) and further investigation is oriented to the *structure system identification*. Once the investigator accepts a particular structure system, he may then define the system by this structure (another inductive step) and focus, if desirable, on the identification of a meaningful systems at higher levels.

3.2. Experimental Investigation

Given an object, purpose of its investigation, and some constraints imposed upon the investigation, systems modeling basically attempts to determine, subject to the given constraints, such properties of the object which are most relevant to the purpose of investigation.

The *object of investigation* is loosely defined as a part of the world which can be identified as a single entry for an appreciable period of time and which is desirable for a particular investigation. The *purpose of investigation* can be viewed as a set of questions regarding the object which the investigator (or his client) wants to answer. *Constraints* associated with an investigation consist of financial and time limitations, available measuring instruments, limited manpower, and various other restrictions imposed upon the investigators.

Objects can almost never be studied in all of their complexities. The first task of the investigator is, therefore, to define a source system on the object. No formal procedure is possible here. It is likely that the better the investigator understands the questions which motivate the investigation, the more meaningful source system will be defined on the object.

Once a source system is defined, the next step in experimental investigation is *data gathering*. Its objective is to collect information about the actual states of the basic variables within the parameter set. The result of data gathering is a data system.

The procedure of experimental investigation is schematically summarized in the first part of the flow diagram in Fig. 1. It is followed by theoretical investigations at various epistemological levels, as discussed in the subsequent sections.

3.3. Generative System Identification

Once a data system is available, further investigation becomes theoretical in nature. It basically involves appropriate processing of the data through which attempts are made to model the given data system by

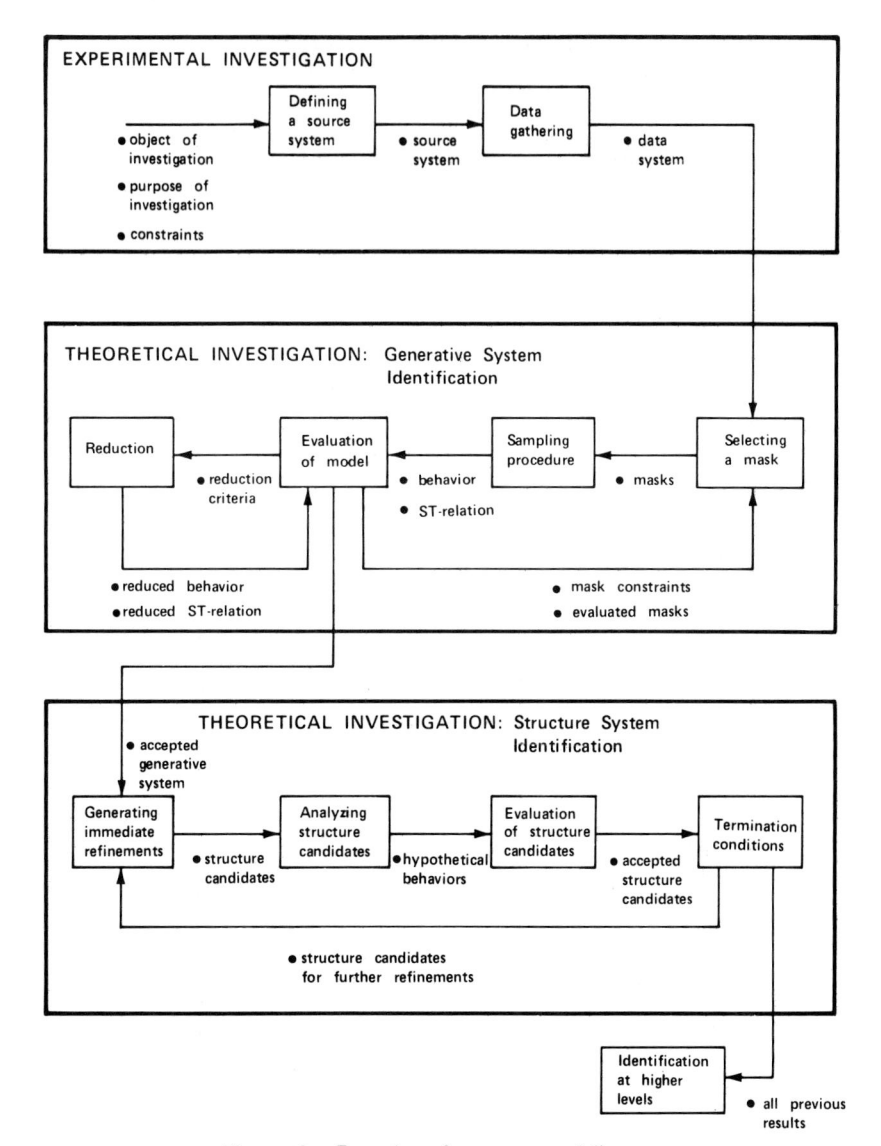

Figure 1. Procedures for systems modeling.

suitable systems at higher epistemological levels. The major objectives of the modeling consist of: (a) representing the data in a parsimonious fashion (*descriptive characteristics* of the model); (b) identifying useful structural properties of the data such as reconstructability of the overall data set from data sets based on subsets of the variables (*explanatory characteristics* of the model); (c) extending the data beyond the limits of the parameter set chosen in the experimental investigation (*predictive characteristics* of the model).

Theoretical investigations start at level 2 of the epistemological hierarchy, where the identification of suitable generative systems is involved. Normally, the objective of this identification is to determine such generative systems (not necessarily only a single system) which best satisfy a compromise between two conflicting criteria, *complexity* and *approximation*, with respect to the given data, both defined in some operational way by the modeler; preference is obviously given to simple generative systems and to those with high degree of approximation (Gaines, 1977).

A computer-aided procedure for the identification of generative systems, which is described in detail in a previous paper (Klir, 1975), is summarized in the second part of the simplified flow chart in Fig. 1. The procedure consists of the following basic steps:

(a) A mask is selected as a paradigm for representing the data.

(b) The data are sampled within the whole parameter set (in an appropriate order) on the basis of the chosen mask. The total numbers of the individual samples are determined and used for the calculation of probabilities representing the basic behavior for the mask (see Section 2.3); other forms of the generative system can easily be derived from the basic behavior when desirable (Klir, 1975; Broekstra, 1976a).

(c) The generative system obtained by the sampling procedure is evaluated by the investigator. If it is satisfactory, the investigation at level 2 terminates and, if desirable, the structure system identification may be initiated. If the results are too complex to provide the modeler with any insight, they can be reduced in many different ways. This may provide the modeler with a spectrum of simplified generative systems, all based on the same data and mask; the modeler can choose reduction criteria from a set of available options.

The whole procedure can be repeated for different masks and the generative systems obtained for the individual masks compared by some subjective or objective criteria. In case of objective criteria, the procedure can be repeated automatically until an optimal mask is reached within given constraints.

The entropy function is used in the procedure for generative system identification as a reasonable measure of *generative uncertainty* through which individual masks can be objectively evaluated and compared with respect to the given data (Klir, 1975). For neutral systems, entropy H_M for mask M is given by the formula

$$H_M = - \sum_{\bar{s}_g} p(\bar{s}_g) \sum_{s_g} p(s_g|\bar{s}_g) \cdot \log_2 p(s_g|\bar{s}_g),$$

where the symbols have the same meaning as defined in Section 2.3. A slightly modified formula must be used for directed systems (Klir, 1975). Normalized *reduction of uncertainty* A_M in the generative process, due to mask M, can be then evaluated by the formula

$$A_M = (H_{max} - H_M)/H_{max},$$

where H_{max} denotes the largest possible entropy for the mask and the source system under consideration. A *quality* Q_M *of mask* M is viewed as the reduction of generative uncertainty per sampling variable of the mask, i.e.,

$$Q_M = A_M/n,$$

where n stands for the number of sampling variables in mask M.

A software package has been developed in which these measures are employed to evaluate and compare submasks of a given maximal acceptable mask. All admissible submasks and their behaviors and/or ST relations can be determined automatically by the procedure; mask M is called admissible iff any mask $M' \neq M$ whose behavior gives a better approximation in accounting for the data contains more sampling variables. Alternatively, masks with the highest quality Q_M can be determined, if this is desirable.

3.4. Structure System Identification

Once a generative system has been accepted, the next objective of systems modeling focuses on the structure system identification. There are basically two alternatives of this problem:

(a) The problem of identifying meaningful decompositions of the generative system in which new coupling variables are introduced.

(b) The problem of identifying structure systems whose elements represent projections of the overall behavior of the given generative system, each associated with a subset of variables of the generative system, and through which the overall behavior can be reconstructed with an acceptable degree of approximation.

In ecosystem modeling, problem (a) has extensively been studied by Overton (1972, 1975, 1977; Overton and White, 1978). A method for solving problem (b) was proposed by Klir (1976) and further developed by Klir and

Uyttenhove (1976a, b, 1977). This method, which has been fully implemented on a computer, is summarized in this section. It is assumed here that the systems under consideration are neutral; modifications to directed systems are rather trivial. An alternative approach to problem (b), based on some concepts of information theory, has been developed by Broekstra (1976b, 1978a, b).

Let V denote the set of variables involved in the given data system and the generative system derived from it. A structure system involved in problem (b) can uniquely be defined by a family of subsets of V and the following conventions: (a) Each subset in the family, say subset E_a, identifies an element of the structure system; (b) the behavior of each element E_a is a projection of the overall behavior of the given generative system into the subspace associated with the submask based only on variables in set E_a; (c) couplings between pairs of elements are defined in terms of intersections between the corresponding subsets in the family.

Let C denote a family of subsets of V which, using the described conventions, uniquely defines a structure system. Let $R_C \subset V \times V$ denote a binary relation such that $(x, y) \in R_C$ iff $(x, y) \in E_a$ for at least one $E_a \in C$; let $R_a \subset V \times V$ denote a binary relation such that $(x, y) \in R_a$ iff $(x, y) \in E_a$, and let Tr(R) denote the transitive closure of a binary relation R. To exclude families of subsets of V which are not meaningful in problem (b) or are, in some sense, redundant, the following conditions are defined:

(α) $\bigcup_{E_a \in C} E_a = V$;

(β) There is no $E_a \in C$ such that $R_C \subseteq \text{Tr}(R_C - R_a)$.

Condition (α) is justified by the fact that the structure system is supposed to approximate the whole data system but nothing else.

Condition (β) guarantees that no relation among the same subset of sampling variables appears in the same structure system in two different forms: (a) as an experimental relation obtained directly from the data system (a relation attached to a particular element); (b) as a theoretical relation obtained through a hypothetical sequence of joins of the experimental relations based on the pattern of couplings in the structure system. This means that, when condition (β) is satisfied, every element of the structure system contains some unique empirical information which is neither included in any of the other elements nor derived through their couplings. Furthermore, condition (β) excludes duplicates of the elements, as well as elements identified by subsets of variables of other elements in the same structure system; such elements do not enrich the structure system by any additional information and are, therefore, of no value to the structure system as far as the structure identification is concerned.

Structure systems which satisfy conditions (α) and (β), as well as the conventions described previously, will be called *structure candidates* of the given generative system. Conditions (α) and (β) will be referred to as *axioms of structure candidates*.

Let C_i and C_j denote two structure candidates based on the same set of variables. Then, we say that C_i is a *refinement* of C_j and write $C_i \leqslant C_j$ iff for each $A \in C_i$ there exists $B \in C_j$ such that $A \subseteq B$. Each set of all structure candidates for a specific set of variables is partially ordered by this relation $\leqslant$. Moreover, the structure candidates together with the relation form a lattice.

Let $\mathscr{L}_n$ denote the *lattice of structure candidates* based on set $V = \{v_1, v_2, \ldots, v_n\}$ of n variables. Then the greatest candidate in $\mathscr{L}_n$, say C_0, consists of one element identified by all variables in V. It is the least refined structure candidate which can be viewed as an interface between the generative system identification (Section 3.3) and the structure system identification. The least candidate in $\mathscr{L}_n$, say C_l, contains n elements, each identified by only one of the variables.

If $C_i \leqslant C_j$ and there does not exist any C_k such that $C_i \leqslant C_k \leqslant C_j$, then C_i is called an *immediate refinement* of C_j. An algorithm for generating all immediate refinements of a given structure candidate was developed by Klir and Uyttenhove (1976a). Although the algorithm is computationally quite simple, its proof, also presented by Klir and Uyttenhove (1976b), is rather complicated. The algorithm enabled the authors to design a computer-aided procedure for structure system identification. The procedure consists of four major steps (see the third part in Fig. 1):

(1) All immediate refinements of a given structure candidate in the appropriate lattice $\mathscr{L}_n$ are generated (Klir and Uyttenhove, 1976); either C_0 or a structure candidate specified by the user is taken as the initial structure candidate.

(2) Each structure candidate generated in Step 1 is analyzed by joining relations associated with its elements according to the pattern of couplings among the elements.[1] Each join of two probabilistic relations can conveniently be performed by multiplying matrices describing the relations (Section 2.3). To preserve states of coupling variables in each matrix product, the first matrix in each pair of multiplied matrices must be made redundant in the sense that states of the involved coupling variables are

---

[1] A join $R * Q$ of two binary relations $R \subset A \times B$ and $Q \subset B \times C$ is a ternary relation defined as $R * Q = \{(a, b, c) | (a, b) \in R$ and $(b, c) \in Q\}$. If probabilities $p(a, b)$ and $p(b, c)$ are associated with elements (a, b) and (b, c) of relations R and Q, respectively, then probabilities $p(a, b, c)$ of elements (a, b, c) of the join $R * Q$ are calculated by the formula $p(a, b, c) = p(a, b) \cdot p(c|b)$.

attached to both rows and columns of the matrix. More detailed description of this way of analyzing structure candidates is included in a previous paper (Klir, 1976).

(3) The relation resulting from the analysis performed in Step 2 is compared with the relation obtained directly from the data, i.e., with the relation associated with candidate C_0. The comparison is based on a *distance* d_k between basic behaviors of candidates C_k and C_0, defined by

$$d_k = \sum_{\mathbf{s}} |p_k(\mathbf{s}) - p_0(\mathbf{s})|, \tag{1}$$

where $p_k(\mathbf{s})$ denotes the probability of sample $\mathbf{s}$ obtained through structure candidate C_k, and $p_0(\mathbf{s})$ stands for the probability of sample $\mathbf{s}$ determined directly from the data system and represented by candidate C_0. Clearly, $0 \leqslant d_k \leqslant 2$ and $d_k = 0$ represents perfect structure candidates. Several other measures for comparing structure candidates have been suggested (Klir, 1976; Klir and Uyttenhove, 1976b; Broekstra, 1978a, b; Cavallo, 1978). While there is no reason to describe these measures here, it is important to emphasize that the actual choice of a measure for a candidate cannot be made outside of the context of the purposes of investigation.

(4) If the smallest distance determined in Step 3 is larger than acceptable and/or the minimal distance increases from one iteration to the next more than acceptable, then the procedure terminates and may be followed up by identification procedures at higher epistemological levels. Otherwise, the procedure is repeated for all structure candidates with the smallest distance and, possibly, those with the distance sufficiently close to the smallest distance.

Further refinement of only structure candidates with the smallest distance or those whose distance is sufficiently close to the smallest one (Step 4) is justified by the following considerations.

When $C_i \leqslant C_j$ and $C_i \neq C_j$, then some experimental relations which conform in C_j exactly to the corresponding relations in C_0 are represented in C_i partially on the experimental basis and partially on the basis of some postulated properties; this does not hold the other way around: All empirical evidence included in C_i is also included in C_j. Hence, if C_i as a whole conforms perfectly to C_0, the conformation of C_j is perfect too. That is to say, if $d_i = 0$, then $d_j = 0$. This means that if the aim of structure identification is to determine the most refined perfect structure system, it is sufficient to consider at each level of refinement only candidates whose distance is zero. If it is not required that the final structure system be perfect, it is likely, but not guaranteed, that $d_i \geqslant d_j$. Extensive experimentation based on computer simulation with full information was performed to determine how well justified the criterion of minimal distance is and to

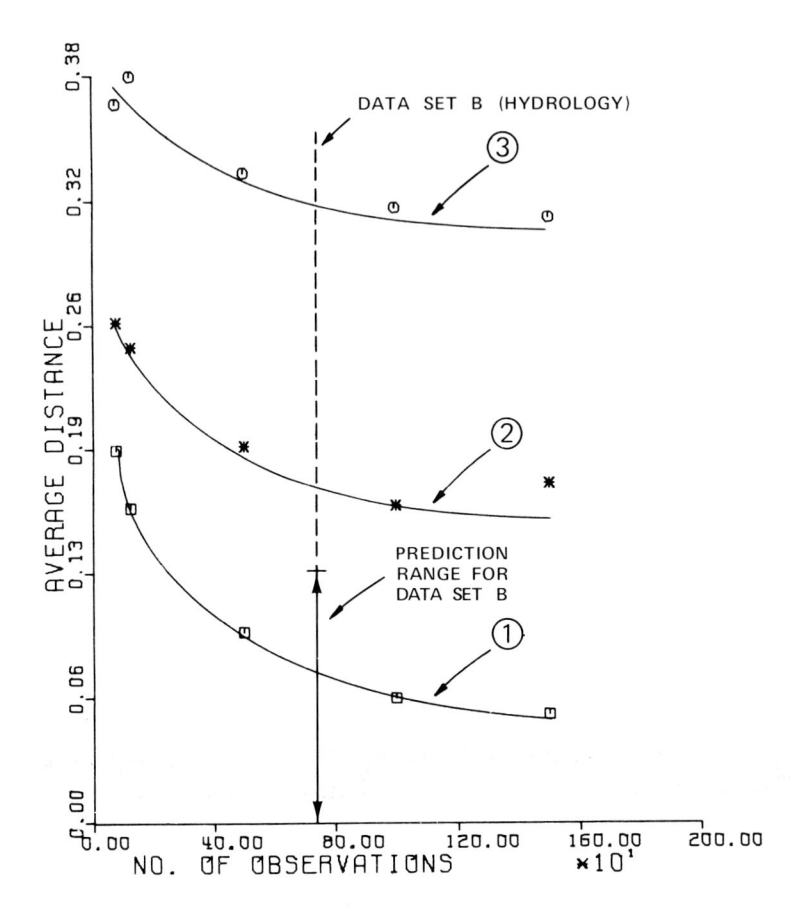

Figure 2. A summary of simulation experiments for the structure system identification.

provide the modeler with some guidelines for proper interpretation of values of the distance, increment in the distance between iterations, distance discrimination among a set of evaluated structure candidates at the same level of refinement, and the dependence of the distance on the size of available data (Klir and Uyttenhove, 1976b, 1977).

Several hundred simulation experiments have been performed in which data sequences containing 1500 observations were generated by a special APL program on the basis of randomly selected structure systems and the structure system identification procedure was then applied to these data sequences as well as their segments of 75, 125, 250, 500, and 1000 of initial observations. For example, the plots in Fig. 2 summarize certain results of the experiments performed for systems with five variables. Plot 1 shows the dependence of the average distance of the correct structure candidates (at the correct level of refinement) on the number of observations in the analyzed data. Plot 2 represents the average distance of the structure candidate with the smallest distance, except for the distance of the correct candidate at the same (correct) level of refinement. Plot 3 shows the average distance of all the structure candidates evaluated at the correct level of refinement except for the correct one.

The experiments, which are fully reported in previous papers (Klir and Uyttenhove, 1976b, 1977) showed that a correctly identified structure system has a strong ability to recover samples which are not included in the available data, although the variables can assume them. For instance, in the experiments for 5 variables, all samples except 10 out of a total of 255 missing samples in the analyzed data were recovered through the identified structure systems. This indicates a powerful mechanism which deserves further study.

4. EXAMPLES OF SYSTEMS MODELING IN ECOLOGY

Procedures involved in systems modeling, as described in Section 3, has been extensively tested on artificial data generated through a simulation program. Their use in real-world applications has thus far been modest and exploratory. The explored areas of applications include social sciences and computer performance evaluation (Klir and Uyttenhove, 1977; Cavallo, 1978).

Two examples which apply the procedures to ecological systems modeling are discussed in this section. They are based on data sets which were provided for this purpose by Efraim Halfon (Canada Centre for Inland Waters, Burlington, Ontario) and W. Scott Overton (Dept. of Forest Science, Oregon State University, Corvallis, Oregon); these data sets are referred to as data sets A and B, respectively.

4.1. Lake Ontario

Data set A consists of measurements taken during a 1-yr period (April 10, 1972 through March 17, 1973) at 32 different locations which encompassed almost all of the area of Lake Ontario. The average of all the locations was calculated over the top 20 m of the lake for each of the following variables.[2]

---

[2] Details regarding the data gathering are included in "Proceedings of the 55th Annual Meeting of the American Geophysical Union: International Field Year for the Great Lakes, April 1974"; Department of Commerce, Rockville, Md., 169 pp.

Variable 1: Temperature (in °C);
Variable 2: Soluble reactive phosphorus (SRP, in µg/liter);
Variable 3: Soluble ammonia (NH_3, in µg/liter);
Variable 4: Total filtered nitrite and nitrate (NO_2/NO_3, in µg/liter);
Variable 5: Chlorophyll a (in µg/liter);
Variable 6: Zooplankton biomass (in µg/liter);
Variable 7: Solar radiation (langleys/day).

To obtain a model of the lake behavior, a five-state resolution level was defined for each variable (Table Ia). The procedures for systems modeling, as described in Sections 3.3 and 3.4, were then applied to test whether it was possible to derive, solely from information contained in the data, the structure of an ecologically realistic model.

4.1.1. Identification of a Four-Variable Model

The investigation was initiated for variables 1, 2, 5, 7. This led to the data matrix in Table Ib and an overall memoryless behavior containing 11 aggregate states of these variables. When analyzed by the structure identification procedure (Section 3.4), the only acceptable structure candidates were found at the first level of refinement. To see the whole situation, all structure candidates generated at levels 1 and 2 are listed in Table II. They are specified in terms of families of subsets of the variables; individual subsets are separated by a slash. For instance, 127/157 specifies a structure candidate whose block diagram is shown in Fig. 3a (candidate C_1); it consists of two elements associated with variables 1, 2, 7 and 1, 5, 7, respectively.

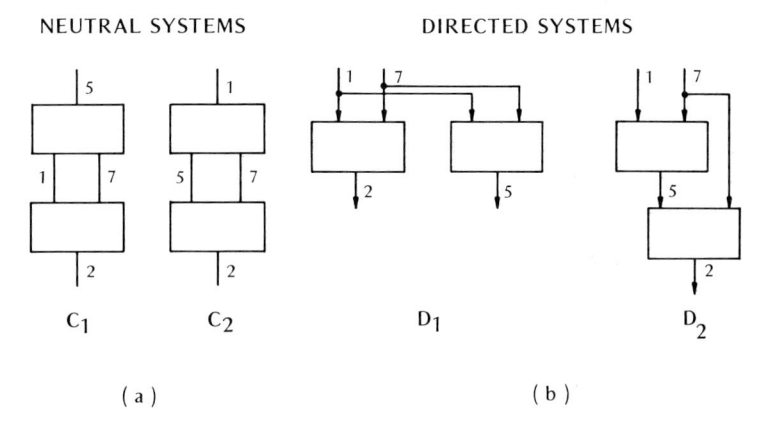

Figure 3. Lake Ontario: Identified structure candidates for four variable data. 1, Temperature; 2, phosphorus; 5, chlorophyll; 7, light.

Table I Resolution Levels for Data A (Lake Ontario)[a]

(a) Variables and Specifications of Their States

Variables	\multicolumn{5}{c}{State identifiers}
	1	2	3	4	5
1. Temperature (°C)	2–4.9	5–7.9	8–11.9	12–14.9	15–18
2. SRP (µg/liter)	0–2.9	3–5.9	6–8.9	9–11.9	12–15
3. NH_3 (µg/liter)	2–5.9	6–9.9	10–13.9	14–17.9	18–20
4. NO_3/NO_2–N (µg/liter)	40–89.9	90–139.9	140–189.9	190–229.9	230–260
5. Chlorophyll (µg/liter)	0–2.9	3–3.9	4–4.9	5–5.9	6–8
6. Zooplankton (µg/liter)	0–14.9	15–50.9	51–119.9	120–184.9	185–250
7. Light (langleys/day)	70–139.9	140–179.9	180–219.9	220–259.9	260–300

(b) Data Matrix Based on the Chosen Resolution Levels for the Variables[b]

Variables	\multicolumn{9}{c}{1972}	\multicolumn{3}{c}{1973}
Month:	A	M	J	J	A*	S	O	N	D*	J	F*	M
1. Temperature	1	1	2	4	5	5	4	3	3	2	2	1
2. SRP	5	4	2	1	1	1	2	3	4	5	5	5
3. NH_3	2	1	2	5	4	4	4	2	2	2	2	2
4. NO_3/NO_2	5	5	3	1	1	1	3	4	4	5	5	5
5. Chlorophyll	2	2	4	5	5	4	2	1	1	1	1	1
6. Zooplankton	1	1	2	5	5	4	3	3	3	2	2	2
7. Light	1	2	2	2	3	4	5	5	3	1	1	1

[a] Measurements for each variable have been classified into five states.
[b] Months marked with asterisks indicate interpolated values; light values are classified up to the saturation light level of 300 langleys/day.

Two structure candidates, labeled as C_1 and C_2 in Table II, are clearly the only acceptable structure systems in this case. While the distance [Eq. (1)] of both C_1 and C_2 is zero, all of the remaining structure candidates at the same level of refinement have quite large distances. Moreover, all structure candidates generated at the second level of refinement and higher levels have prohibitively large distances.

Block diagrams of structure candidates C_1 and C_2 are shown in Fig. 3a. They are neutral structure systems. From the ecological point of view, however, variables 1 (temperature) and 7 (solar radiation) are considered as input variables. When this consideration is incorporated into C_1 and C_2, directed structure systems D_1 and D_2 are obtained, respectively (Fig. 3b).

Although structure system D_1 happens to be in perfect agreement with

Table II Lake Ontario: Structure System Identification for 4 Variables[a]

Refinement level	Generated structure candidates	Distance	Acceptable structure candidates
1	125/127	0.1667	
	125/157	0.3333	
	127/157	0	C_1
	125/257	0.1667	
	125/257	0.1667	
	157/257	0	C_2
2	12/157	0.7778	
	27/157	0.6667	
	127/15	0.6667	
	127/57	0.8333	
	17/257	0.8333	
	15/257	0.9553	
	157/25	0.9667	

[a] 1, Temperature; 2, phosphorus; 5, chlorophyll; 7, solar radiation.

the data, it is not ecologically meaningful. It is based on the assumption that SRP can be determined from temperature and light (solar radiation), without considering the influence of the algal behavior. Structure system D_2 takes this influence into account. Temperature and light effect SRP through the algal biomass (chlorophyll a). In addition, SRP is regulated through algal growth. In the block diagram, this is indicated by the direct influence that light has on SRP. Given any combination of light and temperature, D_2 can be used for determining chlorophyll and SRP.

4.1.2. Identification of a Five-Variable Model

Data regarding variables 1, 2, 5, 6, 7 were analyzed for a one-column mask (memoryless behaviors). The structure system identification procedure identified seven perfect structure candidates (with zero distance) at the first level of refinement, seven perfect candidates at the second level, and only one perfect candidate at the third level. After the third level, all distances become prohibitively large (0.6667 or larger). A block diagram of the most refined perfect structure candidate (the only one at level 3) is shown in Fig. 4a. Viewing variables 1 (temperature) and 7 (solar radiation) as input variables, we obtain the directed system (D_3) specified in Fig. 4b.

Structure system D_3 incorporates structure system D_2 as a subsystem. It is clear from D_3 that zooplankton (the variable by which D_2 is extended into D_3) depends directly on the algae, their rate of growth (through the

NEUTRAL SYSTEM   DIRECTED SYSTEM

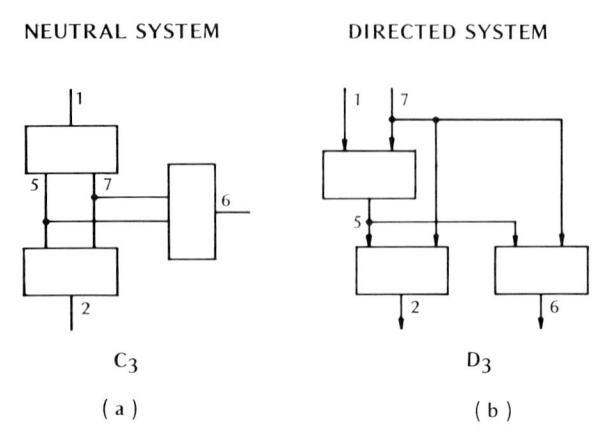

C_3 D_3

(a)   (b)

Figure 4. Lake Ontario: Identified structure candidate for five variable data (1, temperature; 2, phosphorus; 5, chlorophyll; 6, zooplankton; 7, light).

direct light dependence), and indirectly on temperature and light which drive the algae behavior. It is interesting to note here that, given the additional information regarding zooplankton, the structure identification procedure identifies a structure system which is an extension of D_2 while no extension of D_1 is identified. This is exactly what was expected on the basis of previous ecological knowledge.

4.1.3. Identification of a Seven-Variable Model

Given D_3, two additional nitrogen variables were added, NH_3 and NO_3, NO_2 (variables 3 and 4, respectively). The expected result is the directed system (D_4) shown in Fig. 5b. The two nutrients are related to algae in a manner similar to the way SRP is related to it. No transformation of NH_3–N into NO_3–N was expected because information about the organisms which effect this transformation (that is, bacteria) was not included in the data. However, the nutrients are indirectly related to each other through the algae.

The structure identification procedure derived several structure candidates with zero distance at refinement levels 1 through 6. At level 7, structure candidate C_4 (Fig. 5a) was identified as the only candidate with zero distance. When temperature and light are viewed as input variables, directed structure system D_4 (Fig. 5b) is obtained from C_4. This system again conforms to ecological knowledge and contains D_3 (Fig. 4b) as a subsystem. Distances of all structure candidates identified at levels 8 and higher are prohibitively large. Hence, D_4 is the only acceptable directed structure system at the highest level of refinement.

NEUTRAL SYSTEM

(a)

DIRECTED SYSTEM

(b)

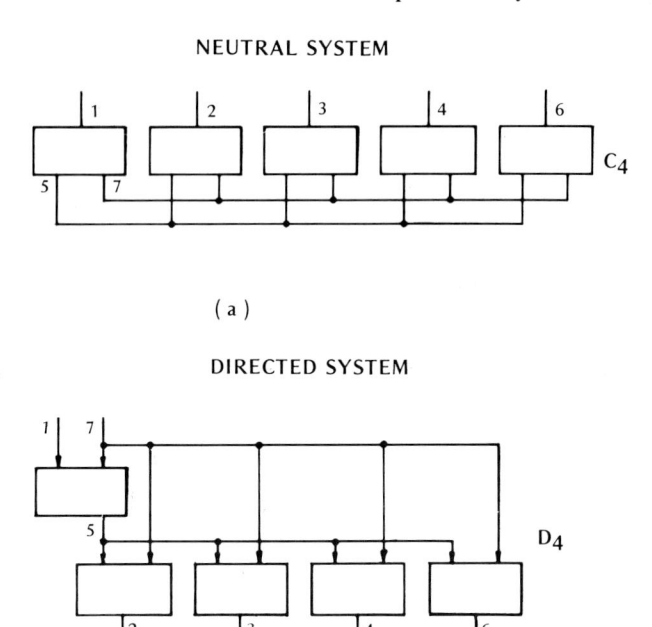

Figure 5. Lake Ontario: Identified structure candidate for seven variable data (1, temperature; 2, phosphorus; 3, ammonia; 4, nitrite and nitrate; 5, chlorophyll; 6, zooplankton; 7, light).

It is important to realize at this point that the structure identification procedure is assumption-free. As such, it does not give any semantic meaning to the variables. The interpretation of the identified structure systems is done after they are identified. The fact that in all of the investigated cases the identified structure system conforms to the one recognized by ecologists indicates a rather profound principle: Information about the structure system is implicitly included in data and can be utilized using an appropriate structure identification procedure for developing models in areas where knowledge regarding the structure is not available.

4.2. Andrews Experimental Forest

Data set B consists of 2196 measurements taken daily in a watershed of the Andrews Experimental Forest in Oregon over a period of 2 yr (exactly 732 days) for the following hydrology oriented variables (Franklin et al., 1972):

318   George J. Klir

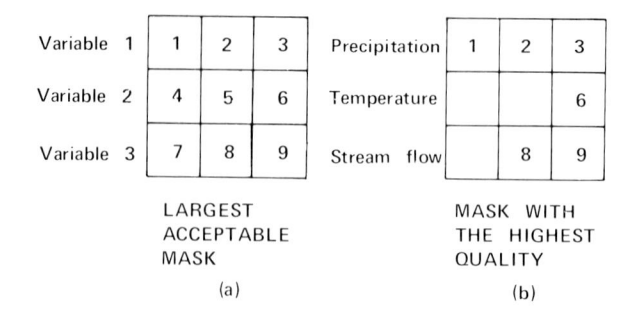

Figure 6. Andrews Experimental Forest: Selection of mask for data set B.

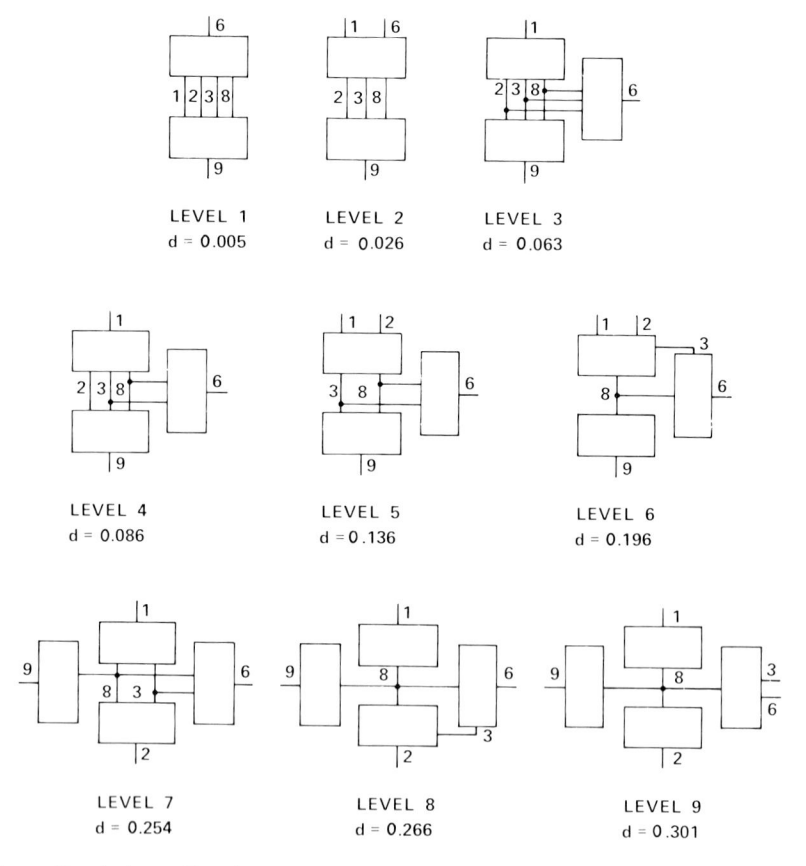

Figure 7. Andrews Experimental Forest: Sequence of structure candidates identified for data set B.

Variable 1: Daily precipitation (in m³/ha);
Variable 2: Daily average air temperature (in °F);
Variable 3: Daily observed stream flow (in m³/ha).

Various modeling studies involving the data have previously been undertaken by Overton (1972, 1975) and Overton and White (1978).

The supporting variable in data set B is obviously time and the number of observations (732) is sufficiently large to provide a good discrimination between structure candidates according to the characteristics in Fig. 2.

Resolution levels of the individual variables were basically chosen by taking logarithms of their actual values. This leads to 4 states distinguished for variable 1, 3 states for variable 2, and 5 states for variable 3; these state sets are described by integers 0 through 3, 2, and 4, respectively.

Since time is linearly ordered, it is meaningful to search for a suitable mask in this case. In the particular study presented here, the largest acceptable mask described in Fig. 6a was chosen as a domain for the search. The mask specified in Fig. 6b was determined as the only mask with the best quality. This mask produced a behavior consisting of 174 samples (out of 4800 possible samples). To simplify the structure system identification, only samples with probabilities greater than 0.0028 were considered.

In the structure system identification, each sampling variable was considered as an independent entity. The procedure was applied only up to the ninth level of refinement where the distance became prohibitively large. Block diagrams and distances of the identified structure candidates are listed in Fig. 7. This sequence of structure candidates provides a broad basis for interpretation which is left to experts in hydrology.

Each identified structure candidate generate some additional samples which are not included in the reduced behavior for which the structure system identification was performed. Some of these additional samples are those which were previously excluded (samples with probabilities smaller than 0.0028) but some are new samples, which do not occur in the data. For instance, at level 1, there is only one new sample: 000210 (variables naturally ordered); at level 2 there are four new samples: 000010, 100010, 100110, 200121; at level 3, there are 10 new samples, etc. If the distance of a candidate which generates some additional samples is sufficiently small in terms of the characteristics in Fig. 2, then the additional samples are likely to exist for the investigated variables and should be viewed as predictions following from the structure systems modeling. The confidence of such predictions depends on the distance of the candidate and the data size evaluated in terms of the characteristics in Fig. 2. The confidence, say c, in predictions based on structure candidate C_k can reasonably be defined as

$$c = \begin{cases} \dfrac{d_{max} - d_k}{d_{max}} & \text{when } d_{max} - d_k \geq 0 \\ 0 & \text{when } d_{max} - d_k < 0, \end{cases}$$

where d_{max} stands for the largest acceptable distance and d_k denotes the distance of candidate C_k. It seems appropriate to choose the largest acceptable distance as equal to the distance corresponding to the middle point between plots 1 and 2 in Fig. 2 for each particular number of observations. This leads to a prediction range of distances for each particular data set. The prediction range for data set B is indicated in Fig. 2. From this range and distances specified in Fig. 7, the values of predictive confidence for new samples generated by the identified structure candidates at levels 1 through 4 are 0.96, 0.8, 0.52, 0.34, respectively; they are 0 for level 5 or higher levels.

Details of the examples discussed here, as well as other examples of systems modeling, are available through the General Systems Depository associated with the *International Journal of General Systems* (Klir and Uyttenhove, 1978).

5. CONCLUSIONS

The approach to systems modeling as described in this article is characterized by the following features:

(1) It is based on a symbiotic man–computer interactive mode of operation and involves two methodological tools (sets of computer programs), one for the generative system identification and one for the structure system identification. These methodological tools are components of a larger package of methodological tools incorporated in the previously mentioned general systems problem solver (Cavallo and Klir, 1978).

(2) It focuses on procedures which are applicable to variables of any scale (quantitative as well as qualitative) and to well-defined and fuzzy variables. In conformity with the nature of experimental data, the procedures are developed for discrete variables.

(3) It recognizes the fact that several models, each reflecting certain aspects of the data, may give the investigator much better insight than any of them could afford alone.

(4) Closely related to features (1) and (3) is the important factor that a given systems modeling undertaking, utilizing the full spectrum of epistemological levels embedded in the general systems problem solver, represents an integrated overall investigation. In this way it encompasses all aspects of the epistemological process, from definition of object system and

observation to the integration of goals and purposes of the investigation. This aspect implies important and fundamental advantages over the isolated use of techniques.

The fact that the structure system identification is preceded by the generative system identification reflects one of the basic principles of systems thinking: Do not *a priori* assume that the whole can be reconstructed from its parts. The generative system identification attempts to deal with the whole data system under consideration; the data system is divided into parts only if its investigation involves prohibitive computational complexity. Once an acceptable generative relation is identified for the whole data system, the structure system identification procedure explores the reconstructability properties of this overall relation. As a result of the procedure, those structure systems are identified at the various levels of structure refinement which are best equipped to reconstruct the overall relation from its appropriate projections.

When repeatedly applied in specific application areas, it is hoped that the structure system identification will eventually lead to the discovery of reconstructability patterns in these areas of inquiry. In areas with persistently poor reconstructability at all levels of structure refinement, systems modeling at higher epistemological levels, allowing changes in the generative and structure systems within the parameter set, may lead to more satisfactory results and should definitely be explored (Uyttenhove, 1978).

One of the most appealing features of the structure system identification is its predictive power. Although a simple suggestion of how to employ it is described in Section 4, this fundamental and important aspect of the structure system identification needs a comprehensive study, closely associated with further simulation experiments such as those summarized in Fig. 2.

There are various other aspects of the generative and structure systems identification which deserve further research. One of them is a study of alternative measures through which generative systems and structure systems are evaluated. Another one is a development of a generalized structure systems identification procedure characterized by weakening axiom (β) of structure candidates into $E_a \not\subset E_b$ for any pair E_a, $E_b \in C$ and allowing to combine projection relations in a structure candidate through various sequences of operations applicable on relations, such as composition, join, extension, intersection, union, complement, restriction, etc.

The examples of systems modeling in ecology, which are discussed in Section 4, are only exploratory and are based on data readily available at the time of writing this chapter. The techniques described in this article, as well as other techniques for systems modeling, can be expected to have some impact on ecology only if a close cooperation develops between ecologists and systems researchers. Such cooperation between specialists in various disciplines and systems researchers is, in a much broader sense, the very essence of the general systems problem solver described by Cavallo and Klir (1978).

ACKNOWLEDGMENTS

The author is very grateful to Efraim Halfon and W. Scott Overton for providing him with ecological data for examples discussed in Section 4. He is also grateful to Hugo Uyttenhove for his assistance in processing the data, and to Roger Cavallo for his valuable comments regarding this article. Finally, SUNY–Binghamton should be acknowledged for providing the author with the necessary computer time.

REFERENCES

Broekstra, G. (1976a). Some comments on the application of informational measures to the processing of activity arrays. *Int. J. Gen. Syst.* **3**, 43–51.

Broekstra, G. (1976b). Constraint analysis and structure identification. *In* "Annals of Systems Research" (B. van Rootselaar, ed.), Vol. 5, pp. 67–80. Nijhoff, The Hague.

Broekstra, G. (1978a). Structure modelling: A constraint (information) analytic approach. *In* "Applied General Systems Research: Recent Developments and Trends" (G. J. Klir, ed.), pp. 117–132. Plenum Press, New York.

Broekstra, G. (1978b). On the representation and identification of structure systems. *Int. J. Syst. Sci.* (in press).

Cavallo, R. (1978). "The Role of Systems Methodology in Social Science Research." Nijhoff, The Hague.

Cavallo, R., and Klir, G. J. (1978). A conceptual foundation for systems problem solving. *Int. J. Syst. Sci.* **9**, 219–236.

Franklin, J. F. *et al.*, eds. (1972). "Proceedings of the Symposium on Research on Coniferous Forest Ecosystem." Pac. Northwest For. Range Exp. Stn., For. Serv., USDA, Portland, Oregon (available from Superintendent of Documents, US Govt. Printing Office, Washington, D.C. 20402; Stock No. 0101-0233).

Gaines, B. R. (1977). System identification, approximation and complexity. *Int. J. Gen. Syst.* **3**, 145–174.

Klir, G. J. (1969). "An Approach to General Systems Theory." Van Nostrand-Reinhold, New York.

Klir, G. J. (1975). On the representation of activity arrays. *Int. J. Gen. Syst.* **2**, 149–168.

Klir, G. J. (1976). Identification of generative structures in empirical data. *Int. J. Gen. Syst.* **3**, 89–104.

Klir, G. J. (1978). General systems concepts. *In* "Cybernetics: A Sourcebook" (R. Trappl, ed.), Halsted, Washington, D.C. (in press).

Klir, G. J., and Uyttenhove, H. J. J. (1976a). Procedure for generating hypothetical structures in the structure identification problem. *In* "Progress in Cybernetics and Systems Research," Vol. 3 (R. Trappl, G. J. Klir, L. Ricciardi, eds.), pp. 19–29. Hemisphere, Washington, D.C.

Klir, G. J., and Uyttenhove, H. J. J. (1976b). Computerized methodology for structure modelling. *In* "Annals of Systems Research" (B. van Rootselaar, ed.), Vol. 5, pp. 29–66. Nijhoff, The Hague.

Klir, G. J., and Uyttenhove, H. J. J. (1977). On the problem of computer-aided structure identification: Some experimental observations and resulting guidelines. *Int. J. Man–Mach. Stud.* **9**, pp. 593–628.

Klir, G. J., and Uyttenhove, H. J. J. (1978). "Examples of Systems Modelling in Ecology," Gen. Syst. Depository No. 1978/1 (associated with *International Journal of General Systems*).

Overton, W. S. (1972). Toward a general model structure for forest ecosystem. *In* "Proceedings of the Symposium on Research on Coniferous Forest Ecosystem" (J. F. Franklin *et al.*, eds.), Pac. Northwest For. Range Exp. Stn., For. Serv., USDA, Portland, Oregon.

Overton, W. S. (1975). The ecosystem modeling approach in the Coniferous Forest Biome. *In* "Systems Analysis and Simulation in Ecology" (B. C. Patten, ed.), Vol. 3, pp. 117–138. Academic Press, New York.

Overton, W. S. (1977). A strategy of model construction. *In* "Ecosystems Modeling in Theory and Practice" (C. A. S. Hall and J. W. Day, Jr., eds.), pp. 50–73. Wiley, New York.

Overton, W. S., and White, C. (1978). Evolution of a hydrology model: An exercise in modelling strategy. *Int. J. Gen. Syst.* **4**, 89–104.

Uyttenhove, H. J. J. (1978). Meta-system identification. *In* "Applied General Systems Research: Recent Developments and Trends" (G. J. Klir, ed.), pp. 147–160. Plenum, New York.

Zeigler, B. P. (1974). A conceptual basis for modelling and simulation. *Int. J. Gen. Syst.* **1**, 213–228.

Zeigler, B. P. (1976a). "Theory of Modelling and Simulation." Wiley, New York.

Zeigler, B. P. (1976b). The hierarchy of system specifications and the problem of structural inference. *In* "PSA 1976" (F. Suppe and P. D. Asquith, eds.), pp. 227–239. Philos. Sci. Assoc., East Lansing, Michigan.

Chapter **13**

IDENTIFICATION OF THE MATHEMATICAL MODEL OF A COMPLEX SYSTEM BY THE SELF-ORGANIZATION METHOD

Aleksej G. Ivakhnenko, Georgij I. Krotov, and Vladimir N. Visotsky

1. Introduction	326
2. Present State of the Theory of Computer-Aided Self-Organization of Mathematical Models	326
2.1 The Principal Shortcoming in Model Development	327
3. Computer-Aided Self-Organization of Models	328
3.1 Criteria for Model Selection	329
3.2 Three Algorithms for Model Sifting	332
3.3 Parameter Estimation	335
4. Discovery of Laws with the Aid of GMDH	336
4.1 Prediction of Predictions	337
4.2 Automatic Control with Prediction Optimization	338
5. Application of GMDH to Environmental Problems	339
5.1 One-Dimensional Problem	340
5.2 Multidimensional Problems	343
5.3 Improvement of the Models Using the Combined Criterion	350
6. Conclusions	350
References	352

1. INTRODUCTION

Deductive and inductive approaches can be used to solve a problem. The deductive approach implies that a solution can be found by processing *a priori* information on a given system with experimental data. The inductive approach, as applied to the identification problem, can be realized by the sifting (enumeration) of a great number of equations on a computer to choose the one which is best, given a set of criteria. The requirement for a deep understanding of the processes which take place in a system is reduced if the inductive (or self-organization) approach is used for the mathematical modeling of a system. In fact, data implicitly contain information about the processes. For example, the complex processes of water self-purification are in some way reflected in the experimental data and this information can be used to develop a mathematical model. Thus, the mathematical model of self-purification could be found only by sifting, without complete knowledge about the mechanisms of the complex processes.

The ideas at the base of the inductive approach are explained by means of two different examples: (a) discovery of the Streeter–Phelps' equation, and (b) identification of a water quality model in a reservoir. In the first example, the problem is one-dimensional and the purpose of the identification is to find the number of time delays to be considered, or prehistory time to be taken into account. In the second example, the problem is two-dimensional, because time and space are considered. The problem here is to choose the best elementary structure (also called "pattern") in the finite-difference equation.

The examples considered show how the physical laws, which are hidden in the noisy experimental data, can be discovered.

2. PRESENT STATE OF THE THEORY OF COMPUTER-AIDED SELF-ORGANIZATION OF MATHEMATICAL MODELS

Since 1969, computer-aided methods of self-organization of mathematical models (GMDH, Group Methods of Data Handling) have been investigated at the Section of Combined Control Systems at the Institute of Cybernetics of the Academy of Sciences of the Ukranian SSR, and at the Department of Engineering and Cybernetics of the Kiev Polytechnical Institute. The principal result of these investigations consists, not so much of the examples of computer-designed models presented below, as of a change in views about cybernetics as a science of model construction in general, and of the role of modern applied mathematics.

The deterministic approach nowadays predominates in science. Its principle is "the more complex a model, the more exact it will be." This is based on the analysis of cause–effect relationships. Another commonly held opinion is that in the man–machine dialogue the predominant role is played by the human operator, whereas the computer is left with the passive role of a "large calculator." Indeed, the opposite is true: In a self-organization algorithm, the role of the human operator is passive; he is no longer required to have a profound knowledge of the system. In fact, he merely gives orders and need possess only a minimal amount of *a priori* information. He must: (1) convey to the computer a criterion of model selection that is very general (e.g., "the prediction must be very accurate," etc.); (2) specify the list of feasible reference functions, such as polynomials or rational fractions, harmonic series, and so on; (3) specify the simulation "environment," that is, a list of possible variables. The most effective set (composition) of variables and the appropriate reference function are selected by the computer itself.

The objective character of the models obtained by self-organization is very important for the resolution of many scientific controversies (Gabor, 1971). The man–machine dialogue is raised to the level of a highly abstract language. Man communicates with the machine, not in the difficult language of details, but in a generalized language of integrated signals (the selection criteria). Self-organization restores the belief that a "cybernetic paradise" on earth is just around the corner, governed by a symbiosis between man (the giver of instructions) and the machine (an intelligent executer of the instructions). The self-organization of models can be regarded as a specific algorithm of computer artificial intelligence.

2.1. The Principal Shortcoming in Model Development

Here we assume that most of mathematical models of ecosystems are developed on the basis of a small set (table) of data (15–20 points in the table). In this development, the principal shortcoming is the so-called "inappropriate choice of the external complement."

According to the well-known incompleteness theorem of Gödel (von Neumann, 1966), it is in principle impossible to find a unique model of an object on the basis of empirical data without using an "external complement." An exact definition of this concept can be found in the well-known book (Beer, 1963). The method of regularization of the solutions of incorrectly formulated problems proposed by Tikhonov and Arsenin (1974) is also based on Gödel's theorem. Hence "external complementation" and "regularization" are synonyms expressing the same concept. When there is no external complement, we enter the domain of the theory of multiplicity of models (Shannon, McCulloch, Pitts, Ashby, etc.). As an example we can take regression analysis. The rms error, determined on the basis of all experimental points, monotonically decreases when the model complexity gradually increases; this error drops to zero when the number n of coefficients of the model equation becomes equal to the number of experimental points N. Every equation which possesses n coefficients can be regarded as absolutely accurate model. This, in fact, has generated the above mentioned theory of multiplicity of models, which asserts that, on basis of a finite number of experimental data, it is not possible in principle to find a unique model. In spite of this, various investigators have found a unique model without stating that they used (consciously or unconsciously) an external complement, necessary in principle for obtaining a unique model.

Hence, none of the investigators has appropriately selected the external complement.

3. COMPUTER-AIDED SELF-ORGANIZATION OF MODELS

Mathematical statistics with concepts such as a normal (or any other) distribution cannot be used if we possess only few experimental points and if the object changes its characteristics. In the absence of probability distributions, we are left only with nonparametric methods. The most important of these is the inductive method of sifting (successive testing) of models whose complexity is gradually increased. It is found that under certain conditions this sifting process is not at all unlimited, and is well within the capability of a computer.

The principle of self-organization can be formulated as follows: When the model complexity gradually increases, certain criteria that have the property of external complementation (called selection criteria) pass through a minimum. By sifting, the computer finds this minimum, and thus indicates to the operator the model of optimum complexity (Fig. 1).

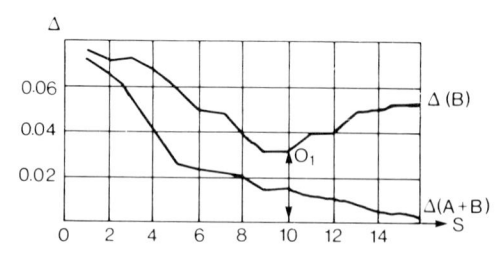

Figure 1. Variation of the rms errors $\Delta(A)$ and $\Delta(A+B)$ for a regression equation of increasing complexity (S); O_1 is the model of optimum complexity.

3.1. Criteria for Model Selection

The global minimum of the selection criterion, reached by sifting all the feasible models, is a measure of model adequacy. If no deep minimum is obtained, then the model has not been found. This occurs in the following cases: (a) The data are too noisy; (b) there are no essential variables among them; (c) the selection criterion is not suitable for the given object of investigation; (d) time delays are not sufficiently taken into account.

In these cases, it is necessary to extend the domain of sifting until we obtain a minimum. We consider both stationary and nonstationary systems. [In the latter case it is often useful to reduce the amount of data used in model development (training set) and use the rest in model verification (testing set).] The allocation of points to the training (N_A) and testing (N_B) sequences ($N = N_A + N_B$) has been examined in various papers, under the heading of "purposeful regularization" (Ivakhnenko, 1975).

As useful external complements in the theory of self-organization we utilize the following criteria:

3.1.1. The Regularity Criterion

This consists of the rms error calculated on the basis of N_B testing points:

$$\Delta(B) = \frac{\sum_{1}^{N_B}(q - q_{\text{table}})^2}{\sum_{1}^{N_B} q_{\text{table}}^2} \cdot 100\% \to \min. \tag{1}$$

where q_{table} is the measured value of the output variable, and q is the simulated value.

3.1.2. The Minimum-of-Bias Criterion

This consists of the rms difference between the outputs of two models developed on the basis of two distinct sets of data A and B:

$$n_{bs} = \frac{\sum_{1}^{N}(q_A - q_B)^2}{\sum_{1}^{N} q_{\text{table}}^2} \cdot 100\% \to \min, \tag{2}$$

where N is the total number of points, q_A is the simulated value of the output variable obtained on the basis of set A, and q_B is the corresponding value based on set B.

Usually the data with the higher values of variance are included into the set A, while those with the smaller variance are put in the set B (when $N_A = N_B$).

When the model is exactly unbiased, then the output variables are equal: $q_A = q_B$ and $n_{bs} = 0$. Therefore, the comparison of the polynomials using the criterion $n_{bs} \to$ min allows us to choose the most unbiased model. The coefficients of all the models upon comparison and selection can be reestimated using the minimum mean square error method applied to the whole data table.

The regularity criterion must be used in the self-organization of prediction models. In the presence of noise, the equations are not a physical model of the system. The optimum prediction model is often "absurd" from the point of view of human logic.

By the minimum-of-bias criterion it is possible to select a model that is not sensitive to the data on which it is based. This criterion requires that the model yield the same results at successive experimental points of A and B. By this criterion it is possible to recover a physical law which is hidden in noisy experimental data. The algorithm for this purpose is described in Section 4.

3.1.3. Combined Criterion for Model Selection

The original idea of selection is incorporated in the method described. For example, when a gardener selects for red roses, he chooses not only the red ones but also those with definite red tints. And we must choose not only the minimum-of-bias models but also those stable models which have the best prediction properties. The combined criterion $k_3 = (n_{bs}^2 + \Delta(C)^2)^{1/2}$ $\to$ min provides the most unbiased stable and accurate predicting models where $\Delta(C)^2$ is the mean square error of prediction calculated using a validating data set C (Section 3.1.1). To calculate the combined criterion, the data table should be divided into three parts A, B, and C (usually in proportion $A = 40\%$, $B = 40\%$, $C = 20\%$ points of table). A and B are used to calculate the minimum-of-bias criterion and part to calculate the prediction error for C. The distribution of the points between A and B is best obtained by minimax optimization of the minimum-of-bias criterion (Visotsky, 1976).

3.1.4. Balance-of-Variables Criterion for Long-Term Predictions

By the regularity criterion it is possible to obtain an exact approximation of a system as well as short-term prediction (for one or two steps ahead) of the processes taking place in it. In the interpolation interval all of the models yield almost the same results (we have the principle of multiplicity of models). In the extrapolation interval the predictions diverge, forming a so-called "fan" of predictions.

13. Identification of the Mathematical Model

The minimum-of-bias criterion yields a narrower fan, and hence a longer prediction time. Prediction is possible for several steps ahead (medium-term prediction). However, the theory of self-organization would not solve the problems to which it is applied unless it yielded examples of exact long-term predictions.

For long-term predictions, a selection criterion has been proposed (the balance-of-variables criterion) which requires simultaneous prediction of several interrelated variables. In many examples, these system variables are constructed artificially (Ivakhnenko and Ivakhnenko, 1974, 1975). With the aid of the minimum-of-bias criterion, the machine finds the laws connecting the correlated system variables. For example, for three variables it is possible to discover the laws:

$$q_1 = f_{1-1}(q_2, q_3), \qquad q_2 = f_{2-2}(q_1, q_3), \qquad q_3 = f_{3-3}(q_1, q_2). \qquad (3)$$

The balance-of-variables criterion requires that these relations between pairs of variables be satisfied not only in the interpolation interval, but also in the extrapolation interval. For this purpose, we construct the differences between so-called "direct" and "inverse" functions. The inverse functions $q_1^* = f_{2-1}(q_2, q_3)$, $q_2^* = f_{3-2}(q_1, q_3)$, and $q_3^* = f_{1-3}(q_1, q_2)$ are computed from the second, third, and first laws (Eq. 3), respectively (Ivakhnenko and Ivankhnenko, 1974, 1975). The balance of variables criterion is then

$$b = \frac{\sum_{1}^{N_{pre}} [f_{1-1}(q_2, q_3) - f_{2-1}(q_2, q_3)]^2}{\sum_{1}^{N_{pre}} [f_{1-1}(q_2, q_3)^2]} \cdot 100\%$$

$$+ \frac{\sum_{1}^{N_{pre}} [f_{2-2}(q_1, q_3) - f_{3-2}(q_1, q_3)]^2}{\sum_{1}^{N_{pre}} [f_{2-2}(q_1, q_3)]^2} \cdot 100\%$$

$$+ \frac{\sum_{1}^{N_{pre}} [f_{3-3}(q_1, q_2) - f_{1-3}(q_1, q_2)]^2}{\sum_{1}^{N_{pre}} [f_{3-3}(q_1, q_2)]^2} \cdot 100\% \to \min, \qquad (4)$$

where N_{pre} is the number of prediction points.

This criterion yields reference points in the future; it requires that a law that was effective up to the present should also continue into the future in the extrapolation interval.

The correctness of the prediction is checked according to the value of the minimum of the balance-of-variables criterion. By gradually increasing the prediction time, we arrive at a prediction time T_{pre} for which it is no' longer possible to find an appropriate trend in the fan of a given reference function. The value of the minimum function begins to increase, which means that an appropriate action must be taken. For example, it may be necessary to change the reference function. For a richer choice of models, it is also recommended to go over from algebraic to finite-difference equations, take other system variables, estimate the coefficients, etc.

3.2. Three Algorithms for Model Sifting

3.2.1. Combinatorial Algorithm GMDH

The coefficients of all degrees of the polynomials are separately equated to zero. Then, we sift the models as long as the selection criterion decreases. In the presence of noise, it is necessary to sift not only the complete polynomials, but also all of the "incomplete" polynomials, obtained by equating to zero certain coefficients of a complete polynomial. For example, the model

$$q = a_0 + a_1 x_1 + a_2 x_1 x_2 \tag{5}$$

may turn out to be more accurate or more unbiased than the total polynomial model

$$q = a_0 + a_1 x_1 + a_2 x_2 + a_3 x_1 x_2 + a_4 x_1^2 + a_5 x_2^2. \tag{6}$$

3.2.2. Multilayer GMDH Threshold Algorithm

This algorithm uses a self-sampling threshold in each selection layer (Fig. 2). The values of the variable of interest q and other data, $x_1, x_2, \ldots, x_n$ are included in the training set. The model to be found is the "complete description," where the output variable is a nonlinear function of all the arguments and their delayed values.

$$q = f(x_{1(0)}, x_{1(-1)}, x_{1(-2)}, \ldots, x_{2(0)}, x_{2(-1)}, x_{2(-2)}, \ldots, x_{n(0)}, x_{n(-1)}, x_{n(-2)}, \ldots), \tag{7}$$

where f is a polynomial of high degree. This "complete description" is found by several layers of approximation. By a renotation of variables the nonlinear polynomial can be substituted by a linear one:

$$u = a_0 + a_1 w_1 + a_2 w_2 + \ldots + a_n w_n, \tag{8}$$

where $w_1, w_2, \ldots, w_n$ are functions of the data and their delayed values.

At the first layer of selection, as it is shown in Fig. 2, the complete description is substituted by some partial descriptions

13. Identification of the Mathematical Model

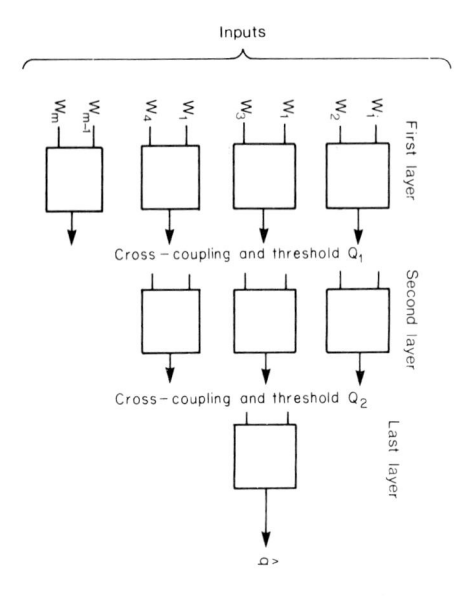

Figure 2. Multilayer threshold structure of GMDH. The output from each layer becomes an input to the next. This process is repeated until the last layer, when the desired output is generated.

$$u_1 = b_0 + b_1 w_1 + b_2 w_2,$$
$$u_1' = b_0' + b_1' w_1 + b_2' w_3,$$
$$u_1'' = b_0'' + b_1'' + b_2'' w_4,$$
$$\ldots$$
$$u_1^{(n)} = b_0^{(n)} + b_1^{(n)} w_{n-1} + b_2^{(n)} w_n. \tag{9}$$

There will be $C_n^2 = (n^2 - n)/2$ such partial descriptions. This is the number of possible combinations of n components by 2. The values of the partial description coefficients can be found by the mean square error method using the training set of data N_A. After this selection the goodness of fit of every description is evaluated and g_1 of them are chosen (usually $g_1 = n$). The error of selected partial descriptions is less than threshold value θ_1 (Fig. 2). *At the second layer of selection* the partial descriptions of the following form are used:

$$v_1 = c_0 + c_1 u_1 + c_2 u_2,$$
$$v_2' = c_0' + c_1' u_1 + c_2' u_3,$$
$$\ldots$$
$$v_2^{(g)} = c_0^{(g)} + c_1^{(g)} u_{g-1} + c_2^{(g)} u_g. \tag{10}$$

The estimation of coefficients (on the training set N_A) and the choice of $g_2 \leq g_1$ best partial descriptions are repeated again.

The number of selection layers increases as long as the lower (or sometimes the average for g_i) value of the criteria is decreasing. Thus the process is continuously repeated with the imposition of ever more rigid thresholds $n \geq g_1 \geq g_2 \geq g_3 \geq \ldots \geq g_p$ so that finally a unique signal is selected as the output on one layer; this signal corresponds to a minimum in the selection criterion.

Selection hypothesis. The above process is the mathematical counterpart of the process used by a gardener in selectively raising various species for the purpose of obtaining a hybrid type that has desired properties. For example, he begins at first with a group of flowers, in which each flower is crossed with another flower of this group; then he collects the obtained seeds and sows them. He studies the new generation of flowers and throws away the types which in his view do not "improve" the sort. This process is equivalent to one layer in the GMDH algorithm. After that, he repeats this process by crossing new plants each time. If the selector uses an appropriate threshold, that is, if he throws away the flowers correctly, he should be able to observe in the obtained generations of flowers an ever-increasing trend toward the presence of more and more of the desired properties. In the end, he obtains a single flower, which is the best that could be obtained by raising several generations of flowers.

Hierarchy of criteria. The principal requirement for a selection criteria is a smooth character of its curve as a function of the complexity of the models. Two criteria (regularity and unbiasedness) considered in Section 3.1 have this property. The smoothness of a criterion makes it possible to let through many models during sifting, and thus to speed up the selection of the model of optimum complexity. There also exist other criteria, with a "sharp" characteristic (e.g., the minimum of bias based on comparison of estimates of the coefficients obtained on the data sequences A and B). It is recommended to use a so-called hierarchy of criteria: The sharp criteria are used together with the smooth criteria. First, we select the region of the models on the basis of a smooth criterion, and then the sharp criterion makes it possible to determine one of them exactly. Very efficient use of a "smooth" and "sharp" selection criterion can be found in Ivakhnenko and Stepashko (1975).

There already exist about 100 algorithms realizing various versions of efficient multilayer selection. The principal aim of these algorithms is neither to omit nor to lose any result that could be obtained by complete sifting. Especially effective are ortogonalized GMDH algorithms (Ivakhnenko, 1975).

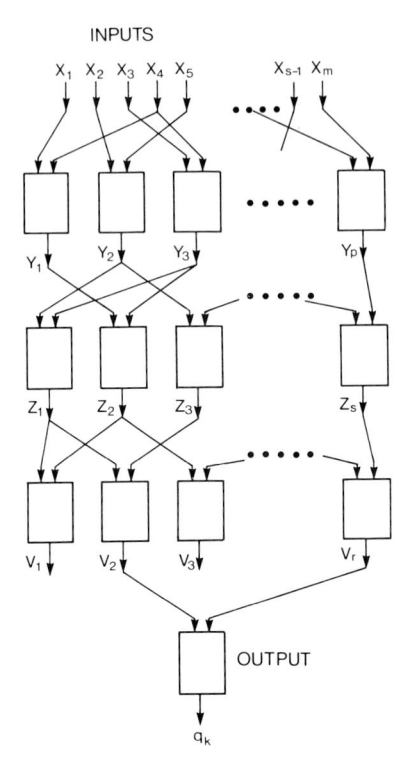

Figure 3. Multilayer GMDH structure (neuron-type net) in which the coefficients are equated to zero (the equation of the elements is $y = a_0 + a_1 x_1 + a_2 x_2 + a_3 x_1 x_2$).

3.2.3. Multilayer GMDH Algorithm

Instead of using a threshold, it is possible to equate to zero the coefficients (as described above for the first algorithm). The obtained structure resembles a neuron net. Such a net was developed by the US firm "Adaptronix" (Barron, 1971; Gilstrap, 1971) for realizing the high-speed computer Hypercomp-80 TM. A net with elements of the form $q = a_0 + a_1 x_1 + a_2 x_2 + a_3 x_1 x_2$ is presented in Fig. 3. Any of the variables obtained in the net that minimize the selection criterion can be taken as the output variable.

3.3. Parameter Estimation

In the simplest form, this stage consists of recalculating (after finding the structure of the models) the coefficients on the basis of all experimental data. After finding the optimal (for a given selection criterion) structure of the model, the estimates of the coefficients are improved to obtain a zero (or almost zero) value of the selection criterion.

Two programs were compiled for this stage, that are included in the Ukranian Republic Library of algorithms and programs: (a) a program of optimization of estimates by random search (compiled by V. K. Svetal'skiy); (b) a program that uses the maximum principle (compiled by V. N. Vysotsky).

4. DISCOVERY OF LAWS WITH THE AID OF GMDH

Identification of system characteristics can be of two kinds.

In the first case, identification involves obtaining an equation that best fits the experimental points. Such identification can be called an approximation in the region in which the experimental data are obtained, that is, the interpolation interval. For solving this problem it is possible to recommend the regularity criterion.

In the second case, it is necessary to filter the noise. On the basis of noisy data we must restore (discover) the law governing the system. For solving this problem it is recommended to use the minimum-of-bias criterion.

For the discovery of laws using the minimum-of-bias criterion, it is not necessary for the human operator to specify the set (ensemble) of independent variables, the input and output variables, the control variables, and the disturbances, etc. All of this is done by the computer.

The GMDH algorithm can be described as follows: Assume that the complete list of variables (which is given with a large margin) contains S variables. In this case, in the first selection layer we estimate by the minimum-of-bias criterion a total of $C_s^2 = (S^2 - S)/2$ models of the form

$$q_1 = f(q_i, q_j), \quad i = 1, 2, \ldots, s; \quad j = 1, 2, \ldots, s. \tag{11}$$

Then we select the g_1 best models.

In the second layer we check models of the form

$$q_2 = f_2(q_i, q_j, q_k), \quad k = 1, 2, \ldots, s, \tag{12}$$

(the number of such models is $C_{g_1}^3$).

In the third layer we estimate models with four independent variables:

$$q_3 = f_3(q_1, q_j, q_k, q_l), \quad l = 1, 2, \ldots, s \tag{13}$$

(the number of such models is $C_{g_1}^4$).

Stopping rule. The selection stops as soon as the unbiasedness criterion begins to increase. By finding a minimum of the selection criterion, we discover a law. The obtained formulas express the laws governing a given system. In this way, we determine the composition of the set of independent variables, the input and output variables, and the structure of the system (Stone et al., 1977).

4.1. Prediction of Predictions

Is it possible to obtain "Cassandra" prediction? Cassandra, the daughter of King Priam of Troy, predicted the downfall of Troy while the city was still winning over the Hellenes. As it is well known, Troy indeed fell and Cassandra's prediction came true.

If we assert that "combinatorics plus regularization" is a language specific to machine (artificial) intelligence, it is necessary to ask whether a machine can produce a conclusion which at first sight appears to only within the capacity of humans. It is also necessary to ask whether, on the basis of monotonically increasing noisy data, a computer can predict a drop in the very near future, and conversely, on the basis of continuously decreasing data, predict that a rise will soon occur. It is found that this is possible. For solving the problem, it suffices to restore the law governing the system. If this law indeed exists, the machine will find the inflection point and predict it not worse than Cassandra did (Akishin and Ivakhnenko, 1975).

Our experience shows that in complex systems, the laws connecting

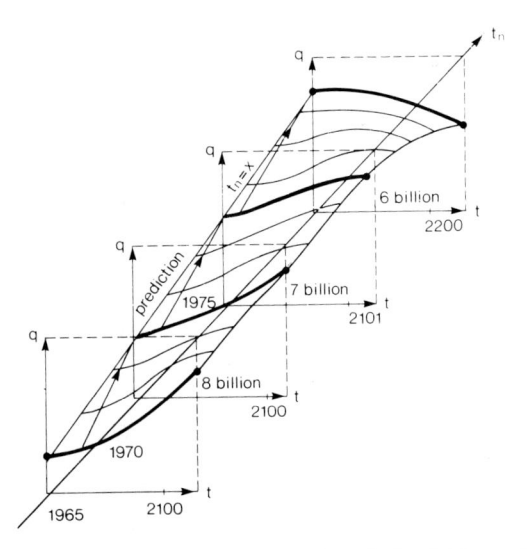

Figure 4. Illustration of the idea of "prediction of predictions" (the prediction is obtained with respect to the line $t_0 = t$).

correlated variables usually do not remain constant. Thus, by going back one interval into the past, we obtain a model for the problem of prediction. For example, Fig. 4 describes the prediction of world population until year 2200.

By using the GMDH identification methods, it is possible to obtain from past data a polynomial model of the form $q = f(t_0, t)$, where t_0 is the time when the prediction begins to be made. This model has time-varying coefficients. The variation of the coefficients must be extrapolated in the future, thus taking into account the law of variation of the coefficients in time. Note that in the problem of prediction of population growth only the method of prediction of predictions makes it possible to determine the year after which the world population begins to decrease. According to this prediction, this will take place in approximately the year 2030 (Ivakhnenko and Ivakhnenko, 1974, 1975).

4.2. Automatic Control with Prediction Optimization

Since it is possible to make a fairly accurate long-term prediction, it is natural to formulate the problem of prediction of a complex system in such a way as to optimize a criterion in the case of a sliding optimization interval:

$$J = \sum_{k}^{k+T_{pre}} f(q, k) \to \min, \quad (14)$$

where k is the present time. The interval from k to $k + T_{pre}$ is sliding because the time is going on, thus the optimization interval is sliding.

There are very few research papers dealing with this type of optimization. This problem has been solved exactly by Parks et al. (1974) for the case of a quadratic criterion and a linear system of differential equations. Moreover, a so-called simplified maximum principle and a method of linearization of a nonlinear performance criterion were developed in the paper. The investigations made it possible to find a so-called asymptotic law. It was found that any (even a very unstable) system controlled with prediction optimization becomes stable when the prediction time T_{pre} is increased.

It is evident that, for the automatic control of enterprises or branches of industry, the economy of a country, or other systems with large delays, it is possible to recommend this principle, which has been checked by computer simulation of control systems [as a plant we used the pulp-and-paper industry (Ivakhnenko et al., 1975) and the British economy (Parks et al., 1974)].

A brief survey of past work shows that, despite the fact that this theory is still in its early stages, it has been possible to overcome the principal difficulties. Note that by solving the problem of long-term prediction in systems with a time-varying law, it is possible to make new decisions and to control complex systems. This heralds the advent of a new era of completely objective computer predictions and optimal control taking into account all of the available information.

5. APPLICATION OF GMDH TO ENVIRONMENTAL PROBLEMS

The problem is how to organize the sifting of the models to find the law hidden in the noisy input data given the assumption that this law exists and remains constant during the time of the data measurement.

Discrete measurements correspond to the discrete form of the model equation

$$q_{+1} = f_1(q_0, q_{-1}, q_{-2}, \ldots, q_{-Lq}, x_0, x_{-1}, x_{-2}, \ldots, x_{-Lx}, t). \quad (15)$$

The current step of measurement is denoted by 0. The variables having indexes -1, -2, -3 are delayed arguments. Using so-called "forward differences," it is possible to write Eq. (15) in the differential form

$$\Delta q = q_{+1} - q_0 = f_2(q_0, q_{-1}, q_{-2}, \ldots, q_{-Lq}, x_0, x_{-1}, x_{-2}, \ldots, x_{-Lx}, t). \quad (16)$$

With the introduction of new variables, Eq. (16) can be transformed into the linear "complete polynomial":

$$q_{+1} = a_0 + a_1 q_0 + a_2 q_{-1} + a_3 q_{-2} + \ldots + a_{Lq+1} q_{-Lq}$$
$$+ b_1 x_0 + b_2 x_{-1} + b_3 x_{-2} + a_{Lx+1} x_{-Lx}. \quad (17)$$

All partial polynomials are obtained by equating to zero the coefficients of this complete polynomial. The complete set of models consists of $n = (2^n - 1)$ equations (where n is the number of coefficients of the complete polynomial).

The coefficients are estimated by the mean square error method. All the equations are compared among themselves using the minimum-of-bias or the combined criteria so that only the best mathematical model is selected. In the one-dimensional example considered below (Section 5.1) the complete polynomial has eight terms. Altogether $(2^8 - 1)$ or 255 regression equations can be obtained by eliminating one or another term of the complete polynomial.

For solving very complex problems, complete polynomials with a considerable number of terms ($n > 20$) should be used. The use of the multilayered selection algorithms of the GMDH is also recommended. In

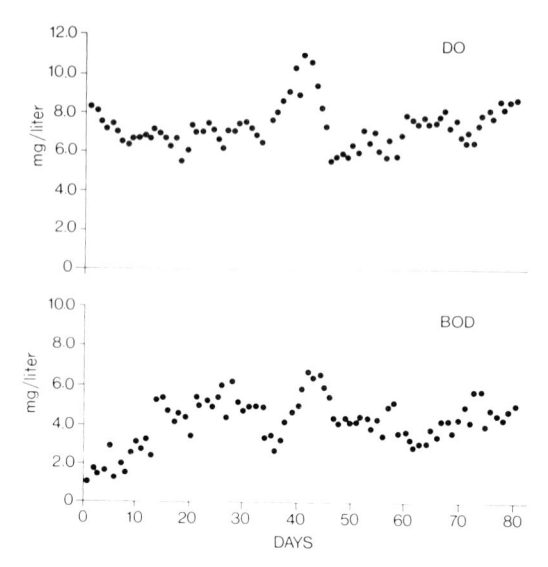

Figure 5. Input data (Beck, 1976).

the second, two-dimensional example, there are 80 terms in the complete polynomials. The multilayered algorithm was used in this case.

5.1. One-Dimensional Problem

When the problem is one-dimensional, model optimization means the choice of the number (L_q and L_z) of time delays:

$$L_q \to \text{opt} \quad \text{and} \quad L_z \to \text{opt}.$$

For the synthesis of the optimal complexity model, L_q and L_z must be gradually increased until the selection criterion decreases. The optimum corresponds to the minimum of the combined criterion. The values of the model coefficients can be found from all data points using the minimum mean square error method.

5.1.1. Example of Solution of One-Dimensional Problem

To illustrate the prediction of a process in an ecosystem we show how the well-known Streeter–Phelps' law can be discovered using experimental data taken from Beck (1976) (Fig. 5). The discrete version for this law is as follows:

$$\begin{aligned} q_{+1} &= k_1 q^{(s)} + (1-k_1)q_0 - k_2 z_0, \\ z_{+1} &= z_0 - k_2 z_0, \end{aligned} \qquad (18)$$

13. Identification of the Mathematical Model

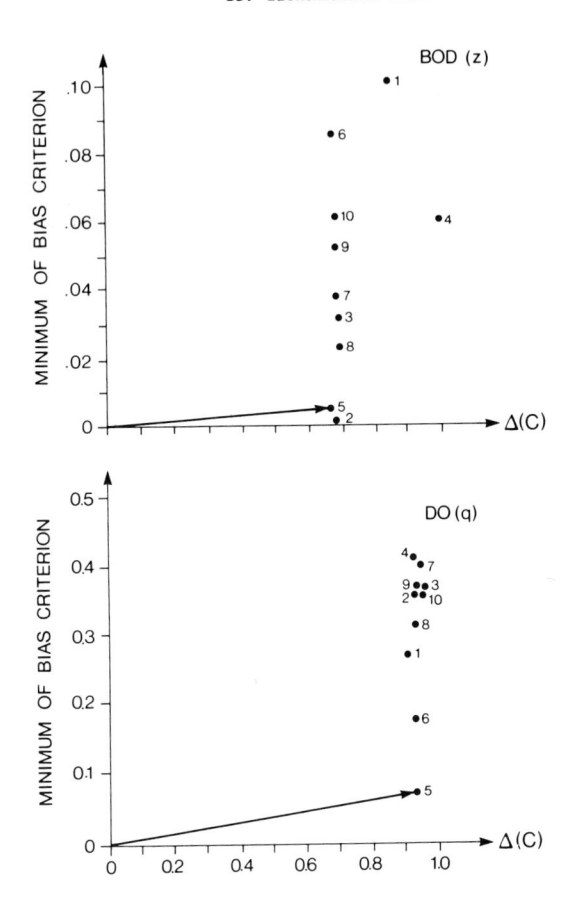

Figure 6. Display of the best models on the $[n_{bs} \div \Delta(C)]$ plane. The model of optimal complexity is No. 5.

where

q_{+1} = predicted concentration of dissolved oxygen (DO) in mg liter^{-1},

q_0 = the present DO concentration,

$q^{(s)}$ = maximum DO concentration (Beck, 1976),

z_{+1} = predicted biochemical oxygen demand (BOD) in mg liter^{-1},

z_0 = the present BOD value,

k_1 = the rate of reaeration (day^{-1}), and

k_2 = the rate of BOD decrease (day^{-1}).

The complete polynomials were taken in the linear form (with $L_q = L_z = 0, 1, 2, 3$):

$$q_{+1} = a_0 + a_1 q_0 + a_2 q_{-1} + a_3 q_{-2} + \ldots + a_{Lq+1} q_{-Lq}$$
$$+ b_1 z_0 + b_2 z_{-1} + b_3 z_{-2} + \ldots + b_{Lz+1} z_{-Lz} \quad (19)$$
$$z_{+1} = c_0 + c_1 q_0 + c_2 q_{-1} + c_3 q_{-2} + \ldots + c_{Lq+1} q_{-Lq}$$
$$+ d_1 z_0 + d_2 z_{-1} + d_3 z_{-2} + \ldots + d_{Lz+1} z_{Lz}.$$

The sifting on the computer of all possible partial linear polynomials able to fit the experimental data using the combined criteria gives the following optimal model:

$$q_{+1} = 1.3350 + 0.8142 q_0 - 0.00001 z_0,$$
$$z_{+1} = z_0 - 0.2545 z_0 + 0.1471 q_{-3}. \quad (20)$$

The model takes into account three arguments. The difference between the theoretical (19) and empirical (20) models can be explained by the fact that physical effects are not considered in Eq. (19). The plane $n_{bs} \div \Delta(C)$ is shown in Fig. 6. The model which is nearer to the origin is the best one. The points in it correspond to the best ten models (from the point of view of the minimum-of-bias criterion). In Fig. 6, No. 5 corresponds to Eqs. (20): This is the best one when we use the combined criteria. The most unbiased models are No. 5 (Fig. 7) for DO and No. 5 for BOD. Extrapolation error for DO is equal to 7% and for BOD 14%.

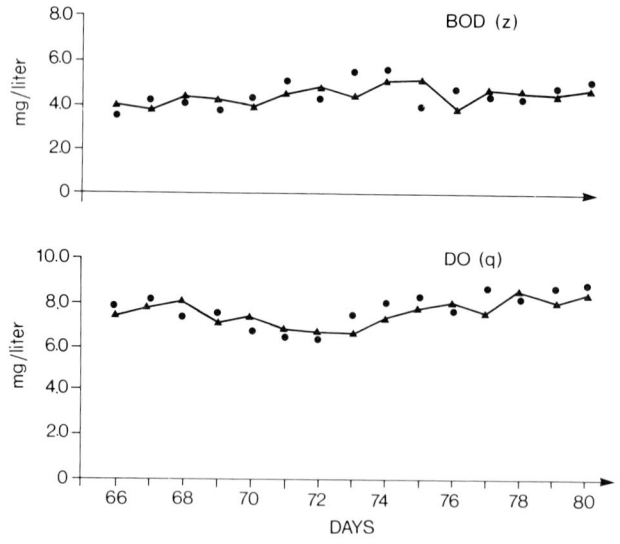

Figure 7. The extrapolation of the processes $q(t)$ and $z(t)$, DO and BOD, respectively, to control the accuracy of the prediction.

5.2. Multidimensional Problems

In discrete mathematics, it is known that every structure of the left hand operator of a finite-difference equation corresponds to a definite "elementary difference scheme" or "pattern" (Fig. 8). The pattern is a graphical space figure which shows the dependence of the value of the output function (in the given point of space at the given moment of time) from its values, measured in all neighboring points in space and time.

The "pattern" describes in mathematical form the process being studied. Two steps along a given axis in the pattern are necessary to build a discrete analog of the second derivative. Therefore, physically this is the diffusion in a given direction. One step makes it possible to build the discrete analog of the first derivative. Thus, this is physical advection.

The so-called "complete pattern" for solution of the three-dimensional problem, presented in Fig. 8c, shows that the diffusion and advection take place along the three coordinates. The patterns for the solution of two-dimensional modeling problems which provide for the analog of the first and the second derivatives, are shown in Fig. 8. This pattern corresponds to the following set of arguments:

$$q_{+1,0} = f(q_{0,0}, q_{0,-1}, q_{0,+1}, q_{-1,0}, q_{-2,0}). \tag{21}$$

The two subscripts indicate delayed arguments in the two dimensions t and x (Section 5). In the case of linear law, this equation can be written in the form (with $q_{-2,0} = 0$):

$$(q_{+1,0} - 2q_{0,0} - q_{-1,0}) + a_1(q_{0,+1} - 2q_{0,0} + q_{0,-1})$$
$$+ a_2(q_{0,0} - q_{-1,0}) + a_3(q_{0,0} - q_{0,-1}) = 0. \tag{22}$$

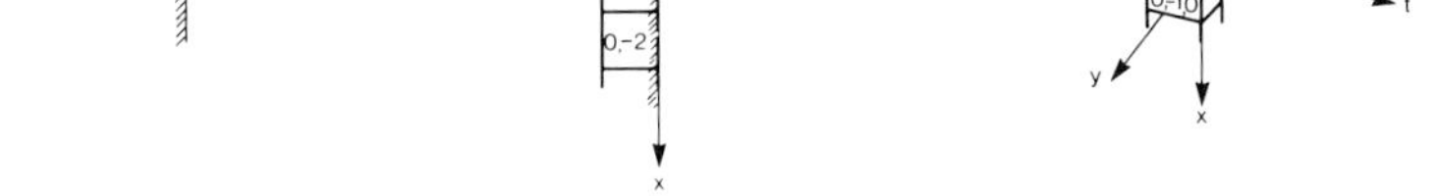

(a)

$q_{+1} = f(q_0, q_{-1}, q_{-2}, \ldots$
$x_0, x_{-1}, x_{-2}, \ldots)$

(b)

$q_{+1,0} = f(q_{0,0}, q_{-1,0}, q_{-2,0}, \ldots$
$q_{0,+1}, q_{0,+2}, \ldots, q_{0,-1}, q_{0,-2}, \ldots$
$x_{0,0}, x_{-1,0}, x_{-2,0}, \ldots$
$x_{0,+1}, x_{0,+2}, \ldots, x_{0,-1}, x_{0,-2}, \ldots)$

(c)

$q_{+1,0,0} = f(q_{0,0,0}, q_{-1,0,0}, q_{-2,0,0}, \ldots$
$q_{0,+1,0}, \ldots q_{0,-1,0}, \ldots q_{0,0,-1}, \ldots q_{0,0,+1},$
$x_{0,0,0}, x_{-1,0,0}, x_{-2,0,0}, \ldots$
$x_{0,+1,0}, \ldots, x_{0,-1,0}, \ldots x_{0,0,-1}, \ldots x_{0,0,+1})$

Figure 8. Patterns for the solution of (a) one-dimensional, (b) two-dimensional, and (c) three-dimensional extrapolation problems.

The continuous analog is

$$\frac{\partial^2 q}{\partial x^2} + a_1 \frac{\partial^2 q}{\partial y^2} + a_2 \frac{\partial q}{\partial x} + a_3 \frac{\partial q}{\partial y} = 0. \tag{23}$$

This describes the advection and diffusion along the x and y axis. This example shows the correspondence between the pattern and the adopted theory. In fact, we think that if a deterministic theory of the system behavior exists, then the model equations should be written according to it.

In real problems, the right-hand side of the equation is not correct and has some error. We call this error "the remainder" or "source" function. For example, in the turbulent diffusion problem we can write down

$$\left[\frac{\partial q}{\partial t} + u \frac{\partial q}{\partial x} - k_x \frac{\partial^2 q}{\partial x^2} \right] = f(x, t), \tag{24}$$

where

$[\cdot]$ = operator of the left-hand side,
$f(x, t)$ = "the remainder."

Usually $f(x, t) = a_0 + a_{01} x + a_{02} t$. The remainder takes into account the side inflows caused by change of derivatives $\partial q / \partial t$ and $\partial q / \partial x$ along the axis x and t.

The optimal pattern and the optimal remainder can be found in two ways: (a) on the basis of *a priori* information about the process physics (deductive or deterministic approach); (b) by sifting all the possible patterns and remainders on the computer using the combined criterion of selection (inductive or model self-organization approach). The choice of the optimal pattern and remainder leads to discovery of the law acting on the complex multidimensional object.

5.2.1. An Example of the Solution of Two-Dimensional Problem

Three variables at three stations of a water reservoir were measured at the depth of $\frac{1}{2}$ m: (1) dissolved oxygen DO $[q(t)]$, (2) biochemical oxygen demand BOD $[z(t)]$, and (3) temperature $[T(t)]$.

The observations were carried out eight times at 4-week intervals starting on the 15th of May. The results are represented in Table I, where $3 \times 8 = 24$ points of measurement are shown.

Two stage interpolation. The number of measurement points was increased up to $16 \times 16 = 256$ points by means of the quadratic interpolation (Table II). This method was adopted because smooth functions can be well approximated within the interpolation interval. But for extrapolation (along the x axis) and prediction (along the t axis), the use of a finite-difference equation obtained by means of GMDH is necessary.

13. Identification of the Mathematical Model

Table I Table of Original Data

Control points (distance in km), x		20	22	24	26	28	30	32	34
		\multicolumn{8}{c	}{Time of measurements (weeks)}
0	$q =$ 9.12	10.26	11.08	10.06	11.06	9.2	10.26	10.12
	$z =$ 0.78	2.07	1.07	1.5	1.32	2.1	1.28	1.4
	$T =$ 12	13.2	14.5	15.5	13.6	13.6	13.5	11.4
16	$q =$ 9.91	9.52	9.5	8.55	8.74	9.67	11.06	10.21
	$z =$ 0.57	1.7	1.6	1.5	1.3	1.8	1.7	1.5
	$T =$ 12	13.7	13.7	17.4	15.1	14	13.8	11.8
24	$q =$ 9.5	9.5	11.2	11.02	11.25	10.88	10.92	10.88
	$z =$ 0.27	0.22	1.5	1.4	1.3	1.36	2.64	1.69
	$T =$ 11.9	10.8	14	13.1	17.7	16.4	13.4	13.4

The prediction problem can be solved with the help of identification of finite-difference equation of the following form (the first index relates to the t axis, the second one to the x axis)

$$q_{+1,0} = f_1(x_{+1,0}, t_{+1,0}) + f_2(q_{0,0}, q_{-1,0}, q_{-2,0}, q_{0,-1}, q_{0,+1},$$
$$z_{0,0}, z_{-1,0}, z_{-2,0}, z_{0,-1}, z_{0,+1}, T). \quad (25)$$

$$z_{+1,0} = f_3(x_{+1,0}, t_{+1,0}) + f_4(q_{0,0}, q_{-1,0}, q_{-2,0}, q_{0,-1}, q_{0,+1},$$
$$z_{0,0}, z_{-1,0}, z_{-2,0}, z_{0,-1}, z_{0,+1}, T).$$

To get the input data table (Table II) the pattern (Fig. 8b) should be moved along the field of numbers of the Table II, step by step, to the right along the t axis. Every position of the pattern gives us one line of Table III. Step-by-step integration of Eqs. (25) allows us to predict the variables $q(t)$ and $z(t)$.

The extrapolation problem can be solved with the help of identification of the following type of finite-difference equation:

$$q_{0,+1} = f_5(x_{0,+1}, t_{0,+1}) + f_6(q_{0,0}, q_{0,-1}, q_{0,-2}, q_{-1,0}, q_{+1,0},$$
$$z_{0,0}, z_{0,-1}, z_{0,-2}, z_{-1,0}, z_{+1,0}, T) \quad (26)$$

$$z_{0,+1} = f_7(x_{0,+1}, t_{0,+1}) + f_8(q_{0,0}, q_{0,-1}, q_{0,-2}, q_{-1,0}, q_{+1,0},$$
$$z_{0,0}, z_{0,-1}, z_{0,-2}, z_{-1,0}, z_{+1,0}, T).$$

To obtain the table of input data, the pattern (Figure 8b) should be moved downward, along the x axis. Step-by-step integration permits the extrapolation of the data given in Table II.

Table II Results of the First Step of Interpolation (by Algebraic Models)

Number of weeks

Distance (in km)		20	21	22	23	24	25	26	27	28	29	30	31	32	33	34
0	$q =$	9.12	9.474	10.26	9.176	11.08	9.083	10.06	9.147	11.06	9.404	9.2	9.902	10.26	9.776	10.12
	$z =$	0.78	1.059	2.07	1.457	1.7	1.671	1.5	1.721	1.32	1.645	2.1	1.718	1.28	1.505	1.4
	$T =$	12.0	12.6	13.2	13.85	14.5	15.0	15.5	14.55	13.6	13.55	13.5	13.55	13.6	12.5	11.4
2	$q =$	9.317	9.193	9.123	9.103	9.135	9.222	9.361	9.5	9.692	9.964	10.288	9.87	9.623	9.392	9.213
	$z =$	0.807	1.063	1.293	1.458	1.585	1.679	1.722	1.743	1.658	1.694	1.689	1.731	1.773	1.522	1.137
	$T =$	12.0	12.63	13.26	13.83	14.4	15.07	15.74	14.76	13.79	13.67	13.56	13.59	13.625	12.54	11.45
4	$q =$	9.14	9.123	9.158	9.24	9.375	9.57	9.819	10.362	10.358	9.935	9.681	9.486	9.343	9.217	9.144
	$z =$	0.81	1.085	1.311	1.458	1.577	1.685	1.715	1.76	1.693	1.702	1.704	1.743	1.781	1.537	1.162
	$T =$	12.0	12.66	13.32	13.81	14.3	15.14	15.97	14.97	13.97	13.8	13.62	13.635	13.65	12.575	11.5
⋮		⋮	⋮	⋮	⋮	⋮	⋮	⋮	⋮	⋮	⋮	⋮	⋮	⋮	⋮	⋮
22	$q =$	10.035	10.255	10.525	10.89	10.508	10.264	10.072	9.976	9.931	9.871	9.864	9.891	9.974	10.137	10.382
	$z =$	0.749	0.721	0.687	1.476	1.505	1.569	1.628	1.76	1.657	1.785	1.869	1.875	1.733	1.708	1.676
	$T =$	11.9	11.715	11.53	12.725	13.92	14.05	14.17	15.61	17.05	16.435	15.8	14.65	13.5	13.25	13.0
24	$q =$	9.5	10.903	9.5	10.266	11.2	9.964	11.02	9.984	11.25	10.026	10.88	10.245	10.42	10.696	10.88
	$z =$	0.27	0.553	0.22	1.064	1.5	1.469	1.4	1.754	1.3	1.69	1.36	1.888	2.64	1.737	1.69
	$T =$	11.9	11.35	10.8	12.4	14.0	13.55	13.1	15.4	17.7	17.05	16.4	14.9	13.4	13.4	13.4

Table III Beginning and the End of the Table of Interpolation Data for Prediction (A Model with Two Delayed Arguments)

Number of points	+1,0	0,0	−1,0	−2,0	0,−1	0,+1	+1,0	0,0	−1,0	−2,0	0,−1	+0,1	T	x	t
1	9.103	9.123	9.193	9.317	10.26	9.158	1.458	1.293	1.063	0.807	2.07	1.311	13.26	2	22
2	9.135	9.103	9.123	9.193	9.176	9.24	1.585	1.458	1.293	1.063	1.457	1.458	13.83	2	23
3	9.222	9.135	9.103	9.123	11.08	9.375	1.679	1.585	1.458	1.293	1.7	1.577	14.4	2	24
⋮	⋮	⋮	⋮	⋮	⋮	⋮	⋮	⋮	⋮	⋮	⋮	⋮	⋮	⋮	⋮
120	9.974	9.891	9.864	9.871	9.75	10.245	1.733	1.875	1.869	1.785	1.888	1.888	14.65	22	32
121	10.137	9.974	9.891	9.864	9.737	10.42	1.708	1.733	1.875	1.869	2.64	2.64	13.5	22	33

347

5.2.2. Identification of Eqs. (20) and (25) Using Multilayered Algorithm of GMDH with Renotation of Variables

When we use a two-dimensional pattern (Fig. 8b), the complete polynomial of the second degree has one "source function": 14 linear terms, 11 square terms, and 55 covariations, in all, $3 + 11 + 11 + 55 = 80$ terms. For example, Eq. (25) gives:

$$q_{+1,0} = (a_0 + a_{01}x_{+1,0} + a_{02}t_{+1,0}) + (a_1 q_{0,0} + a_2 q_{-1,0} + a_3 q_{-2,0}$$
$$+ a_4 q_{0,-1} + a_5 q_{0,+1} + a_6 z_{0,0} + a_7 z_{-1,0} + a_8 z_{-2,0}$$
$$+ a_9 z_{0,-1} + a_{10} z_{0,+1} + a_{11} T + a_{12} q_{0,0}^2 + a_{13} q_{-1,0}^2 + a_{14} q_{-2,0}^2$$
$$+ a_{15} q_{0,-1}^2 + a_{16} q_{0,+1}^2 + a_{17} z_{0,0}^2 + a_{18} z_{-1,0}^2 + a_{19} z_{-2,0}^2$$
$$+ a_{20} z_{0,-1}^2 + a_{21} z_{0,+1}^2 + a_{22} T^2 + a_{23} q_{0,0} q_{-1,0} + a_{24} q_{0,0} q_{-2,0}$$
$$+ \ldots + a_{77} z_{0,+1} T), \quad (27)$$

where $(a_0 + a_{01}x_{+1,0} + a_{0,2}t_{+1,0})$ is the "source function" taken in a simple linear form.

When we change the notation of variables:

$$q_{+1,0} = u; \quad x_{+1,0} = w_{01}; \quad t_{+1,0} = w_{02};$$
$$q_{0,0} = w_1; \quad q_{-1,0} = w_2, \ldots, z_{0,+1} T = w_{77}.$$

We obtain a linear polynomial containing 80 terms

$$u = (a_0 + a_{01} w_{01} + a_0 w_{02}) + (a_1 w_1 + a_2 w_2 + a_3 w_3 + \ldots + a_{77} w_{77}); \quad (28)$$

A similar polynomial can be obtained also for Eqs. (25) and (26).

In the first layer of selection, the following partial description (base function) is used:

$$u_1 = b_0 + b_1 w_i + b_2 w_j; \quad i = j = 0, 01, 02, 1, 2, 3, \ldots, 77. \quad (29)$$

To estimate the model bias, the coefficients b_0, b_1, and b_2 must be estimated twice, using data sets A and B. Altogether we obtain $C_{80}^2 = 3160$ polynomials (or models) from which the most unbiased $g_1 = 80$ polynomials are to be selected (using the minimum-of-bias criterion).

In the second layer of selection, the partial description

$$u_2 = c_0 + c_1 u_{1,i} + c_2 u_{2,j}$$

is used. We obtain 3160 polynomials again, from which we select $g_2 = 80$, the most unbiased polynomials.

In the third layer of selection, the partial description

$$u_3 = d_0 + d_1 u_{2,i} + d_2 u_{2,j}$$

works. We obtain 3160 polynomials again and so on. The building up of the

13. Identification of the Mathematical Model

selection layers continues as long as the selection criterion decreases. In the last layers we have two alternatives:

(a) to choose one of the most unbiased models using the criterion $n_{bs} \to \min$, or

(b) to choose $g_3 = 20$ better models in order to find the best one among them using the minimax criterion

$$n_{bs} \max \to \min \qquad (21)$$

by varying the contents of points in the sets A and B, as explained earlier. The second way is better and is recommended to be used when the input data are noisy.

When the unique best model is found, the coefficients are calculated again using the mean square method applied to all data points.

The results of modeling. The following models are obtained with the help of the GMDH algorithm described above (using $n_{bs} \to \min$ criterion):

—For extrapolation:

$q_{0,+1} = 9.6915 + 2.9860 \times 10^{-2} z_{0,0} + 3.4797 \times 10^{-2} z_{+1,0} z_{-2,0}$,

$\qquad n_{bs} \min = 2.085 \times 10^{-3}, \qquad \Delta(B) \min = 0.043;$

$z_{0,+1} = 0.6470 - 0.1632 z_{0,-2} + 1.0349 \times 10^{-4} z_{0,0}^2$

$\qquad - 3.2834 \times 10^{-4} z_{0,0} z_{0,-2} - 2.6637 \times 10^{-4} z_{0,0} z_{-1,0}$

$\qquad + 7.6496 \times 10^{-2} z_{0,-1} z_{0,-2} - 1.3309 \times 10^{-3} z_{-1,0} z_{0,-1}$

$\qquad + 4.5817 \times 10^{-2} z_{0,-2}^2 - 3.2259 \times 10^{-6} z_{0,-2} z_{+1,0}$

$\qquad + 0.3309 z_{-1,0}^2 + 4.6004 \times 10^{-4} z_{-1,0} z_{1,0}$,

$\qquad n_{bs} \min = 7.245 \times 10^{-4}, \qquad \Delta(B) \min = 0.076.$

—For prediction:

$q_{+1,0} = 6.3700 + 0.1958 z_{0,0} q_{0,+1} + 5.6351 \times 10^{-4} q_{0,+1} z_{0,-1}$,

$\qquad n_{bs} \min = 3.054 \times 10^{-4}, \qquad \Delta(B) \min = 0.040;$

$z_{+1,0} = 1.4958 - 6.2680 \times 10^{-4} z_{0,0} + 2.6289 \times 10^{-4} z_{0,0} z_{-1,0}$

$\qquad + 1.6642 \times 10^{-2} z_{0,0} q_{-2,0} - 1.5704 \times 10^{-5} z_{0,0} q_{0,-1}$

$\qquad + 1.1195 \times 10^{-2} z_{-1,0} z_{0,-1} + 1.4887 \times 10^{-5} q_{0,0} z_{0,+1}$

$\qquad - 4.4883 \times 10^{-3} q_{-2,0} q_{0,-1} + 3.5511 \times 10^{-5} z_{0,-1} z_{0,+1}$,

$\qquad n_{bs} \min = 4.059 \times 10^{-4}, \qquad \Delta(B) \min = 0.070.$

We can see that not all the arguments represented in the pattern remain in the models. The model is, so to say, the shortened pattern.

5.3. Improvement of the Models Using the Combined Criterion

The models mentioned above give high accuracy of prediction and extrapolation with prediction lead for two or three steps (the error is not over 8% when calculated using the data set). The 20 most unbiased models should be chosen for long-term prediction, to find the best one among them we use the combined criteria (Section 3.1.3).

$$k_3 = (n_{bs}^2 + \Delta^2(C))^{1/2} \to \min.$$

The following models were obtained using the combined criterion.
—For extrapolation:

$q_{0,+1} = 9.5237 + 0.0937 z_{0,0} z_{0,-2} + 0.0031 z_{0,0} T_{0,0} + 0.2500 z_{0,-1} z_{0,-2}$
$\quad - 0.1403 z_{0,-1} z_{+1,0} - 0.0271 z_{0,-2}^2 + 0.0392 z_{0,-2} q_{-1,0}$
$\quad - 0.1782 z_{0,-2} z_{-1,0} - 0.0288 z_{-1,0}^2 - 0.1046 z_{-1,0} z_{1,0},$
$k_3 \min = 1.547 \times 10^{-3}, \quad n_{bs} = 6.950 \times 10^{-6}, \quad \Delta(C) = 1.551 \times 10^{-3}, \quad \Delta(B) \min = 0.038;$

$z_{0,+1} = -0.1798 + 1.1573 z_{0,0} - 0.1124 z_{0,0}^2 + 9.6622 \times 10^{-3} z_{-1,0} T_{0,0},$
$k_3 \min = 7.512 \times 10^{-4}, \quad n_{bs} = 5.987 \times 10^{-4}, \quad \Delta(C) = 4.586 \times 10^{-4}, \quad \Delta(B) \min = 0.050.$

—For prediction:

$q_{+1,0} = 12.4306 - 4.6477 z_{0,0} + 0.1615 q_{-2,0} z_{0,0} + 0.8896 z_{0,0}^2$
$\quad - 0.0035 z_{0,0} z_{+1,0} + 0.0004 t_{+1,0},$
$k_3 \min = 1.182 \times 10^{-3}, \quad n_{bs} = 2.025 \times 10^{-4}, \quad \Delta(C) = 1.165 \times 10^{-3}, \quad \Delta(B) \min = 0.037;$

$z_{+1,0} = 1.1149 + 0.0200 z_{-1,0} T_{0,0} + 0.0042 z_{0,+1} T_{0,0},$
$k_3 \min = 2.969 \times 10^{-3}, \quad n_{bs} = 5.402 \times 10^{-4}, \quad \Delta(C) = 3.937 \times 10^{-3}, \quad \Delta(B) \min = 0.064.$

The accuracy of the step-by-step integration for these models is lower in case of short-term prediction (the error is 15%) but considerably higher in case of long-term prediction or extrapolation for 10 to 20 steps ahead (the error is not over 20%).

6. CONCLUSIONS

Both theoretical development and applications of system identification have a variety of possibilities to be implemented. Theoretical developments should involve the study of uniqueness of the identified model. System identification techniques for nonlinear models and for systems described by partial differential equations should be improved. Applied mathematicians should simplify algorithms and computer programs to reduce costs and increase benefits.

Future system identification procedures should develop self-organizing models. With use of the deterministic approach a system of equations must be invented, which must describe all the elements of a complex system. No one element can be left without mathematical description. With such "forced" invention the more mistakes that are made, the more the object is complex and nonunderstandable. The natural researchers' tendency to the simple linear relationships may cause a low accuracy of the model, whereas exact models should have an optimum number of arguments and should take into account the past values of arguments, that is, equations with time delays.

Such complex equations cannot be invented with a deterministic approach. The equations of optimum complexity can be obtained only as a result of a self-organizing method using a computer. Such a self-organizing approach is completely open for introducing any reliable *a priori* information. Thus, it is enough to show the computer only the means of the problem solution, experimental data, as well as the goals (criteria) of the solution of the problem. The goal can be a predictive or an extrapolation model. If the data are not too variable, the computer itself can find the best unique model for prediction or the best one exhibiting cause and effect relationships. By application of the self-organizing theory, the computer will be able to discover the natural law that objectively exists in the observed object. A number of algorithms based on the self-organization principle have been developed. In these algorithms, the freedom of choice is provided for by the fact that a number of decisions of foregoing choices is produced at the next point of choice and no unique decision is made. It turns out that the random processes which should not be suitable for being forecasted from our point of view actually can be forecasted exactly enough.

When these concepts are regarded in the perspective of ecological theory, it can be assumed that the development of system identification methods may lead to laws governing ecosystems, with some help from computers. Such evolution is still some time away, but the prospect indicates why system identification should be an active field of research for ecologists.

REFERENCES

Akishin, B. A., and Ivakhnenko, A. G. (1975). Possibility of extrapolation (prediction) on the basis of monotonically varying noisy data. *Automatika* **4**, 22–30; *Sov. Autom. Contr. (Engl. Transl.)* **8**, July–Aug., 17–23 (1975).

Barron, R. L. (1971). Adaptive transformation networks for modeling, prediction, and control. *Annu. Symp. IEEE Syst., Man Cybernet. Group, 1971* pp. 254–263.

Beck, M. B. (1976). Modelling of dissolved oxygen in a non-tidal stream. *In* "The Use of Mathematical Models in Water Pollution Control" (A. James, ed.), pp. 1–38. Wiley, New York.

Beer, S. T. (1963). "Cybernetics and Management" (transl. from English). Fizmatgiz, Moscow.

Gabor, D. (1971). Cybernetics and the future of industrial civilization. *J. Cybernet.* **1**, 1–4.

Gilstrap, L. O. (1971). Keys to developing machines with high level artificial intelligence. *Proc. ASME Des. Eng. Conf., 1971* Paper 71-DE-21, pp. 1–16.

Ivakhnenko, A. G. (1975). "Long Term Prediction and Control of Complex Systems" (in Russian). Tekhnika, Kiev.

Ivakhnenko, A. G., and Ivakhnenko, N. A. (1974). Long term prediction of random processes by GMDH algorithms using the unbiasedness criterion and balance-of-variables criterion. Part I. *Automatika* **4**, 52–59; *Sov. Autom. Control (Engl. Transl.)* **7**, July–Aug., 40–45 (1974).

Ivakhnenko, A. G., and Ivakhnenko, N. A. (1975). Long term prediction of random processes by GMDH algorithms using the unbiasedness criterion and balance-of-variables criterion. Part II. *Automatika* **4**, 31–47; *Sov. Autom. Control (Engl. Transl.)* **8**, July–Aug., 24–38 (1975).

Ivakhnenko, A. G., and Stepashko, V. S. (1975). Self-organization of models and long term prediction of river runoff by the balance criterion. *Automatika* **5**, 34–41; *Sov. Autom. Control (Engl. Transl.)* **8**, Sept.–Oct., 27–33 (1975).

Ivakhnenko, A. G., Tolokhnyanenko, V. A., and Yaremenko, A. G. (1974). Model self-organization and control with prediction optimization as a means of transition from automated to automatic control systems. *Automatika* **5**, 22–31; *Sov. Autom. Control (Engl. Transl.)* **7**, Sept.–Oct., 16–23 (1974).

Parks, P., Ivakhnenko, A. G., Boychuk, L. M., and Svetal'skiy, V. K. (1974). Self-organization of a model of the British economy by balance-of-variables criterion for control with prediction optimization. *Automatika* **6**, 30–52; *Sov. Autom. Control (Engl. Transl.)* **7**, Nov.–Dec., 25–43 (1974).

Stone, R., Ivakhnenko, A. G., Visotsky, V. N., and Semina, L. P. (1977). Discovery of laws of complex system in a case when output variables are not indicated. *Automatika* **6**, 27–40; *Sov. Autom. Control (Engl. Transl.)* **10**, Nov.–Dec., 10–21 (1977).

Tikhonov, A. I., and Arsenin, Y. A. (1974). "Methods of Solution of Incorrectly Formulated Problems." Nauka, Moscow.

Visotsky, V. N. (1976). Optimum partitioning of experimental data on GMDH algorithms. *Automatika* **3**, 71–73; *Sov. Autom. Control (Engl. Transl.)* **9**, May–June, 62–65 (1976).

von Neumann, J. (1966). "Theory of Self-Reproducing Automata." Univ. of Illinois Press, Urbana.

Part IV
MODEL ANALYSIS, CONTROL THEORY, AND STABILITY

Chapter **14**

AN ANALYSIS OF TURNOVER TIMES IN A LAKE ECOSYSTEM AND SOME IMPLICATIONS FOR SYSTEM PROPERTIES

Vicki Watson and Orie L. Loucks

1. Introduction	356
1.1 Nutrient Cycling and Turnover Times	356
1.2 Hypotheses	359
2. Methods	361
2.1 The Lake Wingra Ecosystem Model (WINGRA III)	361
2.2 Calculation of Turnover Times	364
2.3 Relative Retention and Recycling of Nutrients	366
3. Results	368
3.1 Compartment Turnover Times	368
3.2 System Turnover Times	372
4. Discussion	374
4.1 Comparison of Results with Observed Data and Other Simulated Data	374
4.2 Role of Compartments in System Nutrient Dynamics	375
4.3 Compartment and System Variability	377
4.4 Material Cycling and System Properties	378
5. Summary and Conclusions	380
References	381

1. INTRODUCTION

As Aristotle suggested, a complete understanding of the parts of a whole may not be necessary and certainly is not sufficient to achieve a complete understanding of the properties of the whole. System properties are those properties which are not obvious from an analysis of the system's parts but which are evident only from a study made at the system's level of organization. The investigation of such system properties is the principal goal of this study.

A system's properties are generally held to be a product of the couplings between its parts (Klir, 1969). Thus, if all the interactions within a system were understood, perhaps the system and all its properties would stand revealed. However, to grasp the functioning of a system, one must reduce its tremendous complexity to a minimum set of structural components which are believed to be functionally distinct with regard to that aspect of the system's functioning which is of interest (see Zeigler, Chapter 1, this volume). The challenge to the system's ecologist is to achieve an understandable, yet meaningful, simplification and to attach biological meaning to the behavior of such analogs.

The system or level of organization investigated here is the ecosystem. It is conceptualized as being composed of functional subunits termed *compartments*. The quantities of nutrients in these compartments are here referred to as *pools*. The system attribute of interest is nutrient retention, approached via an investigation of ecosystem nutrient dynamics. The recycling of nutrients is proposed as an explanation of retention, and evidence for retention and recycling is sought through an analysis of turnover times and flux rates in a lake ecosystem model.

Pool sizes, flux rates, and turnover times are available for the organic matter, nitrogen (N), and phosphorus (P) pools of the Lake Wingra, Wisconsin ecosystem. These values may be simulated with the aid of the Lake Wingra pelagic zone ecosystem model WINGRA III, developed in conjunction with the IBP studies (O'Neill, 1975). An examination of these quantities should add to the understanding of the relative importance of recycling in the various element cycles in an aquatic ecosystem, as well as adding to the understanding of the roles of system components in system nutrient dynamics.

1.1. Nutrient Cycling and Turnover Times

The system processes that have yielded the most information on ecosystem properties to date are energy flow and nutrient cycling. Of the two, the analysis of nutrient cycling may reveal more about the importance

14. An Analysis of Turnover Times 357

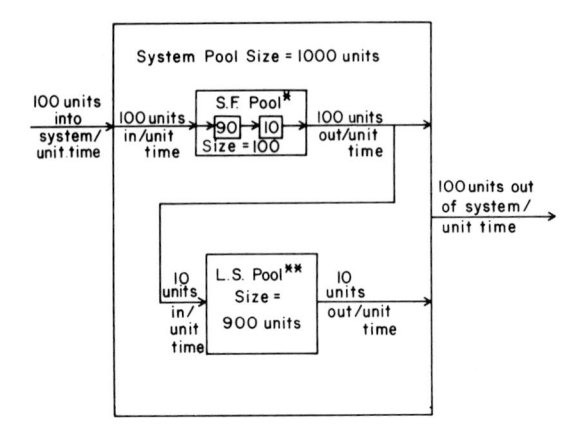

```
                    ┌─────────────────────────────────────┐
                    │  System Pool Size = 1000 units      │
                    │         ┌─────────────┐             │
  100 units         │         │  S.F. Pool* │             │
  ──into──→  100 units  →     │ [90]→[10]   │ → 100 units │
  system/     in/unit         │  Size = 100 │   out/unit  │
  unit time    time           └─────────────┘    time     │
                    │                                     │ → 100 units out
                    │                                     │   of system/
                    │                                     │   unit time
                    │    10      ┌─────────────┐   10     │
                    │  units →   │ L.S. Pool** │ → units  │
                    │   in/      │   Size =    │  out/unit│
                    │   unit     │  900 units  │   time   │
                    │   time     └─────────────┘          │
                    └─────────────────────────────────────┘
```

Figure 1. Hypothetical system of 1000 units which gains and loses 100 units for every unit of time.

of system couplings since it emphasizes feedback (as opposed to the one-way flow of energy).

With the realization that static studies of systems are inadequate to the elucidation of many system properties, interest has shifted to a dynamic approach which emphasizes rates and time constants of processes acting on the static structure of the system. The turnover times[1] of a system's components have come to be regarded as significant aspects of the system's functioning (Pomeroy, 1970; Waide et al., 1974). In fact, the turnover time of the system itself appears to be largely a function of the turnover time of its parts and the nature of the couplings between them.

To illustrate, if a hypothetical *system* were composed of 1000 units of relatively inert matter which gained and lost 10 units of matter every unit of time, this *system* would completely turn over in 100 units of time. However, if only 10% of that *system* (100 units of matter) were composed of labile matter which gained and lost 100 units every unit of time (Fig. 1), that 10% would turn over in one unit of time and the *system* in 10 units.

The small, *fast* compartment in an ecosystem model may be viewed as being composed of living matter (which respires, excretes, dies, etc.) and of labile detritus (which is subject to microbial action). The large, inert compartment is analogous to detritus which is not subject to bacterial decomposition.

The turnover time of the system illustrated in Fig. 1 seems likely to be short relative to the turnover time of the large, slow pools which make up a

---

[1] For a formal definition of turnover time, see Section 2.2.

Figure 2

A

- System Pool Size = 1000
- S.F. Pool Size = 100
- 90 into system → 90 in → 100 in → [90] → [10] → 100 out → 90 out → 90 out of system
- 10 recycled
- 10 in → L.S. Pool Size = 900 → 10 out

System Turnover Time = $\frac{1000}{90} = 11$

B

- System Pool Size = 1000
- S.F. Pool Size = 100
- 10 into system → 10 in → 100 in → [90] → [10] → 100 out → 10 out → 10 out of system
- 90 recycled
- 10 in → L.S. Pool Size = 900 → 10 out

System Turnover Time = $\frac{1000}{10} = 100$

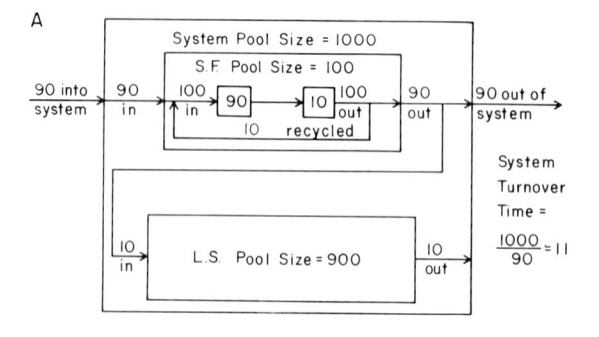

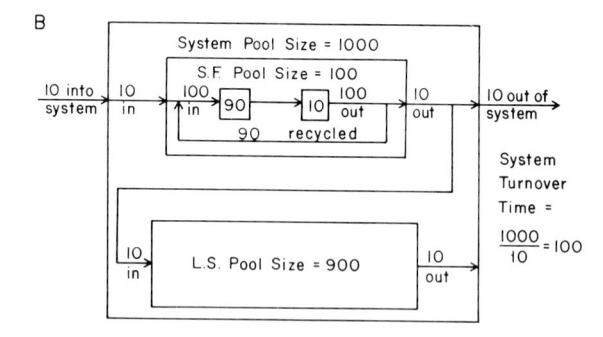

Figure 2. System similar to the one shown in Fig. 1 but which has: (A) recycling, (B) a greater degree of recycling.

substantial portion of the total system. Small, fast pools which move a great deal of material through the system rapidly should have a great influence on the turnover time of the system.

The effect of recycling within an ecosystem should be to increase the total system turnover time for the recycled element. Figure 2A shows a system similar to that in Fig. 1 in structure, pool size, and flux rate but to which recycling has been added. One-tenth of the output of the biotic populations and labile detritus which make up the small, fast pool is recycled within this pool between its components. As a result of this recycling, the system's turnover time is increased from 10 units of time to 11 units. Notice that the individual components of the small, fast pool are turning over at the same rate as those in the system without recycling.

Since most elements are recycled to some extent within the ecosystem, Fig. 2B is included to demonstrate the effect of a *greater* degree of recycling. In this model, 90% of the output of the small, fast pool is recycled instead of

10%. As a result, the total system turnover time is increased to 100 units of time, nearly a tenfold increase over that of the system in Fig. 2A. Once again, the individual components of the system are turning over at the same rate as in the previous example.

From these simplified models, it is possible to see that system turnover time depends on interactions between the components of a system which affect overall system flux rates. A comparison of Figs. 2A and B reveals that two systems with similar structures and internal flux rates could have different total system flux rates as a consequence of differences in the interactions between the system parts. It should be obvious that if the compartments in the hypothetical systems had had different turnover times as well, the system turnover times would have been altered further.

1.2. Hypotheses

The hypotheses to be examined here are formulated to answer the following questions concerning the rate of turnover of organic matter and nutrients in an ecosystem:

(a) What differences in turnover times may be expected between the different compartments of a lake ecosystem model? What implications do these differences have for the roles of the biotic populations and abiotic components which correspond to these compartments?

(b) Can the turnover times of the system be explained by the behavior of the individual parts? What differences in turnover times may be expected between the essential elements at the ecosystem level? What emergent system properties may be deduced from the answers to these questions?

As a first step toward answering these questions, the following hypotheses are posed as testable constructs derived in part from principles appearing in recent literature.

Hypothesis *a*. Turnover times and functional differences between compartments: Compartments with dissimilar functions in the system exhibit substantially different turnover times; moreover, the relative magnitudes of these turnover times can be predicted from compartment characteristics.

Earlier, system compartments were defined as those parts of the system which differ in function in some meaningful way. Bastin (1969) postulates that the parts of a whole that differ in function may be distinguished by different characteristic time constants. Assuming for the moment that the ecosystem has been correctly conceptualized (i.e., divided into parts which do have substantially different functions), one may expect these compartments to exhibit substantially different turnover times.

Webster et al. (1974) suggested that small compartments with rapid cycling of nutrients (i.e., short turnover times) contribute to system resilience (ability to "bounce back" after disturbance). Large compartments with long turnover times, on the other hand, seem to increase the system's resistance to perturbation. Consequently, the compartments of an ecosystem would be expected to exhibit a wide range of turnover times as a result of the "adaptiveness" of achieving a compromise between resistance and resilience. Harwell et al. (1977) have since objected to the measures of resistance and resilience used by Webster et al.; however, the intuitive arguments are still attractive.

Given certain compartment characteristics which affect flux rates, one should be able to predict the relative magnitudes of the turnover times of the various compartments. For example, the shorter life span and higher metabolic rates of the smaller organisms at the base of the aquatic food chain should cause the compartments in which they predominate to turn over more rapidly than the compartments dominated by larger, longer-lived organisms with lower metabolic rates which occupy the top of the chain (Odum, 1967).

Hypothesis b. Properties of ecosystems: The behavior of an ecosystem cannot be explained simply from the behavior of its parts. With particular reference to the cycling of essential elements, the turnover time of the total system pool of an element cannot be explained from information about the turnover time of its compartments.

Recycling is postulated as a mechanism capable of inducing intensive nutrient retention manifested by system turnover times which cannot be explained by the turnover times of the individual system parts. Carbon is predicted to have a shorter system turnover time than expected from the turnover time of the slowest compartments, as a consequence of the rapid flux through the small labile compartments (as in Fig. 2A). Nitrogen and phosphorus are predicted to have very long system turnover times relative to what might be expected from an analysis of compartment turnover times (as in Fig. 2B) as a consequence of the greater degree of recycling postulated for limiting nutrients (Odum, 1975). Differences in the degree or extent of recycling (obtained by comparing system fluxes and turnover times to those of the system compartments) may provide one quantifiable expression of the Aristotelian maxim: A whole is more than the sum of its parts.

Another ecosystem property which may be investigated through an inquiry into nutrient pool turnover times is system stability. Here stability is defined simply as the variability in pool size, flux rate, and turnover time.

The variability in these values for the system and its parts may be compared to determine if there is evidence for buffering of these values at the ecosystem level. Analysis of other stability measures appears elsewhere in this volume (see Šiljak, Harte, Goh, and Jeffries).

2. METHODS

The nominal run of the Lake Wingra pelagic zone ecosystem model, WINGRA III, provided the daily pool sizes and flux rates for one growing season. Turnover times were calculated from these model-generated data. WINGRA III has been given extensive testing and has been shown to reproduce reasonably well the dynamics of Lake Wingra, a small, shallow, eutrophic lake in a partly urbanized watershed in Madison, Wisconsin (Huff et al., 1973; Loucks and Weiler, 1979).

2.1. The Lake Wingra Ecosystem Model (WINGRA III)

The approach taken by the WINGRA III modelers emphasizes the quantity of energy and materials in system compartments. Each compartment is modeled by a theoretically derived, mechanistic model. The overall structure of the model appears in Fig. 3. The physical and biological processes which result in transfers between compartments are represented by letters in Fig. 3 and are briefly defined in Table I.

The lake system is conceptualized as being divisible into compart-

Table I Summary of Symbols Used for Biomass and Nutrient Equations in the WINGRA III Model

Symbol	Brief definition	Symbol	Brief definition
A	Nutrient incorporation transfers—assimilation	L	Littoral zone inputs
		M	Detrital resuspension transfers—mixing
B b	Biomass pools		
C	Consumption transfers	N n	Nitrogen pools
D	Death or mortality transfers	O	Hydrologic outfall transfer
DF	Dryfall deposition input	P p	Phosphorus pools
DEN	Denitrification	R	Respiration transfers
E	Emergence of benthic insects	S	Sinking rate terms
F	Egestion transfers	U	Excretion transfers
G	Transfers in fish reproduction	V	Microbial process transfers
I	Hydrologic inputs	W	Algal nutrient uptake terms
		Z	Depth

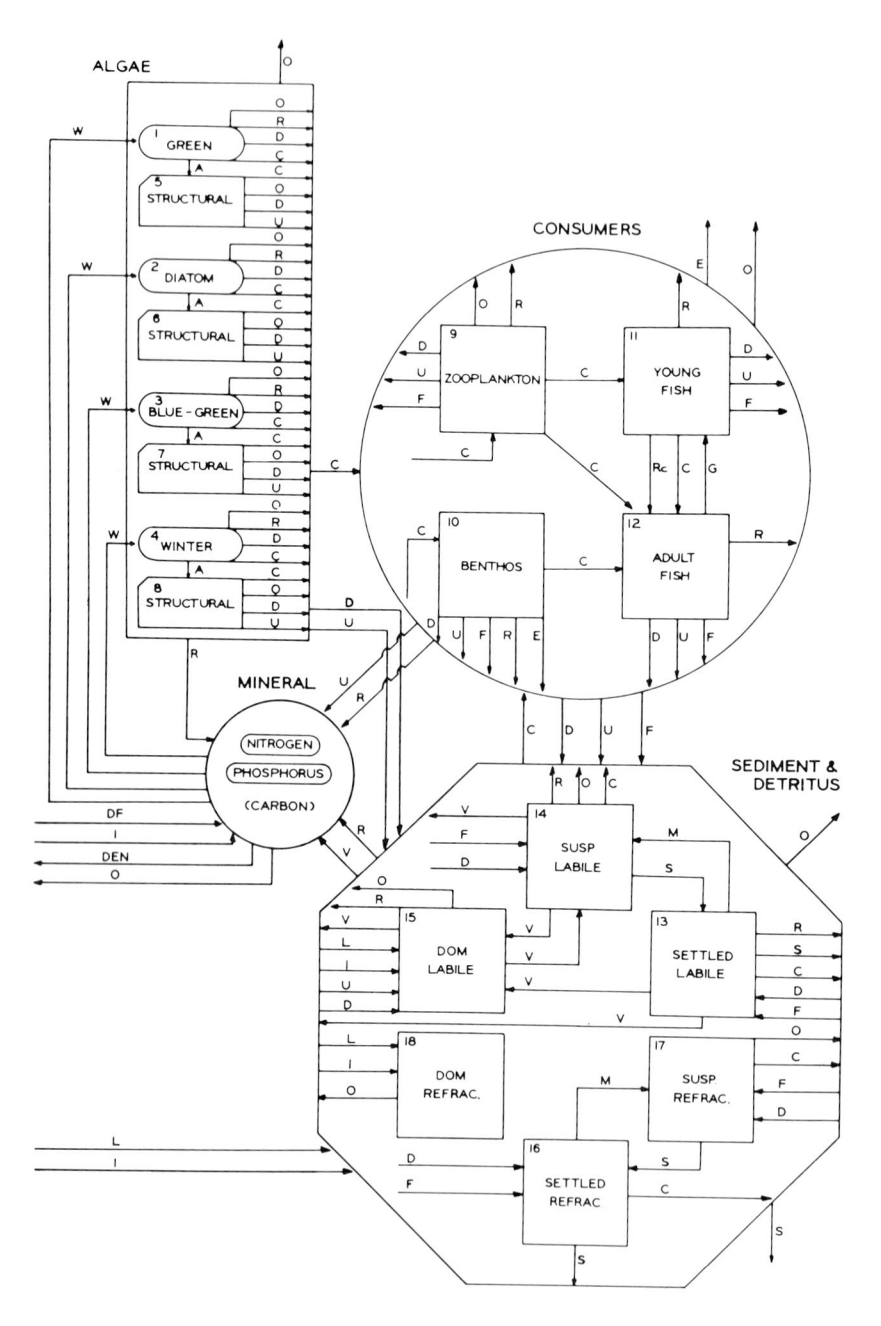

Figure 3. Structure of the Lake Wingra ecosystem model.

ments corresponding to the principal primary producers, consumers, detrital pools, and inorganic nutrient pools in a square meter column of lake water and in the surface sediments below it. Each compartment has a pool of organic matter, nitrogen, and phosphorus, except dissolved inorganic nitrogen (DIN) and phosphorus (DIP) which have no corresponding carbon pool in this model. Inorganic carbon is assumed to be constant and undepletable by algae.

The algal compartments include four groups of algae with substantially different physiological responses to the environment (greens, diatoms, blue-greens, and a winter association of algae). Each of these is divided into two subcompartments: (1) recently photosynthesized carbohydrates and unassimilated nutrients, and (2) assimilated carbohydrates and nutrients built into structural biomass. The latter is not as labile with respect to nutrient exchange as is the former.

The consumer compartments include zooplankton, benthos (principally, chironomid larvae), and two age/size classes of fish (principally the blue gill sunfish) aged up to 2 yr and from 2 to 4 yr.

Detritus is divided into suspended and settled particulates and dissolved organic matter (DOM). Settled detritus is defined as the top 5 cm of bottom sediment which is subject to resuspension and where most benthic activity takes place. These three components are subdivided into labile and refractory fractions (only the former is subject to microbial decomposition). The N and P in detritus are not divided into these fractions. Although there is a dissolved organic N (DON) pool corresponding to the DOM pool, there is no dissolved organic P (DOP) pool in this version).

Inorganic nutrient compartments include dissolved inorganic nitrogen (DIN), that is, ammonia, nitrates, and nitrites, and dissolved inorganic phosphorus (DIP), that is, reactive orthophosphates.

The WINGRA III model is represented mathematically by a system of differential equations. For each pool a first-order differential equation states that the rate of change in that pool is equal to inputs minus outputs or:

$$\frac{dx_i}{dt} = \sum_{j=1}^{18} f(x_i, x_j) - \sum_{j=1}^{18} g(x_j, x_i) + IN_i - OUT_i,$$

where x_i is the ith organic pool; t is time; $f(x_i, x_j)$ is the transfer rate into the ith pool from the jth pool; $g(x_j, x_i)$ is the loss rate from the ith pool to the jth pool; IN_i is the rate of input from sources outside the system; and OUT_i is the output rate to sinks outside the system. Similar equations represent the rate of change in N and P pools.

The transfers specified in the above equation which are nonzero are simulated by deterministic process submodels, represented by equations using maximum rate parameters multiplied by reduction factors to account for effects of nonoptimal conditions. Since some of these terms are nonlinear, the overall model is nonlinear.

Two important assumptions inherent in the model—conservation of mass and constant C:N:P ratios through time—add to the realism of the model. Conservation of mass is inherent in the model's treatment of biomass, N and P. This assumption demands that any loss from a pool must be accounted for by a gain in another pool or a loss from the system, while any gain to a pool must be accounted for by a loss from another pool or a gain to the system. Such an assumption represents a system constraint which serves to tie together the parts into an organic whole.

The other premise involves assuming that the amount of N and P in the algal structural biomass and in the consumer biomass is proportional to the amount of that biomass (since the C:N:P ratios in biomass are fairly uniform through time). This stoichiometrically realistic assumption allows the change in these nutrient pools to be simulated by the simple expression:

$$\frac{dN_i}{dt} = \eta_i \frac{dx_i}{dt} \quad \text{for} \quad i = 5,\ldots,12,$$

where N_i is the nitrogen in compartment i and η_i is the nitrogen yield coefficient for compartment i and is equal to N_i/x_i. A similar equation determines the change in the phosphorus pools in the algal structural biomass and the consumer biomass.

The N and P pools which obey the above equation are *implicit variables* (pool size is known implicitly when the corresponding biomass pool size is known). The other pools, represented by independent functions, are *explicit variables*. Although the assumption of stoichiometry simplifies the implicit equations, the equations which specify the corresponding explicit biomass pools must be quite complicated to maintain constant C:N:P ratios while obeying the conservation of mass assumption. For details of this problem, see Loucks and Weiler (1979).

The differential equations which describe the behavior of the 49 state variables are solved simultaneously by the Madison Academic Computing Center's UNIVAC 1110 digital computer, using a fifth-order, variable size integration step Runge Kutta algorithm.

2.2. Calculation of Turnover Times

Turnover time may be defined as the time required for the amount of a substance transferred into or out of a compartment to be equal to the amount present in the compartment (Robertson, 1957). A common method of calculating turnover time involves dividing pool size by either the gain rate or the loss rate of the pool. This calculation is made with the assumption that the pool or the system is in steady state, that is, gains equal losses over the time of observation. To provide a more meaningful estimate of turnover time of a pool which is not at steady state, turnover time may be assumed to be equal to the pool size divided by the rate of flux through the pool. Flux rate is here defined as the average of the input and output rates of the pool. Thus:

$$\begin{pmatrix} \text{Turnover time} \\ \text{of pool } i \end{pmatrix} = \frac{\text{pool size of pool } i \cdot 2}{\begin{pmatrix} \text{input rate} \\ \text{of pool } i \end{pmatrix} + \begin{pmatrix} \text{output rate} \\ \text{of pool } i \end{pmatrix}}.$$

Mean turnover times for some period of time may be obtained in a number of ways. Instantaneous pool sizes may be divided by instantaneous flux rates to obtain instantaneous turnover times at some point in time. Alternatively, the average pool size over some period (such as a day) may be divided by the total flux for that period to obtain an averaged daily turnover time. A series of instantaneous turnover times or daily turnover times might be averaged to obtain a mean value for some longer period. If the pool size and flux rate are random, independent variables, the average of discrete daily turnover times over a year will be similar in value to the single yearly value obtained by averaging the pool size for a year and dividing it by the average daily flux. However, the greater the covariance between pool size and flux rate, the greater will be the difference between the two *averages* of turnover time. In other words,

$$E(PS/FR) = E(PS) \cdot E(1/FR) + \text{Cov}(PS, 1/FR),$$

where E is the mean value operator, PS is pool size, FR is flux rate, and $\text{Cov}(PS, 1/FR)$ is the covariance between PS and $1/FR$.[2]

The compartment turnover times used in this study were obtained for each day of the growing season (defined to be April 10 to September 15) by dividing the pool size averaged over each day by the flux rate averaged over each day. This provided a picture of the seasonal change in turnover time as well as a measure of the variability. These daily values were averaged over the spring (April 10 to July 15) and over the summer (July 15 to September 15). The lake system exhibits very different behavior during these two seasons; consequently, an average over the entire growing season was not felt to be meaningful.

---

[2] Moreover, $\text{Cov}(PS, 1/FR) = \sigma_{PS}\sigma_{1/FR}\rho(PS, 1/FR)$, where σ_{PS} is the variance in PS, $\sigma_{1/FR}$ is the variance in $1/FR$ and $\rho(PS, 1/FR)$ is the correlation between PS and $1/FR$. Thus, as any of these factors increase, the covariance increases and the difference between the two averages of turnover time increases.

In addition to averaging daily turnover times, seasonal turnover times were obtained for each compartment for the spring and summer by dividing the pool size averaged over each season by the flux rate averaged over the season.

The turnover time of the total system was obtained in a similar fashion except that, instead of using the average pool sizes and flux rates over the day, instantaneous values at the end of the day were used, these being the only values available. Consequently, the system turnover times presented are the instantaneous values at the end of the day. Because of the greater algal activity in light and the higher respiration rates during the day, this end-of-the-day turnover time is likely to be shorter than the average-over-the-day value.

It should be noted that while the inputs and outputs to the compartments used in these calculations represent model output, the inputs to the lake system are observed values.

2.3. Relative Retention and Recycling of Nutrients

Comparing the turnover times of the individual compartments of an ecosystem should shed light on the roles of these compartments in the cycling of nutrients. In this context, it should be possible to discern whether a compartment acts as a *holder* or a *mover* of nutrients within the system (Pomeroy, 1974). Comparing the turnover time of the individual compartments with the turnover times of the total system should give some insight into the extent of nutrient retention in that system. The ratio of the system turnover time to the time required for an atom to take the longest path through that system may be used to assess the extent of nutrient retention. This ratio, which may be called the relative retention (RRT) of the system, may be determined for various essential elements and used to elucidate differences in the degree of retention of these elements at the ecosystem level.

It should be noted that retention of nutrients may be accomplished by recycling nutrients or by the gradual accumulation of nutrients in large, slow pools within the system. In addition, the effect of small, fast compartments on the turnover time of the system may obscure the effect of recycling (recall Fig. 1). That is, the turnover time of the system might be equal to the sum of the turnover times of the compartments even if recycling is occurring. Consequently, an analysis of system and compartment fluxes may be more revealing than a comparison of system and compartment turnover times.

In a catenary system (i.e., one in which there are no couplings between

Figure 4

A

```
90 in  →  ▭  →  90 out
10 in  →  ▭  →  out
```
$\dfrac{TCF}{SF} = \dfrac{100}{100} = 1$

B

```
90 in → 90 in ▭ 90 out → 80 out
         10 in ▭ 10 out → 10 out
```
$\dfrac{TCF}{SF} = \dfrac{100}{90} = 1.1$

C

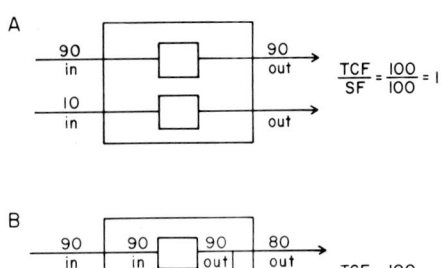

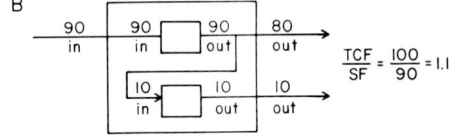

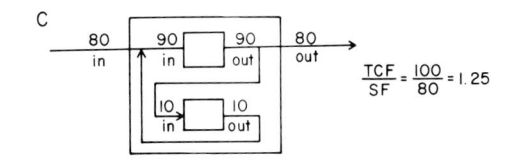

```
80 in → 90 in ▭ 90 out → 80 out
         10 in ▭ 10 out
```
$\dfrac{TCF}{SF} = \dfrac{100}{80} = 1.25$

Figure 4. (A) System with no couplings (flux through system equals flux through compartments). (B) System with coupling resulting in cycling of materials. (C) System with cycling and recycling of materials.

the compartments), the flux through the system would be equal to the total flux through all the compartments (Fig. 4A). In such a system the amount of material entering and leaving the system over some unit of time should equal the total amount of material entering and leaving all the system's parts over that period. As the connectivity of the system increases, the amount of material which is *cycling* (i.e., passing through more than one compartment) may increase (Fig. 4B). As a result of such cycling, the ratio of the total compartment flux (TCF) to the system flux (SF) increases. Couplings between system compartments which result in *recycling* (i.e., feedback of nutrients), as in Fig. 4C, result in still greater ratios of TCF to SF.

Obviously, the ratio of these quantities is affected by the lumping of compartments which exchange materials with one another. As a result, the ratio provides a means of assessing the degree of complexity and connectivity occurring within a system.[3] To some extent, the complexity

---

[3] Complexity is used here as an assessment of the number of compartments in a system (or model of a system) while connectivity is an assessment of the number of couplings or interactions between these compartments. Obviously, both terms are meaningful only when *comparing* systems.

and connectivity in a model are a function of the modeler's taste and purpose. However, it is also true that certain systems are more complex and highly connected in meaningful ways than are others.

When comparing two systems with a similar number of compartments and a similar amount of connectivity, qualitative differences in cycling and recycling disappear. Thus, differences in the ratio of TCF to SF indicate the relative amount of material which is cycling and recycling within the two systems. That is, a system in which most material passes through several compartments before leaving the system will have a greater ratio of TCF to SF than a structurally similar system in which most material passes through one compartment before leaving the system.

The ratio TCF/SF is dimensionless and independent of pool measurement units (mg N versus gram dry weight of organic matter). Hence, it may be used to compare the degree of recycling for different elements in the same system or to compare the system to itself at different times of the year and under different management regimes. The ratio of TCF to SF might be called relative recycling (RRC) and used to assess recycling where appropriate, as described above.

3. RESULTS

The results of the analysis of the pool sizes, flux rates, and turnover times obtained with the aid of the WINGRA III model are summarized in Tables II through V. The mean, standard deviation, and coefficient of variation of the daily turnover times for each pool and for the total system during the spring and summer seasons appear in Tables II, III, and V. The seasonal turnover times for these periods appear in Tables IV and V. These two methods of averaging turnover times produce similar results in most cases. The maximum percent difference between turnover times obtained by these methods is less than 100% in all cases except for the system turnover time of P. Consequently, the mean daily turnover time and the seasonal turnover time differ by less than an order of magnitude.

3.1. Compartment Turnover Times

Generally, the pools of unassimilated carbohydrates in the algae turn over in approximately 1 day in both spring and summer (Tables II and IV). Algal structural biomass turns over somewhat more slowly. Because diatoms and winter algae almost disappear in the summer, their computed turnover times are probably not reliable during this period and are not reported. Notice that the differences between the mean turnover times of the

Table II Analysis of the Daily Turnover Times for the Organic Matter Pools in the Compartments of WINGRA III[a]

Organic matter pools in:	Mean (days)	Standard deviation (days)	Coefficient of variation
Unassimilated carbohydrates			
Green algae	0.41 (0.33)	0.20 (0.063)	0.48 (0.19)
Diatoms	1.3	0.98	0.75
Blue-green algae	0.57 (0.44)	0.29 (0.068)	0.51 (0.15)
Winter algae	2.6	0.75	0.29
Structural biomass			
Green algae	1.8 (1.3)	0.88 (0.12)	0.49 (0.092)
Diatoms	4.5	1.9	0.42
Blue-green algae	2.5 (1.6)	1.5 (0.17)	0.62 (0.10)
Winter algae	8.0	2.1	0.26
Consumer biomass			
Zooplankton	31 (20)	10 (2.9)	0.33 (0.14)
Benthos	37 (24)	21 (7.6)	0.56 (0.32)
Young fish	180 (270)	110 (32)	0.58 (0.12)
Adult fish	260 (310)	140 (47)	0.53 (0.15)
Labile detritus			
Settled	11 (10)	3.0 (0.40)	0.28 (0.040)
Suspended	9.1 (5.7)	4.8 (0.63)	0.53 (0.11)
DOM	9.4 (8.5)	2.3 (0.59)	0.24 (0.069)
Refractory detritus			
Settled	2100 (2700)	440 (700)	0.21 (0.26)
Suspended	12 (6.7)	6.5 (0.75)	0.56 (0.11)
DOM	490 (940)	280 (400)	0.58 (0.43)

[a] Obtained by averaging daily turnover times over the spring and summer seasons. Spring values appear first followed by summer values in parentheses. No summer values are reported for diatoms and winter algae because these pools drop to zero in summer.

unassimilated carbohydrates and the structural biomass within an algal group appears to be greater than the differences between the algal groups.

As for the consumers, zooplankton and benthos turn over in 1 month in the spring, young fish in 5 to 6 months, and adult fish in 7 to 9 months (Tables II and IV). In the summer, zooplankton and benthos turn over more rapidly while fish turn over more slowly. The assertion that turnover time increases with higher position in the trophic structure (Odum, 1967) appears to be supported by these results.

Labile detritus, along with the refractory suspended detritus, exhibits

Table III Analysis of the Daily Turnover Times for the Nitrogen and Phosphorus Pools in the Compartments of WINGRA III[a]

	Mean (days)	Standard deviation (days)	Coefficient of variation
Nitrogen pools			
Unassimilated nutrients			
Green algae	0.72 (0.46)	0.41 (0.057)	0.57 (0.12)
Diatoms	3.3	2.6	0.80
Blue-green algae	1.1 (0.69)	0.63 (0.095)	0.58 (0.14)
Winter algae	7.4	2.6	0.35
Detritus			
Settled	1800 (2200)	430 (380)	0.24 (0.17)
Suspended	14 (8.5)	7.2 (0.76)	0.51 (0.089)
DOM	100 (180)	33 (11)	0.33 (0.061)
Dissolved inorganic N	11 (11)	5.8 (1.6)	0.52 (0.14)
Phosphorus pools			
Unassimilated nutrients			
Green algae	0.72 (0.54)	0.25 (0.042)	0.35 (0.078)
Diatoms	2.9	2.1	0.72
Blue-green algae	1.3 (0.88)	0.56 (0.093)	0.43 (0.11)
Winter algae	6.7	2.4	0.36
Detritus			
Settled	1500 (1400)	370 (150)	0.25 (0.11)
Suspended	10 (6.8)	4.8 (0.58)	0.48 (0.085)
Dissolved inorganic P	0.59 (0.52)	0.51 (0.085)	0.86 (0.16)

[a] Obtained by averaging daily turnover times over the spring and summer seasons. Spring values appear first followed by summer values in parentheses. No summer values are reported for diatoms and winter algae because these pools drop to zero in summer.

turnover times on the order of 1 to 2 weeks. Refractory settled detritus and dissolved organic matter, the two largest pools in the system, appear to be the "slowest" pools, turning over at a rate of 6 yr and 1 yr, respectively, in the spring; and 7 and 3 yr, respectively, in the summer.

Most N and P pools exhibit turnover times similar or identical to those exhibited by the organic matter pools of which they are a part (Tables III and IV). Turnover times of implicit N and P pools are identical to corresponding organic matter pools by definition; consequently, these were not reported. Dissolved inorganic nutrients, DIN and DIP, turn over in 11 and 0.5 days, respectively. Apparently, inorganic nutrients in the water and

Table IV Seasonal Turnover Times (in days) for the Nutrient and Organic Matter Pools in the Compartments of WINGRA III[a]

	Organic matter	Nitrogen	Phosphorus
Unassimilated nutrients			
Green algae	0.43 (0.33)	0.73 (0.47)	0.67 (0.55)
Diatoms	0.72	1.4	1.5
Blue-green algae	0.45 (0.45)	0.69 (0.69)	0.82 (0.88)
Winter algae	2.0	3.7	3.5
Structural biomass			
Green algae	1.8 (1.3)	Same as organic matter	
Diatoms	3.4		
Blue-green algae	1.5 (1.6)		
Winter algae	7.5		
Consumers			
Zooplankton	29 (20)	Same as organic matter	
Benthos	28 (17)		
Young fish	140 (270)		
Adult fish	220 (300)		
Detritus			
Settled, labile/refractory	10 (10)/2100 (2500)	1800 (2100)	1500 (1400)
Suspended, labile/refractory	8.1 (5.7)/10 (6.7)	14 (8.5)	9.8 (6.8)
DOM, labile/refractory	9.0 (8.5)/320 (540)	98 (180)	
Dissolved inorganic nutrients		10 (11)	0.50 (0.50)

[a] Obtained by dividing pool size averaged over the season by flux rate averaged over the season. Spring values appear first followed by summer values in parentheses. No summer values are reported for diatoms and winter algae because these pools drop to zero in summer.

unassimilated nutrients in the algae represent the "smallest, fastest" pools. From the fastest pools to the slowest pools, turnover times vary over three orders of magnitude in this model of a lake ecosystem.

Although the mean values of the turnover times of each compartment are not greatly different from spring to summer, their variance is. All compartment turnover times, except for that of refractory settled detritus, are substantially more variable in the spring than in the summer (Tables II and III). This is also true of the pool sizes and flux rates of many compartments (Watson, 1976). Exceptions to this include diatoms, winter algae, benthos, refractory settled detritus, and refractory DOM.

Table V Analysis of the Spring and Summer Pool Sizes, Flux Rates, and Turnover Times of the Total System Pools of Organic Matter, N, and P in Lake Wingra[a]

	Mean	Standard deviation	Coefficient of variation
Total organic matter			
Pool size (g dry wt m^{-2})	1100 (1100)	11 (3.7)	0.01 (0.0034)
Flux rate (g dry wt m^{-2} day^{-1})	3.8 (3.5)	2.0 (0.51)	0.51 (0.14)
Turnover time (days)			
Daily average[b]	390 (310)	280 (42)	0.72 (0.14)
Seasonal value[c]	290 (310)		
Total system nitrogen			
Pool size (g N m^{-2})	41 (41)	0.18 (0.13)	0.0044 (0.0032)
Flux rate (g N m^{-2} day^{-1})	$3 \cdot 10^{-2}$ ($1 \cdot 10^{-2}$)	$1 \cdot 10^{-2}$ ($8 \cdot 10^{-3}$)	0.48 (0.53)
Turnover time (days)			
Daily average[b]	1700 (3400)	490 (1300)	0.29 (0.39)
Seasonal value[c]	1500 (2900)		
Total system phosphorus			
Pool size (g P m^{-2})	2.3 (2.3)	0.02 (0.01)	0.01 (0.0028)
Flux rate (g P m^{-2} day^{-1})	$8 \cdot 10^{-4}$ ($5 \cdot 10^{-4}$)	$2 \cdot 10^{-3}$ ($7 \cdot 10^{-4}$)	2.1 (1.4)
Turnover time (days)			
Daily average[b]	7300 (10,300)	4700 (5800)	0.64 (0.56)
Seasonal value[c]	2700 (5000)		

[a] Spring values are followed by summer values in parentheses.
[b] Average of turnover times calculated each day.
[c] Average system pool size for the season divided by average system flux rate for the season.

3.2. System Turnover Times

Analysis of daily system pools and fluxes indicates that organic matter turns over for the Lake Wingra system in approximately 1 yr on the average. N turns over in 5 to 9 yr and P in 20 to 28 yr (Table V). Seasonal turnover times (based on the quotient of seasonal averages of system pools and fluxes) are somewhat shorter: less than 1 yr for organic matter, 4 to 5 yr for N, and 7 to 9 yr for P (Table V).

One may compare these system turnover times to the time required for material to take the longest one-way path through the system.[4] This is the RRT discussed in Section 2.3. For organic matter the longest one-way path requires 7 to 9 yr; for N, 6 to 8 yr; and for P, 5 yrs (Table VI). Hence, organic matter turns over in less time than is required for material to follow

---
[4] The longest path through the system in this case is approximately equal to the sum of the compartment turnover times.

Table VI Comparison of System Turnover Time (STT) to Total Compartment Turnover Time (TCTT)[a]

	Organic matter		Nitrogen		Phosphorus	
Total compartment turnover time						
Daily average	2700	(3400)	2400	(2800)	2000	(2000)
Seasonal value	2600	(3100)	2300	(2700)	1900	(2000)
System turnover time						
Daily average	390	(310)	1700	(3400)	7300	(10300)
Seasonal value	290	(310)	1500	(2900)	2700	(5000)
Relative retention (STT/TCTT)						
Daily average	0.14 (0.091)		0.71 (1.2)		3.65 (5.15)	
Seasonal value	0.11 (0.10)		0.65 (1.1)		1.4 (2.5)	

[a] Turnover times are in days.

the longest path through the system, N turns over in about the same length of time, and P turns over in a longer time. This suggests that these materials are retained and recycled to different extents in this system. However, a comparison of system flux and total compartment flux allows a more quantitative assessment of the magnitude of the difference in recycling.

The average daily total compartment flux (TCF) for organic matter (Table VII) is 10 g dry wt m^{-2} day^{-1} in the spring and 9.8 g dry wt m^{-2} day^{-1} in the summer (individual compartment fluxes appear in Watson, 1976). Similar values for N and P are 260 mg N m^{-2} day^{-1} (spring), 240 mg N m^{-2} day^{-1} (summer), and 22 mg P m^{-2} day^{-1} (spring and summer). The ratio of these TCFs and the system fluxes is the relative recycling (RRC) discussed in Section 2.3.

Table VII Comparison of System Flux (SF) to Total Compartment Flux (TCF)

Material	System flux	Total compartment flux	Relative recycling (TCF/SF)
Organic matter (g dry wt m^{-2} day^{-1})	3.8 (3.5)	10 (9.8)	2.6 (2.8)
Nitrogen (mg N m^{-2} day^{-1})	27 (15)	260 (240)	9.6 (16)
Phosphorus (mg P m^{-2} day^{-1})	0.87 (0.49)	22 (22)	25 (45)

Since the organic matter, N, and P models are essentially templates of one another with a similar number of compartments and couplings, the ratio of their RRCs is a measure of the relative amount of cycling and recycling of these elements. The RRC of N is four times that of organic matter in the spring and six times that of the latter in the summer (Table VII). The RRC of P is three times that of N.

The assessment of relative variability (coefficient of variation) indicates that system pool sizes are much less variable through the growing season than are most compartment pools. Only refractory settled detritus and DOM have similarly low variability in pool size (Watson, 1976). However, system turnover times are no less variable than are compartment turnover times (see Tables II, III, and V). As was true of the individual compartments, the system pools, fluxes, and turnover times of organic matter and P show more variability in spring than in summer (Table V); however, the reverse is true for the system flux and turnover time of N.

4. DISCUSSION

4.1. Comparison of Results with Observed Data and Other Simulated Data

Few measurements of turnover time have been made in the field and those that have been are not always directly comparable to the values in this study. Hutchinson (1957) calculated turnover times of 40 to 180 days for P in the sediments of a number of lakes. In WINGRA III, P in settled detritus exhibits a turnover time of 1400 to 1500 days; however, these sediment pools may not be completely analogous (e.g., activity may not have been assessed at the same depth).

Turnover times for DIP in lakes range from hours (Pomeroy, 1960) to minutes (Schindler *et al.*, 1975). This is in general agreement with the average turnover time for DIP produced by WINGRA III (12 hr, with a range from 2.5 hr to 2.5 days).

The turnover time for ammonia seems to be on the order of 24 hr (subarctic lakes—Alexander, 1970) to 40 hr (tropical marine surface waters—Dugdale, 1969). Since nitrate is used only in the absence of ammonia (Pomeroy, 1970), DIN, which contains both of these inorganic forms, might be expected to turn over more slowly than this, as it does in the WINGRA III simulation.

A model of seasonal biomass and nutrient dynamics for a cove in Lake Texoma (Patten *et al.*, 1975) makes possible the calculation of turnover times for several biomass pools in that system, by taking the reciprocal of the maximum turnover rates. Generally, the Lake Texoma model produced algal turnover times which were considerably longer than those produced by WINGRA III; however, the turnover times of most other compartments (zooplankton, fish) appear to be similar (Watson, 1976).

Turnover times also were calculated (Watson, 1976) from measurements of pool sizes, loss rates, and gain rates for dissolved organic carbon (DOC) and particulate organic carbon (POC) reported for Lawrence Lake, Michigan (Wetzel and Rich), 1973). Turnover times of 280 and 16 days for DOC and POC, respectively, appear to agree reasonably well with those of refractory DOM (490 to 940 days) and suspended (particulate) organic matter (7 to 12 days) in Lake Wingra.

Total system turnover times for Lake Wingra cannot be compared to those of other systems because of gross differences in structure and function. However, the relative magnitudes of the turnover times of the elements can be compared. Since biomass is assumed to be 40% C on the average in WINGRA III, the rate of turnover of organic matter gives an indication of the rate of turnover of C relative to N and P. Organic matter (and presumably C) turns over faster than N in Lake Wingra as in several terrestrial systems (Witkamp, 1971; Henderson and Harris, 1975). Contrary to Witkamp's prediction, however, the turnover time of P is longer than that of either N or C.

4.2. Role of Compartments in System Nutrient Dynamics

The analysis of the pool sizes and turnover times of the compartments in the Lake Wingra ecosystem suggests the roles of the various compartments in ecosystem function. Those pools which experience a rapid flux of certain elements, or which store large amounts of these elements for long periods, are of greater significance for certain aspects of ecosystem function (i.e., cycling) than are pools which do not (Pomeroy, 1974).

As part of their theoretical treatment of ecosystems, O'Neill et al. (1974) hypothesized that the evolution of ecosystems toward those of maximum persistent biomass is accomplished by selection toward those with *large, slow* pools. These pools act to conserve essential elements over a time span exceeding that of any living pool. Such conservation is possible because these pools are capable of retaining elements without maintenance energy input. In addition to their importance as conservation mechanisms, these pools are often relatively slow in response to short-term environmental change.

Stability analysis of model systems suggests that large storage pools tend to increase the system's resistance to perturbation (Webster *et al.*,

1974) perhaps by attenuating input signals from the environment (Waide *et al.*, 1974). This buffering of the nutrient dynamics of an ecosystem against environmental perturbations makes large, slow pools an important part of a strategy for persistence.

Nutrient pool buffers in terrestrial systems include litter and soil organic matter and the structural biomass of large autotrophs and herbivores (Burns, 1970; Jordan *et al.*, 1971). In aquatic systems, such as Lake Wingra, the inert fractions of the particulate and dissolved organic matter appear to be the largest, slowest pools and consequently may be expected to function as buffers.

O'Neill *et al.* (1974) also hypothesized that *small, fast* pools are of special significance in the strategy of persistence in that they allow the rapid recycling of limited nutrients. Rapid transfer of nutrients is important in maintaining a high level of productivity when certain essential elements are limiting. In addition, small, fast pools allow the system greater flexibility to take advantage of optimal conditions to achieve high levels of productivity. The O'Neill paper postulates that heterotrophs serve this function in terrestrial systems while minimizing losses of nutrients from the system. In Lake Wingra, dissolved inorganic nutrients in the water and in the algae provide readily available nutrients for the production of biomass. It is not clear that these pools minimize system losses; however, minimization of nutrient losses, while apparently of prime importance to terrestrial systems (Waide *et al.*, 1974), would seem to be of less importance in a eutrophic lake.

The possession of large, slow pools and small, fast pools by ecosystems may explain how these systems balance two mutually exclusive aspects of stability—resistance and resilience—as discussed by Webster *et al.* (1974). Resistance to perturbation was suggested to be related to large, slow storage compartments and a large amount of recycling while resilience was related to smaller, faster pools and a high rate of recycling. The variation in average compartment turnover times of over three orders of magnitude from the slowest to the fastest pools exhibited by this lake model demonstrates how an ecosystem may have "the best of both worlds."

Comparisons of large, slow pools and small, fast pools of different systems could suggest whether these systems place a greater premium on buffering against perturbations or on maximizing productivity and resilience. Odum (1974) asserts that an element is stored in a system only when external sources are irregular and in short supply. Costs involved in storage cause ecosystems and their components to avoid storing elements which are in large or steady external supply. Odum predicted nutrients in such short supply will have the largest, slowest pools in the system, while Pomeroy (1970) claimed that such nutrients will have the smallest, fastest pools. Pomeroy stressed concentration rather than regularity of supply, but pointed out that nutrients in low concentration were not necessarily limiting in the Liebig sense. A nutrient in low concentration can be recycled rapidly and thus not limit the existing system at its present equilibrium.

In Lake Wingra, DIP turns over more rapidly than does DIN, indicating that it may be limiting the system in the sense described above. Recent work by Schindler (1977) seems to substantiate this. On the other hand, the largest, slowest N pool (that in settled detritus) is larger and slower than the largest, slowest P pool. This implies that the system has a less regular external supply of N compared to P, if one accepts Odum's premise. However, the main source of P to the lake is storm runoff while the main source of N is groundwater, suggesting that P is in less regular supply. Consequently, the results from Lake Wingra do not seem to support Odum's hypothesis.

4.3. Compartment and System Variability

After a discussion of the importance of various pools to system stability, it is appropriate to note that the variation in system pool size was indeed low relative to that of most compartments (Watson, 1976). Only the large, slow detrital pools exhibited similarly stable pool sizes. However, system turnover times were not any less variable than were compartment turnover times. It should be realized that in the face of changing inputs unchanging pool sizes and turnover times are mutually exclusive. To maintain a constant pool size, losses must equal gains. As gains and losses increase, turnover time decreases.

The observation concerning differences between spring and summer nutrient dynamics noted in the results may have implications for ecosystem regulation. The nutrient pools of all compartment and system pools (except system N) exhibited much less variability in turnover times in the summer than in the spring. Most also exhibited less variability in pool size and flux rate in the summer (Watson, 1976).

From these observations one might conclude that the nutrient dynamics of most compartments and of the total system are dominated by different kinds of processes in the spring and summer. In the spring, increasing temperature, day length, and nutrient concentrations result in an algal bloom which, in turn, contributes to an increase in zooplankton. In addition, the precipitation record suggests that spring storms introduce stochastic perturbations into the system, causing spring dynamics to be highly variable. Summer, on the other hand, is characterized by less extreme environmental fluctuation and long periods without storms. In the physically stable summer system, lake populations stabilize around an equilibrium point or carrying capacity. Since inputs are low, in-system regeneration becomes an important source of nutrients. The increased importance of recycling adds to the stability of the system by increasing feedback. Pomeroy (1970) pointed out the importance of bioregulation to systems existing in physically stable environments.

One might expect then that in the spring the lake system is driven by population-density-independent forces from outside the system. In the summer, however, the biotic populations, and subsequently the system, appear to be regulated by density-dependent regulation.

4.4. Material Cycling and System Properties

Despite the widely recognized importance of material cycling to such ecosystem properties as productivity and stability (Odum, 1969; Pomeroy, 1970, 1975; Jordan *et al.*, 1972), this concept has no widely accepted operational definition. Consequently, comparisons of the amount of recycling calculated for different systems by different workers are of limited value. In addition, because of differences in pool size and units of measure for different elements, attempts to compare recycling of different elements in the same system are not appropriate without a measure of recycling which normalizes these differences.

The measure of relative recycling (RRC), introduced in Section 2.3., may serve as one means of filling this gap. The dimensionless value obtained by dividing total compartment flux (TCF) by system flux (SF) normalizes the difference in pool sizes and units for the different elements in a system, allowing the comparison of their degree of cycling and recycling. An element recycled to a greater extent will have a greater RRC.

A similar concept was developed with some rigor by Hannon (1973) and extended by Finn (1976). Average path length (APL_i) was defined as the average number of compartments through which a given inflow passes. A weighted mean of all inflow paths gives an average path length for the system ($\overline{APL}$). $\overline{APL}$ was shown to be equal in magnitude to the ratio of the total system throughflow to system inflow. The quantity, total system throughflow (TST), appears to be essentially the same as total compartment flux in this study, and in a steady-state system, system inflows are the same as the system flux used here. Consequently, in a system in steady state, relative recycling and average path length are similar conceptually and have the same magnitude.

Like RRC, mean path length ($\overline{APL}$) gives an idea of the relative amount of recycling in different systems. Like TST and TCF, this quantity depends on the number of compartments in the model. As a consequence, it would not be appropriate to compare APL (RRC) or TST (TCF) of models which are different in structure (i.e., differ in number of compartments). However, comparing a system to itself at different times of the year or under different management regimes would be appropriate, as would comparing different elements in the same system.

Earlier, a comparison of RRCs indicated that P undergoes three times the amount of cycling and/or recycling that is experienced by N, and that N is cycled/recycled four to six times as much as is organic matter in general. One may also compare the RRCs obtained for the highly variable spring to those of the more stable summer (Table VII). One finds that in each case the RRC is greater in the summer. Thus, while recycling is apparently higher in the summer, nutrient input is lower (Prentki *et al.*, 1977). Jordan *et al.* (1972) predicted a higher degree of stability for systems with a high ratio of recycled flux to influx than for systems with a low ratio. Waide *et al.* (1974) concurred and went on to assert that bioregulation takes on greater importance in such systems. Reasons for greater significance of bioregulation in the Lake Wingra summer system as opposed to its spring system were discussed in Section 4.3.

It should be realized that, in addition to nutrient cycling, storage of nutrients in a growing pool of organic matter can have a profound effect on system turnover times. Vitousek and Reiners (1975) argue that, while the organic matter pool in an ecosystem is increasing, there will be nutrient retention, where nutrient retention is defined as occurring when nutrient output from the system is less than nutrient input. Once a system reaches a steady state and the pool of organic matter is no longer increasing, nutrient inputs essentially will equal nutrient outputs.

There is no net increase in the system pool of organic matter in the model over the growing season. However, total N increases somewhat and total P increases still more (relative to its initial pool size). This increase in pool size over the growing season may explain part of the differential retention of these elements. However, it appears from the relative recycling values summarized in Table VII that a large quantity of material is fluxing through the system compartments, relative to the flux through the system, rather than remaining in a few large storage compartments. Also, the big, slow pools of N and P are not slower than those of organic matter, and yet the RRC and system turnover times for these elements are much longer than for organic matter.

Thus, the progressively *higher* rate of internal flux (TCF) relative to system flux (SF) for elements with progressively *longer* system turnover times provides some support for the postulated differential recycling of elements as an explanation for some, if not all, of the difference in turnover times in system organic matter, N and P, in Lake Wingra.

5. SUMMARY AND CONCLUSIONS

Hypotheses concerning ecosystem properties and the role of system parts were investigated through an analysis of the simulated nutrient dynamics of a lake ecosystem. Pool sizes, flux rates, and turnover times were computed for the organic matter, N, and P pools of the Lake Wingra ecosystem, as represented by the WINGRA III ecosystem model. Results indicate that

(a) Turnover times vary substantially between different compartments of the ecosystem (the range is over three orders of magnitude). Specifically, turnover times increase from lower to higher trophic levels, from labile to refractory detritus, and from inorganic nutrient pools to organic pools.

(b) The variability in turnover times, flux rates, and pool sizes was observed to be much greater in the spring relative to the summer; variability in pool sizes was much less for system pools and large, slow compartments than for small, fast compartments; system pools and compartments exhibited similar variability in turnover times.

(c) The ratio of the system turnover time to the total compartment turnover times was taken to give an indication of the relative retention (RRT) of nutrients, while the ratio of total flux through all compartments to system flux was taken to give an indication of the relative amount of recycling (RRC) in the system. Both ratios were found to be greater for P than for N and greater for N than for organic matter in general (i.e., P > N > organic matter). RRC was found to be greater in the summer than in the spring for all pools while RRT was greater in the summer for N and P only.

From these results it is concluded that

(a) Differences in turnover times between compartments suggest that these compartments serve different functions in the ecosystem. Small, fast components serve as *movers* and *cyclers* of elements while large, slow components act as *holders* of elements and *buffers* against perturbation. The wide spread in turnover times within a system suggests how an ecosystem may *balance* resilience and resistance.

(b) Nutrient dynamics are more stable, retention appears greater, and cycling/recycling occurs to a greater extent in the summer than in the spring, suggesting that the nutrient dynamics of this lake ecosystem are influences greatly by stochastic forces external to the system in spring and are more influenced by internal regulation in summer.

(c) Higher degrees of retention and cycling/recycling of P relative to N and of N relative to organic matter in general are indicated. This, coupled with the shorter turnover time of available P relative to available N, suggests that P may be limiting the productivity of the system more often than is N.

ACKNOWLEDGMENTS

We would sincerely like to thank the following for their time, advice, criticisms, and ideas: T. F. H. Allen, P. Allen, P. R. Weiler, D. Rogers, B. Reynolds, R. M. Friedman, and R. T. Prentki.

This research was supported by National Science Foundation Grant No. DEB76-11776, and by the Eastern Deciduous Forest Biome, US-IBP, funded by the National Science Foundation under Interagency Agreement AG-199, and under BMS76-00761 with the Energy Research and Development Administration—Oak Ridge National Laboratory.

Contribution No. 302 from the Eastern Deciduous Forest Biome, Oak Ridge National Laboratory, Oak Ridge, Tennessee.

REFERENCES

Alexander, V. (1970). Relationships between turnover rates in the biological nitrogen cycle and algal productivity. *Eng. Bull. Purdue Univ., Eng. Ext. Ser.* **137**, 1–7.

Bastin, T. (1969). A general property of hierarchies. *In* "Towards a Theoretical Biology. II. Sketches" (C. H. Waddington, ed.), IUBS Symposium, pp. 252–267. Aldine Publ. Co., Chicago.

Burns, L. A. (1970). Analog simulation of a rain forest with high-low pass filters and a programmatic spring pulse. *In* "A Tropical Rain Forest: A Study of Irradiation and Ecology at El Verde, Puerto Rico" (H. T. Odum and R. F. Pigeon, eds.), Appendix D, Vol. I, pp. 284–289. USAEC Tech. Inf. Ext., Oak Ridge, Tennessee.

Dugdale, R. C. (1969). The nitrogen cycle in the sea. *In* "Biology and Ecology of Nitrogen," pp. 16–18. Natl. Acad. Sci., Washington, D.C.

Finn, J. T. (1976). Measures of ecosystem structure and function derived from analysis of flows. *J. Theor. Biol.* **56**, 363–380.

Hannon, B. (1973). The structure of ecosystems. *J. Theor. Biol.* **41**, 535–546.

Harwell, M. A., Cropper, W. P., Jr., and Ragsdale, H. L. (1977). Nutrient recycling and stability: A re-evaluation. *Ecology* **58**, 660–666.

Henderson, G. S., and Harris, W. F. (1975). An ecosystem approach to characterization of the nitrogen cycle in a deciduous forest watershed. *In* "Forest Soils and Forest Land Management" (B. Bernier and C. H. Winget, eds.), pp. 179–193. Laval Univ. Press, Quebec.

Huff, D. D., Koonce, J. F., Ivarson, W. R., Weiler, P. R., Dettmann, E. H., and Harris, R. F. (1973). Simulation of urban runoff, nutrient loading and biotic response of a shallow eutrophic lake. *In* "Modeling the Eutrophication Process" (E. J. Middlebrooks, D. H. Falkenborg, and T. E. Maloney, eds.), pp. 33–55. Ann Arbor Sci. Publ., Ann Arbor, Michigan.

Hutchinson, G. E. (1957). "A Treatise on Limnology. Geography, Physics and Chemistry," Vol. I. Wiley, New York.

Jordan, C. F., Kline, J. R., and Sasscer, D. S. (1972). Relative stability of mineral cycles in forest ecosystems. *Am. Nat.* **106**, 237–253.

Jordan, P. A., Botkin, D. B., and Wolfe, M. L. (1971). Biomass dynamics in a moose population. *Ecology* **52**, 147–152.

Klir, G. J. (1969). "An Approach to General Systems Theory." Van Nostrand-Reinhold, New York.

Loucks, O. L., and Weiler, P. R. (1979). Description and evaluation of the WINGRA III ecosystem models. I. Objectives and form of the WINGRA III ecosystem model. In preparation.

Odum, E. P. (1969). The strategy of ecosystem development. *Science* **164**, 262–270.

Odum, H. T. (1967). Biological circuits and the marine systems of Texas. *In* "Pollution and Marine Ecology" (T. A. Olson and F. J. Burgess, eds.), pp. 99–157. Wiley, Interscience, New York.

Odum, H. T. (1974). Combining energy laws and corollaries of the maximum power principle with visual systems mathematics. *In* "Ecosystem: Analysis and Prediction" (S. A. Levin, ed.), pp. 239–263. SIAM Inst. Math. Soc., Philadelphia.

O'Neill, R. V. (1975). Modeling in the eastern deciduous forest biome. *In* "Systems Analysis and Simulation in Ecology" (B. C. Patten, ed.), Vol. 3, pp. 49–72. Academic Press, New York.

O'Neill, R. V., Harris, W. F., Ausmus, B. S., and Reichle, D. E. (1974). A theoretical basis for ecosystem analysis with particular reference to element cycling. *EDFB Memo Rep.* **74-7**, 1–21.

Patten, B. C., Egloff, D. A., Richardson, T. H., and 38 other co-authors. (1975). Total ecosystem model for a cove in Lake Texoma. *In* "Systems Analysis and Simulation in Ecology" (B. C. Patten, ed.), Vol. 3, pp. 205–421. Academic Press, New York.

Pomeroy, L. R. (1960). Residence time of dissolved phosphate in natural waters. *Science* **131**, 1731–1732.

Pomeroy, L. R. (1970). The strategy of mineral cycling. *Annu. Rev. Ecol. Syst.* **1**, 171–190.

Pomeroy, L. R., ed. (1974). "Cyclings of Essential Elements." Dowden, Hutchinson & Ross, Inc., Stroudsburg, Pennsylvania.

Pomeroy, L. R. (1975). Mineral cycling in marine ecosystems. *In* "Mineral Cycling in Southeastern Ecosystems" (F. G. Howell, J. B. Gentry, and M. H. Smith, eds.), pp. 209–223. ERDA Symposium Series. Technical Information Center, 1975.

Prentki, R. T., Rogers, D. S., Watson, V. J., Weiler, P. R., and Loucks, O. L. (1977). Summary tables of Lake Wingra basin data. *IES Report No. 85*. Center for Biotic Systems. Institute of Environmental Studies, University of Wisconsin, Madison.

Robertson, J. S. (1957). Theory and use of tracers in determining transfer rates in biological systems. *Physiol. Rev.* **37**, 133–154.

Schindler, D. W. (1977). Evolution of phosphorus limitation in lakes. *Science* **195**, 260–262.

Schindler, D. W., Lean, D. R. S., and Fee, E. J. (1975). Nutrient cycling in freshwater ecosystems. *In* "Productivity of World Ecosystems" (D. E. Reichle, J. F. Franklin, and D. W. Goodall, eds.), pp. 96–105. Natl. Acad. Sci., Washington, D.C.

Vitousek, P. M., and Reiners, W. A. (1975). Ecosystem succession and nutrient retention: A hypothesis. *BioScience* **25**, 376–381.

Waide, J. B., Krebs, J. E., Clarkson, S. P., and Setzler, E. M. (1974). A linear systems analysis of the calcium cycle in a forested watershed ecosystem. *Prog. Theor. Biol.* **3**, 261–345.

Watson, V. (1976). An analysis of turnover times in a lake ecosystem and some implications for emergent properties. Master's Thesis. University of Wisconsin, Madison.

Webster, J. R., Waide, J. B., and Patten, B. C. (1974). Nutrient recycling and stability. *In* "Mineral Cycling in Southeastern Ecosystems" (F. G. Howell, J. B. Gentry, and M. H. Smith, eds.), pp. 1–27. ERDA Symposium Series. Technical Information Center, 1975.

Wetzel, R. G., and Rich, P. H. (1973). Carbon in freshwater ecosystems. *In* "Carbon and the Biosphere" (G. M. Woodwell and E. V. Pecan, eds.), pp. 241–263. *24th Brookhaven Symposium in Biology*, AEC Publication, Springfield, Va.

Witkamp, M. (1971). Soils as components of ecosystems. *Annu. Rev. Ecol. Syst.* **2**, 85–110.

Chapter **15**

THE USEFULNESS OF OPTIMAL CONTROL THEORY TO ECOLOGICAL PROBLEMS

B. S. Goh

1. Introduction	385
2. Discrete-Time Optimal Control	387
2.1 Introduction	387
2.2 Nonlinear Programming	388
2.3 Standard Discrete-Time Optimal Control Problem	389
2.4. Optimal Size Limit for a Fishery	391
3. Continuous-Time Optimal Control	394
3.1 Introduction	394
3.2 Standard Continuous-Time Optimal Control Problem	394
3.3 Optimal Harvesting of a Fish Population	396
4. Conclusions	398
References	398

1. INTRODUCTION

The main purpose of management is to make good decisions subject to realistic constraints. This leads naturally to a mathematical optimization problem. Ecosystems are dynamical systems. It follows that the appropriate tool for formulating optimal policies is optimal control theory. But there are two great difficulties in applying optimal control theory in the management of an ecosystem. First, it requires an adequate model of the ecosystem. Second, realistic models of ecosystems would have many state variables. This leads to considerable computational difficulties. For these reasons the potential usefulness of optimal control in the management of ecosystems is demonstrated by applying it to two simple problems in the management of fisheries.

Optimal control theory is the modern version of calculus of variations, its origin going back to ancient Greece. The prototype necessary conditions for optimality in the calculus of variations were obtained by Euler, Legendre, Jacobi, and Weierstrass in 1744, 1786, 1837, and 1879, respectively (see Bliss, 1925, 1946). Most of the necessary conditions for optimality in a continuous-time optimal control problem which are needed in practice were available by 1940. These are described in the classic book on calculus of variations by Bliss (1946).

In the early 1950s Bellman (1957) formulated the principle of optimality for dynamic programming and applied it to many problems. The applications of dynamic programming to problems in automatic control motivated Pontryagin in 1956 to develop the Maximum Principle (see Pontryagin *et al.*, 1962). Independent of these developments, a small group of scientists in the early 1950s were using the classical calculus of variations to compute optimal rocket trajectories (see Lawden, 1963; Leitmann, 1965). It soon became clear that the Pontryagin Maximum Principle provided only a more lucid version of some necessary conditions in the classical calculus of variations (Berkovitz, 1961). It is unfortunate that many books on optimal control theory fail to explain that the Pontryagin Maximum Principle contains only first-order necessary conditions. It does not include second-order conditions like the focal point condition (Bliss, 1946). When the optimal control vector belongs to the interior of its admissible set, the focal point condition provides the criterion which states when a sequence of short-term optimal policies fails to be a long-term optimal policy. When the control vector is on the boundary of its admissible set, long-term optimality is given by the switching conditions in the Pontryagin Maximum Principle.

One important theoretical development in the 1960s was the derivation of new necessary conditions for singular control (Kelley *et al.*, 1967; Goh, 1966, 1973; Bell and Jacobson, 1975). Singular control usually appears in an optimal control problem in which one or more of the control variables appear linearly in the system dynamics and the objective function. Interesting problems of this kind first appeared in rocket problems (Lawden, 1963). This is because the thrust control variable appears linearly in rocket dynamics. Similarly, singular control is important in the management of ecosystems because control variables often occur linearly in ecosystem dynamics.

Applications of optimal control theory to ecological problems appeared only in the late 1960s. It appears that several authors, independently of each other, began to apply optimal control theory to ecological problems at about the same time (Watt, 1968; Goh, 1969/1970; Becker, 1970; Clark, 1971). Recently Clark (1976), Conway (1977), and Wickwire (1977) have reviewed the applications of optimal control theory in resource management and in the control of epidemics.

In this volume Brewer (Chapter 16) discusses a method to overcome the severe computational difficulties of optimal control theory. This method approximates control programs by control impulses. This approach converts a dynamic optimal control problem into a static optimization problem which is then amenable to effective numerical optimization techniques. Also in this volume, Chapter 17, Singh discusses hierarchical methods for solving optimization and control problems for large-scale interconnected dynamical systems which can be described by a system of linear differential equations or a system of linear difference equations.

2. DISCRETE-TIME OPTIMAL CONTROL

2.1. Introduction

There is a growing realization that difference equations can provide realistic models of ecosystems when differential equations fail to do so (Innis, 1974; May, 1976). The optimization technique for formulating optimal policies in the management of an ecosystem described by difference equations is the discrete-time optimal control theory (Goh, 1977; Singh, this volume, Chapter 17). Around 1960 it became obvious to several control theorists that a discrete-time optimal control problem is none other than a nonlinear programming problem with a special structure. This means that numerical algorithms which have been developed for problems in nonlinear programming can be applied to a discrete-time optimal control problem (Tabak and Kuo, 1971; Canon et al., 1970). For this reason we shall first describe briefly some relevant results in nonlinear programming. Unfortunately, in this approach a low-order optimal control problem with a long planning horizon is converted into a nonlinear programming problem with a large number of variables (see Canon et al., 1970). This leads to computational difficulties. There is a need for further research in this area.

2.2. Nonlinear Programming

The standard nonlinear programming problem is as follows:

Minimize $f(x_1, x_2, \ldots, x_n)$
subject to $g_s(x_1, x_2, \ldots, x_n) = 0$, $\quad s = 1, 2, \ldots, k$,
$h_r(x_1, x_2, \ldots, x_n) \geq 0$, $\quad r = 1, 2, \ldots, m$.

Let $p_1, p_2, \ldots, p_k, w_1, w_2, \ldots, w_m$ be a set of constants. The augmented scalar function is

$$F(\mathbf{x}, \mathbf{p}, \mathbf{w}) = f(\mathbf{x}) + \sum_{s=1}^{k} p_s g_s(\mathbf{x}) - \sum_{r=1}^{m} w_r h_r(\mathbf{x}).$$

By definition a vector $\mathbf{x}$ is admissible if it satisfies the constraints in the problem. For easy reference we shall state some of the mathematical results as theorems.

Theorem 1. A necessary condition for $f(\mathbf{x})$ to be optimal at an admissible point $\mathbf{x}^*$ is that there exists a set of multipliers $p_1, p_2, \ldots, p_k$, $w_1, w_2, \ldots, w_m$ such that

$$w_r \geq 0, \; w_r h_r(\mathbf{x}^*) = 0, \quad r = 1, 2, \ldots, m,$$
$$\partial F(\mathbf{x}^*, \mathbf{p}, \mathbf{w})/\partial x_i = 0, \quad i = 1, 2, \ldots, n.$$

A lucid proof of this theorem is given in the book by Fiacco and McCormick (1968). It characterizes an optimal solution but it does not provide a constructive method for getting an optimum solution. It requires that the gradient of the augmented function $\partial F(\mathbf{x}^*, \mathbf{p}, \mathbf{w})/\partial \mathbf{x}$ must be a zero vector at an optimum solution. In practice the optimum solution must first be determined by means of one of the many numerical algorithms for solving a nonlinear programming problem which are described in the book by Himmelblau (1972).

Consider a nonlinear programming problem with a special structure which applies directly to an important class of discrete-time optimal control problems. The problem is as follows:

Minimize $f(y_1, y_2, \ldots, y_n, u_1, u_2, \ldots, u_m)$
subject to $g_s(y_1, y_2, \ldots, y_n, u_1, u_2, \ldots, u_m) = 0$, $\quad s = 1, 2, \ldots, k$,
$a_r \leq u_r \leq b_r$, $\quad r = 1, 2, \ldots, m$. (1)

Let $p_1, p_2, \ldots, p_k$ be constants. The scalar function

$$G(\mathbf{y}, \mathbf{u}, \mathbf{p}) = f(\mathbf{y}, \mathbf{u}) + \sum_{s=1}^{k} p_s g_s(\mathbf{y}, \mathbf{u}).$$

Theorem 2. A necessary condition for $f(\mathbf{y}, \mathbf{u})$ to be optimal at an admissible point $(\mathbf{y}^*, \mathbf{u}^*)$ is that there exists a set of multipliers $p_1, p_2, \ldots, p_k$ such that for $i = 1, 2, \ldots, n$ and $r = 1, 2, \ldots, m$

$\partial G(\mathbf{y}^*, \mathbf{u}^*, \mathbf{p})/\partial y_i = 0$,
$u_r^* = a_r$ only if $\partial G(\mathbf{y}^*, \mathbf{u}^*, \mathbf{p})/\partial u_r \geq 0$,
$a_r < u_r^* < b_r$ only if $\partial G(\mathbf{y}^*, \mathbf{u}^*, \mathbf{p})/\partial u_r = 0$,
$u_r^* = b_r$ only if $\partial G(\mathbf{y}^*, \mathbf{u}^*, \mathbf{p})\partial u_r \leq 0$.

Proof. The inequality in Eq. (1) is equivalent to

$$u_r - a_r \geq 0, \qquad b_r - u_r \geq 0, \qquad r = 1, 2, \ldots, m.$$

Let $w_1, w_2, \ldots, w_m, v_1, v_2, \ldots, v_m$ be constants and

$$F = f + \sum_{s=1}^{k} p_s g_s - \sum_{r=1}^{m} w_r(u_r - a_r) - \sum_{r=1}^{m} v_r(b_r - u_r).$$

By Theorem 1 necessary conditions for optimality are

$\partial F/\partial y_i = \partial G/\partial y_i = 0$, $i = 1, 2, \ldots, n$,
$w_r \geq 0, w_r(u_r^* - a_r) = 0$, $r = 1, 2, \ldots, m$,
$v_r \geq 0, v_r(b_r - u_r^*) = 0$, $r = 1, 2, \ldots, m$,
$\partial F/\partial u_r = \partial G/\partial u_r - w_r + v_r = 0$, $r = 1, 2, \ldots, m$.

If $u_r^* = a_r$, these conditions imply that $v_r = 0$ and

$$\partial G/\partial u_r = w_r \geq 0.$$

Similarly if $u_r^* = b_r$, we have $\partial G/\partial u_r \leq 0$. This completes the proof.

2.3. Standard Discrete-Time Optimal Control Problem

Let $\mathbf{x} = (x_1, x_2, \ldots, x_n)^T$ be the state vector of a system, and $\mathbf{u} = (u_1, u_2, \ldots, u_m)^T$ be the control vector which can be manipulated within certain constraints. For $r = 1, 2, \ldots, m$, let a_r and b_r be constants. Assume that the state at time $t = 0$ can be measured and the environment is deterministic. If the environment is not deterministic the control policy should be updated periodically. This means that the control policy should be implemented in a feedback manner. Suppose the ecosystem dynamics is described by a set of difference equations.

The problem is to choose a control sequence $\mathbf{u}(0), \mathbf{u}(1), \ldots, \mathbf{u}(T-1)$ which will drive the system from an observed state at time $t = 0$ to a state at time $t = T$, which satisfies a set of terminal conditions, and such that a given objective function is minimized. Briefly, the standard discrete-time optimal control problem for the management of ecosystems is as follows:

System: $x_i(t+1) = f_i[\mathbf{x}(t), \mathbf{u}(t)], \quad i = 1, 2, \ldots, n;$
Initially: $x_i(0) = x_{i0};$
Terminally: $\phi^s[\mathbf{x}(T)] = 0, \quad s = 1, 2, \ldots, k;$
Constraints: $a_r \leqslant u_r(t) \leqslant b_r, \quad r = 1, 2, \ldots, m;$
Objective: $\min g[\mathbf{x}(T)] + \sum_{t=0}^{T-1} L[\mathbf{x}(t), \mathbf{u}(t)].$

Here g and L are scalar functions.

By definition the Hamiltonian function at time t is

$$H[\mathbf{x}(t), \mathbf{u}(t), \mathbf{p}(t+1)] = L[\mathbf{x}(t), \mathbf{u}(t)] + \sum_{i=1}^{n} p_i(t+1) f_i[\mathbf{x}(t), \mathbf{u}(t)],$$

where $\mathbf{p}(t+1)$ is a vector at time $t+1$. The vector $\mathbf{p}(t)$ is called the costate vector. For convenience we use $H(t)$ in place of $H[\mathbf{x}(t), \mathbf{u}(t), \mathbf{p}(t+1)]$.

Theorem 3. If $\mathbf{u}(0), \mathbf{u}(1), \ldots, \mathbf{u}(T-1)$ is an optimal sequence and $\mathbf{x}(0), \mathbf{x}(1), \ldots, \mathbf{x}(T-1)$ is an optimal trajectory, then there exists costate vectors $\mathbf{p}(0), \mathbf{p}(1), \ldots, \mathbf{p}(T)$ and scalar multipliers $\lambda_1, \lambda_2, \ldots, \lambda_k$ such that

$$p_i(t) = \partial H/\partial x_i(t), \qquad i = 1, 2, \ldots, n, \qquad (2)$$

For

$$r = 1, 2, \ldots, m, a_r < u_r < b_r \quad \text{only if} \quad \partial H/\partial u_r = 0, \qquad (3a)$$
$$u_r = a_r \quad \text{only if} \quad \partial H/\partial u_r \geqslant 0, \qquad (3b)$$
$$u_r = b_r \quad \text{only if} \quad \partial H/\partial u_r \leqslant 0, \qquad (3c)$$
$$p_i(T) = \frac{\partial g}{\partial x_i(T)} + \sum_{s=1}^{k} \lambda_s \frac{\partial \phi^s}{\partial x_i(T)}, \qquad i = 1, 2, \ldots, n. \qquad (4)$$

Proof. By definition the augmented function

$$\begin{aligned}
G &= g[\mathbf{x}(T)] + \sum_{t=0}^{T-1} \Bigg[L\{\mathbf{x}(t), \mathbf{u}(t)\} \\
&\quad + \sum_{i=1}^{n} p_i(t+1)\{f_i(\mathbf{x}(t), \mathbf{u}(t)) - x_i(t+1)\} \Bigg] \\
&\quad + \sum_{s=1}^{k} \lambda_s \phi^s[\mathbf{x}(T)] \\
&= g[\mathbf{x}(T)] + \sum_{t=0}^{T-1} H(t) - \sum_{t=1}^{T-1} \sum_{i=1}^{n} p_i(t) x_i(t) \\
&\quad - \sum_{i=1}^{n} p_i(T) x_i(T) + \sum_{s=1}^{k} \lambda_s \phi^s[\mathbf{x}(T)].
\end{aligned}$$

By Theorem 2 necessary conditions for optimality are

$$\partial G/\partial x_i(t) = \partial H(t)/\partial x_i(t) - p_i(t) = 0,$$

$$\partial G/\partial x_i(T) = \partial g/\partial x_i(T) - p_i(T) + \sum_{s=1}^{k} \lambda_s \partial \phi^s/\partial x_i(T) = 0,$$

$$u_r = a_r \quad \text{only if} \quad \partial G/\partial u_r = \partial H/\partial u_r \geqslant 0,$$
$$a_r < u_r < b_r \quad \text{only if} \quad \partial G/\partial u_r = \partial H/\partial u_r = 0,$$
$$u_r = b_r \quad \text{only if} \quad \partial G/\partial u_r = \partial H/\partial u_r \leqslant 0,$$

where $i = 1, 2, \ldots, n$ and $r = 1, 2, \ldots, m$. This completes the proof.

From this proof it should be clear that the optimal control of a discrete-time system with time delays is no more difficult than that for a discrete-time system without any time delay. For this reason, in the optimal control of a biological system with time delays, the discrete-time optimal control theory should be used in preference to the continuous-time optimal control theory.

One of the most successful areas in the applications of mathematics to biological systems is in the management of fisheries. We shall illustrate the use of the discrete-time optimal control theory to a central problem in the management of a fishery. In this problem we seek the policy which will maximize the biomass yield from a fishery which contains fish of different ages.

2.4. Optimal Size Limit for a Fishery

Theorem 3 can be used to derive a criterion for computing the optimal size limit for a fishery with a limited season. First, we develop the optimal policy for harvesting a fishery with a single year-class. We then establish that a fishery with many year-classes can be exploited optimally if an optimal size limit is prescribed.

By definition a year-class is the cohort of fish born in a given year. At time t let the age of a year-class be t and let the number of fish in the year-class be $x(t)$. Let the limited fishing season take place during the subinterval $[t, t+\Delta t]$ in the unit time interval $[t, t+1]$. Let $s = 1/\Delta t$. During an initial period the dynamics of a year-class is usually very variable and is highly dependent on external factors. The dynamics of the year-class from the age of recruitment onward can be adequately described by a linear equation (Beverton and Holt, 1957). Let T be the lifespan of the year-class.

In the absence of fishing,

$$x(t+\Delta t) = (1-m)x(t), \tag{5}$$

where m is the natural mortality coefficient in the interval $[t, t+\Delta t]$. It follows that

$$x(t+1) = (1-m)^{s-1}x(t+\Delta t).$$

If $u(t)$ is the fishing mortality coefficient during the fishing season $[t, t+\Delta t]$, then

$$x(t+\Delta t) = [1 - m - u(t)]x(t).$$

Combining these equations

$$x(t+1) = f[u(t)]x(t), \tag{6}$$

where $f[u(t)] = (1-m)^{s-1}[1-m-u(t)]$.

Initially

$$x(t_0) = x_0,$$

and at time T,

$$x(T) > 0.$$

In practice the fishing mortality is limited because the number of boats and the number of hours that they are in service are limited. This implies that

$$0 \leqslant u(t) \leqslant b,$$

where b is a constant.

Let $w(t+\theta\Delta t)$ be the average weight of a fish in the year-class at age $t+\theta\Delta t$ where $\theta = 0.5$. Suppose the management problem is to choose a sequence of fishing levels $u(t_0), u(t_0+1), \ldots, u(T-1)$ so as to maximize the total biomass yield from the year-class. The total biomass yield is

$$Y = \sum_{t=t_0}^{T-1} [u(t)x(t)w(t+\theta\Delta t)].$$

We shall minimize $-Y$ because in the standard problem a function is minimized. A maximization problem can be converted into a minimization problem by using the identity $\max Y = -\min(-Y)$.

The Hamiltonian function is

$$H[x, u, p] = p(t+1)f[u(t)]x(t) - u(t)x(t)w(t+\theta\Delta t). \tag{7}$$

Equations (2) and (6) give

$$p(t) = \partial H/\partial x(t) = p(t+1)f[u(t)] - u(t)w(t+\theta\Delta t)$$
$$\Rightarrow p(t)x(t) - p(t+1)x(t+1) = -u(t)w(t+\theta\Delta t)x(t) \tag{8}$$
$$\Rightarrow p(t)x(t) - p(T)x(T) = -\sum_{\tau=t}^{T-1} u(\tau)w(\tau+\theta\Delta t)x(\tau). \tag{9}$$

At the terminal time, $x(T) > 0$ but otherwise it is unconstrained. Hence by Eq. (4), $p(T) = 0$. Therefore Eq. (9) becomes

$$p(t)x(t) = -\sum_{\tau=t}^{T-1} u(\tau)w(\tau + \theta\Delta t)x(\tau). \tag{10}$$

The optimal control program seeks the balance between gain in weight from growth of individual fish in the year-class and loss in biomass yield from natural mortalities. This observation suggests that there should be a single switch in the optimal control program. Hence there exists t^* such that $u^* = 0$ for $t = t_0, t_0+1, \ldots, t^*-1$, and $u^* = b$ for $t = t^*, t^*+1, t^*+2, \ldots, T-1$. The purpose of the optimal control analysis is to get a criterion which will determine t^* quantitatively.

Equation (7) implies

$$\partial H/\partial u(t) = -(1-m)^{s-1}p(t+1)x(t) - w(t+\theta\Delta t)x(t). \tag{11}$$

Equation (3) implies that $u(t) = 0$ only if $\partial H/\partial u(t) \geq 0$, and $u(t) = b$ only if $\partial H/\partial u(t) \leq 0$. Equations (8) and (10) imply

$$p(t+1)x(t) = -\frac{1}{f[u(t)]} \sum_{\tau=t+1}^{T-1} u(\tau)w(\tau + \theta\Delta t)x(\tau). \tag{12}$$

Equations (11) and (12) give

$$\frac{\partial H}{\partial u(t)} = -w(t+\theta\Delta t)x(t) + \frac{(1-m)^{s-1}}{f[u(t)]} \sum_{\tau=t+1}^{T-1} u(\tau)w(\tau + \theta\Delta t)x(\tau). \tag{13}$$

Equations (3), (6), and (13) imply

$$w(t+\theta\Delta t) > (1-m)^{s-1} \sum_{\tau=t+1}^{T-1} bw(\tau + \theta\Delta t)[f(b)]^{\tau-t-1}$$

whenever $t \geq t^*$. This inequality criterion enables the determination of t^* numerically.

Suppose that the fishery contains many year-classes. If nets are used it is not possible to exploit each year-class separately. Thus the problem has a very intricate constraint. Assume for a moment that it is possible to exploit each year-class separately. This means that the nonselectivity constraint is relaxed and hence the set of admissible options is enlarged. For a collection of year-classes which can be exploited separately, the optimal policy is to exploit each year-class optimally. The crux of the argument is that this relaxed optimal policy can be implemented when the fishery contains many different year-classes which cannot be exploited separately. It can be implemented by specifying the optimum size limit which is the length of a fish at age t^*. This policy gives the maximum total biomass yield for a fixed fishing effort.

3. CONTINUOUS-TIME OPTIMAL CONTROL

3.1. Introduction

For ecosystem models described by differential equations, the appropriate tool for formulating optimal policies is the continuous-time optimal control theory.

In some ways the continuous-time optimal control theory is more deeply developed than the discrete-time optimal control theory. For example, in a continuous-time optimal control problem, the Hamiltonian function is minimized relative to the control variables whenever the optimal control vector belongs to the interior of its admissible set. Without convexity assumptions this condition is not necessary in discrete-time optimal control theory. In continuous-time optimal control theory the generalized Legendre conditions for a singular control have been very effective in establishing optimality or nonoptimality of the singular control in some important problems. In discrete-time optimal control, the state and control variables have the same status. For this reason it is difficult to develop a useful theory of second-order necessary conditions for singular control in discrete-time optimal control.

The main advantages of the discrete-time optimal control theory relative to the continuous-time optimal control theory are (i) time delays do not pose any additional difficulties and (ii) numerical methods developed for nonlinear programming problems can be used directly.

3.2. Standard Continuous-Time Optimal Control Problem

In brief, the standard continuous-time optimal control problem for managing an ecosystem is as follows:

$$
\begin{aligned}
&\text{System:} && \dot{x}_i = f_i(\mathbf{x}, \mathbf{u}), && i = 1, 2, \ldots, n; \\
&\text{Initially:} && x_i(0) = x_{i0}; \\
&\text{Terminally:} && \phi^s[T, \mathbf{x}(T)] = 0, && s = 1, 2, \ldots, k; \\
&\text{Constraints:} && a_r \leqslant u_r \leqslant b_r, && r = 1, 2, \ldots, m; \\
&\text{Objective:} && \min g[T, \mathbf{x}(T)] + \int_0^T L(\mathbf{x}, \mathbf{u})\, dt.
\end{aligned}
\tag{14}
$$

Here g and L denote scalar functions. For convenience the set of admissible control vectors is denoted by U.

The correct proofs of necessary conditions in continuous-time optimal control theory are very difficult. Therefore we shall state without proof some useful sets of necessary conditions. By definition the Hamiltonian function

$$H(\mathbf{x}, \mathbf{u}, \mathbf{p}) = L(\mathbf{x}, \mathbf{u}) + \sum_{i=1}^{n} p_i f_i(\mathbf{x}, \mathbf{u}).$$

Theorem 4. If $(\mathbf{x}^*, \mathbf{u}^*)$ is a normal optimal set, then there exists a costate vector $\mathbf{p}(t)$ and constant multipliers $\lambda_1, \lambda_2, \ldots, \lambda_k$ such that for $i = 1, 2, \ldots, n$ and $r = 1, 2, \ldots, m$

$$\dot{p}_i = -\partial H/\partial x_i,$$

$$p_i(T) = \partial g/\partial x_i(T) + \sum_{s=1}^{k} \lambda_s \partial \phi^s/\partial x_i(T),$$

$$H(T) + \partial g/\partial T + \sum_{s=1}^{k} \lambda_s \partial \phi^s/\partial T = 0,$$

$u_r^* = a_r \qquad$ only if $\partial H/\partial u_r > 0$
$u_r^* = b_r \qquad$ only if $\partial H/\partial u_r < 0$
$a_r < u_r^* < b_r \qquad$ only if $\partial H/\partial u_r = 0$.

If $\mathbf{u}^*$ belongs to the interior, necessary conditions for optimality are (i) $\partial H/\partial u_r = 0$ for $r = 1, 2, \ldots, m$ and (ii) the matrix $(\partial^2 H/\partial u_r \partial u_s)$ must be positive semidefinite.

By definition the control vector $\mathbf{u}^*$ is singular if the matrix $(\partial^2 H(\mathbf{x}^*, \mathbf{u}^*, \mathbf{p})/\partial u_r \partial u_j)$ is singular. This usually occurs when one or more control variables appear linearly in the Hamiltonian. By definition $\mathbf{x}^*(t)$ is called an extremal if $(\mathbf{x}^*, \mathbf{u}^*, \mathbf{p})$ satisfies Theorem 4.

Let $\mathbf{v}$ and $\mathbf{w}$ be subvectors of $\mathbf{u}$. Suppose $\mathbf{v}$ appears nonlinearly in $H(\mathbf{x}, \mathbf{v}, \mathbf{w}, \mathbf{p})$ and $\mathbf{w}$ appears linearly in H. We shall denote partial differentiations relative to the set $(\mathbf{x}, \mathbf{p}, \mathbf{v}, \mathbf{w}, \dot{\mathbf{v}}, \ddot{\mathbf{v}}, \ldots)$ by subscripts. Total differentiation with respect to time is denoted by a dot or by D. For example, $D^2 H = d^2 H/dt^2$. The next theorem describes some necessary conditions for a singular extremal to be optimal. More general conditions are discussed elsewhere (Goh, 1973).

Theorem 5. Along a singular extremal, if $H_{ww} = 0$ and $H_{wv} = 0$, then necessary conditions for optimality are

(i) $[DH_w]_w = 0$,
(ii) if (i) is satisfied, the matrix

$$\begin{bmatrix} H_{vv} & -[DH_w]_v^T \\ -[DH_w]_v & -[D^2 H_w]_w \end{bmatrix}$$

must be positive semidefinite.

In the management of a renewable resource singular control often forms the major portion of a long-term optimal program. We demonstrate this importance of singular control by developing the long-term feedback optimal policy in the harvesting of a fish population.

3.3. Optimal Harvesting of a Fish Population

Let x be the density of a fish population and $u(t)$ be the rate of harvesting. Let a and u_{max} be positive constants. The planning period is $[0, T]$, with the nonextinction constraint $x(T) \geqslant a$ and the quota constraint $0 \leqslant u \leqslant u_{max}$. The problem is to prescribe a harvesting policy so as to maximize the total yield for a given initial density x_0. The optimal harvesting policy is not sensitive to the model used for describing the unexploited population. The logistic model is therefore used.

In brief the problem is as follows:

$$\text{System:} \quad \dot{x} = (r/k)x(k-x) - u \tag{15}$$
$$\text{Initially:} \quad x(0) = x_0$$
$$\text{Terminally:} \quad T - c = 0, \quad x(T) \geqslant a, \tag{16}$$
$$\text{Constraint:} \quad 0 \leqslant u \leqslant u_{max}$$
$$\text{Objective:} \quad \min -\int_0^T u\,dt.$$

Here r, k and c are positive constants. In order to use the standard continuous-time optimal control problem directly the maximization problem is converted into a minimization problem.

The terminal condition in (16) is not of the same form as that in Eq. (14). It must be decomposed into cases, namely, (i) $x(T) - a = 0$ and (ii) $x(T) > a$, but otherwise $x(T)$ is unconstrained. The Hamiltonian function is

$$H = -u + p(r/k)[x(k-x) - u].$$

By Theorem 4 necessary conditions for optimality are

$$\dot{p} = -p(r/k)(k - 2x) \tag{17}$$
$$u = 0 \qquad \text{only if} \quad \partial H/\partial u = -1 - p > 0, \tag{18}$$
$$u = u_{max} \qquad \text{only if} \quad \partial H/\partial u = -1 - p < 0, \tag{19}$$
$$0 < u < u_{max} \qquad \text{only if} \quad \partial H/\partial u = -1 - p = 0. \tag{20}$$

Equations (17) and (20) imply

$$DH_u = -\dot{p} = p(r/k)(k - 2x) = 0$$
$$\Rightarrow x = k/2.$$

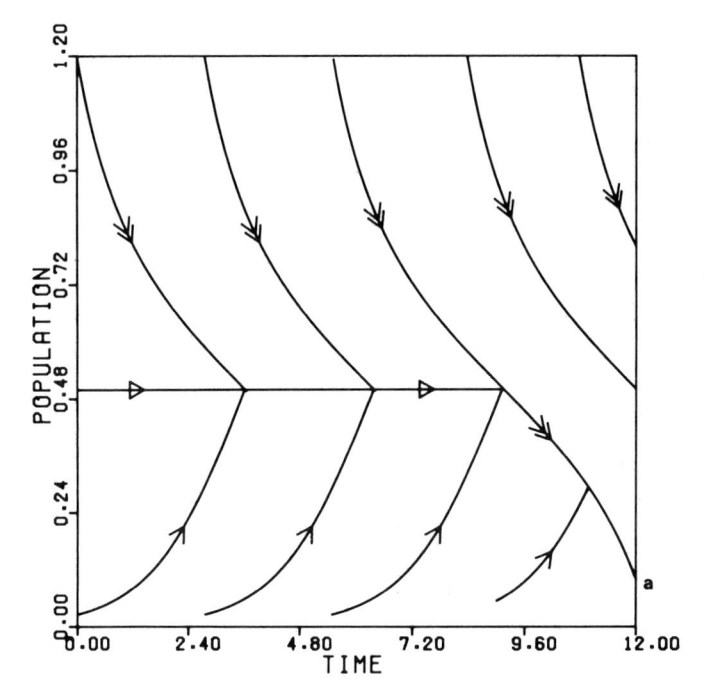

Figure 1. Feedback optimal control policy, → $u = 0$; →→ $u = u_{max}$; ⇾ singular control, $u = rk/4$.

Along the singular extremal $x = k/2$, Eq. (15) gives $u = rk/4$. Along this extremal, $p = 1$ and

$$D^2 H_u = 2(r/k)\dot{x}$$
$$\Rightarrow [D^2 H_u]_u = -2(r/k) < 0.$$

Hence the singular solution $(x^*, u^*) = (k/2, rk/4)$ satisfies Theorem 5.

The final job of putting together the optimal controls, $u^* = 0$, $u^* = u_{max}$, and $u^* = rk/4$, in the right sequence has to be done numerically. The optimal feedback control policy for the case when $u_{max} > rk/4$ is displayed in Fig. 1. Biologically the optimal control policy is very plausible. The singular control $u = rk/4$ creates an equilibrium at $x = k/2$, where the net natural growth rate is a maximum. In fact, the singular control policy is none other than the classical "maximum sustainable yield" policy used by fishery scientists. Clearly it is desirable (see Fig. 1) to attain this equilibrium as soon as possible from other initial states.

The constant quota harvest policy $u = rk/4$ creates an unstable equilibrium. It is better to use the corresponding constant effort policy

$v = r/2$. In this case the exploited model $\dot{x} = (r/k)x(k-x) - vx$ is globally stable. The optimal control policy for effort harvesting has been developed by Cliff and Vincent (1973).

By an application of the focal point condition to the singular control trajectory it can be shown that the feedback control policy displayed in Fig. 1 is indeed the long-term optimal policy for harvesting the fish population.

4. CONCLUSIONS

This chapter gives a brief description of some of the more useful theorems of optimal control theory. This mathematical tool has been successfully applied to a number of problems in the management of renewable resources, principally fisheries, pest control, and the management of epidemics. The success of these applications is to a large extent due to the fact the dynamics of a system in these applications could be adequately described by a nonlinear model with one or two variables. The computational difficulties in the solution of a nonlinear optimal control problem rises rapidly with the number of variables. In general it is not possible to apply optimal control theory to a nonlinear model with many variables.

REFERENCES

Becker, N. G. (1970). Control of a pest population. *Biometrics* **26**, 365–375.
Bell, D. J., and Jacobson, D. H. (1975). "Singular Optimal Control Problems." Academic Press, New York.
Bellman, R. E. (1957). "Dynamic Programming." Princeton Univ. Press, Princeton, New Jersey.
Berkovitz, L. A. (1961). Variational methods of control and programming. *J. Math. Anal. Appl.* **3**, 145–169.
Beverton, R. J. H., and Holt, S. J. (1957). On the dynamics of exploited fish populations. *Fish. Invest., London* **19**, 1–533.
Bliss, G. A. (1925). "Calculus of Variations." Open Court Co., London.
Bliss, G. A. (1946). "Lectures on the Calculus of Variations." Univ. of Chicago Press, Chicago, Illinois.
Canon, M. D., Cullum, C. D., and Polak, E. (1970). "Theory of Optimal Control and Mathematical Programming." McGraw-Hill, New York.
Clark, C. W. (1971). Economically optimal policies for the utilization of biologically renewable resources. *Math. Biosci.* **12**, 245–260.
Clark, C. W. (1976). "Mathematical Bioeconomics: The Optimal Management of Renewable Resources." Wiley, New York.
Cliff, E. M., and Vincent, T. L. (1973). An optimal policy for a fish harvest. *J. Optim. Theor. Appl.* **12**, 485–496.
Conway, G. R. (1977). Mathematical models in applied ecology. *Nature (London)* **269**, 291–297.

Fiacco, A. V., and McCormick, G. P. (1968). "Nonlinear Programming: Sequential Unconstrained Minimization Techniques." Wiley, New York.

Goh, B. S. (1966). Necessary conditions for singular extremals involving multiple control variables. *SIAM J. Control* **4**, 715–731.

Goh, B. S. (1969/1970). Optimal control of a fish resource. *Malayan Sci.* **5**, 65–70.

Goh, B. S. (1973). Compact forms of the generalized Legendre conditions and the derivation of singular extremals. *Proc. Hawaii Conf. Syst. Sci., 6th, 1973*, 115–117.

Goh, B. S. (1977). Optimum size limit for a fishery with a limited fishing season. *Ecol. Modell.* **3**, 3–15.

Himmelblau, D. M. (1972). "Applied Nonlinear Programming." McGraw-Hill, New York.

Innis, G. (1974). Dynamics analysis in "soft science" studies: in defence of difference equations. *In* "Mathematical Problems in Biology" (P. van den Driessche, ed.), pp. 102–122. Springer-Verlag, Berlin and New York.

Kelley, H. J., Kopp, R. E. and Moyer, H. G. (1967). Singular extremals. *In* "Topics in Optimization" (G. Leitmann, ed.), pp. 63–101. Academic Press, New York.

Lawden, D. F. (1963). "Optimal Trajectories for Space Navigation." Butterworth, London.

Leitmann, G. (1965). Rocket trajectory optimization. *Appl. Mech. Rev.* **18**, 345–350.

May, R. M. (1976). Simple mathematical models with very complicated dynamics. *Nature (London)* **261**, 459–467.

Pontryagin, L. S., Boltyanskii, V. G., Gamkrelidze, R. V., and Mischenko, E. F. (1962). "The Mathematical Theory of Optimal Processes." Wiley (Interscience), New York.

Tabak, D., and Kuo, B. C. (1971). "Optimal Control by Mathematical Programming." Prentice-Hall, Englewood Cliffs, New Jersey.

Watt, K. E. F. (1968). "Ecology and Resource Management." McGraw-Hill, New York.

Wickshire, K. (1977). Mathematical models for the control of pests and infectious diseases: A survey. *Theor. Popul. Biol.* **11**, 182–238.

Chapter **16**

TOWARD OPTIMAL IMPULSIVE CONTROL OF AGROECOSYSTEMS

John W. Brewer

1. Introduction. 401
2. Optimal, Single-Impulse Control of S-Shaped Growth . . 403
   2.1 Control Cost Plus Cumulative Damage 406
   2.2 Control Cost Plus Terminal Penalty 409
3. Numerical Experiments 412
4. Comments on the Optimal Impulse Control of J-Shaped Growth 414
5. Concluding Comments. 416
   References 416

1. INTRODUCTION

The movement, transformation, and impact of biocides within the environment is a set of highly complex phenomena. Nisbet (1975a–e) has summarized anecdotal evidence of the impact of biocides. Haque and Freed (1975) edited a set of tutorial papers on the physical mechanisms of the movement and transformation of biocides in the environment. Solow (1971) and Rausser and Howitt (1975) describe the economic and institutional problems of the control of biocide use.

The one certain conclusion is that biocides should be used as efficiently as possible; that is to say, as little biocide as possible should be employed in agroecosystems. Of course, the damage done by pest species requires that some biocide be used sometimes. The challenge is to determine both the timing and the minimum amounts of biocide applications in order to properly balance environmental costs against pest damage.

System engineers have recently been responding to the challenge. Mitchiner *et al.* (1975) and Vincent *et al.* (1975) provide bibliographies which are partial lists of these efforts. The tendency has been to apply mathematical optimization techniques to highly idealized models of agroecosystems. The results presented in this paper may be described in the same way. The hope is that idealized studies will evolve to the point that practical results will be obtained.

The main feature of the analysis presented here is recognition of the pulselike nature of biocide applications. In spite of the low volatility and solubility of many pesticides, there are physical–chemical mechanisms which cause these biocides to quickly disperse from the area to which they are applied (Nisbet, 1975a; Haque and Freed, 1975). (This dispersal is the reason that the impacts of biocide use are so widespread and the reason that the impacts are economic "externalities.") In mathematical terms, the biocide application will be described here, in a highly ideal manner, by the Dirac delta function. That is to say, applications will be idealized as "impulses."

The use of the impulse idealization provides mixed analytical benefits. The solution of certain "bilinear" equations of ecosystem dynamics is thereby made easier; however, the standard derivations of the calculus of variations and of the Minimum Principle do not apply to such unbounded inputs. Lee and Markus (1967) demonstrate the extraordinary methods required to develop necessary conditions when unbounded variations are feasible. Gilbert and Harasty (1971) consider the particular case of impulsive variations, although their results apply only to systems described by linear equations in which the input appears in an additive fashion (as opposed to the multiplicative or "bilinear" fashion of inputs in ecosystem equations).

No attempt was made in this study to extend variational calculus to the problem at hand, although such extensions would be a very important contribution.

The chapter is organized in the manner now described. Section 2 is a complete solution of the optimal single-impulse control of S-shaped growth. The results may be applicable to the control of microbial pests. The analysis is based upon the Riccati transformation of the logistic equation (Brewer, 1976) which is used to convert the problem to one of the minimization of an algebraic function. Some analytical evidence is provided to support the conjecture that the optimal impulsive control is the optimal single-impulse control derived here. Section 3 describes dynamic programming experiments which substantiate the analytical results of Section 2 and which lend further credence to the conjectures about optimal controls. Also, Section 3 contains some suggestions for future research into the optimal control of S-shaped growth. Section 4 contains some comments on the optimal impulsive control of the J-shaped growth associated with insect pests. The problem is outlined and the difficulties of the solution are highlighted. Finally, concluding comments are presented in Section 5.

2. OPTIMAL, SINGLE-IMPULSE CONTROL OF S-SHAPED GROWTH

In this section, the optimal impulsive control of a pest described by the time-varying logistic equation

$$\frac{dp}{dt} = \alpha p \left(1 - \frac{p}{P_0 v(t)}\right) - \alpha k p b(t) \tag{1}$$

is studied (Brewer, 1976). Here,

$$p = \text{population level}, \tag{2}$$

$P_0 v(t) =$ time varying "capacity" (maximum potential population), and (3)

$$b(t) = \text{biocide concentration.} \tag{4}$$

The parameters α, k, and P_0 are taken to be constants. The capacity will often depend upon the level of a commercial species.

Brewer (1976) introduced the "Riccati" transformation of variables

$$t = \tau/\alpha, \tag{5}$$

$$x(t) = P_0/p(t), \tag{6}$$

$$u(t) = kb(t)/P_0, \tag{7}$$

so that Eq. (1) becomes, in terms of *reciprocal population*,

$$dx/d\tau = -(1-u(t))x + (1/v(t)), \tag{8}$$

which is a *bilinear* differential equation (Mohler, 1973).

The study of the optimal impulsive control of a system described by Eq. (8) may provide some insight into the solution of the much more difficult bilinear impulsive control problem described in Section 4. Additionally, the solution of a problem associated with Eq. (8) may be useful in its own right since this equation may describe the growth of bacterial or fungus pest species for which age distribution effects are not as important.

The performance index for this study is taken to be

$$I = w_0 \frac{k}{P_0} \int_0^T b(t)dt + w_0 \frac{w_1}{P_0} \int_0^T p(t)dt + w_0 \frac{w_2}{P_0} p(T^+)$$

which is to be minimized. Minimizing I is equivalent to minimizing

$$J \triangleq \frac{I}{w_0} = \int_0^{\tau_f} u(\tau)d\tau + w_1 \int_0^{\tau_f} \frac{d\tau}{x(\tau)} + \frac{w_2}{x(\tau_f^+)}, \qquad \tau_f = \alpha T. \tag{9}$$

This index reflects a desire to minimize cost of control, cumulative damage by the pest, and a final value of pest at known harvest time, T. The constants w_0, w_1, and w_2 are weighting factors which provide for the proper combination of the agroecosystem costs. The "+" superscript means that the pest level is to be evaluated a short time after a possible impulse at harvest time.

In what follows, it is assumed that

$$v(t) \equiv 1.0 \tag{10}$$

although the analysis with time-varying capacity is possible. For instance the analysis for the case of a logistic growth of the commercial (that is, the pest's food) species was provided by Brewer (1976).

It is easily verified that the solution of Eqs. (8) and (10) is the one provided by Brewer (1976):

$$x(\tau) = e^{-(\tau-\tau_0)} \exp\left[\int_{\tau_0}^{\tau} u(\xi)d\xi\right]$$
$$\times \left\{x(\tau_0) + \int_{\tau_0}^{\tau} e^{(\xi-\tau_0)} \exp\left[-\int_{\tau_0}^{\xi} u(\eta)d\eta\right] d\xi\right\}. \tag{11}$$

In particular, consider a single impulse which occurs slightly after time t_i:

$$b(t) = B_i \delta(t - t_i^+),$$

where t_i is specified and the constant B_i is to be determined. In terms of transformed variables

$$u(\tau) = U_i \delta(\tau - \tau_i^+), \tag{12}$$

where

$$U_i = (k/P_0)B_i > 0, \tag{13}$$

$$\tau_i = \alpha t_i. \tag{14}$$

16. Toward Optimal Impulsive Control of Agroecosystems

The solution of Eq. (11) becomes for $\tau_0 = \tau_i$,

$$x(\tau) = e^{-(\tau-\tau_i)}e^{U_i}x(\tau_i) + 1 - e^{-(\tau-\tau_i)}. \tag{15}$$

Notice that in *bilinear* systems the effect of an impulse is a discontinuous change in the effective initial condition from $x(\tau_i)$ to $e^{U_i}x(\tau_i)$.

When Eq. (15), for the reciprocal population, is substituted into Eq. (9), the result is

$$J = U_i + w_1[\tau_f - \tau_i + \ln(1 + a_i e^{-\tau_f}) - \ln(1 + a_i e^{-\tau_i})]$$
$$+ \frac{w_2}{(1+a_i e^{-\tau_f})} + w_1 \int_0^{\tau_i} \frac{d\tau}{x(\tau)} \tag{16}$$

since the integral term in Eq. (9), from $\tau = 0$ to $\tau = \tau_i$, does not depend upon U_i. Also

$$a_i = (e^{U_i}x(\tau_i) - 1)e^{\tau_i}. \tag{17}$$

U_i is chosen to minimize the right-hand side of Eq. (16). Thus, the optimal control problem has been reduced to an algebraic minimization problem.

First notice that

$$da_i/dU_i = a_i + e^{\tau_i} \tag{18}$$

so that

$$\frac{dJ}{dU_i} = 1 + w_1 \left\{ \frac{(a_i+e^{\tau_i})e^{-\tau_f}}{1+a_i e^{-\tau_f}} - 1 \right\} - w_2 \frac{(a_i+e^{\tau_i})e^{-\tau_f}}{(1+a_i e^{-\tau_f})^2} \tag{19}$$

and

$$\frac{d^2J}{dU_i^2} = w_1 \frac{(a_1+e^{\tau_i})}{(a_1+e^{\tau_f})^2}(e^{\tau_f}-e^{\tau_i}) + w_2 \frac{(a_i+e^{\tau_i})e^{\tau_f}}{(a_i+e^{\tau_f})^2}\left[\frac{a_i+2e^{\tau_i}-e^{\tau_f}}{a_i+e^{\tau_f}}\right]. \tag{20}$$

Suppose U_* is such that

$$dJ(U_*)/dU_i = 0 \tag{21}$$

and

$$d^2J(U_*)/d^2U_i > 0 \tag{22}$$

Then U_* is the optimal value of U_i.

2.1. Control Cost Plus Cumulative Damage

Consider the special case

$$w_2 = 0. \tag{23}$$

Condition (21) becomes

$$1 + w_1(e^{\tau_i} - e^{\tau_f})/(a_i + e^{\tau_f}) = 0, \tag{24}$$

which may be transformed to

$$x(\tau_i)e^{U_*}e^{\tau_i} - e^{\tau_i} + e^{\tau_f} + w_1(e^{\tau_i} - e^{\tau_f}) = 0 \tag{25}$$

and has the solution

$$U_* = \ln\{(w_1 - 1)(1 - e^{-(\tau_f - \tau_i)})/x(\tau_i)e^{-(\tau_f - \tau_i)}\}. \tag{26}$$

For U_* to be positive, the argument of the logarithm must not be less than one. This condition is assured if

$$e^{-(\tau_f - \tau_i)} \leqslant (w_1 - 1)/(w_1 - 1 + x(\tau_i)) \tag{27}$$

or, equivalently, if

$$x(\tau_i) \leqslant (w_1 - 1)(e^{\tau_f - \tau_i} - 1). \tag{28}$$

Notice that this condition can only be fulfilled if $w_1 > 1$.

Define an *application function*

$$s(\tau) = (w_1 - 1)(e^{\tau_f - \tau} - 1) \tag{29a}$$

so that condition (28) may be rewritten

$$x(\tau_i) \leqslant s(\tau_i). \tag{29b}$$

This relation is interpreted in Fig. 1.

Equation (15) indicates that just after application of the biocide

$$x(\tau_i^+) = e^{U_*}x(\tau_i).$$

Combine Eqs. (26) and (28) with this result to show

$$x(\tau_i^+) = s(\tau_i) \tag{30}$$

independent of $x(\tau_i)$. That is to say, whenever biocide is applied in an optimal manner, the reciprocal population is immediately increased to the current value of the application function.

Combine Eqs. (15) and (30) to show that after application of U_*,

$$x(\tau) - s(\tau) = s(\tau_i)e^{-(\tau - \tau_i)} + 1 - e^{-(\tau - \tau_i)} - (w_1 - 1)(e^{(\tau_f - \tau)} - 1)$$

or

$$x(\tau) - s(\tau) = w_1(1 - e^{-(\tau - \tau_i)}) \geqslant 0. \tag{31}$$

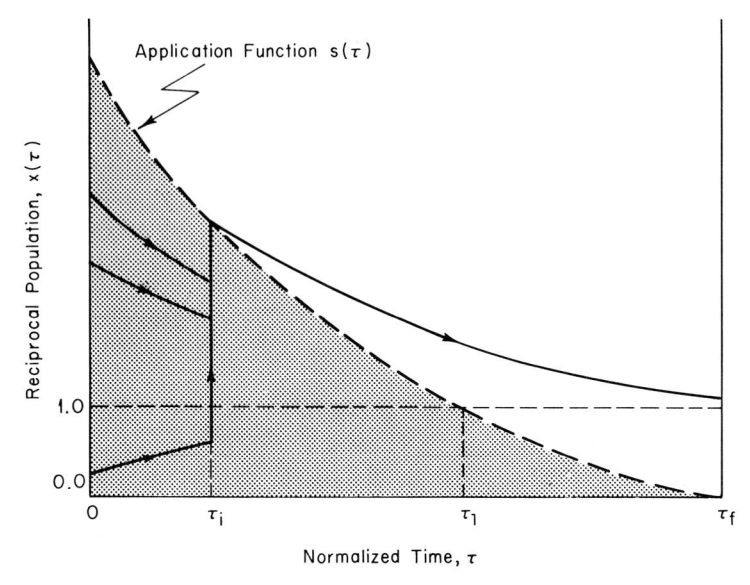

16. Toward Optimal Impulsive Control of Agroecosystems 407

Figure 1. Optimal single-impulse control of S-shaped growth (control cost plus cumulative damage): An optimal single impulse is positive (feasible) only if the state lies in the shaded region bounded by the *application function*, $s(\tau)$. The time τ_1 is that time when the application function is 1.0. The solid lines with arrows are optimal time histories for three different initial values, $x(0)$.

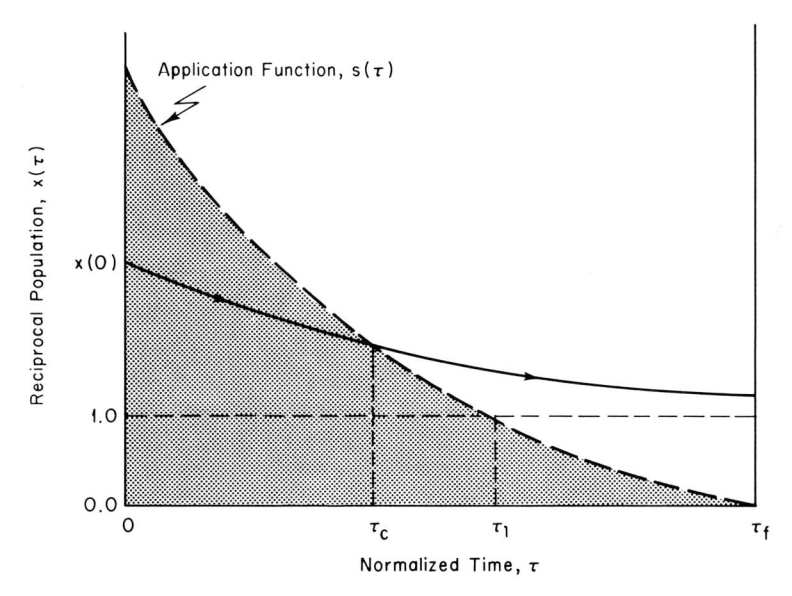

Figure 2. An uncontrolled time history, $x(\tau)$, leaves the feasible region at $\tau_c < \tau_1$ if $x(0) \geq 1$.

This means that state trajectories starting on $s(\tau)$ will never again satisfy relation (29). This fact is illustrated in Fig. 2 and encourages the conjecture that the optimal control may be the optimal single-impulse control.

Consider the sufficiency condition (22); it follows from Eqs. (17), (20), (23), and (31) that

$$\frac{d^2 J}{dU_i^2}(U_*) = w_1 s(\tau_i) e^{\tau_i} \frac{(e^{\tau_f} - e^{\tau_i})}{(a_i + e^{\tau_f})^2}$$

which is clearly positive if $w_1 > 0$ and $\tau_i < \tau_f$ so that (26) is indeed a minimizing solution.

The remaining question is: When should the impulse be applied? After much algebra, it can be shown that the optimal value of the performance index

$$J_* \triangleq J(U_*, \tau_i)$$
$$= w_1[\tau_f - \ln\{x(0)\}] + (1 - w_1)\ln\{s(\tau_i)\} + w_1 \ln\{w_1(1 - e^{\tau_i - \tau_f})\}.$$

Thus

$$\frac{dJ_*}{d\tau_i} = \frac{(w_1 - 1)e^{\tau_f - \tau_i} - w_1}{e^{\tau_f - \tau_i} - 1}. \tag{32}$$

It can then be shown that

$$\frac{dJ_*}{d\tau_i} \geqslant 0 \quad \text{for} \quad w_1 > 1 \tag{33}$$

and

$$0 \leqslant \tau_i \leqslant \tau_1, \tag{34}$$

where τ_1 is the crossover time illustrated in Fig. 1. The equality in relation (33) occurs when

$$\tau_i = \tau_1 = \tau_f - \ln\{w_1/(w_1 - 1)\}. \tag{35}$$

An uncontrolled trajectory for $x(0) \geqslant 1$ (i.e., for the important range of $p(0) \leqslant P_0$) is illustrated in Fig. 2. Condition (28) is satisfied only if

$$0 \leqslant \tau_i \leqslant \tau_c \leqslant \tau_1. \tag{36}$$

It follows from relations (32) and (34) that for the feasible range of application times (for $x(0) \geqslant 1$), J_* increases with τ_i. Thus *if τ_i is arbitrary, it should be as early in the season as possible*. Notice that this conclusion does not necessarily apply if $x(0) < 1$.

The above results can now be summarized and rephrased in terms of nonnormalized variables. *The single impulse control of S-shaped growth, is optimized ($w_2 = 0$) with an impulse of magnitude*

$$B_i = \frac{P_0}{k} \ln\{(w_1 - 1)(e^{\alpha(T-t_i)} - 1)\} + \frac{P_0}{k} \ln\left\{\frac{p(t_i)}{P_0}\right\} \qquad (37)$$

and t_i taken to be as early as possible.

Notice that Eq. (37) is a nonlinear, time-varying feedback control law.

One can imagine other uses for Eq. (37). Once the impulse has been applied, criterion (28)

$$p(t) \geq P_0/(w_1 - 1)(e^{\alpha(T-t)} - 1) \qquad (38)$$

will never be satisfied again if no "disturbances" (unforeseen events) affect the growth of the pest. However, such disturbances are inevitable. One could imagine a feedback scheme in which condition (38) is checked periodically during the same growing season. When it is found to be satisfied, impulse magnitudes of Eq. (37) could then be applied.

It must be emphasized that only the optimal single-impulse control has been obtained. This control must not be confused with the optimal control which may be more than a single impulse. However, relations (33) and (31) do much to fuel the suspicion that the optimal single-impulse control is the optimal control. It is worth noting that Neustadt (Edelbaum, 1967) has demonstrated that the optimal impulsive control of a linear system is never constituted of more impulses than the order of the system. It may well be that his result can be extended to the case of bilinear systems.

2.2. Control Cost Plus Terminal Penalty

Consider the special case

$$w_1 = 0. \qquad (39)$$

A hypothetical situation for which Eq. (39) might be a reasonable assumption is one in which the pest population creates an unsightly or unappetizing commercial population but creates no cumulative damage.

For this case, Eq. (19) and (21) become

$$1 - w_2(a_i + e^{\tau_i})e^{-\tau_f}/(a_i + e^{\tau_f})^2 = 0. \qquad (40)$$

This equation is a quadratic equation in e^{U_i} which has the two solutions

$$\ln\left\{\frac{w_2 e^{\tau_f} - 2(e^{\tau_f} - e^{\tau_i}) \pm (w_2 e^{\tau_f}(w_2 e^{\tau_f} - 4(e^{\tau_f} - e^{\tau_i})))^{1/2}}{2e^{\tau_i}x(\tau_i)}\right\}. \qquad (41)$$

It can be shown that the inequality

$$e^{\tau_i} \geq \tfrac{1}{4}(4 - w_2)e^{\tau_f} \qquad (42)$$

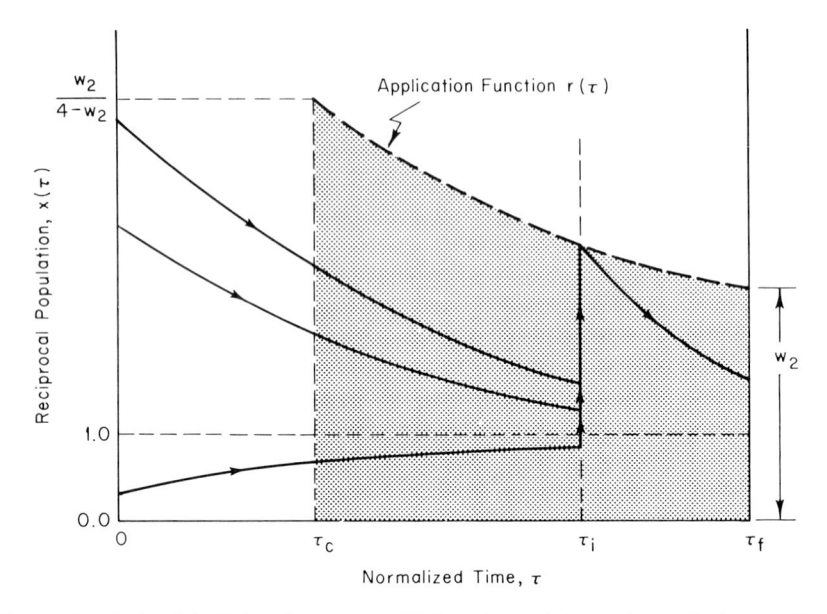

Figure 3. Optimal single-impulse control of S-shaped growth (control cost plus harvest time penalty): The optimal control is real only if $\tau_i > \tau_c = \tau_f + \ln((4-w_2/4)$ and positive only if $x(\tau_i)$ lies within the region below the application function, $r(\tau)$. The application function illustrated here is characteristic of the case $w_2 < 4$.

is precisely the condition that ensures that the argument of the logarithm is real and is more than sufficient to ensure that the same argument is nonnegative for either sign. This condition can be satisfied *late in the season* ($\tau_i \to \tau_f$) for any positive value of w_2 and can be satisfied for any value of $\tau_i \leqslant \tau_f$ if $w_2 \geqslant 4$.

Equation (20) becomes

$$\frac{d^2 J}{dU_i^2} = \frac{w_2 e^{U_i} x(\tau_i) e^{\tau_i} e^{\tau_f}}{(a_i + e^{\tau_f})^2} \left[\frac{e^{U_i} x(\tau_i) e^{\tau_i} - (e^{\tau_f} - e^{\tau_i})}{e^{U_i} x(\tau_i) e^{\tau_i} + (e^{\tau_f} - e^{\tau_i})} \right]. \tag{43}$$

Notice that the term outside the brackets is positive as is the denominator of the term in brackets (for $\tau_i \leqslant \tau_f$). Thus, condition (22) becomes

$$e^{U_*} x(\tau_i) e^{\tau_i} > e^{\tau_f} - e^{\tau_i}. \tag{44}$$

Substitute Eq. (41) into (44) to find that the sufficiency condition becomes

$$1 \pm (1 - (4/w_2)(1 - e^{-(\tau_f - \tau_i)}))^{1/2} > (4/w_2)(1 - e^{-(\tau_i - \tau_1)}). \tag{45}$$

It is not difficult to show that only the positive sign will satisfy this condition for the range of τ_i defined by relation (42). (The negative sign is associated with a local maximum.) Thus, the optimal impulse magnitude

$$U_* = \ln\left\{\frac{w_2 e^{\tau_f} - 2(e^{\tau_f} - e^{\tau_i}) + (w_2 e^{\tau_f}(w_2 e^{\tau_f} - 4(e^{\tau_f} - e^{\tau_i})))^{1/2}}{2e^{\tau_i} x(\tau_i)}\right\}. \qquad (46)$$

Since U_* must be nonnegative, it is necessary to require that

$$x(\tau_i) \leq r(\tau_i), \qquad (47)$$

where the *application function*

$$r(\tau) = \frac{1}{2}\left\{w_2 e^{(\tau_f - \tau)}\left[1 + \left(1 - \frac{4}{w_2}(1 - e^{-(\tau_f - \tau)})\right)^{1/2}\right] - 2(e^{\tau_f - \tau} - 1)\right\}. \qquad (48)$$

Condition (47) can never be satisfied if $x(0) \geq 1$ and $w_2 < 1$.

Condition (47) is illustrated in Fig. 3.

It is not difficult to show that after the application of the impulse at τ_i,

$$\begin{aligned}x(\tau_i^+) &= e^{U_*} x(\tau_i) \\ &= r(\tau_i).\end{aligned} \qquad (49)$$

Use Eqs. (48) and (49) to show that

$$x(\tau) - r(\tau) = \tfrac{1}{2} w_2 e^{(\tau_f - \tau)}\left\{\left(1 - \frac{4}{w_2}(1 - e^{-(\tau_f - \tau_i)})\right)^{1/2} - \left(1 - \frac{4}{w_2}(1 - e^{-(\tau_f - \tau)})\right)^{1/2}\right\} \leq 0$$

if

$$\tau_i \leq \tau \leq \tau_f. \qquad (50)$$

Thus, if Eq. (47) is satisfied for some τ_i and only impulses of magnitude (46) are applied, relation (47) will be satisfied for all succeeding times up to harvest. This is quite the opposite of the result obtained in Section 2.1.

It can be shown that

$$\begin{aligned}J_* &\triangleq J(U_*, \tau_i) \\ &= \ln\left\{\frac{r(\tau_i)}{1 + [x(0) - 1]e^{-\tau_i}}\right\} + \frac{2}{1 + \left(1 - \dfrac{4}{w_2}(1 - e^{-(\tau_f - \tau_i)})\right)^{1/2}}.\end{aligned} \qquad (51)$$

The value of τ_i can be selected by finding that value which minimizes J_*. A relation similar to (33) was not easily obtained for this case. Numerical investigations were resorted to. In all cases, *it was noted that J_* diminished as τ_i was taken to be later in the season.* It was noted that J_* seemed to be a local maximum at the minimum value of τ_i that satisfies inequality (42).

In summary: *The single impulse control of S-shaped growth is optimized* ($w_1 = 0$) *with an impulse of magnitude*

$$B_i = \frac{P_0}{k} \ln \left\{ \frac{w_2}{2} e^{\alpha(T-t_i)} \left[1 + \left(1 - \frac{4}{w_2}(1 - e^{\alpha(t_i - T)}) \right)^{1/2} \right] + (1 - e^{\alpha(T-t_i)}) \right\}$$
$$+ \frac{P_0}{k} \ln \frac{p(t_i)}{P_0}. \tag{52}$$

Generally, the best choice is

$$t_i = T. \tag{53}$$

At the application time, an impulse is applied only if

$$p(t_i) \geq \frac{P_0}{\frac{w_2 e^{\alpha(T-t_i)}}{2} \left[1 + \left(1 - \frac{4}{w_2}(1 - e^{\alpha(t_i - T)}) \right)^{1/2} \right] + (1 - e^{\alpha(T-t_i)})}. \tag{54}$$

Once again, a nonlinear feedback law is obtained.

3. NUMERICAL EXPERIMENTS

Dynamic programming (Bellman and Dreyfus, 1962) is a numerical method that can be used to obtain optimal controls for low-order system models and for specific parameter values. This method was applied to the problem treated analytically in Section 2.

Analytical solutions are always superior to numerical solutions because, among other things, conditions for existence are obtained. For instance, refer to relations (29b), (42), and (47). The single advantage provided by dynamic programming in the present case is that the optimal impulsive control (rather than the optimal single-impulse control) is obtained.

In any case, dynamic programming solutions provide a convenient check on analytical results. Also, they can be used to indicate possible extensions of analytical results.

The result of one dynamic simulation experiment is provided in Fig. 4. This is a "control matrix" (i.e., optimal values of impulse magnitude at discrete times and at discrete reciprocal population levels) for $w_2 = 0$. Also shown in this figure is the theoretical application function (Eq. 29a) and the dynamic programming solution for the optimal trajectory for $x(0) = 2$. If these impulse numbers are compared with Eq. (26), agreements to two and three significant figures are obtained. Clearly, for this specific case, the optimal multipulse control is the optimal single-impulse control derived in Section 2.1.

Similarly, agreement between the optimal multipulse control ($w_1 = 0$)

16. Toward Optimal Impulsive Control of Agroecosystems   413

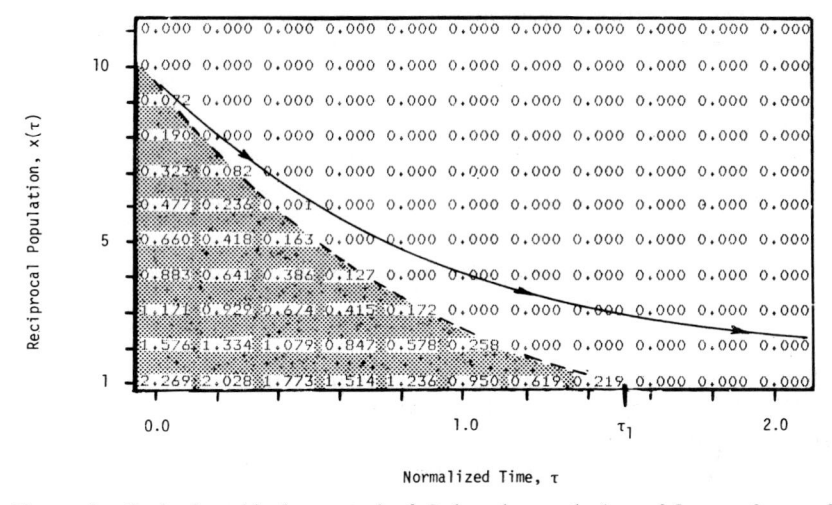

Figure 4. Optimal multipulse control of S-shaped growth ($w_1 = 2.5$, $w_2 = 0$, $\tau_f = 2$): Dynamic programming was used to determine this matrix of optimal impulse magnitudes at various times and reciprocal population levels. The optimal time history (solid curve) is for $x(0) = 2$. The analytical optimal single-impulse control, obtained in Section 2.1, is equal to this optimal multipulse control in every respect. The dashed curve is the theoretical application function, $s(\tau)$ (see Fig. 1).

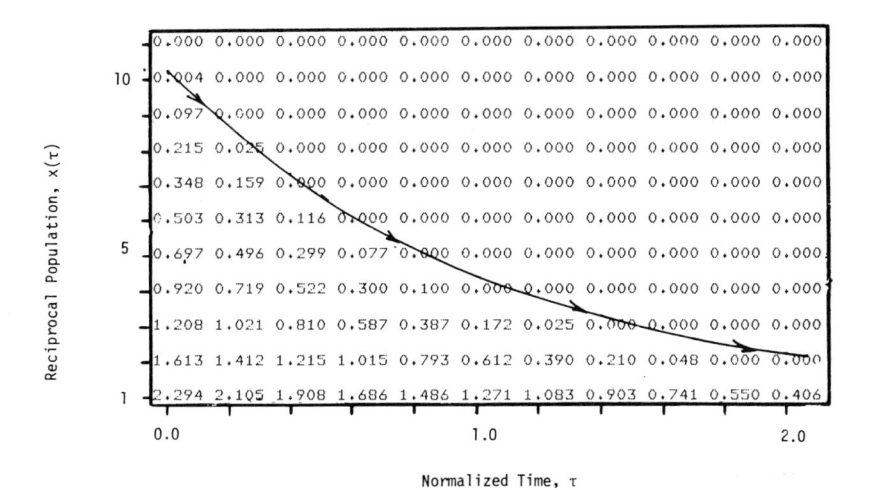

Figure 5. Optimal multipulse control of S-shaped growth ($w_1 = 1.5$, $w_2 = 1.5$, $\tau_f = 2.0$). Dynamic programming was used to determine the optimal control matrix for this general form of the performance index. The optimal time history is for $x(0) = 2$. The optimal control is very much like that for the case of optimal single-impulse control for control cost plus cumulative damage only.

and the theory derived in Section 2.2 was obtained for several specific cases. In these experiments, it was found that only the last column of the control matrix was nonzero which means that the optimum multipulse control is a single impulse at harvest time. Agreement with Eq. (46) was obtained.

A third experiment yielded a surprise. Both w_1 and w_2 were given nonzero values. The resulting optimal control matrix is provided in Fig. 5. It was expected that the optimal multipulse control might be a combination of the controls derived in Sections 2.1 and 2.2. Instead, *the optimal control was a single impulse at the beginning of the season.* As is indicated in Fig. 5, there seems to be an application function-like effect. Perhaps the results of Section 2.1 can be extended to this more general case. Of course, this result applies only to specific values of parameters and weighting factors.

These are many extensions of the results of Section 2 that would hasten practical applications. One improvement would be to recognize that the pest's food species is also accumulating biomass during the growing season. This would correspond to a time-varying $v(t)$ in Eq. (1) (see Brewer, 1976). Another improvement would be to account for the inherent stochastic nature of biocide applications. That is to say, the effective impulse magnitude is related to the actual applied amount in a stochastic manner because of meteorological disturbances.

4. COMMENTS ON THE OPTIMAL IMPULSE CONTROL OF J-SHAPED GROWTH

Insect pests often display J-shaped growth; that is to say, growth in which the number of individuals in a given age class vary in an oscillatory manner during the growing season. Apparently, this type of growth is an age–time phenomenon.

Oster *et al.* (1976) demonstrate an impressive agreement of J-shaped growth data with theory based upon the *von Foerster partial differential equation.*

$$\frac{\partial \rho}{\partial t} + \frac{\partial \rho}{\partial a} = -\mu(a,t)\rho, \tag{55}$$

where a is the age, $\rho(a,t)$ is the proportion of total population in age class $a - \frac{1}{2}da$ to $a + \frac{1}{2}da$ at time t, and $\mu(a,t)$ is the deaths per individual at age a at time t and per unit time.

Obviously

$$\int_0^\infty \rho(a,t)da = 1. \tag{56}$$

16. Toward Optimal Impulsive Control of Agroecosystems

The *intrinsic death rate*

$$D(t) \triangleq \int_0^\infty p(a,t)\mu(a,t)\,da \tag{57}$$

and the *intrinsic birth rate*

$$B(t) \triangleq \int_0^\infty p(a,t)b(a,t)\,da, \tag{58}$$

where $b(a,t)$ is the number of eggs laid by an individual at age a at time t and per unit time.

The evolution of the total population, $p(t)$, is then given by

$$dp/dt = \{B(t) - D(t)\}p. \tag{59}$$

The total cumulative damage of the pest is given by

$$\int_0^T p(t) \int_0^\infty w(a)p(a,t)\,da\,dt. \tag{60}$$

$w(a)$ is a weighting function which is used to indicate that not all age classes damage the commercial population equally; for instance, it may be that only the larva do damage.

The death rate term

$$\mu(a,t) = \mu_0(a,t) + u(a,t), \tag{61}$$

where μ_0 is the natural (uncontrolled, age specific) death rate and u is the controlled rate.

The control is a function of age because biocides only affect some of the age classes and not others.

The impulse idealization is

$$u(a,t) = f(a)\delta(t - t_i). \tag{62}$$

Notice the bilinear nature of Eqs. (55) *and* (61).

The performance index will be a combination of Eq. (60) with a control cost term. This optimal control problem is associated with partial differential equation (55) and ordinary differential equation (59) as side conditions. [The famous method of characteristics can be used to convert the problem of solving (55) to the problem of solving an integral equation in the unknown egg production rate $p(0,t)$.]

The solution of the optimization problem outlined above is made even more formidable when unbounded inputs (62) are considered feasible.

5. CONCLUDING COMMENTS

Edelbaum (1967) and Gobetz and Doll (1969) reviewed the historical development of another theory of optimal impulse control: the impulse control of artificial satellites. The need to minimize fuel consumption during satellite control is reminiscent of the need to minimize biocide during agroecosystem control.

None of the specific results of the control of satellites is directly applicable to control of agroecosystems because in the former case, the system model is that of a pure inertia in an inverse-square-law gravitational field. The lessons in method may be useful however. In the early development of this theory, analysts optimized single-impulse control. In short, the theory began with an analysis similar to that presented in this paper.

The "bilinear" theme is noted in both the discussion of S- and J-shaped growth. The theoretical developments of bilinear analysis (Mohler, 1973; Bruni et al., 1974) were inspired by practical applications in the control of nuclear reactors. It may well be that the thorough study of the literature of satellite and reactor control will lead to the optimal impulse control of agroecosystems.

A final comment deals with the impulse idealization. Some will argue that the actual application is a bounded pulse—so why not recognize this fact at the beginning so that standard variational methods can be brought to bear on the optimization problem? It may well be that the trade of analytical simplicity for the use of standard variational principles is a good one. Only time and much research will provide the answer.

I suspect that the impulse idealization will prove to be the favored route. One method for obtaining the true nature of the biocide pulse is to add "cumulative biocide" to the list of dynamic state variables (e.g., see Mitchner et al., 1975). However, when this dynamic equation is added, one discovers that the input to the system is the *rate of application* of the biocide. The impulse is an even better approximation of this type of input than it is for the biocide concentration taken as the input in this paper.

REFERENCES

Bellman, R. E., and Dreyfus, S. E. (1962). "Applied Dynamic Programming." Princeton Univ. Press, Princeton, New Jersey.

Brewer, J. W. (1976). The analytical solution of a time-varying logistic growth equation. *IEEE Trans. Syst., Man, and Cybernet.* **smc-6**, 384–386.

Bruni, C., Dipillo, G., and Koch, G. (1974). Bilinear systems: An appealing class of "nearly linear" systems in theory and applications. *IEEE Trans. Autom. Control.* **ac-19**, 334–348.

Edelbaum, T. N. (1967). How many impulses. *Astronaut. Aeronaut.* **5**, 64–69.
Gilbert, E. C., and Harasty, G. A. (1971). A class of fixed-time fuel-optimal impulsive control problems and an efficient algorithm for their solution. *IEEE Trans. Autom. Control.* **ac-16**, 1–11.
Gobetz, F. W., and Doll, J. R. (1969). A survey of impulsive trajectories. *AIAA J.* **7**, 801–836.
Haque, R., and Freed, V. H., eds. (1975). "Environmental Dynamics of Pesticides." Plenum, New York.
Lee, E. G., and Markus, L. (1967). "Foundations of Optimal Control Theory." Wiley, New York.
Mitchiner, J. L., Kennish, W. J., and Brewer, J. W. (1975). Application of optimal control and optimal regulator theory to the "integrated" control of insect pests. *IEEE Trans. Syst., Man, Cybernet.* **smc-5**, 111–116.
Mohler, R. R. (1973). "Bilinear Control Processes." Academic Press, New York.
Nisbet, I. C. T. (1975a). Global pollutants. *Technol. Rev.* 6–7.
Nisbet, I. C. T. (1975b). Some unsolved puzzles of chlorinated hydrocarbons. *Technol. Rev.* p. 13.
Nisbet, I. C. T. (1975c). Ecological magnification. *Technol. Rev.* p. 6.
Nisbet, I. C. T. (1975d). Persistent pesticides and wildlife. *Technol. Rev.* 8–9.
Nisbet, I. C. T. (1975e). Pesticides and breeding failures in birds. *Technol. Rev.* 8–9.
Oster, G. F., Auslander, D. M., and Allen, T. T. (1976). Deterministic and stochastic effects in population dynamics. *ASME J. Dyn. Syst., Meas., Control.* V. 98, Series G, N.1, 44–48.
Rausser, G. C., and Howitt, R. (1975). Stochastic control of environmental externalities. *Ann. Econ. Social. Meas.* **4**, 271–292.
Solow, R. M. (1971). The economist's approach to pollution and its control. *Science* **173**, 498–503.
Vincent, T. L., Lee, C. S., Pulliam, H. R., and Everett, L. G. (1975). Applications of optimal control to the modeling and management of ecosystems. *Simulation*, pp. 65–72.

Chapter **17**

HIERARCHICAL METHODS IN RIVER POLLUTION CONTROL

Madan G. Singh

1. Introduction	420
2. Problem Formulation	421
2.1 The Goal Coordination Approach	422
2.2 Remarks	425
3. The Three-Level Method of Tamura	426
3.1 The Goal Coordination Method for Discrete Dynamical Systems	427
3.2 The Modification of Tamura	428
3.3 Remarks	430
4. The Time Delay Algorithm of Tamura	430
4.1 Remarks	434
5. The Interaction Prediction Approach	434
5.1 Remarks	436
6. River Pollution Control	437
6.1 The Model	437
6.2 Steady-State Considerations	438
6.3 Multiple Effluent Inputs	439
6.4 Hierarchical Solution Using the Prediction Principle	441
6.5 Simulation Results	442
6.6 Remarks	442
7. Hierarchical Feedback Control for Linear Quadratic Problems	444
7.1 Remarks	446
7.2 The Closed-Loop Controller	446
8. Extension to the Servomechanism Case	448
8.1 Control of the Three-Reach Distributed Delay Model	449
9. Conclusions	450
References	451

1. INTRODUCTION

Hierarchical methods have been used in the past few years to solve optimization and control problems for large-scale systems (Singh, 1977; Mesarovic et al., 1970; Singh and Tamura, 1974). Many ecological systems can be described by large-scale models (see chapter by Patten and Finn in this volume) and it may be possible to use these methods for solving control and optimization problems in ecology. In this chapter a brief description of some important hierarchical techniques is given and the techniques are illustrated for a particular ecological problem, that is, river pollution control.

The particular river pollution problem is the control of treatment plants in a multipolluter stream for the purpose of reaching a balance between the cost of treatment and the social cost to the community of having a polluted stream. This problem can be formulated as a large-scale dynamic optimization problem (Singh, 1977; Tamura, 1974; Beck, 1973) having a particular structure. Large-scale dynamic optimization problems are difficult to solve using standard methods owing to Bellman's (1957) "curse of dimensionality." This arises from the fact that as the order of a system increases, the computational burden for its dynamic optimization increases much faster. Such problems can be solved using hierarchical techniques.

The overall system can often be split up into interacting compartments which can be considered separately if certain variables, called coordination variables, are supplied by a higher level, the coordinator. The coordinator carries out an iterative exchange of information with the subproblems; as a result, each iteration of the optimization procedure on

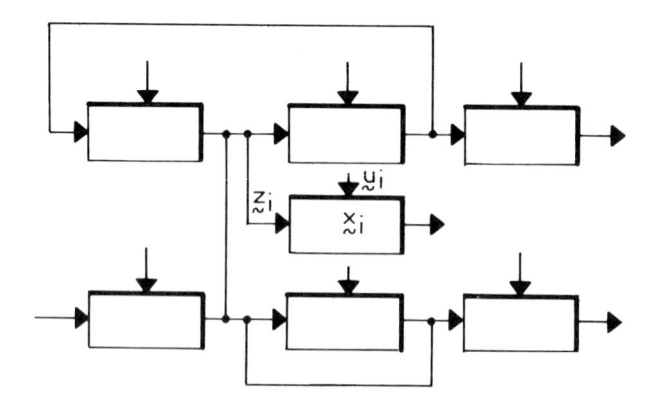

Figure 1. An interconnected dynamical system.

the subproblems improves the solution of the integrated global problem. In the context of the river pollution control problem the compartments are the "reaches" of a river, where a "reach" is a convenient length of the stream having a treatment facility.

It is assumed that a model of the system behavior is available. In general such models are represented by nonlinear differential or difference equations. However, in river pollution control, as well as in many other ecological problems, it is quite adequate to describe the dynamic behavior of the system using a linear differential or difference equation (Patten, 1975). This class of models will be treated here. The reader concerned with a particular nonlinear problem is referred to Singh's book (1977).

The rest of this chapter is divided into two main parts. In the first part the hierarchical approaches to the optimization and control of linear interconnected dynamical systems are described; in the second part the various approaches to the river pollution control problem are illustrated and the numerical efficiency of the various methods is compared.

2. PROBLEM FORMULATION

It is assumed that the overall system is a collection of N interconnected compartments as shown in Fig. 1. This subdivision can be made on the basis of ecological knowledge or on a theoretical basis (see Zeigler, this volume). For the ith compartment, $\mathbf{x}_i$ is an n_i-dimensional state vector, $\mathbf{u}_i$ is an m_i-dimensional control vector, and $\mathbf{z}_i$ is an r_i-dimensional vector of inputs from the other compartments. The compartment dynamics are assumed to be linear and can be represented by the following state space equations:

$$\dot{\mathbf{x}}_i(t) = A_i \mathbf{x}_i(t) + B_i \mathbf{u}_i(t) + C_i \mathbf{z}_i(t) \quad \text{with} \quad \mathbf{x}_i(0) = \mathbf{x}_{i0}. \tag{1}$$

It is assumed that the vector of inputs, $\mathbf{z}_i$, is a linear combination of the states of the N compartments, i.e.,

$$\mathbf{z}_i = \sum_{j=1}^{N} L_{ij} \mathbf{x}_j. \tag{2}$$

In Eq. (2), note that $i = j$ may be a possibility, which represents simple feedback around the subsystem. Note that in Fig. 1 there are two input vectors into each subsystem, that is, $\mathbf{u}_i$ and $\mathbf{z}_i$. $\mathbf{u}_i$ is the subsystem control and $\mathbf{z}_i$ represents all interactions with other subsystems.

It is desirable to choose the controls $\mathbf{u}_1, \ldots, \mathbf{u}_N$ in order to minimize the cost function

$$J = \sum_{i=1}^{N} \left(\tfrac{1}{2}\|x_i(T)\|_{Q_i}^2 + \int_0^T \tfrac{1}{2}[\|x_i(t)\|_{Q_i}^2 + \|u_i\|_{R_i}^2 + \|z_i\|_{S_i}^2]\, dt \right) \qquad (3)$$

subject to the constraints (1) and (2), where Q_i are positive semidefinite matrices and R_i, S_i are positive definite $\|c\|_L^2 = c^T L c$.

This cost function implies that (a) there is a quadratic cost function for each compartment and the overall function in Eq. (3) is a sum of these functions; (b) since it is possible to eliminate $\|z_i\|_S^2$ using Eq. (2), the cost function for each compartment allows, by a suitable choice of Q_i, R_i to maintain each state x_i around its desired value while not utilizing an excessive amount of energy; (c) it should be noted that the term $\|z_i\|_{S_i}^2$ is necessary for technical reasons (Singh, 1977), and it will be possible to eliminate it in certain methods.

If the interconnection relationship (2) is substituted back into Eq. (1) it is possible to obtain a standard overall description of the form

$$\dot{x} = Ax + Bu, \qquad (4)$$

where

$$x = \begin{bmatrix} x_1 \\ x_2 \\ \vdots \\ x_N \end{bmatrix}, \quad u = \begin{bmatrix} u_1 \\ u_2 \\ \vdots \\ u_N \end{bmatrix},$$

A and B are full matrices.

2.1. The Goal Coordination Approach

The basis of this approach is that it is possible to convert the original minimization problem into a simpler maximization problem and then solve this problem using a two-level iterative calculation structure.

To do this, define a dual function $\Phi(\lambda)$, where

$$\Phi(\lambda) = \min_{x,u,z} \{L^*(x, u, z, \lambda)\} \quad \text{(subject to Eq. (1))}, \qquad (5)$$

where

$$L^*(x, u, z, \lambda) = \sum_{i=1}^{N} \left\{ \tfrac{1}{2}\|x_i(T)\|_{Q_i}^2 + \int_0^T \left(\tfrac{1}{2}\|x_i\|_{Q_i}^2 + \tfrac{1}{2}\|u_i\|_{R_i}^2 + \tfrac{1}{2}\|z_i\|_{S_i}^2 \right. \right.$$

$$\left. \left. + \lambda^T \left(z_i - \sum_{j=1}^{N} L_{ij} x_j \right) \right) dt \right\}, \qquad (6)$$

where λ is an r-dimensional vector of Lagrange multipliers. Note that the Lagrangian L^* is a scalar. The theorem of strong Lagrange duality (Geffrion, 1971) asserts that for cases like the one considered here, where all the constraints are convex,

$$\max_{\lambda} \Phi(\lambda) = \min_{\mathbf{u}} J. \tag{7}$$

That is, the problem of minimizing J in Eq. (3) subject to the linear equality constraints given by (1) and (2) is equivalent to maximizing the dual function $\Phi(\lambda)$ with respect to λ. This can be done within a two-level structure since from Eq. (6) for a given $\lambda = \lambda^*$,

$$\begin{aligned} L(\mathbf{x}, \mathbf{u}, \mathbf{z}, \lambda^*) &= \sum_{i=1}^{N} \left\{ \tfrac{1}{2} \|\mathbf{x}_i(T)\|_{Q_i}^2 + \int_0^T \left(\tfrac{1}{2} \|\mathbf{x}_i\|_{Q_i}^2 + \tfrac{1}{2} \|\mathbf{u}_i\|_{R_i}^2 \right. \right. \\ &\quad \left. \left. + \tfrac{1}{2} \|\mathbf{z}_i\|_{S_i}^2 + \lambda_i^{*T} \mathbf{z}_i - \sum_{j=1}^{N} \lambda_j^{*T} L_{ji} \mathbf{x}_i \right) dt \right\} \\ &= \sum_{i=1}^{N} L_i, \end{aligned} \tag{8}$$

where L_i is a scalar, that is, the Lagrangian L is additively separable and can be decomposed into N independent sub-Lagrangians, one for each subsystem. Thus, for a given $\lambda = \lambda^*$ which is treated as a known trajectory, it is possible to minimize the sub-Lagrangian

$$\begin{aligned} L_i &= \tfrac{1}{2} \|\mathbf{x}_i(T)\|_{Q_i}^2 + \int_0^T \left(\tfrac{1}{2} \|\mathbf{x}_i\|_{Q_i}^2 + \tfrac{1}{2} \|\mathbf{u}_i\|_{R_i}^2 \right. \\ &\quad \left. + \tfrac{1}{2} \|\mathbf{z}_i\|_{S_i}^2 + \lambda_i^{*T} \mathbf{z}_i - \sum_{j=1}^{N} \lambda_j^{*T} L_{ji} \mathbf{x}_i \right) dt \end{aligned} \tag{8a}$$

independently for the N subsystems. Each subsystem's minimization is subject to that subsystem's dynamic constraints given by Eq. (1). By guessing $\lambda = \lambda^*$, a first value of $\Phi(\lambda^*)$ in Eq. (5) can be obtained, and the $\Phi(\lambda^*)$ can be improved successively by an iterative exchange of information with a second level, which improves $\Phi(\lambda^*)$ using the N-independent first-level minimizations. The actual mechanism for the improvement of $\Phi(\lambda^*)$, in order to maximize it, relies on the fact that it is possible to write a simple expression for the gradient of $\Phi(\lambda^*)$ in terms of the solutions of the first-level minimizations. In fact, the gradient is given by the error in the interconnection relationship, that is,

$$\nabla \Phi(\lambda)|_{\lambda = \lambda^*} = \mathbf{z}_i - \sum_{j=1}^{N} L_{ij} \mathbf{x}_j = \mathbf{e}_i, \quad i = 1, 2, \ldots, N. \tag{9}$$

424   Madan G. Singh

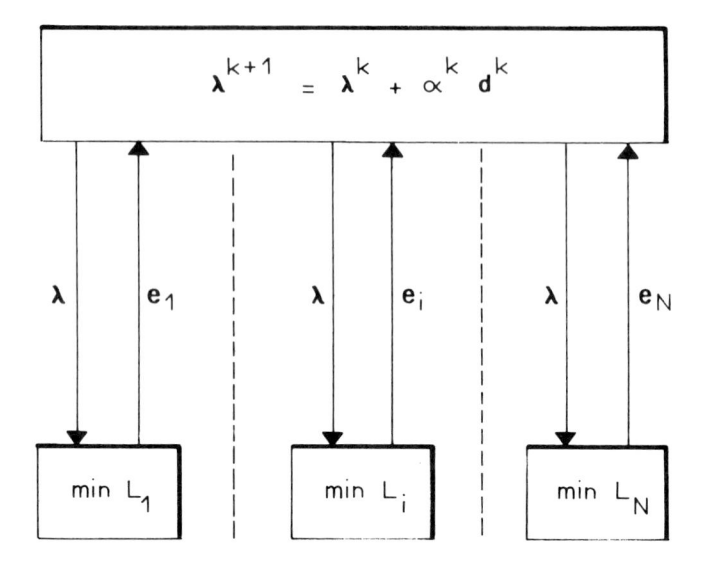

Figure 2. The two-level goal coordination.

It is thus possible to envisage a two-level hierarchical algorithm, as shown in Fig. 2, where on level 1 for a given $\lambda = \lambda^*$, supplied by the second level, L_i is minimized subject to the subsystem dynamic constraints and the resulting $\mathbf{x}_i$, $\mathbf{u}_i$ are sent back to level 2. At level 2, these vectors are collated and substituted into Eq. (9) to form the interconnection error

$$\mathbf{e} = \begin{bmatrix} \mathbf{e}_1 \\ \mathbf{e}_2 \\ \vdots \\ \mathbf{e}_N \end{bmatrix}.$$

This error vector is used in a gradient procedure to produce a new λ^*. For example, from iteration k to $k+1$

$$\lambda^{k+1}(t) = \lambda^k(t) + \alpha^k \mathbf{d}^k(t), \qquad 0 \leq t \leq T, \tag{10}$$

where α is the step length and $\mathbf{d}^k$ is the search direction. If the steepest ascent method is used, then $\mathbf{d}^k(t) = \mathbf{e}^k(t)$ ($0 \leq t \leq T$).

Instead, if the conjugate gradient method is used then

$$\mathbf{d}^{k+1}(t) = \mathbf{e}^{k+1}(t) + \beta^{k+1} \mathbf{d}^k, \qquad 0 \leq t \leq T, \tag{11}$$

where
$$\beta^{k+1} = \int_0^T (\mathbf{e}^{k+1})'(\mathbf{e}^{k+1})\,dt \Big/ \int_0^T (\mathbf{e}^k)'(\mathbf{e}^k)\,dt$$

with $\mathbf{d}^0 = \mathbf{e}^0$, where ' denotes the transpose of the vector. The overall optimum is achieved when $\mathbf{e}(t)$ ($0 \le t \le T$) is sufficiently close to zero.[1]

2.2. Remarks

The above method is called the Goal Coordination Method because coordination of the subproblems is performed via the Lagrange multipliers which enter into each subsystem cost function. Mesarovic et al. (1970) call it the Interaction Balance Approach because at the optimum the interactions balance, that is, $\mathbf{z}_i = \sum_{j=1}^N L_{ij}\mathbf{x}_j$. The algorithm is very elegant and has been tested by Pearson (1971) on a 12th-order example. Its main strengths are the following.

(a) It is possible to tackle large scale "linear-quadratic" problems with this approach since the computer storage requirements are no longer prohibitive. If parallel processors are used for the first-level computation, truly large problems can be solved.

(b) It is possible, at least in principle, to include inequality constraints on the states and controls of the subsystems since only low-order subproblems are solved.

(c) It is possible to show that the two-level controller will converge uniformly to the optimal solution (Varaiya, 1969).

Against the above advantages, the method has certain drawbacks and these are perhaps the principal reason why the method has not, at least in this form, been extensively used to solve practical large-scale problems. The main disadvantages of the method are as follows.

(a) Although the second-level algorithm is attractive in principle, since the gradient is easy to calculate, the choice of the step length causes some problems. One can either use a constant one or attempt to find an α at each iteration which gives the largest increase in $\Phi(\lambda)$. In the former case no α is, in general, suitable for all the convergence process. A large one may be desirable in the beginning while a much smaller one may be the most

---

[1] In principle one cannot be sure that the maximum of a function has been achieved merely because the gradient is zero. However, we know that (a) the dual function to be maximized is concave thus having a unique maximum and (b) the gradient algorithm ensures that this maximum is approached at each iteration since only the gradient direction which increases $\Phi(\lambda)$ is used. Thus in this case, if the error becomes zero, the original problem has indeed been solved.

appropriate nearer to the optimum. If a one-dimensional search is made for the best α, a lot of additional computation is required since it will be necessary to solve the lower-level problem a number of times to obtain the best step length. Given that one requires doing all of this at each iteration, and since the whole convergence process may require many iterations, the overall calculation, although yielding substantial savings in computer storage (since only low-order problems are solved), is unlikely to yield a saving in computation time if a single computer is used to perform the calculations of the two levels sequentially. However, this is not a real drawback since in future applications it is quite likely that parallel minicomputers will be used at the first level thus improving the computation time.

(b) The large computation time on a single computer of the present approach could perhaps be reduced if one were willing to accept good suboptimal control by terminating the iterations before the optimum was reached. However, this is not possible because the method does not give a practical control except at the optimum. It is only at this point that all the system and interconnection constraints are satisfied.

(c) Another disadvantage of this approach is that the inclusion of the term $\frac{1}{2}\|z_i\|_{S_i}^2$ in the cost function does not correspond to a realistic physical situation and has been added to ensure that singular solutions do not arise at the first level (Singh, 1977).

Now, although singular solutions are perfectly valid solutions to optimization problems, they are certainly undesirable in the iterative hierarchical scheme. They complicate the lower-level calculation enormously while one of the main justifications of hierarchical optimization is the ease of calculation achieved by decentralization. Thus, although the goal coordination method is attractive, it has not been extensively used for two main reasons: (1) the second-level calculation gets complicated because of the need for finding a good step length and (2) terms which are not physically meaningful must be introduced in order to avoid singularities. Nevertheless, the method is significant because of its conceptual simplicity. Indeed, Tamura (1974) used a discrete time version of this method to solve river pollution control problems.

3. THE THREE-LEVEL METHOD OF TAMURA

This method (Tamura, 1974) treats the problem in discrete time because the discretization is necessary for the lowest level. For the sake of completeness, the goal coordination method for discrete dynamical systems is described followed by the modification of Tamura.

3.1. The Goal Coordination Method for Discrete Dynamical Systems

The problem is to minimize

$$J = \sum_{i=1}^{N} \left[\tfrac{1}{2}\|\mathbf{x}_i(K)\|_{Q_{i(K)}}^2 + \sum_{k=0}^{K-1} \tfrac{1}{2}\Big(\|\mathbf{x}_i(k)\|_{Q_{i(k)}}^2 \right. $$
$$\left. + \tfrac{1}{2}\|\mathbf{z}_i\|_{S_i}^2 + \|\mathbf{u}_i(k)\|_{R_{i(k)}}^2 \Big)\right], \tag{12}$$

where $\tfrac{1}{2}\|\mathbf{x}_i(K)\|_{Q_{i(k)}}^2$ is the cost for the terminal interval K and the terms within the inner summation represent the cost over the rest of the optimization sequence, that is, from $k = 0$ to $K - 1$. This cost function is the discrete time equivalent of Eq. (3) and was chosen for the same reason.

This minimization is to be performed subject to the subsystem dynamic constraints, that is,

$$\mathbf{x}_i(k+1) = A_i\mathbf{x}_i(k) + B_i\mathbf{u}_i(k) + C_i\mathbf{z}_i(k),$$
$$i = 1, 2, \ldots, N, \quad k = 0, 1, 2, \ldots, K-1, \tag{13}$$

with the assumption that the initial state is known,

$$\mathbf{x}_i(0) = \mathbf{x}_{i0}. \tag{14}$$

As in the continuous time case, $\mathbf{z}_i$ is the vector of interaction inputs from the other subsystems,

$$\mathbf{z}_i(k) = \sum_{j=1}^{N} L_{ij}\mathbf{x}_j(k), \quad k = 0, 1, \ldots, K-1, \quad i = 1, 2, \ldots, N; \tag{15}$$

To solve this problem it is necessary, as in the continuous time case, to maximize a dual function $\phi(\lambda)$ with respect to λ, where

$$\phi(\lambda) = \min_{\mathbf{x},\mathbf{u}} L^*(\mathbf{x}, \mathbf{u}, \lambda) \tag{16}$$

subject to Eqs. (13) and (14), where

$$L^*(\mathbf{x}, \mathbf{u}, \lambda) = \sum_{i=1}^{N} \left\{ \tfrac{1}{2}\|\mathbf{x}_i(K)\|_{Q_i}^2 + \sum_{k=0}^{K-1} \tfrac{1}{2}\|\mathbf{x}_i(k)\|_{Q_i}^2 \right.$$
$$\left. + \tfrac{1}{2}\|\mathbf{u}_i(k)\|_{R_i}^2 + \tfrac{1}{2}\|\mathbf{z}_i\|_{S_i}^2 + \lambda_i^T \mathbf{z}_i - \sum_{j=1}^{N} \lambda_j^T L_{ji}\mathbf{x}_i(k) \right\}$$
$$= \sum_{i=1}^{N} L_i, \tag{17}$$

where

$$L_i = \tfrac{1}{2}\|\mathbf{x}_i(K)\|^2_{Q_i(K)} + \sum_{k=0}^{K-1} \tfrac{1}{2}\|\mathbf{x}_i(k)\|^2_{Q_i(k)} + \tfrac{1}{2}\|\mathbf{u}_i(k)\|^2_{R_i(k)}$$
$$+ \lambda_i^T \mathbf{z}_i - \sum_{j=1}^{N} \lambda_j^T L_{ji} \mathbf{x}_i(k) + \tfrac{1}{2}\|\mathbf{z}_i\|^2_{S_i}. \tag{18}$$

Note again that L_i is a scalar since L^* is a scalar. Thus as in the continuous time case, it is possible to separate the problem of minimizing the Lagrangian L^* into minimizing N independent sub-Lagrangians L_i for given sequences $\lambda = \lambda^*$ supplied by a second level, each subject to Eqs. (13) and (14). The Lagrange multiplier vector sequences can be improved at the second level by using a gradient-type algorithm since

$$\nabla \phi(\lambda)|_{\lambda = \lambda^*} = \mathbf{z}_i(k) - \sum_{j=1}^{N} L_{ij} \mathbf{x}_j(k),$$
$$i = 1, 2, \ldots, N; \quad k = 0, 1, \ldots, K-1. \tag{19}$$

3.2. The Modification of Tamura

The basis of this approach is the observation that for a given trajectory $\lambda^*(k)$, $k = 0, 1, \ldots, K-1$, the first-level problem of minimizing L_i subject to Eqs. (13) and (14) can itself be treated by duality and decomposition. The Lagrangian is not decomposed into the sub-Lagrangians for each subsystem, but the subsystem Lagrangian itself is decomposed by the index k leading at the lowest level to a *parametric* as opposed to a *functional* optimization.

Define the dual problem of minimizing L_i in Eq. (18) subject to Eqs. (13) and (14). Maximize $M(\mathbf{p})$ where

$$M(\mathbf{p}) = \min_{\mathbf{x},\mathbf{u}} \Big\{ \tfrac{1}{2}\|\mathbf{x}_i(K)\|^2_{Q_i} + \sum_{k=0}^{K-1} \Big(\tfrac{1}{2}\|\mathbf{x}_i(k)\|^2_{Q_i} + \tfrac{1}{2}\|\mathbf{u}_i(k)\|^2_{R_i}$$
$$+ \mathbf{p}_i(k)^T [A_i \mathbf{x}_i(k) + B_i \mathbf{u}_i(k) + C_i \mathbf{z}_i(k) - \mathbf{x}_i(k+1)]$$
$$+ \lambda_i^{*T} \mathbf{z}_i - \sum_{j=1}^{N} \lambda_j^{*T} L_{ji} \mathbf{x}_i + \tfrac{1}{2}\|\mathbf{z}_i(k)\|^2_{S_i} \Big) \Big\}. \tag{20}$$

To solve this dual problem numerically, it is necessary to compute the value of the dual function $M(\mathbf{p})$ for given $\mathbf{p} = \mathbf{p}^*$ and then to maximize $M(\mathbf{p})$ by some gradient technique. The gradient of $M(\mathbf{p})$ is given by

$$\nabla M(\mathbf{p})|_{\mathbf{p} = \mathbf{p}^*} = -\mathbf{x}_i(k+1) + A_i \mathbf{x}_i(k) + B_i \mathbf{u}_i(k) + C_i \mathbf{z}_i(k),$$
$$k = 0, 1, \ldots K-1; \quad i = 1, 2, \ldots N, \tag{20a}$$

where $\mathbf{x}_i$, $\mathbf{u}_i$ are the solutions obtained after minimization of L_i subject to Eq. (13) for a given $\mathbf{p} = \mathbf{p}^*$. The computation of $M(\mathbf{p})$ for a fixed $\mathbf{p} = \mathbf{p}^*$ and $\lambda = \lambda^*$ can be performed by minimizing the function independently for each time index k as follows: Define the Hamiltonian of the ith compartment by

$$H_i(\mathbf{x}_i(k), \mathbf{u}_i(k), k) = \tfrac{1}{2}\|\mathbf{x}_i(k)\|_{Q_i}^2 + \tfrac{1}{2}\|\mathbf{u}_i(k)\|_{R_i}^2 + \lambda_i^{*T}\mathbf{z}_i$$

$$+ \tfrac{1}{2}\|\mathbf{z}_i\|_{S_i}^2 - \sum_{j=1}^N \lambda_j^{*T} L_{ji}\mathbf{x}_i + \mathbf{p}_i^{*T}(k)$$

$$\times [-\mathbf{x}_i(k+1) + A_i\mathbf{x}_i(k) + B_i\mathbf{u}_i(k) + C_i\mathbf{z}_i(k)],$$

$$k = 0, 1, \ldots, K-1; \quad i = 1, 2, \ldots, N. \qquad (21)$$

Then, using Eq. (21)

$$M(\mathbf{p}) = \tfrac{1}{2}\|\mathbf{x}_i(K)\|_{Q_{i(K)}}^2 - \mathbf{p}_i^{*T}(K-1)\mathbf{x}_i(K) + \sum_{k=0}^{K-1}(H_i(\mathbf{x}_i(k), \mathbf{u}_i(k), k)$$

$$- \mathbf{p}_i^*(k-1)^T\mathbf{x}_i(k)),$$

where $\mathbf{p}(-1)$ is defined to be zero. The minimization problem for a fixed $\mathbf{p} = \mathbf{p}^*$ then becomes:

For $k = 0$. Minimize with respect to

$$\mathbf{u}_i(0), \mathbf{z}_i(0): \{H_i(\mathbf{x}_i(0), \mathbf{u}_i(0)) \text{ subject to } \mathbf{x}_i(0) = \mathbf{x}_{i0}\}.$$

It is possible to obtain an explicit solution by setting the partial derivative of H_i with respect to $\mathbf{u}_i(0)$ equal to zero to yield

$$\mathbf{u}_i(0) = -R_i^{-1}B_i^T\mathbf{p}_i^*(0), \quad \mathbf{z}_i(0) = -S_i^{-1}(C_i^T\mathbf{p}^*(0) + \lambda_i^*(0)). \qquad (22)$$

For $k = 1, 2, \ldots, K-1$. Minimize

$$\{H_i(\mathbf{x}_i(k), \mathbf{u}_i(k), k) - \mathbf{p}_i^*(k-1)^T\mathbf{x}_i(k)\}$$

with respect to $\mathbf{x}_i(k)$, $\mathbf{u}_i(k)$, $\mathbf{z}_i(k)$. The explicit solution in this case is

$$\mathbf{x}_i(k) = -Q_i(k)^{-1}\left[A_i^T\mathbf{p}_i^*(k) + \mathbf{p}_i^*(k-1) + \sum_{j=1}^N [\lambda_j^{*T}L_{ji}]^T\right],$$

$$\mathbf{u}_i(k) = -R_i^{-1}B_i^T\mathbf{p}_i^*(k), \qquad (23)$$

$$\mathbf{z}_i(k) = -S_i^{-1}(C_i^T\mathbf{P}_i^*(k) + \lambda_i^*(k)).$$

For $k = K$. Minimize with respect to $\mathbf{x}_i(K)$

$$\tfrac{1}{2}\|\mathbf{x}_i(K)\|_{Q_{i(K)}}^2 - \mathbf{p}_i^{*T}(K-1)\mathbf{x}_i(K)$$

which gives

$$\mathbf{x}_i(K) = \mathbf{p}_i^*(K-1). \qquad (24)$$

Thus the integrated problem of minimizing J in Eq. (17), subject to the dynamics given by Eqs. (13)–(16), can be solved by a three-level algorithm, where on level 1 for a given $\lambda^*(k)$, $\mathbf{p}^*(k)$ sequences, it is merely necessary to substitute into the explicit solutions given by Eqs. (22)–(24) in order to obtain the optimal x, u which can be used at level 2 to calculate the gradient of $M(\mathbf{p})$ from Eq. (20a) and thus to improve $\mathbf{p}$ to maximize $M(\mathbf{p})$. On the third level, the optimal $\mathbf{p}$ obtained from the second-level optimization, can be used to iteratively improve $\phi(\lambda)$ to maximize it. The overall optimum is achieved when both $\phi(\lambda)$ and $M(\mathbf{p})$ go to zero. Figure 3 shows the optimization structure.

3.3. Remarks

(a) This method is attractive because an explicit solution is obtained at the lowest level, which makes the level 1 problem trivial and ensures that complicated Riccati equations need not be solved.

(b) It is still necessary as in the goal coordination method (Section 2.1) to introduce the term $\frac{1}{2}\|\mathbf{z}_i\|_{S_i}^2$ in the cost function in order to avoid a singular solution.

In the next section we describe how Tamura used this concept of three-level hierarchy to produce one of the most powerful and useful algorithms in hierarchical optimization practice, that is, the Time Delay Algorithm.

4. THE TIME DELAY ALGORITHM OF TAMURA

This algorithm solves a class of problems which are of great practical importance. The overall system has multiple pure time delays in the state and control variables. In addition, the states and controls are bounded by inequality constraints.

The system dynamics are assumed to be represented by a high-order difference equation of the form

$$\mathbf{x}(k+1) = A_0\mathbf{x}(k) + A_1\mathbf{x}(k-1) + \cdots + A_\theta\mathbf{x}(k-\theta) + B_0\mathbf{u}(k) \\ + B_1\mathbf{u}(k-1) + \cdots + B_\theta\mathbf{u}(k-\theta), \qquad (25)$$

where A_i ($i = 0, 1, \ldots \theta$) are $n \times n$ matrices, $\mathbf{x}$ is a $1 \times n$ vector, $\mathbf{u}$ is a $1 \times r$ vector, and B_i ($i = 0, 1, \ldots \theta$) are $n \times r$ matrices. In addition it is assumed that

$$\mathbf{x}(k) = 0, \qquad \mathbf{u}(k) = 0 \quad \text{for } k < 0 \qquad \text{and} \qquad \mathbf{x}(0) = \mathbf{x}_0. \qquad (26)$$

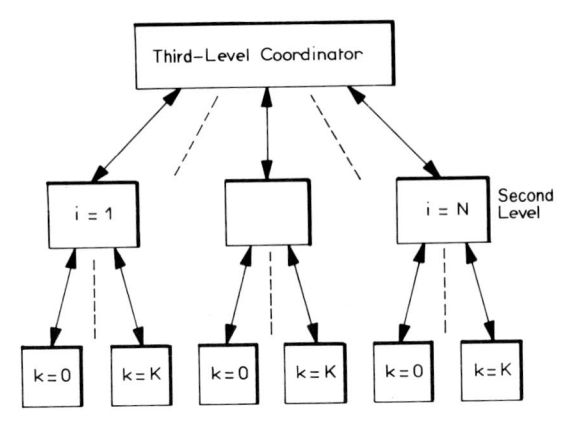

Figure 3. The three-level method of Tamura.

Equation (26) can be interpreted as follows. The system is assumed to be at some steady-state operating point up to the instant $k = 0$ when it receives an unknown disturbance which takes the state of the system to a known value $\mathbf{x}_0$. Although the steady-state operating point is assumed to be zero, it is just as easy to consider a nonzero operating point by interpreting the sequences $\mathbf{x}(k)$, $\mathbf{u}(k)$, $k >$ to be varying about this actual steady state.

The states and controls can be bounded by the inequality constraints

$$\mathbf{x}_{\min} \leqslant \mathbf{x}(k) \leqslant \mathbf{x}_{\max}, \quad k = 0, 1, \ldots, K, \tag{27}$$
$$\mathbf{u}_{\min} \leqslant \mathbf{u}(k) \leqslant \mathbf{u}_{\max}, \quad k = 0, 1, \ldots, K-1,$$

and it is desired to minimize

$$J = \tfrac{1}{2}\|\mathbf{x}(K)\|^2_{Q(K)} + \sum_{k=0}^{K-1} \tfrac{1}{2}(\|\mathbf{x}(k)\|^2_{Q(k)} + \|\mathbf{u}(k)\|^2_{R(k)}), \tag{28}$$

where Q and R are assumed to be positive definite block-diagonal matrices.

It is very difficult to solve this problem for large-scale systems mainly because of the existence of the multiple time delays as well as the inequality constraints on the states and controls. The delays are usually eliminated by augmenting the state space and introducing additional variables for the delays. But this solution is undesirable for large-scale systems. In fact, the problem even without the augmentation already has a dimension too large to be solved by standard techniques. The inequality constraints prove even more of a problem if functional optimization techniques are used. In the Tamura method, both of these difficulties are circumvented if the basic concept of the three-level hierarchy is used (Section 3), where at the lowest level a quadratic programming problem is solved.

Write the "Hamiltonian"[2] of the overall system as

$$H(\mathbf{x}(k), \mathbf{u}(k), \mathbf{p}(k), k) = \tfrac{1}{2}(\|\mathbf{x}(k)\|^2_{Q(k)} + \|\mathbf{u}(k)\|^2_{R(k)}) + \sum_{j=0}^{\theta} \mathbf{p}(k+j)^T \\ \times (A_j \mathbf{x}(k) + B_j \mathbf{u}(k)), \quad k = 0, 1, \ldots, K-1, \quad (29)$$

where $\mathbf{p}(k)$ is defined to be zero for $k \geq K$.

For a fixed $\mathbf{p} = \mathbf{p}^* = [\mathbf{p}(0)^*, \ldots, \mathbf{p}(K-1)^*]$ it is possible to write the Lagrangian as

$$L(\mathbf{x}, \mathbf{u}, \mathbf{p}^*, k) = \tfrac{1}{2}\|\mathbf{x}(K)\|^2_{Q(K)} - \mathbf{p}^{*T}(K-1)\mathbf{x}(K) + \sum_{k=0}^{K-1} \{H(\mathbf{x}, \mathbf{u}, \mathbf{p}^*, k) \\ - \mathbf{p}^*(k-1)\mathbf{x}(k)\} \quad (30)$$

subject to

$$\mathbf{x}_{\min} \leq \mathbf{x}(k) \leq \mathbf{x}_{\max}, \quad k = 0, 1, 2, \ldots, K,$$
$$\mathbf{u}_{\min} \leq \mathbf{u}(k) \leq \mathbf{u}_{\max}, \quad k = 0, 1, 2, \ldots, K-1.$$

As in the previous sections, in order to obtain the optimal control, it is necessary to maximize the Lagrangian $L(\mathbf{x}, \mathbf{u}, \mathbf{p})$ with respect to $\mathbf{p}$ and minimize it with respect to $\mathbf{x}, \mathbf{u}$. One of the attractions of this formulation is that the Lagrange multiplier vector $\mathbf{p}$ is of the same dimension as $\mathbf{x}$ despite the existence of the delays.

As in the three-level algorithm, the Lagrangian [Eq. (30)] can be decomposed into the following $(K+1)$ independent minimization problems for a fixed $\mathbf{p}$.

(a) *For $k = 0$.* From Eq. (30) using the definition of H from Eq. (29), the first problem is

$$\min_{\mathbf{u}(0)} H(\mathbf{x}(0), \mathbf{u}(0), \mathbf{p}(0)) = \min_{\mathbf{u}(0)} \tfrac{1}{2}(\|\mathbf{x}(0)\|^2_{Q(0)} + \|\mathbf{u}(0)\|^2_{R(0)}) \\ + \sum_{j=0}^{\theta} \mathbf{p}^{*T}(j)(A_j \mathbf{x}(0) + B_j \mathbf{u}(0))$$

subject to

$$\mathbf{x}(0) = \mathbf{x}_0, \quad \mathbf{u}_{\min} \leq \mathbf{u}(0) \leq \mathbf{u}_{\max}.$$

The solution of this parametric optimization problem is further simplified by the fact that here $R(0)$ has been assumed to be a diagonal matrix so that the minimization problem reduces to a set of r independent one-variable

---

[2] The "Hamiltonian" here is defined for convenience and it corresponds to the conventional Hamiltonian of the discrete maximum principle if $\theta = 0$, i.e., for the no-delay case.

minimizations. For a single-variable minimization it is, of course, easy to include the inequality constraints. The explicit solution is thus given by

$$\mathbf{u}^*(0) = Sat_2\left[-R^{-1}(0)\sum_{j=0}^{\theta} B_j^T \mathbf{p}^*(j)\right] \quad (31)$$

where for $i = 1, 2, \ldots r$, the ith element of $Sat_2(\eta)$ is given by

$$Sat_2(\eta) = \begin{cases} u_{max,i}, & \text{if } \eta_i > u_{max,i}, \\ \eta_i, & \text{if } u_{min,i} \leq \eta_i \leq u_{max,i}, \\ u_{min,i}, & \text{if } \eta_i < u_{min,i}. \end{cases} \quad (32)$$

The solution given in Eq. (32) is obtained from a trivial manipulation by setting $\partial H/\partial \mathbf{u}(0) = \mathbf{0}$ and then noting that since R is a diagonal matrix, this can be viewed as a set of r independent one-variable solutions bounded by the limits.

(b) For $k = 1, 2, \ldots, K-1$. Similarly, for this case the minimization problem is Minimize with respect to $\mathbf{x}(k), \mathbf{u}(k)$

$$H(\mathbf{x}(k), \mathbf{u}(k), \mathbf{p}^*(k), k) - \mathbf{p}^*(k-1)^T \mathbf{x}(k)$$

subject to

$$\mathbf{x}_{min} \leq \mathbf{x}(k) \leq \mathbf{x}_{max}, \quad \mathbf{u}_{min} \leq \mathbf{u}(k) \leq \mathbf{u}_{max}.$$

Again, since $Q(k), R(k)$ are assumed to be diagonal matrices, this becomes a set of $(n+r)$ independent one-variable minimizations with the solution given by

$$\begin{aligned}\mathbf{x}^*(k) &= Sat_1\left\{-Q^{-1}(k)\left[-\mathbf{p}^*(k-1) + \sum_{j=0}^{\theta} A_j \mathbf{p}^*(k+j)\right]\right\}, \\ \mathbf{u}^*(k) &= Sat_2\left\{-R^{-1}(k)\left[\sum_{j=0}^{\theta} B_j \mathbf{p}^*(k+j)\right]\right\},\end{aligned} \quad (33)$$

where for $i = 1, 2, \ldots, n$, the ith element of $Sat_1(\xi)$ is

$$Sat_1(\xi) = \begin{cases} x_{max,i}, & \text{if } \xi_i > x_{max,i}, \\ \xi_i, & \text{if } x_{min,i} \leq \xi_i \leq x_{max,i}, \\ x_{min,i}, & \text{if } \xi_i < x_{min,i}. \end{cases} \quad (34)$$

(c) For $k = K$. Minimize $\mathbf{x}(K)$

$$\tfrac{1}{2}\|\mathbf{x}(K)\|_{Q(K)}^2 - \mathbf{p}^{*T}(K-1)\mathbf{x}(K)$$

subject to

$$\mathbf{x}_{min} \leq \mathbf{x}(K) \leq \mathbf{x}_{max}.$$

The solution is
$$\mathbf{x}^*(K) = \mathrm{Sat}_1[Q^{-1}\mathbf{p}^*(K-1)]. \tag{35}$$

The solutions given by Eqs. (31)–(35) effectively allow the analytical computation of the minimum of the Lagrangian for a given $\mathbf{p} = \mathbf{p}^*$. On the second level it is necessary to improve $\mathbf{p}^*$ in order to maximize the dual function. The gradient is given by the error in the system equation [i.e., right-hand side of Eq. (25) minus $\mathbf{x}\,(k+1)$] so that a gradient method (Section 2) could be used.

4.1. Remarks

The time delay algorithm of Tamura is virtually the only goal-coordination-type algorithm which has so far been used to solve practical problems. The main reasons for this are (a) there is a substantial computational saving since it is not necessary to augment the state space to account for the delays; (b) the inequality constraints are treated in a very simple way; (c) the cost function is meaningful because no additional terms have been introduced to avoid the singular solutions which occur in the standard goal coordination solution.

Thus the method is able to get around most of the disadvantages of the standard goal coordination approach (except that it is still necessary to perform a linear search for the second-level gradient algorithm). However, the most important practical advantage of the approach is that time delay systems have in general relatively slow dynamics (because of the delays). This means that given some disturbance before the instant $k = 0$ which takes the system state to some known state $\mathbf{x}_0$ (or a state $\mathbf{x}_0$ which can be obtained by suitable measurement), it may well be possible to calculate the control sequences rapidly enough before the initial state changes significantly. This is one of the few cases where open-loop control methods of the kind developed here could well be used for on-line control.

All of the methods considered here suffer from the disadvantage that a linear search is necessary for the second level and this requires time-consuming multiple evaluations of the first level. The interaction prediction approach avoids this difficulty and provides a simple and computationally attractive algorithm.

5. THE INTERACTION PREDICTION APPROACH[3]

The overall system as usual consists of N linear interconnected subsystems described by

---
[3] Takahara, 1965; Singh, 1977.

17. Hierarchical Methods in River Pollution Control

$$\dot{\mathbf{x}}_i = A_i \mathbf{x}_i + B_i \mathbf{u}_i + C_i \mathbf{z}_i, \qquad i = 1, 2, \ldots, N, \tag{36}$$

where

$$\mathbf{z}_i = \sum_{j=1}^{N} L_{ij} \mathbf{x}_j, \tag{37}$$

and it is desired to minimize

$$J = \sum_{i=1}^{N} \tfrac{1}{2}\|\mathbf{x}_i(t)\|^2_{Q_i(T)} + \int_0^T (\tfrac{1}{2}\|\mathbf{x}_i(t)\|^2_{Q_i} + \tfrac{1}{2}\|\mathbf{u}_i(t)\|^2_{R_i}) \, dt. \tag{38}$$

Note that the cost function in Eq. (38) does not contain the terms $\|\mathbf{z}_i\|^2_{S_i}$ as in Eq. (3). This is more realistic as discussed previously in Section 2.2.

The Lagrangian is

$$L = \sum_{i=1}^{N} \left\{ \tfrac{1}{2}\|\mathbf{x}_i(T)\|^2_{Q_i(T)} + \int_0^T \left(\tfrac{1}{2}\|\mathbf{x}_i(t)\|^2_{Q_i(t)} + \tfrac{1}{2}\|\mathbf{u}_i(t)\|^2_{R_i(t)} \right. \right.$$
$$\left. \left. + \boldsymbol{\lambda}_i^T \left[\mathbf{z}_i - \sum_{j=1}^{N} L_{ij}\mathbf{x}_j \right] + \mathbf{p}_i^T [-\dot{\mathbf{x}}_i + A_i \mathbf{x}_i + B_i \mathbf{u}_i + C_i \mathbf{z}_i] \right) dt \right\}, \tag{39}$$

where $\mathbf{p}_i$ is the n_i-dimensional adjoint vector and $\boldsymbol{\lambda}_i$ is the r_i-dimensional vector of Lagrange multipliers, and these have been introduced as before to ensure satisfaction of the constraints (36) and (37). Now for a given $\boldsymbol{\lambda}_i = \boldsymbol{\lambda}_i^*$, $\mathbf{z}_i = \mathbf{z}_i^*$, L in Eq. (39) is additively separable, that is,

$$L = \sum_{i=1}^{N} L_i = \sum_{i=1}^{N} \tfrac{1}{2}\|\mathbf{x}_i(T)\|^2_{Q_i(T)} + \int_0^T \left(\tfrac{1}{2}\|\mathbf{x}_i(t)\|^2_{Q_i(t)} + \tfrac{1}{2}\|\mathbf{u}_i(t)\|^2_{R_i(t)} \right.$$
$$\left. + \boldsymbol{\lambda}_i^{*T}\mathbf{z}_i^* - \sum_{j=1}^{N} \boldsymbol{\lambda}_j^{*T} L_{ji}\mathbf{x}_i + \mathbf{p}_i^T [-\dot{\mathbf{x}}_i + A_i\mathbf{x}_i + B_i\mathbf{u}_i + C_i\mathbf{z}_i^*] \right) dt.$$

By regrouping the terms in the index i[4] where

$$L_i = \tfrac{1}{2}\|\mathbf{x}_i(T)\|^2_{Q_i(T)} + \int_0^T \left(\tfrac{1}{2}\|\mathbf{x}_i(t)\|^2_{Q_i} + \tfrac{1}{2}\|\mathbf{u}_i\|^2_{R_i} + \boldsymbol{\lambda}_i^T * \mathbf{z}_i^* \right.$$
$$\left. - \boldsymbol{\lambda}_j^T * L_{ji}\mathbf{x}_i + \mathbf{p}_i^T [A_i \mathbf{x}_i + B_i \mathbf{u}_i + C_i \mathbf{z}_i^* - \dot{\mathbf{x}}_i] \right) dt. \tag{40}$$

Here, unlike in the goal coordination case where the coordination vector was only the Lagrange multiplier vector, the coordination vector is $[\begin{smallmatrix}\lambda\\z\end{smallmatrix}]$. This is of higher dimension than the coordination vector for the goal coordination case. However, the second-level algorithm is exceedingly

---
[4] This regrouping of the terms in the index i provides the term $\sum \boldsymbol{\lambda}_j^* L_{ji} \mathbf{x}_i$.

simple and this ensures that there is no disadvantage in using the more complex coordination vector. The second-level algorithm provides an improvement for the coordination vector by reinjecting the value of the vector from the previous iteration into the stationarity conditions, that is, from iteration k to $k+1$

$$\begin{bmatrix} \lambda^{*k+1} \\ z^{*k+1} \end{bmatrix} = \begin{bmatrix} \lambda^*(\mathbf{x}^k, \mathbf{u}^k, \mathbf{p}^k) \\ z^*(\mathbf{x}^k, \mathbf{u}^k, \mathbf{p}^k) \end{bmatrix}, \qquad (41)$$

where the expression on the right-hand side of Eq. (41) is obtained by setting

$$\partial L/\partial z_i^* = \mathbf{0} \quad \text{and} \quad \partial L/\partial \lambda_i^* = \mathbf{0},$$

that is,

$$\lambda_i^* = C_i^T \mathbf{p}_i \quad \text{and} \quad z_i^* = \sum_{j=1}^{N} L_{ij} \mathbf{x}_j,$$

thus making the coordination rule

$$\begin{bmatrix} \lambda_i^* \\ z_i^* \end{bmatrix}^{k+1} = \begin{bmatrix} C_i^T \mathbf{p}_i \\ \sum_{j=1}^{N} L_{ij} \mathbf{x}_j \end{bmatrix}^k. \qquad (42)$$

The method therefore involves the minimization of the N independant sub-Lagrangians L_i for given λ_i^*, z_i^* and then using the resultant $\mathbf{p}$ and $\mathbf{x}$ to calculate the new prediction for λ_i^*, z_i^* by substituting into the right-hand side of Eq. (42).

5.1. Remarks

The above method is attractive for many reasons.

(a) The second-level algorithm is very simple. It is not necessary to do the inefficient linear search as in the goal coordination method.

(b) The problem formulation is more meaningful since a quadratic term in z need not be included to avoid singularities as in the goal coordination approach.

(c) Experience (Singh and Hassan, 1976a) shows that the method has extremely fast convergence.

These approaches are now illustrated in the river pollution control problem. A discrete version of this problem has been treated by Tamura (1974). The method has also been used by Fallside and Perry (1975) to solve a water resource problem, and by Singh and Tamura (1974) to solve traffic control problems. Here we will consider the more practical prediction principle approach.

6. RIVER POLLUTION CONTROL

In recent years there has been much interest in regulating the levels of pollution in rivers (Beck, 1973; Singh, 1975; Singh and Hassan, 1976b; Singh *et al.*, 1976). A good measure of the "quality" of a stream can be obtained from two main factors, that is, (a) the instream biochemical oxygen demand (BOD)[5] and (b) the dissolved oxygen (DO) in the stream. If the DO falls below certain levels or the BOD rises above certain levels, fish die.

The ecological balance of the river is often disturbed by unknown perturbations and it becomes necessary to vary the BOD content of the sewage (by increasing or decreasing the treatment levels) in order to bring the river quality back to the desired values. It is the problem of on-line regulation of sewage discharge BOD from multiple sewage works on a polluted river that we treat in this section.

6.1. The Model

The reach of a river is defined as a stretch of a river of some convenient length which receives one major controlled effluent discharge from a sewage treatment facility. Beck (1973) has developed a second-order state space equation which describes the BOD–DO relationship at some average point in the reach. Each reach is thought of as an ideal stirred tank reactor, as shown in Fig. 4a, so that the parameters and variables are uniform throughout the reach and the output concentrations of BOD and DO are equal to those in the reach. Then, from mass balance considerations, the following equations can be written

$$\text{BOD} \quad \dot{z}_i = -K_{1i}z_i + \frac{Q_{i-1}}{V_i}z_{i-1} - \frac{Q_i+Q_E}{V_i}z_i + \frac{m_i Q_E}{V_i},$$

$$\text{DO} \quad \dot{q}_i = K_{2i}(q_i^s - q_i) - \frac{Q_{i-1}}{V_i}q_{i-1} - \frac{Q_i+Q_E}{V_i}q_i - K_{1i}z_i - \frac{\eta_i}{V_i},$$

(43)

where V_i is the volume of water in reach i in cubic meters; Q_E, the flow rate of the effluent in reach i in cubic meters; z_i, z_{i-1}, the concentrations of BOD in reaches i and $i-1$ in mg liter^{-1}; q_i, q_{i-1}, the concentrations of DO in reaches i and $i-1$ in mg liter^{-1}; K_{1i}, the BOD decay rate day^{-1} in reach i;

---

[5] The biochemical oxygen demand is a measure of the rate of oxygen consumption by decomposing organic matter.

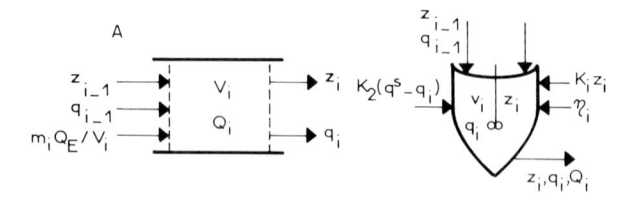

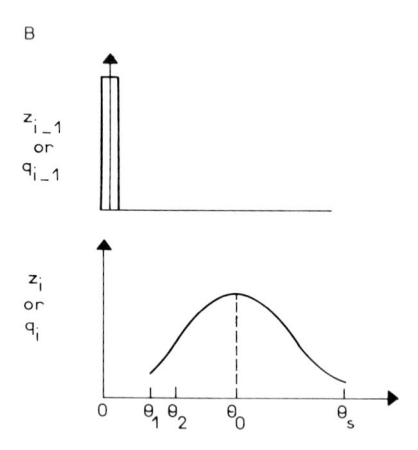

Figure 4. (a) An ideal stirred tank reactor model for a reach of a river. (b) The distributed delay phenomenon.

K_{2i}, the DO reaeration rate day^{-1} in reach i; Q_1, Q_{i-1}, are the stream flow rates in reaches i and $i-1$ in m^3 day^{-1}; q_i^s, the DO saturation level for the ith reach (mg liter^{-1}); η_i/V_i, the removal of DO due to bottom sludge requirements (mg liter^{-1} day^{-1}); and m_i, the concentration of BOD in the effluent in mg liter^{-1}.

For a section of the river Cam near Cambridge in England Beck (1973) found that the following values for the coefficients in Eq. (43) were appropriate: $K_{1i}=0.32$ day^{-1}, $K_{2i}=0.2$ day^{-1}, $\eta_i/V_i=0.1$ mg liter^{-1} day^{-1}, $q_i^s = 10$ mg liter^{-1}, $Q_E/V = 0.1$, $Q/V = 0.9$. Thus for the ith reach, Eq. (43) could be rewritten as:

$$\frac{d}{dt}\begin{bmatrix} z_i \\ q_i \end{bmatrix} = \begin{bmatrix} -1.32 & 0 \\ -0.32 & -1.2 \end{bmatrix}\begin{bmatrix} z_i \\ q_i \end{bmatrix} + \begin{bmatrix} 0.1 \\ 0 \end{bmatrix} m_i + \begin{bmatrix} 0.9 z_{i-1} \\ 0.9 q_{i-1} + 1.9 \end{bmatrix}. \qquad (44)$$

6.2. Steady-State Considerations

Before considering the control problem, it is necessary to define the "desired" values of the controlled variables which any controller should try

Table I Steady-State Values for Reaches

Reach	Desired BOD (mg liter^{-1})	Desired DO (mg liter^{-1})	Resulting effluent discharge (mg liter^{-1})
0	0	10	0
1	4.06	8	53.5
2	5.94	6	41.9
3	5.237	4.69	15.91

to maintain. These desired values should clearly be consistent with the dynamics of the system; otherwise it will not be possible for the controller to attain them. Appropriate desired values are therefore the steady-state values of the system.

In the steady state

$$\frac{d}{dt}\begin{bmatrix} z \\ q \end{bmatrix} \to 0.$$

From Eq. (44) therefore, if z_0^*, z_1^*, z_2^*, z_3^*, q_0^*, q_1^*, q_2^*, q_3^*, m_1^*, m_2^*, m_3^* are the desired values in reaches 0, 1, 2, 3, then

$$-1.32z_i^* + 0.9z_{i-1}^* + 0.1m_i^* = 0,$$
$$-0.32z_i^* - 1.2q_i^* + 0.9q_{i-1}^* + 1.9 = 0. \tag{45}$$

Let it be assumed that the reach 0 is always very "clean" so that z_0^* = 0 mg liter^{-1}, q_0^* = 10 mg liter^{-1}. Let the desired values of DO in reaches 1, 2, and 3 be 8, 6 and 4.69 mg liter^{-1}, respectively. Then from Eq. (45) z_1^* = 4.06, z_2^* = 5.94, z_3^* = 5.237, m_1^* = 53.5, m_2^* = 41.9, m_3^* = 15.91. Table 1 gives the desired values of z^*, q^*, m^* for the three reaches and these are the values which will be used in the control studies in the following sections.

6.3. Multiple Effluent Inputs

Beck's model (1973) is based on a single reach of the river Cam which has only one effluent input. The interesting problems, however, are the ones with multiple inputs. Since no additional data are available, we assume that there exists a system of many reaches, each one having the properties of Beck's model.

The most realistic description of such a stream with multiple polluters was given by Tamura (1974) who assumed that each reach was separated from the next by a distributed delay. This model is able to account for the dispersion of pollutants which actually occurs in rivers. In this model, for $j = 1, 2, \ldots, s$, a fraction a_j of BOD and DO in the $(i-1)$th reach at time $(t - \theta_j)$ arrives in the ith reach at time t, that is, the transport delays are distributed in time between θ_1 and θ_s. Thus in Eq. (46), z_{i-1}, q_{i-1} are given by

$$z_{i-1}(t) = \sum_{j=1}^{s} a_j z_{i-1}(t-\theta_j),$$
$$q_{i-1}(t) = \sum_{j=1}^{s} a_j q_{i-1}(t-\theta_j),$$
(46)

$\sum_{j=1}^{m} a_j = 1$, mean of $\theta_j = \theta_0$, $\theta_1 < \theta_2 < \cdots < \theta_s$. Figure 4b shows the distributed delay phenomenon whereby BOD (or DO) is discharged at time $t = 0$ in the $(i-1)$th reach and fractions a_j arrive at $t = \tau_j$, $j = 1, 2, \ldots, s$ in the ith reach. Thus, it is possible to write the state equations for a three-reach system with distributed delays as

$$\dot{z}_1 = -1.32 z_1 + 0.1 m_1 + 0.9 z_0 + 5.35, \tag{47}$$

$$\dot{q}_1 = -0.32 z_1 - 1.2 q_1 + 0.9_0 + 1.9, \tag{48}$$

$$\dot{z}_2 = 0.9 \sum_{j=1}^{s} a_j z_1(t-\tau_j) - 1.32 z_2 + 0.1 m_2 + 4.19, \tag{49}$$

$$\dot{q}_2 = 0.9 \sum_{j=1}^{s} a_j q_1(t-\tau_j) - 0.32 z_2 - 1.29 q_2 + 1.9, \tag{50}$$

$$\dot{z}_3 = 0.9 \sum_{j=1}^{s} a_j z_2(t-\tau_j) - 1.32 z_3 + 0.1 m_3 + 4.19, \tag{51}$$

$$\dot{q}_3 = 0.9 \sum_{j=1}^{s} a_j q_2(t-\tau_j) - 0.32 z_3 - 1.29 q_3 + 1.9. \tag{52}$$

Tamura gives the following values for s, τ, a, etc., for each of the distributed delays: $s = 3$, $\tau_1 = 0$, $\tau_2 = \frac{1}{2}$ day, $\tau_3 = 1$ day, $z_0 = 0$, $q_0 = 10$, $a_1 = 0.15$, $a_2 = 0.7$, $a_3 = 0.15$.

The system, as described by Eqs. (47)–(52), is nominally of infinite dimension in the state space. It is possible to obtain a good finite-dimensional approximation by expanding the delayed terms in a Taylor series and taking the first two terms. For example, Eq. (49) can be rewritten as

$$\dot{z}_2 = 0.9[0.15 z_1(t) + 0.7 z_1(t-0.5) + 0.15 z_1(t-1)]$$
$$- 1.32 z_2 + 0.1 m_2 + 4.19.$$

Here there are two delays so it is necessary to introduce four additional states. Let these be given by z_4, z_5, z_6, z_7. Let $z_4(t) = z_1(t-0.5)$; then $z_4(s) = z_1(s) e^{-0.5s}$. Now

$$z_1(s) e^{-0.5s} = z_1(s)[1 + 0.5s + (0.25/2)s^2 + \cdots]^{-1}.$$

Taking only the first three terms

$$z_1(t) = z_4(t) + 0.5\dot{z}_4(t) + 0.125\ddot{z}_4(t),$$

let

$$\dot{z}_4 = z_5; \tag{53}$$

then

$$\dot{z}_5 = 8z_1 - 8z_4 - 4z_5. \tag{54}$$

Similarly for the other delay

$$z_1(t-1) = z_6;$$

then

$$\dot{z}_6 = z_7, \tag{55}$$
$$\dot{z}_7 = 2z_1 - 2z_6 - 2z_7. \tag{56}$$

Thus Eq. (49) can be rewritten as

$$\dot{z}_2 = 0.135z_1 + 0.63z_4 + 0.135z_6 - 1.32z_2 + 0.1m_2 + 4.19. \tag{57}$$

Similarly four additional variables each can be introduced for the delays in Eqs. (50)–(52). This makes the overall system of order 22.

6.4. Hierarchical Solution Using the Prediction Principle

In order to examine the application of the prediction principle to this problem we decompose the system into three subsystems, where subsystem 1 consists of Eqs. (47)–(48), subsystem 2 consists of Eqs. (49) and (50) which ultimately yields 10 equations once the approximation to the delay is defined. Subsystem 3 consists of Eqs. (51) and (52) which again gives a 10th-order system once the delays are approximated by a second-order Taylor's series. Thus the system is decomposed into three parts of orders 2, 10, and 10, respectively. Clearly many other decompositions are possible but the present one is the most convenient since it retains the subsystem structure. Other decompositions which consider, for example, 8, 7, 7 variables are almost certainly more efficient computationally but these decompositions have no clear physical meaning. Moreover, when the interaction prediction approach is used for systems with many more reaches, this decomposition becomes computationally still more attractive.

Assume that the problem is to minimize a function of the form

$$J = \int_0^8 [2(z_1 - z_1^*)^2 + (q_1 - q_1^*)^2 + 2(z_2 - z_2^*)^2 + (q_2 - q_2^*)^2$$
$$+ 2(z_3 - z_3^*)^2 + (q_3 - q_3^*)^2 + (m_1 - m_1^*)^2 + (m_2 - m_2^*)^2$$
$$+ (m_3 - m_3^*)^2] \, dt. \tag{58}$$

Minimization of this function ensures that if before time zero an unknown disturbance takes the system state to $\mathbf{x}_0$, then it will be possible to bring the system back to the steady-state values of $z_1{}^*$, $q_1{}^*$, $z_2{}^*$, $q_2{}^*$, $z_3{}^*$, $q_3{}^*$. The controls can also be taken back to the steady-state m_1, m_2, m_3, with the guarantee that there are not unacceptably large deviations from the steady-state BOD, DO, or controls. Note that both positive and negative deviations are penalized. In the case of the control, this is important because the treatment process is biological and it costs money to modify it.[6] We can interpret the quadratic form as follows: Let us assume that there are bands $(z_1{}^* \pm z_1^{**})$ and $(q_1{}^* \pm q_1^{**})$ ($i = 1, 2, 3$) within which the BODs and DOs should be. Then the quadratic forms will ensure that this is so, provided that adequate weights are used in the cost function. After extensive experimentation the weights given in Eq. (58) have proven adequate.

6.5. Simulation Results

The problem of minimizing J in Eq. (58) subject to the system dynamics given in Eqs. (47)–(52) was solved using a hierarchical interaction prediction principle structure on the IBM 370/165 digital computer at the LAAS, Toulouse. The initial conditions for the BODs and DOs were taken to be $z_1(0) = 10$ mg/liter, $z_2(0) = 5.94$ mg/liter, $z_3(0) = 5.237$ mg/liter, $q_1(0) = 7$ mg/liter, $q_2(0) = 6$ mg/liter, $q_3(0) = 4.69$ mg/liter.

This initial condition implies that before $t = 0$, some disturbance affected reach one and polluted it while reaches 2 and 3 in their desired steady state.

Convergence to the optimum took place in eleven second-level iterations. Figures 5–7 show the resulting state and control trajectories. These figures show how the effect of the pollution load is gradually attenuated down the river. Note that the results are quite realistic with the distributed delay since the reaches downstream are only gradually affected as the pollutants are dispersed.

6.6. Remarks

In this section the applicability of the prediction principle approach was demonstrated by solving a realistic large-scale system problem. Even for this very large system, convergence of the two-level algorithm is very

---

[6] The control variables m_1, m_2, m_3 are the rates of discharge of BOD in the treated sewage into the stream. The implementation of such a control requires the possibility of changing the BOD content of the sewage by treatment prior to discharge into the stream. One possible way of doing this may be by detention and then by controlled discharge of the treated effluent.

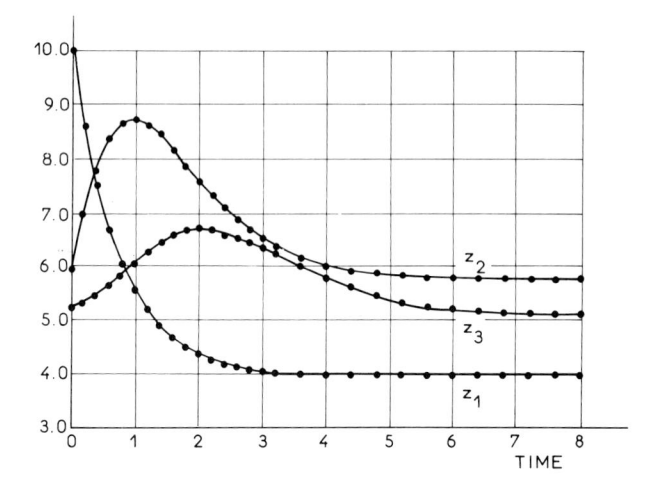

Figure 5. Optimal BOD trajectories for the three reaches.

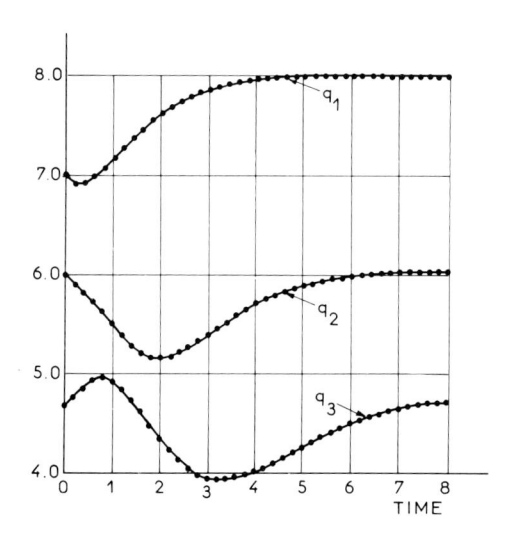

Figure 6. Optimal DO trajectories for the three reaches.

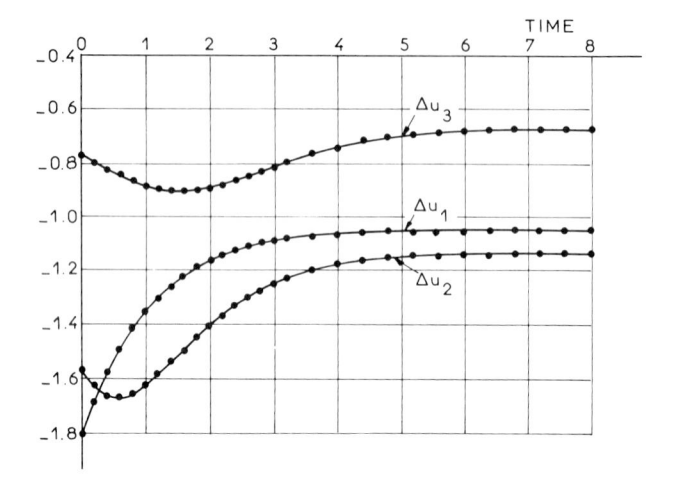

Figure 7. Optimal control deviations for the three reaches.

rapid even though the integration interval is very long. This rapid convergence property, as well as the ease of programming, makes this one of the most powerful approaches to hierarchical optimization.

7. HIERARCHICAL FEEDBACK CONTROL FOR LINEAR QUADRATIC PROBLEMS

In the previous section it was necessary to calculate some control over the fixed period $(t_f - t_0)$ to bring the system back to steady-state $\mathbf{x}_s$ while minimizing some integral function of the states and controls. In that case, it was assumed that the hierarchical calculation could be performed rapidly and the control applied before the state had significantly changed. However, this is not a realistic situation unless the system has particularly slow dynamics (as, for example, in the case of the river) or if the initial value $\mathbf{x}_0$ can be precisely predicted *a priori*. This is, for example, the case of the start up or shut down of plants. In the general case, the initial state $\mathbf{x}_0$ will almost certainly have changed by the time the control is calculated. Thus, if the open-loop control is applied for the actual state $\mathbf{x}$ instead of for the initial state $\mathbf{x}_0$, for which the hierarchical calculation was done, the results could prove to be disastrous. It is therefore highly desirable to be able to calculate a feedback control that is independent of initial conditions. Moreover, it would also not be necessary to calculate the controls repeatedly for differing initial conditions because the same control law would be valid for all initial conditions.

In order to develop this closed-loop control, consider the lower-level problem for the prediction principle. From Eq. (40) L_i need be minimized which leads to the minimization of the Hamiltonian

$$H_i = \tfrac{1}{2}\|\mathbf{x}_i\|_{Q_i}^2 + \tfrac{1}{2}\|\mathbf{u}_i\|_{R_i}^2 + \lambda_i^T \mathbf{z}_i - \sum_{j=1}^{N} \lambda_j^T L_{ji} \mathbf{x}_i$$
$$+ \mathbf{p}_i^T [A_i \mathbf{x}_i + B_i \mathbf{u}_i + C_i \mathbf{z}_i]. \tag{59}$$

Then from the necessary conditions

$$\dot{\mathbf{p}}_i = -Q_i \mathbf{x}_i - A_i^T \mathbf{p}_i + \sum_{j=1}^{N} [\lambda_j^T L_{ji}]^T \tag{60}$$

with

$$\mathbf{p}_i(t_f) = \mathbf{0}, \tag{61}$$
$$\mathbf{u}_i = -R_i^{-1} B_i^T \mathbf{p}_i. \tag{62}$$

Let

$$\mathbf{p}_i = K_i \mathbf{x}_i + \mathbf{s}_i, \tag{63}$$

where $\mathbf{s}_i$ is the open-loop compensation vector. Then

$$\dot{\mathbf{p}}_i = K_i \dot{\mathbf{x}}_i + \dot{K}_i \mathbf{x}_i + \dot{\mathbf{s}}_i \tag{64}$$

substituting into Eq. (1) gives

$$\dot{\mathbf{x}}_i = A_i \mathbf{x}_i - B_i R_i^{-1} B_i^T K_i \mathbf{x}_i - B_i R_i^{-1} B_i^T \mathbf{s}_i + C_i \mathbf{z}_i. \tag{65}$$

Using Eq. (64)

$$[\dot{K}_i + A_i^T K_i + K_i A_i - K_i B_i R_i^{-1} B_i^T K_i + Q_i] \mathbf{x}_i$$
$$+ \left[\dot{\mathbf{s}}_i + A_i^T \mathbf{s}_i - K_i B_i R_i^{-1} B_i^T \mathbf{s}_i + K_i C_i \mathbf{z}_i - \sum_{j=1}^{N} [\lambda_j^T L_{ji}]^T \right] = \mathbf{0}. \tag{66}$$

Since this equation is valid for arbitrary $\mathbf{x}_i$,

$$\dot{K}_i + K_i A_i + A_i^T K_i - K_i B_i R_i^{-1} B_i^T K_i + Q_i = 0 \tag{67}$$

with $K_i(t_f) = 0$ and

$$\dot{\mathbf{s}}_i = [K_i B_i R_i^{-1} B_i^T - A_i^T] \mathbf{s}_i - K_i C_i \mathbf{z}_i + \sum_{j=1}^{N} [\lambda_j^T L_{ji}]^T \tag{68}$$

with $\mathbf{s}_i(t_f) = \mathbf{0}$ and the local control $\mathbf{u}_i$ is given by

$$\mathbf{u}_i = -R_i^{-1} B_i^T K_i \mathbf{x}_i - R_i^{-1} B_i^T \mathbf{s}_i. \tag{69}$$

7.1. Remarks

(i) The K_i in Eq. (67) is independent of the initial state $\mathbf{x}(0)$. Thus the N matrix Riccati equations each involving $n_i \times (n_i + 1)/2$ nonlinear differential equations can be solved independently from the given final condition $K_i(t_f) = 0$. These equations give a partial feedback control. This feedback provides some degree of stabilization against small disturbances and moreover allows the correction of the control based on the current state.

(ii) The $\mathbf{s}_i$ in Eq. (68) is *not* independent of the initial state $\mathbf{x}_i(t_0)$. Thus the second term in Eq. (69) provides open-loop compensation. This can be proved using Eq. (68); At the optimum, $\mathbf{s}_i$ can be written

$$\dot{\mathbf{s}}_i = [-A_i^T + K_i B_i R_i^{-1} B_i^T] \mathbf{s}_i - K_i C_i \sum_{j=1}^{N} L_{ij} \mathbf{x}_j$$
$$+ \sum_{j=1}^{N} L_{ji}^T [-C_j^T K_j \mathbf{x}_j - C_j^T \mathbf{s}_j], \tag{70}$$

that is, $\mathbf{s}_i$ is a function of the states of all the other subsystems, and is thus dependent on the initial state $\mathbf{x}(t_0)$ of the overall system.

7.2. The Closed-Loop Controller

Let $\mathbf{x}$ be the overall state vector of the system, let $\mathbf{u}$ be the overall control vector, let $\mathbf{s}$ be the overall open-loop part of the compensator, and let $A, B, C, L, Q, R, K, \ldots$, be the matrices for the overall system. Then Singh et al. (1976) have shown that the open-loop compensation vector $\mathbf{s}$ and the state $\mathbf{x}$ are related by a transformation Y type

$$\mathbf{s} = Y\mathbf{x}, \tag{71}$$

where Y is an $n \times n$ matrix.

For the infinite time regulator, $Y(t_f, t)$ is time invariant.

7.2.1. The Regulator Solution

As seen in Eq. (70), $\mathbf{s}$ provides the open-loop part of the controller and the above remark shows that $\mathbf{s}$ is related very simply to the overall state vector.

Now, for the infinite time regulator, Y is particularly easy to compute since near $t = 0$, Y is constant whereas $\mathbf{x}$ and $\mathbf{s}$ are not. Thus, if the values of $\mathbf{x}$ and $\mathbf{s}$ are recorded at the first $n = \sum_{i=1}^{N} n_i$ time points, very close to $t = t_0$, Y can be determined as follows.

Form the matrix

$$S = [\mathbf{s}(t_0), \mathbf{s}(t_1), \ldots, \mathbf{s}(t_n)] \quad \text{and} \quad X = [\mathbf{x}(t_0), \mathbf{x}(t_1), \ldots, \mathbf{x}(t_n)]$$

then

$$S = YX \quad \text{or} \quad Y = SX^{-1}.$$

This inversion of X should not pose much of a problem for even large systems since it is to be done off-line.[7]

If it is desired to calculate the time varying Y (i.e., if the horizon must be considered finite) it is possible to do so by solving the problem n times for n different initial conditions and then forming the $n \times n$ matrices

$$S = [\mathbf{s}^1(t), \mathbf{s}^2(t), \ldots, \mathbf{s}^n(t)],$$
$$X = [\mathbf{x}^1(t), \mathbf{x}^2(t), \ldots, \mathbf{x}^n(t)]$$

for each integration point and determining each value of Y by the relationship

$$Y = SX^{-1}.$$

7.2.2. Remarks

1. The above method allows the solution of the large interconnected systems regulator problem within a completely decentralized calculation structure.

2. The resulting gains are independent of the initial conditions. In fact, using Eq. (61) in the composite case

$$\mathbf{u} = -R^{-1}B^{TK}\mathbf{x} - R^{-1}B^T\mathbf{s}, \tag{72}$$

where K is block-diagonal and substituting $\mathbf{s}$ from Eq. (71)

$$\mathbf{u} = -R^{-1}B^T K\mathbf{x} - R^{-1}B^T Y\mathbf{x} = -[R^{-1}B^T K + R^{-1}B^T Y]\mathbf{x} = -G\mathbf{x}. \tag{73}$$

[7] It should be noted that the approach hinges on one's ability to invert X which basically depends on the linear independence of the chosen record. In practical cases, if a large enough perturbation is given, it will be possible to obtain an invertible X. Otherwise, one can solve the problem off-line n times successively for the initial conditions

$$\mathbf{x}(t_0) = \begin{bmatrix} 1 \\ 0 \\ \vdots \\ 0 \end{bmatrix}, \begin{bmatrix} 0 \\ 1 \\ \vdots \\ 0 \end{bmatrix}, \ldots, \begin{bmatrix} 0 \\ 0 \\ \vdots \\ 1 \end{bmatrix}.$$

Then $Y = S$ and it is not necessary to invert X. This of course requires more computation but this is off-line and decentralized.

None of the terms in the gain matrix G are dependent on $\mathbf{x}_0$. Thus, this gain will bring the system back to the steady state optimally from any initial condition.

3. Against the above advantages, there is the difficulty that a large amount of off-line calculation needs to be done for the finite time case. However, even here, all of this off-line computation is within a decentralized structure so that its storage requirements are minimal. Also, for large systems, the case of most practical interest is the one where the period of optimization is infinite and for this important case, the off-line computational requirements are very small.

8. EXTENSION TO THE SERVOMECHANISM CASE

Consider now the problem

$$\min J = \sum_{i=1}^{N} \tfrac{1}{2} \int_{t_0}^{t_f} [\|\mathbf{x}_i - \mathbf{x}_i^*\|_{Q_i}^2 + \|\mathbf{u}_i\|_{R_i}^2] \, dt. \tag{74}$$

Note that the cost function (74) is more general than the one in Eq. (38) in that the two become identical on setting $\mathbf{x}_i^* = \mathbf{0}$. This cost function takes into account the possibility of having a nonzero desired state $\mathbf{x}_i^*$.

We wish to minimize J, subject to

$$\dot{\mathbf{x}}_i = A_i \mathbf{x}_i + B_i \mathbf{u}_i + C_i \mathbf{z}_i + \mathbf{D}_i \tag{75}$$

and

$$\mathbf{z}_i = \sum_{j=1}^{N} L_{ij} \mathbf{x}_j. \tag{76}$$

Here $\mathbf{x}_i^*$ is a constant known desired trajectory for the ith subsystem and $\mathbf{D}_i$ is a vector of constant known inputs which come into the ith subsystem.

Using a similar development (as in Section 7.2) it is easy to show that the control is given by

$$\mathbf{u} = -R^{-1}B^T K \mathbf{x} - R^{-1} B^T \xi, \tag{77}$$

where R, B, K are block-diagonal matrices. Similarly it is also easy to show that

$$\xi = T_1 \mathbf{x} + \pi, \tag{78}$$

where for the infinite time case for A, B, C, D, etc., the time invariant T_1 is an $\sum_{i=1}^{N} n_i \times \sum_{i=1}^{N} n_i$ constant matrix and π is a $\sum_{i=1}^{N} n_i$-dimensional constant vector.

Thus

$$\mathbf{u} = -[R^{-1}B^TK + R^{-1}B^TT_1]\mathbf{x} - R^{-1}B^T\pi \quad \text{or} \quad \mathbf{u} = -G\mathbf{x} - R^{-1}B^T\pi. \quad (79)$$

Here again, as for the regulator case, for $t_f - t_0 \to \infty$, T_1 and π can be obtained from the first $(\sum_{i=1}^{N} n_i + 1)$ time points by inverting a $(\sum_{i=1}^{N} n_i \times \sum_{i=1}^{N} n_i)$ matrix, where the points are obtained from an off-line decentralized calculation using the interaction prediction principle. This algorithm is now applied to the three-reach river problem.

8.1. Control of the Three-Reach Distributed Delay Model

Here, as before the system is of order 22 which can be split up into three subsystems of orders 2, 10, and 10.

The equations and parameter values are identical to those in Section 6.

8.1.1. Results

For the off-line calculation, convergence took place in 11 sec on an IBM 370/165. From this, the control gain matrix was calculated (Table II).

Table II The Gain Matrix G for the Three Reaches with Distributed Delays

First row
 +0.099062085 −0.010958791 +0.022846639 +0.009379625 +0.002637625 +0.004595816
 +0.002281010 −0.004275560 −0.001736045 −0.000522673 −0.000974357 −0.000555217
 +0.003174365 +0.001376390 +0.000480294 +0.000990987 +0.000697553 −0.000700831
 −0.000285327 −0.000102699 −0.000220478 −0.000178099

Second row
 +0.022842050 −0.005539298 +0.093502045 +0.022300422 +0.004415929 +0.007374823
 +0.002398074 −0.009108484 −0.003236697 −0.000795007 −0.001329780 −0.000566363
 +0.015544653 +0.006787241 +0.001954496 +0.003401399 +0.001715660 −0.002267599
 −0.001000464 −0.000320613 −0.000610054 −0.000370383

Third row
 +0.003168851 −0.001074106 +0.015543908 +0.003468424 +0.000657767 +0.001095563
 +0.000330418 −0.003199846 −0.000876218 −0.000188380 −0.000311464 −0.000111490
 +0.076797992 +0.017195195 +0.003240436 +0.005426317 +0.001640469 −0.005270451
 +0.003240436 +0.005426317 −0.000784665 −0.000327677 −0.001944035 −0.000475317

Disturbance vector
 −0.588996768
 −0.410924196
 −0.112484366

The initial conditions used for the calculation were $z_1(0) = 10$ mg/liter, $z_2(0) = 5.94$ mg/liter, $z_3(0) = 5.237$ mg/liter, $q_1(0) = 7$ mg/liter, $q_2(0) = 6$ mg/liter, $q_3(0) = 4.69$ mg/liter.

Physically, these conditions imply that the second and third reach are in the steady state when a large pollution load comes into reach 1. Figures 5 and 6 show the attenuation of this pollution down the river. Figure 7 shows the control trajectory.

9. CONCLUSIONS

In this chapter a review has been given of some recent results in hierarchical optimization and control, and these results have been illustrated on a river pollution control problem. The case of linear dynamical constraints and quadratic cost functions has been exclusively treated since this case covers many ecological problems of practical interest including river pollution control. Both open- and closed-loop methods have been considered. River pollution problems, as well as many other ecological problems, are unique in the sense that either approach is suitable because of the relatively slow dynamic response of the system (compared to the speed of computation in modern computers) although, as is usual with control systems, closed-loop control is more attractive. This attractiveness lies in the fact that with the open-loop methods (described in Sections 2 to 5) each control trajectory is valid only for one given initial state, while the closed-loop control provides a gain matrix, the elements of which can be multiplied on-line to provide optimal control for any initial condition. This makes the control more flexible, and feedback also allows one to reduce to some extent the effects of small parameter uncertainty, etc. Against these, the open-loop control calculation is certainly a little easier.

Of the open-loop methods considered, the prediction approach is computationally very attractive but it is not able to take into account inequality constraints on the states and controls. The method of Tamura does this but is only applicable to discrete time systems and in practice the sampling interval for this method needs to be rather large.

The methods treated in this chapter deal, however, only with the deterministic case, and they assume that accurate models are available and also that perfect measurement of the state vector is possible. In practice, models are often quite crude and only corrupted measurements of a part of the state vector may be available. For example, in the case of river pollution control, BOD requires 5 days to measure while control action needs to be taken for a shorter period of time than that. It would therefore be necessary to reconstruct the BOD states and filter the DO states from noise-corrupted measurements of DO using modern state estimation techniques. Hassan *et al.* (1978) and Rinaldi *et al.* (1976) have recently developed techniques for performing this state estimation.

REFERENCES

Beck, M. B. (1973). The application of control and systems theory to problems of river pollution control. Ph.D. Thesis, University of Cambridge, Cambridge, England.

Bellman, R. (1957). "Dynamic Programming." Princeton Univ. Press, Princeton, New Jersey.

Fallside, F., and Perry, P. (1975). Hierarchical optimisation of a water supply network. *Proc. Inst. Electr. Eng.* **22**, 202–208.

Gefferion, A. M. (1971). Duality in non-linear programming. *SIAM Rev.* **13**, 1–37.

Hassan, M., Hurteau, R., Singh, M. G., and Titli, A. (1977). Stochastic hierarchical control of a large scale river system. *Proc. IFAC World Congr., 7th, 1978*.

Hassan, M., Salut, G., Singh, M. G., and Titli, A. (1978). A hierarchical filter structure for the global Kalman filter. *IEEE Trans. Autom. Control* **AC23**, 262–267.

Mesarović, M. D., Maco, D., and Takahara, Y. (1970). "Theory of Hierarchical Multi-Level Systems." Academic Press, New York.

Patten, B. (1975). Ecosystem linearization: An evolutionary design problem. *Am. Nat.* **969**, 529–539.

Pearson, J. D. (1971). Dynamic decomposition techniques. *In* "Optimization Methods for Large Scale Systems" (D. Wismer, ed.), McGraw-Hill, New York.

Rinaldi, S., Romano, P., and Soncini-Sesta, T. (1976). Parameter estimation of a Streeter–Phelps type water pollution model. *Proc. Int. Conf. Ident. Syst. Parameter Estimation, 4th, 1976* Paper 5-1, pp. 98–111.

Singh, M. G. (1975). River pollution control. *Int. J. Syst. Sci.* **6**, 9–21.

Singh, M. G. (1977). "Dynamical Hierarchical Control." North-Holland Publ., Amsterdam.

Singh, M. G., and Hassan, M. (1976a). A comparison of two hierarchical optimisation methods. *Int. J. Syst. Sci.* **7**, 603–611.

Singh, M. G., and Hassan, M. (1976b). A closed loop solution for the river pollution control problem. *Automatica* **12**, 261–264.

Singh, M. G., and Tamura, H. (1974). Modelling and hierarchical optimisation for oversaturated urban road traffic networks. *Int. J. Control* **20**, 913–934.

Singh, M. G., Hassan, M., and Titli, A. (1976). A multi level controller for interconnected dynamical systems using the prediction principle. *IEEE Trans. Syst., Man, Cybernet.* **SMC-6**, 233–239.

Takahara, Y. (1965). M.S. Thesis, Case Western Reserve University, Cleveland, Ohio.

Tamura, H. (1974). A discrete dynamical model with distributed transport delays and its hierarchical optimisation to preserve stram quality. *IEEE Trans. Syst., Man, Cybernet.* **SMC-4**, 424–429.

Varaiya, P. (1969). A decomposition technique for non-linear programming. University of California, Berkeley (unpublished report).

Chapter **18**

ECOSYSTEM STABILITY AND THE DISTRIBUTION OF COMMUNITY MATRIX EIGENVALUES

John Harte

1. Introduction	453
2. A Practical Measure of Stability	455
3. Analysis of the Stability Measure	456
3.1 Mathematical Preliminaries	456
3.2 An Averaging Procedure	458
3.3 Diagonal Community Matrices	459
3.4 The Two-Component System	460
3.5 The Arbitrary Community Matrix	461
4. Discussion	462
Appendix	463
References	465

1. INTRODUCTION

One of the central problems in ecology is to learn how to predict what will happen to a disturbed ecosystem. What one would like is to be able to measure certain properties (ideally, a few simple ones) of ecosystems before they are perturbed, and then on the basis of the results of the measurements, to be able to predict how gently or violently the system will respond to a disturbance. For this reason, much effort has been directed toward understanding what properties of ecosystems tend to enhance stability against external stresses. We call such properties stability indicators. A wide range of definitions of stability (May, 1973; Holling, 1973; Botkin and Sobel, 1974; Orians, 1974) and an even wider range of possible stability indicators for ecosystems have been considered by ecosystem theorists. Among the former are resistance, resilience, neighborhood asymptotic stability, global asymptotic stability, and structural stability, while the latter include species diversity (Odum, 1969; May, 1973; Goodman, 1975), nutrient pathway diversity (MacArthur, 1955), nutrient transit-time diversity (Harte and Morowitz, 1975), time lags (May, 1973), spatial heterogeneity (Andrewartha and Birch, 1954; Horn and MacArthur, 1972), grazing strategies (Steele, 1974), density dependence (May, 1973), stochasticity (May, 1973; Roughgarden, 1975), and successional maturity (Margalef, 1975).

A shortcoming of much of the theoretical work in ecology is that results are often not expressed or expressible as relations among readily measureable quantities. A familiar example is the often-quoted result that a necessary and sufficient condition for asymptotic stability of a system described by a community matrix is the negativity of the real parts of all the eigenvalues of that matrix. While mathematically rigorous, this result unfortunately is not very useful in situations of practical concern such as environmental impact prediction or assessment. Suppose an ecosystem is disturbed by some activity. For example, effluent is dumped in a lake, or a watershed is partially timbered. On the one hand, asymptotic stability is too stringent a condition in the sense that the eventual return of a system to precisely its predisturbed state is neither likely (even if the stress is removed) nor essential to environmental acceptability. On the other hand, the condition may be too lenient if the system takes a very long time to return to predisturbed conditions, or is displaced quite far from its original state prior to its eventual return. Asymptotic stability is a mathematically convenient notion because there is a precise mathematical prescription for determining whether or not a system is asymptotically stable. But the relevant question in much environmental impact assessment work is not one of *whether* or not a system is stable, but rather of *how* stable it is.

This work is motivated by our perceived need to bring theoretical stability analysis more in line with experimental constraints and practical needs. Our objective is, first, to introduce a definition of stability which offers a useful quantitative measure of ecosystem (or more generally any system) response to stress. Although this definition of stability *appears* to be mathematically ungainly, it has the advantage, we argue, of quantifying what is often of most interest to those concerned with the responses of ecosystems to environmental stress. We then explore, through rigorous analytical methods, how the value of this stability measure depends upon some particular ecosystem parameters (such as averaged properties of the distribution of eigenvalues of the community matrix). We conclude with several speculations concerning the use of tracers to empirically determine the values of the stability indicators suggested by our analysis. Because measurement of these ecosystem parameters may be far easier than measurement of all the community matrix eigenvalues, further work along these lines could enhance the empirical relevance of theoretical analysis and eventually allow a greater predictive understanding of how ecosystems respond to external stress.

2. A PRACTICAL MEASURE OF STABILITY

All notions of stability in ecology pertain to the relation between a stressed state of a system and the state the system would have been in had there been no stress. Where they differ is in the aspects of that relation which are selected as being especially important. No single measure of ecosystem stability can be suitable for all purposes and satisfy the interests of all ecologists. The notion of *resistance* to stress is very useful if one's interest is in the maximum *extent* of the *deviation* between the stressed and unstressed system. *Resilience* is of relevance to those concerned with the *rate* at which a system returns to prestress conditions. *Asymptotic stability* is a useful concept for those concerned with whether or not a system will *eventually return* to its prestressed state. Other concepts of stability likewise single out certain aspects of perturbed behavior for emphasis.

The measure of stability analyzed here incorporates both resistance and resilience in a single integrated measure of the deviation between the stressed and unstressed states of the system. Suppose that in the absence of an external disturbance to an ecosystem, the state of the system is characterized by the state variables, $x_i(t)$, with $i = 1, 2, \ldots, n$. The vector $\mathbf{x}$ describes the unperturbed components of the system; its components are usually most conveniently taken to refer to the carbon contents of these compartments although that is not of importance to us here. Let the vector function $\Delta\mathbf{x}(t)$ be a measure of the magnitude of the effect on the system resulting from some perturbation. The Δx_i's are usually taken to describe the same property of the compartments that the state variables describe, although they could describe as well the levels of a toxic substance or any other measure of system response. Assume that the perturbation begins at time $t = 0$. Then let

$$s^{-1} = \frac{1}{n} \sum_i w_i \int_0^\infty dt\, f(t) \left[\frac{\Delta x_i(t)}{x_i(t)}\right]^2 \tag{1}$$

be our measure of stability. A large value of s corresponds to high stability. Note that this definition of stability incorporates a measure of both the magnitude of the deviation of the perturbed system from equilibrium and the rapidity with which it returns, if at all. The w_i may all be chosen to be equal to unity if each compartment is deemed to be of equal weight or importance, although a subjective judgment might favor choosing the w's corresponding to some species to be larger than others. A time-weighting function $f(t)$ under the integral sign can be chosen to be different from a constant in order to reflect the observer's judgment about the relative importance of short- or long-term effects. In order to compare the intrinsic stability of different systems, the integrals in Eq. (1) must be averaged over a set of initial perturbations of specified magnitude, as discussed further in the following section.

While Eq. (1) provides an index or measure of stability which is empirically convenient [or can be made so by appropriate choices of the w_i and $f(t)$] and appears to have the flexibility to reflect well many of the realities of environmental impact concerns, it does not connect in any transparent way with methods of mathematical analysis of theoretical models other than computer simulation. And yet if it is to provide more than just another empirical property of stressed ecosystems and is to be related in some way to properties of the prestressed system, such as diversity, then methods of theoretical analysis of this measure would be useful. By means of such methods, one could identify, for purposes of ultimate verification, interesting candidates for stabilizing or destabilizing properties of ecosystems. The rest of this chapter will develop an analytic approach to the study of this stability index, and present some preliminary results about some possible candidates for stability indicators.

3. ANALYSIS OF THE STABILITY MEASURE

3.1. Mathematical Preliminaries

The value of s for a particular ecosystem subjected to a particular disturbance will depend upon how large the Δx_i grow and how rapidly, if at all, they damp out to zero. In order to see what generalities can be deduced, consider a general model consisting of coupled, nonlinear, first-order differential equations. Assume, now, that the ratios $\Delta x_i/x_i$ are small. Then a linearized matrix equation

18. Ecosystem Stability and Community Matrix Eigenvalues

$$\frac{d}{dt}\Delta x_i(t) = \sum_j A_{ij}\Delta x_j(t) \tag{2}$$

is obtained. A is commonly referred to as the community matrix; its eigenvalues will be denoted by $-\lambda_k$. (Note the minus sign in front of λ_k introduced for future convenience.) We hereafter assume the x_i are *time independent* and the original nonlinear equations have *no explicit time dependence* (so that the λ_i are time independent). Then, *in the absence of eigenvalue degeneracy* the time evolution of the Δx_i's can be easily computed to be

$$\Delta x_i(t) = \sum_k a_{ik} e^{-\lambda_k t}, \tag{3}$$

where the a_{ik} depend upon the matrix elements of A and the initial conditions, $\Delta x_i(0)$. *If the eigenvalues of A all have* negative real parts, as we hereafter assume, then the Δx_i's are guaranteed to damp to zero as $t \to \infty$. The real parts of the eigenvalues $-\lambda_k$, which determine the damping times, are related to the residence times of the compartments of the systems; in general, the damping time constants, or inverse negative real parts of the eigenvalues denoted by $Re^{-1}(\lambda_k)$, are large if the residence times are all long.

The qualitative dependence of integrals $\int dt(\Delta x_j(t))^2$ on time constants, $Re^{-1}(\lambda_k)$, of the system can be conjectured. Dimension counting would suggest that when the time constants are small, the integrals are small. Therefore, small $Re^{-1}(\lambda_k)$ should be an indication of high stability. Moreover, equitability, or small dispersion, of the time constants tends to prevent a situation in which a subset of terms in Eq. (3) with, say, positive coefficients dominate the integrand at early times just subsequent to the perturbation, while the remaining terms with negative coefficients dominate at later times. Therefore, if the sum of the $Re^{-1}(\lambda_k)$ is fixed, the more equitably they are distributed, the less the Δx_i should grow, and the larger s should be. Below, we prove several theorems which establish rigorous connections between s and the distribution of eigenvalues.

Explicitly displaying the dependence of the $\Delta x_i(t)$ on the initial conditions $\Delta x_i(0)$, rewrite Eq. (3) as

$$\Delta x_i(t) = \sum_{j=1}^{N} \sum_{k=1}^{N} \Delta x_j(0) C_{ji}^k e^{-\lambda_k t}. \tag{4}$$

Here, C_{ji}^k is of the form:

$$C_{ji}^k = D_{jk}^{-1} D_{ki}, \tag{5}$$

(no sum intended over repeated subscript) where D is a matrix composed of eigenvectors of A, and D^{-1} is the inverse to D. For future reference note that the C_{ji}^k satisfy

$$\sum_k C_{ji}^k = \delta_{ji}, \tag{6}$$

where δ_{ji} is the Kronecker delta ($\delta_{ji} = 0$ if $i \neq j$ and $\delta_{ii} = 1$) and

$$\sum_i C_{ii}^k = 1, \tag{7}$$

both of which follow from Eq. (5). For the rest of this discussion we will assume that the function $f(t)$ in Eq. (1) is equal to unity, that the w_i are all equal to some constant, w, and that the eigenvalues are real and negative. This will simplify the algebra. It will also influence our results; other cases can be treated, however, by the methods we develop here.

Substituting Eq. (4) into Eq. (1) and performing the integration, we obtain

$$s^{-1} = \frac{w}{n} \sum_{i=1}^n \sum_{j=1}^n \sum_{l=1}^n \sum_{k=1}^n \sum_{m=1}^n \frac{\Delta x_j(0) \Delta x_l(0) C_{ji}^k C_{li}^m}{x_i^2 (\lambda_k + \lambda_m)}. \tag{8}$$

3.2. An Averaging Procedure

In order to proceed we must make an assumption about the $\Delta x_i(0)$. One possible initial condition would be to take a particular $\Delta x_i(0)$ different from zero and the others equal to zero. While in some situations this is a realistic initial condition and is, in fact, the most commonly considered one in theoretical studies, we choose instead to evaluate s by averaging over a set of many initial displacements. We adopt this procedure because we are interested in general, not specific, stability properties of ecosystems. That is, we do not want to know the value of s for a particular initial condition, but rather its value for a random average over a wide range of possible initial conditions. Our averaging procedure will be to take an average of s^{-1}, denoted $\langle s^{-1} \rangle$, over initial displacements which are randomly distributed subject to:

$$\langle \Delta x_i(0) \Delta x_j(0) \rangle = \delta_{ij} x_i^2 / w, \tag{9}$$

where w is the constant appearing in Eq. (8) and is assumed to be sufficiently greater than 1 so that the linearity assumption implicit in Eq. (2) is valid.

The "absence-of-correlation" assumption made here for the initial displacements is based simply on the absence of more detailed information.

The methods we introduce here for analyzing s could in principle be applied to other averaging procedures, with different results generally obtained.

Using Eq. (9) to evaluate $\langle s^{-1} \rangle$, we obtain from Eq. (8)

$$\langle s^{-1} \rangle = \frac{1}{n} \sum_{i=1}^{n} \sum_{j=1}^{n} \sum_{k=1}^{n} \sum_{m=1}^{n} \frac{C_{ji}^{k} C_{ji}^{m} x_j^2}{(\lambda_k + \lambda_m) x_i^2}. \tag{10}$$

3.3. Diagonal Community Matrices

Consider, first, the case in which the community matrix is diagonal:

$$A = \begin{pmatrix} -\lambda_1 & & & 0 \\ & -\lambda_2 & & \\ & & \ddots & \\ 0 & & & -\lambda_n \end{pmatrix}. \tag{11}$$

Letting $-\lambda^*$ be the mean eigenvalue

$$\lambda^* = \frac{1}{n} \sum_{k=1}^{n} \lambda_k, \tag{12}$$

we write

$$\lambda_k = \lambda^* + \varepsilon_k, \tag{13}$$

where

$$\lambda^* > 0, \quad \sum_k \varepsilon_k = 0 \quad \text{and} \quad \varepsilon_k > -\lambda^*. \tag{14}$$

It is then straightforward to show that

$$\langle s^{-1} \rangle = \frac{1}{2\lambda^*} - \frac{1}{n} \sum_k \frac{\varepsilon_k}{2\lambda^*(\lambda^* + \varepsilon_k)} \tag{15}$$

whether or not any degeneracies occur among the λ_k. This result follows from the fact that for this case, $C_{ji}^{k} = \delta_{kj}\delta_{ji}$.

Equations (14) and (15) lead directly to the result that $\langle s^{-1} \rangle$ is minimum (corresponding to maximum stability) when all $\varepsilon_k = 0$. Moreover, for fixed values of the ε_k, $\langle s^{-1} \rangle$ decreases as the mean negative eigenvalue λ^* increases. Thus stability is enhanced when the mean eigenvalue is most negative and when the eigenvalues are most nearly equal in value, as suggested by the qualitative arguments given above. It is natural to inquire at this point as to what happens if the mean eigenvalue diminishes in absolute value *and* the dispersion, $\Sigma_k \varepsilon_k^2/n$, also decreases. It is easy to see from Eq. (15) that for *small* values of ε_k/λ^*, it is the ratio $\Sigma_k \varepsilon_k^2/n\lambda^{*2}$ which determines the dependence of s on the ε_k.

3.4. The Two-Component System

In order to analyze the case of an arbitrary (not necessarily diagonal) community matrix, we will need to use the relations given in Eqs. (6) and (7) to simplify Eq. (10). A two component system will illustrate the techniques we use for nondiagonal A. We denote the two eigenvalues by $-\lambda_1$ and $-\lambda_2$, and we set

$$\lambda_1 = \lambda^* + \lambda', \qquad \lambda_2 = \lambda^* - \lambda', \tag{16}$$

where

$$\lambda^* > 0, \qquad |\lambda'| < \lambda^*. \tag{17}$$

Substituting Eq. (16) into Eq. (10) we obtain

$$\langle s^{-1} \rangle = \frac{1}{2} \sum_{i=1}^{2} \sum_{j=1}^{2} \frac{x_j^2}{x_i^2} \left(\frac{C_{ji}^1 C_{ji}^1}{2\lambda^* + 2\lambda'} + \frac{2C_{ji}^1 C_{ji}^2}{2\lambda^*} + \frac{C_{ji}^2 C_{ji}^2}{2\lambda^* - 2\lambda'} \right). \tag{18}$$

By adding and subtracting appropriate terms this becomes

$$\langle s^{-1} \rangle = \frac{1}{2} \sum_{i=1}^{2} \sum_{j=1}^{2} \frac{x_j^2}{x_i^2} \left[\sum_{k=1}^{2} \sum_{m=1}^{2} \frac{C_{ji}^k C_{ji}^m}{2\lambda^*} + C_{ji}^1 C_{ji}^1 \left(\frac{1}{2\lambda^* + 2\lambda'} - \frac{1}{2\lambda^*} \right) \right.$$
$$\left. + C_{ji}^2 C_{ji}^2 \left(\frac{1}{2\lambda^* - 2\lambda'} - \frac{1}{2\lambda^*} \right) \right]. \tag{19}$$

Using Eq. (6) we can simplify the first term on the right-hand side of the above expression:

$$\frac{1}{2} \sum_{i=1}^{2} \sum_{j=1}^{2} \frac{x_j^2}{x_i^2} \sum_{k=1}^{2} \sum_{m=1}^{2} \frac{C_{ji}^k C_{ji}^m}{2\lambda^*} = \frac{1}{2} \sum_{i=1}^{2} \sum_{j=1}^{2} \frac{x_j^2}{x_i^2} \frac{\delta_{ji} \delta_{ji}}{2\lambda^*}$$
$$= \frac{1}{2} \sum_{i} \frac{\delta_{ii}}{2\lambda^*} = \frac{1}{2\lambda^*}. \tag{20}$$

Then, after a little algebraic rearrangement of the remaining terms, Eq. (19) becomes

$$\langle s^{-1} \rangle = \frac{1}{2\lambda^*} + \frac{1}{4\lambda^{*2}} \sum_{i=1}^{2} \sum_{j=1}^{2} \frac{x_j^2}{x_i^2} ((C_{ji}^1)^2 - (C_{ji}^2)^2)$$
$$+ \frac{\lambda'^2}{4\lambda^{*2}} \sum_{i=1}^{2} \sum_{j=1}^{2} \frac{x_j^2}{x_i^2} \left(\frac{(C_{ji}^1)^2}{\lambda^* + \lambda'} + \frac{(C_{ji}^2)^2}{\lambda^* - \lambda'} \right). \tag{21}$$

The second term on the right-hand side of Eq. (21) can be shown to vanish.

18. Ecosystem Stability and Community Matrix Eigenvalues

This follows directly from Eqs. (6) and (7):

$$\sum_{i=1}^{2}\sum_{j=1}^{2}\frac{x_j^2}{x_i^2}[(C_{ji}^1)^2-(C_{ji}^2)^2]$$

$$=\sum_{i=1}^{2}\sum_{j=1}^{2}\frac{x_j^2}{x_i^2}(C_{ji}^1-C_{ji}^2)(C_{ji}^1+C_{ji}^2)$$

$$=\sum_{i=1}^{2}\sum_{j=1}^{2}\frac{x_j^2}{x_i^2}(C_{ji}^1-C_{ji}^2)\delta_{ij}$$

$$=\sum_{i=1}^{2}(C_{ii}^1-C_{ii}^2)=1-1=0. \qquad (22)$$

Moreover, from Eq. (17), it follows that the last term in Eq. (21) is nonnegative, and thus $\langle s^{-1}\rangle = 1/2\lambda^*$ plus terms which are nonnegative and do not vanish if $\lambda' \neq 0$.

3.5. The Arbitrary Community Matrix

This result generalizes to an arbitrary community matrix. Let

$$\lambda_k = \lambda^* + \varepsilon_k, \qquad (23)$$

where

$$\lambda^* > 0, \qquad \sum_k \varepsilon_k = 0, \quad \text{and} \quad \varepsilon_k > -\lambda^*. \qquad (24)$$

Then following the same procedures as in Eqs. (19) and (20),

$$\langle s^{-1}\rangle = \frac{1}{2\lambda^*} - \frac{1}{2\lambda^* n}\sum_i\sum_j\frac{x_j^2}{x_i^2}\sum_k\sum_m\frac{C_{ji}^k C_{ji}^m(\varepsilon_k+\varepsilon_m)}{2\lambda^*+\varepsilon_k+\varepsilon_m}. \qquad (25)$$

In the Appendix to this chapter it is shown that even in this general case, $\langle s^{-1}\rangle \geq 1/2\lambda^*$.

It should be pointed out that often in the limit of degenerate eigenvalues, Eq. (3) is no longer a correct form for Δx_i and the C_{ji}^k develop singularities behaving like, for example, $1/\varepsilon_k$ as the ε_k approach zero. In such a degenerate limit, our proofs fail. By continuity, we still expect that $\langle s^{-1}\rangle \geq 1/2\lambda^*$ even in such a case. What we do not expect is that $\langle s^{-1}\rangle = 1/2\lambda^*$ in the limit $\varepsilon_k \to 0$, should such singularities develop. If the community matrix has all zero matrix elements either above or below the diagonal, then such singularities do not arise, $C_{ji}^k = \delta_{jk}\delta_{ki}$, Eq. (15) is exact, and $\langle s^{-1}\rangle \to 1/2\lambda^*$ as the $\varepsilon_k \to 0$. In general, for fixed diagonal elements, the more sparse the off-diagonal terms of the community matrix, the more likely that the C_{ji}^k are nonsingular in the degenerate limit and that Eq. (15) is exact.

From these results we draw the following conclusions: (a) $\langle s^{-1}\rangle$ has a minimum value $1/2\lambda^*$, where $-\lambda^*$ is the mean eigenvalue; (b) this minimum can only be reached if the dispersion, $(1/n)(\Sigma \varepsilon_k^2)$ of the eigenvalues vanishes (and thus each $\varepsilon_k = 0$); and (c) this minimum *is* reached if the dispersion vanishes and the C_{ji}^k develop no singularities in that limit.

Additionally, the following qualitative results are suggested by the preceding arguments: (d) for a fixed value of the eigenvalue mean and dispersion, stability is enhanced by community matrices which are either zero above or below the diagonal or are sparse with respect to off-diagonal elements. In an energy representation, the former is likely to be approximately valid because the retention coefficients characterizing the ratio of the energy gained by a predator to the energy lost by a prey in a trophic interaction are typically very small. No matter what representation is used, our results suggest that pathway simplicity (sparseness of off-diagonal community matrix) enhances stability for a fixed value of the eigenvalue mean and small eigenvalue dispersion. (e) the ratio of the eigenvalue dispersion to the square of the eigenvalue mean, $(1/n\lambda^{*2})(\Sigma \varepsilon_k^2)$ provides a parameter roughly characterizing the dependence of $\langle s^{-1}\rangle$ on the dispersion of the eigenvalues. Stability is enhanced as this parameter decreases.

4. DISCUSSION

Our results follow from relatively model-independent arguments. Nevertheless, it is useful to review the assumptions we have made in deriving Eq. (10) from Eq. (1). They are (a) the $\Delta x_i/x_i$ are small so that a *linear* differential equation for Δx_i holds; (b) the x_i are constants; (c) all of the eigenvalues of A are negative and real; (d) the w_i and $f(t)$ in Eq. (1) are equal to unity; and (e) the initial conditions are as specified in Eq. (9).

If these assumptions are relaxed, methods may still exist for analyzing the implications of Eq. (1). Complex eigenvalues can be accommodated by our methods; for example, we have examined the case of diagonal community matrix with complex eigenvalues. There, it is straightforward to show that for fixed values of the real parts of the eigenvalues, $\langle s^{-1}\rangle$ is greatest when the imaginary parts vanish; as all the imaginary parts go to $\pm\infty$, $\langle s^{-1}\rangle$ approaches zero. If $f(t)$ is not a constant but some function of time weighing the future unevenly, then the λ-dependence of the time integrals in Eq. (1) will no longer look like $1/(\lambda_m + \lambda_k)$. Whatever dependence on the λ's emerges from the integrals, the constraints of Eqs. (6) and (7) will still hold and our methods will give *some* result. Determination of the consequences of various forms for $f(t)$ and analysis of what happens when other assumptions in the above list are relaxed will be explored in future studies. It would also be of interest to determine how our results depend on the specific functional dependence of the integrand in Eq. (1) on the $|\Delta x_i|$. All our assumptions were chosen, in part, to simplify the mathematics, but only assumptions (a) and (b) above appear difficult to relax if analytical methods rather than numerical simulation are to be used to explore all these issues. However, because of the importance of understanding time-dependent equilibria and nonlinear behavior far from equilibrium in conjection with our stability measure, extensions of our analytical methods are now being explored.

We conclude with a brief mention of the problem of measurement of an approximate eigenvalue distribution. Clearly, measurement of all the individual λ_k is impossible. In a previous report (Harte and Morowitz, 1975) we have proposed a tracer experiment which measures the transit-time distribution for a trace substance such as ^{14}C or ^{15}N to make a round trip through an ecosystem. This transit-time distribution provides a measure of the spread of the time constants characterizing the pathways for nutrient flow in the system. For a system in which there are only a small number of pathways with rapid flow rates, the distribution would be narrow and peaked at early times. For a system with many pathways and with greatly differing flow rates, the distribution would be spread out in time.

Simple heuristic arguments were used to suggest that the distribution of transit times and the distribution of time constants given by the inverses of the real parts of the community matrix eigenvalues were qualitatively similar. Thus, we conjecture that further exploration of tracer or other techniques for measurement of transit-time distributions for ecosystems could be of great value in verifying the ideas described here.

APPENDIX

In this Appendix we prove that

$$J \equiv -\sum_i \sum_j \frac{x_j^2}{x_i^2} \sum_k \sum_m \frac{C_{ji}^k C_{ji}^m (\varepsilon_k + \varepsilon_m)}{2\lambda^* + \varepsilon_k + \varepsilon_m} \geq 0 \tag{A1}$$

subject to the constraints

$$\sum_i C_{ii}^k = 1, \qquad \sum_k C_{ji}^k = \delta_{ji}, \qquad \sum_k \varepsilon_k = 0, \qquad \varepsilon_k > -\lambda^*. \tag{A2}$$

Using Eq. (25), this will then establish that $\langle s^{-1}\rangle \geq 1/2\lambda^*$ in the most general case.

We first use the identity

$$\frac{\varepsilon_k + \varepsilon_m}{2\lambda^* + \varepsilon_k + \varepsilon_m} \equiv \int_0^\infty d\sigma \left(-\frac{d}{d\sigma} - 2\lambda^*\right) e^{-\sigma(2\lambda^* + \varepsilon_k + \varepsilon_m)} \quad (A3)$$

and the fact that

$$\sum_k \sum_m x_k x_m \equiv \left(\sum_k x_k\right)^2 \quad (A4)$$

for arbitrary x_k, to rewrite J in the form:

$$J = \int_0^\infty d\sigma \left(\frac{d}{d\sigma} + 2\lambda^*\right) \sum_i \sum_j \frac{x_j^2}{x_i^2} \left[\sum_k C_{ji}^k e^{-\sigma(\lambda^* + \varepsilon_k)}\right]^2. \quad (A5)$$

The $d/d\sigma$ term easily simplifies, with the aid of Eq. (A2) as follows:

$$\int_0^\infty d\sigma \frac{d}{d\sigma} \sum_i \sum_j \frac{x_j^2}{x_i^2} \left[\sum_k C_{ji}^k e^{-\sigma(\lambda^* + \varepsilon_k)}\right]^2$$

$$= -\sum_i \sum_j \frac{x_j^2}{x_i^2} \left[\sum_k C_{ji}^k\right]^2 = -\sum_i \sum_j \frac{x_j^2}{x_i^2} \delta_{ij}^2 = -\sum_i 1 = -n. \quad (A6)$$

We now use the fact that the double sum over i and j is bounded below by the sum over $i = j$, and the general inequality

$$\sum_{i=1}^n x_i^2 \geq \frac{1}{n}\left(\sum_{i=1}^n x_i\right)^2 \quad (A7)$$

to get

$$J \geq -n + 2\lambda^* \int_0^\infty d\sigma \frac{1}{n} \left[\sum_i \sum_k C_{ii}^k e^{-\sigma(\lambda^* + \varepsilon_k)}\right]^2. \quad (A8)$$

Using Eqs. (A2) and (A4), this becomes

$$J \geq -n + \frac{2\lambda^*}{n} \sum_k \sum_m \frac{1}{2\lambda^* + \varepsilon_k + \varepsilon_m}. \quad (A9)$$

We next show that

$$K \equiv \sum_k \sum_m \frac{1}{2\lambda^* + \varepsilon_k + \varepsilon_m} \geq \frac{n^2}{2\lambda^*}. \quad (A10)$$

We begin by writing

$$K = \sum_k \sum_m \int_0^\infty d\sigma e^{-\sigma(2\lambda^* + \varepsilon_k + \varepsilon_m)} = \int_0^\infty d\sigma e^{-2\lambda^* \sigma} \sum_k \sum_m e^{-\sigma(\varepsilon_k + \varepsilon_m)}. \quad (A11)$$

But
$$e^{-x} \geqslant 1-x, \qquad (A12)$$
and so
$$K \geqslant \int_0^\infty d\sigma e^{-2\lambda*\sigma} \sum_k \sum_m (1 - \sigma e_k - \sigma \varepsilon_m) = \frac{1}{2\lambda*} \sum_k \sum_m 1 = \frac{n^2}{2\lambda*}. \qquad (A13)$$

Thus, $J \geqslant 0$, and $\langle s^{-1} \rangle \geqslant 1/2\lambda*$.

ACKNOWLEDGMENT

The proof in the Appendix is due to Mark Davidson; I am extremely grateful to him. This work was supported by the Department of Energy and the Electric Power Research Institute.

REFERENCES

Andrewartha, H. G., and Birch, L. C. (1954). "The Distribution and Abundance of Animals." Univ. of Chicago Press, Chicago, Illinois.

Botkin, D., and Sobel, M. (1974). The complexity of ecosystem stability. In "Ecosystem: Analysis and Prediction" (S. A. Levin, ed.), pp. 144–150. SIAM Inst. Math. Soc.

Goodman, D. (1975). The theory of diversity-stability relationships in ecology. Q. Rev. Biol. **50**, 237–266.

Harte, J., and Morowitz, H. (1975). Nutrient transit-time diversity: A measure of ecological organization and stability. Lawrence Berkeley Lab., Berkeley, California LBL 4441, pp. 1–10.

Holling, C. S. (1973). Resilience and stability of ecological systems. Ann. Rev. Ecol. Syst. **4**, 1–23.

Horn, D., and MacArthur, R. H. (1972). On competition in a diverse and patchy environment. Ecology **53**, 749–752.

MacArthur, R. H. (1955). Fluctuations of animal populations, and a measure of community stability. Ecology **36**, 533–536.

Margalef, R. (1974). Diversity, stability, and maturity in natural ecosystems. In "Unifying Concepts in Ecology" (W. H. van Dobben and R. H. Lowe McConnell, eds.), pp. 151–160. Junk Publ., The Hague.

May, R. (1973). "Stability and Complexity in Model Ecosystems." Princeton Univ. Press, Princeton, New Jersey.

Odum, E. (1969). The strategy of ecosystem development. Science **164**, 262–270.

Orians, G. (1974). Diversity, stability and maturity in natural ecosystems. In "Unifying Concepts in Ecology" (W. H. van Dobben and R. H. Lowe McConnell, eds.), pp. 139–150. Junk Publ., The Hague.

Roughgarden, J. (1975). A simple model for population dynamics in stochastic environments. Am. Nat. **109**, 713–736.

Steele, J. (1974). "The Structure of Marine Ecosystems." Harvard Univ. Press, Cambridge, Massachusetts.

Chapter **19**

ROBUST STABILITY CONCEPTS FOR ECOSYSTEM MODELS

B. S. Goh

1. Introduction 467
2. Global and Finite Stability 469
   2.1 Liapunov Functions. 469
   2.2 A Class of Nonlinear Models 471
   2.3 Complex Ecosystems 473
   2.4 Exploited Complex Ecosystems 475
   2.5 Connective Stability. 475
3. Nonvulnerability......................... 478
   3.1 The Concept and Liapunov-Like Functions. 478
   3.2 An Algebraic Method. 480
   3.3 A Numerical Example 481
4. Sector Stability......................... 482
   4.1 Introduction 482
   4.2 A Class of Nonlinear Models 483
   4.3 A Numerical Example 485
5. Conclusion 486
   References............................ 486

1. INTRODUCTION

The standard test for stability in an ecosystem model which is described by a set of differential equations requires that the real part of all the eigenvalues of a matrix be negative (May, 1974). In a nonlinear model this test can only establish stability relative to small perturbations of the initial state from an equilibrium. For this reason this type of stability is called local stability. By itself local stability is of no practical significance to a nonlinear model. When an eigenvalue analysis establishes that an equilibrium is locally stable an estimate of its region of attraction should be made.

The word "stability" is used for many purposes in ecology, mathematics, and engineering. Recently, Holling (1973) and May (1974) reviewed the various concepts of stability which are used in ecology. In ecology the meaning of a stability concept can be made precise by applying it to a prototype mathematical model with specifications of (a) the class of admissible perturbations, (b) the set of admissible initial states, and (c) a set of system responses which characterizes desirable or undesirable behavior. Each set of specifications defines a concept of stability. There are at least three types of perturbations that can be applied to a model: Perturbations of the initial state, external disturbances acting in a continual manner on the system dynamics, and variations in the system parameters.

The most powerful analytical method for studying stability relative to realistic perturbations of the initial state of a ecosystem model is the direct method of Liapunov. The method was discovered in 1892 by the Russian mathematician A. M. Liapunov (see Liapunov, 1966), but was ignored by mathematicians for a long time. In the West it became a popular tool for the analysis of nonlinear systems in the 1950s.

The direct method of Liapunov requires the construction of certain functions called *Liapunov functions*. For physical systems the direct method of Liapunov generalizes the principle that a system, which continuously dissipates energy until it attains equilibrium, is stable.

In an ecosystem the density of each species must be nonnegative. This requires that the concept of global stability in the standard stability theory (e.g., Willems, 1970) be modified. By definition an ecosystem model is globally stable if every trajectory of the system which initiates from a positive state remains in the positive orthant and moves into a decreasing, small neighborhood of a feasible equilibrium. The fact that population densities must be nonnegative requires that in ecological theory a two-sided energy principle for constructing a Liapunov function is more appropriate (Goh, 1977a). A viable single-species population must have net energy dissipation when its density is very high, and it must have net energy absorption when its density is very low. In a complex ecosystem the population of each species should have the behavior of a viable single-species population at extreme densities, and there must be a balance between the rates of net energy exchanges between the species in the ecosystem and also between the species and their environments.

An ecosystem is said to be robust if it is stable relative to large perturbations of its initial state and if the qualitative behavior of its dynamics is not sensitive to small variations in the system parameters. An ecosystem model is robust if it is a collection of subsystems each of which is self-regulating at extreme population densities, and is such that from the total system point of view, the interactions between subsystems are weaker than the self-regulating interactions of the subsystems. The idea that a complex robust ecosystem model is robust if it is a collection of subsystems with some limitations on their interactions has also been discussed by Šiljak (1975; Chapter 7, this volume). However, it is not the purpose of this chapter to prove that nature uses relatively weak intercompartmental interactions.

2. GLOBAL AND FINITE STABILITY

2.1. Liapunov Functions

Let N_i be the population of the ith species in an ecosystem. By definition the positive orthant is the set, $\Omega = \{\mathbf{N} | N_i > 0, i = 1, 2, \ldots, m\}$. A model with m species is

$$\dot{N}_i = N_i F_i(N_1, N_2, \ldots, N_m), \qquad i = 1, 2, \ldots, m. \tag{1}$$

By definition $\mathbf{N}^*$ is a feasible equilibrium if

$$N_1^* > 0, N_2^* > 0, \ldots, N_m^* > 0$$

and

$$F_1(\mathbf{N}^*) = 0 = F_2(\mathbf{N}^*) = \ldots = F_m(\mathbf{N}^*).$$

In order that a given function $V(\mathbf{N})$ is a Liapunov function for Eq. (1) in the positive orthant it must satisfy the following conditions:

(a) $V(\mathbf{N}) > 0$ for all $\mathbf{N} \in \Omega$ such that $\mathbf{N} \neq \mathbf{N}^*$ and $V(\mathbf{N}^*) = 0$.
(b) For $i = 1, 2, \ldots, m$, $V(\mathbf{N}) \to \infty$ as $N_i \to 0+$ or as $N_i \to \infty$ along any trajectory which remains in the positive orthant.
(c) Along solutions of Eq. (1)

$$\dot{V}(\mathbf{N}) = \sum_{i=1}^{m} (\partial V / \partial N_i) N_i F_i(\mathbf{N}) \leq 0 \tag{2}$$

and $\dot{V}(\mathbf{N})$ does not vanish identically along any solution of Eq. (1) other than the solution $\mathbf{N} = \mathbf{N}^*$.

The quadratic function

$$V = \sum_{i=1}^{m}\sum_{j=1}^{m}(N_i - N_i^*)P_{ij}(N_j - N_j^*),$$

where $P = (P_{ij})$ is a positive definite matrix, is often used as a Liapunov function for an engineering system (see Willems, 1970). But it cannot be directly used to establish global stability in the positive orthant because it does not tend to infinity as $N_i \to 0+$.

In economics this function is often incorrectly used to establish global stability of a competitive equilibrium (e.g., Arrow et al., 1959; Takayama, 1974). To use this function correctly, it is necessary to show separately that N^* is the unique limit point in the nonnegative orthant and every trajectory that begins in the positive orthant remains in it.

A Liapunov function in the standard stability theory can be converted into a Liapunov function for Eq. (1) by subjecting it to a transformation which maps the positive orthant into the whole state space. The following example demonstrates this method of constructing a Liapunov function for Eq. (1).

Example 1. Let n denote the population of a single species. Consider

$$\dot{n} = nf(n). \tag{3}$$

Suppose it has a feasible equilibrium at n^*.

The function $V = x^2$ is often used as a Liapunov function in the standard stability theory. Under the transformation $x = \ln(n/n^*)$, it becomes $V = [\ln(n/n^*)]^2$. Along solutions of Eq. (3) we have

$$\dot{V}(n) = 2[\ln(n/n^*)]f(n).$$

It follows that $\dot{V}(n) < 0$ for $n > 0$ and $n \ne n^*$ if (i) $f(n) > 0$ for $n^* > n > 0$ and (ii) $f(n) < 0$ for $n > n^*$.

Another Liapunov function for Eq. (2) is

$$V(n) = n - n^* - n^* \ln(n/n^*). \tag{4}$$

Along solutions of Eq. (3) we have

$$\dot{V}(n) = (n - n^*)f(n).$$

This implies that Eq. (3) is globally stable if $f(n) > 0$ for $n^* > n > 0$ and $f(n) < 0$ for $n > n^*$.

When n/n^* is small the dominant term in the function $V(n)$ is $n^* \ln(n/n^*)$. This is a measure of the energy embodied in the standing biomass. The negative sign in front of this term implies that the population is absorbing energy from an external source if $\dot{V}(n)$ is negative for small values of n/n^*. When n/n^* is large the dominant term in $V(n)$ is n. Therefore for such values of n the value of the function $V(n)$ is again a measure of the energy embodied in the standing biomass. It follows that the population is dissipating energy if $\dot{V}(n)$ is negative at high population densities.

Similarly if we employ the Liapunov functions $V(n) = [\ln(n/n^*)]^2$, $V(n) = n - 2n^* + (n^*)^2/n$, or any other Liapunov function $V(n)$ which tends to infinity as $n \to \infty$ and as $n \to 0+$, then we can conclude that on balance a viable single-species population must absorb energy at low densities and must dissipate energy at high densities.

Let $d_1, d_2, \ldots, d_m$ be positive constants. Suppose $W_i(N_i)$ is a Liapunov function for a single-species population. If the energy flows in a complex model are well balanced, a good candidate to act as a Liapunov function is

$$V(\mathbf{N}) = d_1 W_1(N_1) + d_2 W_2(N_2) + \cdots + d_m W_m(N_m).$$

Using this principle for constructing a Liapunov function for a model of a complex ecosystem, we get

$$V(\mathbf{N}) = \sum_{i=1}^{m} d_i [N_i - N_i^* - N_i^* \ln(N_i/N_i^*)]. \tag{5}$$

At present this $V(\mathbf{N})$ function is the best candidate to act as a Liapunov function for Eq. (1).

Finally note that a Liapunov function for a differential equation model may sometimes be used as a Liapunov function for a difference equation model of a community with the same number of species (Goh, 1977b; Šiljak, Chapter 7, this volume).

2.2. A Class of Nonlinear Models

Some sufficient conditions are here established for global stability in a class of ecosystem models (see also Goh, 1976, 1977a; Jeffries, 1976, Chapter 20, this volume). But generally a model of an ecosystem which is described by a set of differential equations will not provide a valid description of the dynamics of an ecosystem when the density of any component species is very low or very high. Therefore for practical purposes it should be assumed that global stability in a model ensures only a finite region of attraction for the equilibrium of the ecosystem in the real world.

Theorem 1. Let $V(\mathbf{N})$ be the function in Eq. (5). The ecosystem model in Eq. (1) is globally stable if (i) it has a feasible equilibrium at $\mathbf{N}^*$ (ii) there exist positive constants $d_1, d_2, \ldots, d_m$ such that

$$\dot{V}(\mathbf{N}) = \sum_{i=1}^{m} d_i(N_i - N_i^*)F_i(\mathbf{N}) \leq 0 \tag{6}$$

at all points in the positive orthant, and (iii) $\dot{V}(\mathbf{N})$ does not vanish identically along a solution of Eq. (1) except for the equilibrium solution $\mathbf{N} = \mathbf{N}^*$.

Proof. If the conditions of the theorem are satisfied by definition the function $V(\mathbf{N})$ is a suitable Liapunov function. Hence by LaSalle's extension (LaSalle and Lefschetz, 1961) of the direct method of Liapunov the model in Eq. (1) is globally stable in the positive orthant.

One method for choosing the positive constants $d_1, d_2, \ldots, d_m$ is to linearize the $F_1, F_2, \ldots, F_m$ in Eq. (1) and then use Goh's (1977a) algorithm for computing such constants for Lotka–Volterra models. For a low order model it is simpler to use the trial and error method.

Let $b = (b_i)$ and $A = (a_{ij})$ be constant matrices. Consider the generalized Lotka–Volterra model

$$\dot{N}_i = N_i \left[b_i + \sum_{j=1}^{m} a_{ij} N_j \right], \quad i = 1, 2, \ldots, m, \tag{7}$$

where N_i is the population of the ith species. At a feasible equilibrium $\mathbf{N}^*$ we have $\mathbf{b} + A\mathbf{N}^* = 0$. Substituting this into Eq. (7), we get

$$\dot{N}_i = N_i \sum_{j=1}^{m} a_{ij}(N_j - N_j^*), \quad i = 1, 2, \ldots, m. \tag{8}$$

Equations (6) and (8) imply that

$$\dot{V}(\mathbf{N}) = \tfrac{1}{2}(\mathbf{N} - \mathbf{N}^*)^T (DA + A^T D)(\mathbf{N} - \mathbf{N}^*). \tag{9}$$

Therefore model (7) is globally stable if (i) it has a feasible equilibrium at $\mathbf{N}^*$, (ii) there exists a positive diagonal matrix D such that $DA + A^T D$ is negative semidefinite, and (iii) the function $\dot{V}(\mathbf{N})$ in Eq. (9) does not vanish identically along a solution of Eq. (8) except for the solution $\mathbf{N} = \mathbf{N}^*$ (see also Goh, 1977a).

Example 2. Let N_1, N_2 be the population of a prey and its predator. Let b, d, a_{12}, a_{22}, and e be positive constants. The following model illustrates how a predator may provide a robust control of a pest population which is an unstable population in the absence of the predator.

The prey–predator model is

$$\dot{N}_1 = N_1[b - a_{12}N_2], \quad \dot{N}_2 = N_2[-d + ea_{12}N_1 - a_{22}N_2].$$

This model has a feasible equilibrium at $\mathbf{N}^* = [(da_{12} + ba_{22})/ea_{12}^2, b/a_{12}]$. The interaction matrix is

$$A = \begin{bmatrix} 0 & -a_{12} \\ ea_{12} & -a_{22} \end{bmatrix}.$$

If $D = \text{diag}(e, 1)$, the matrix $DA + A^T D = \text{diag}(0, -2a_{22})$; clearly it is negative semidefinite. The function $\dot{V}(N) = -(a_{22})(N_2 - b/a_{12})^2$ does not vanish identically along a solution of the model other than $\mathbf{N} = \mathbf{N}^*$. Hence the model is globally stable in the positive orthant.

2.3. Complex Ecosystems

If a model ecosystem is more nonlinear than a Lotka–Volterra model, it is very difficult to use Theorem 1 for establishing global stability. For this reason an algebraic criterion for global stability in an ecosystem model is developed. Unfortunately it is a very conservative condition for global stability and it is difficult to apply it to a specific nonlinear model of an ecosystem. Its main value is that it provides a deep insight into conditions for global stability in a complex ecosystem. It also identifies a subset of Lotka–Volterra models which have qualitatively the same stability behavior as a large class of nonlinear ecosystem models.

Theorem 2. Suppose Eq. (1) has a feasible equilibrium at $\mathbf{N}^*$. In the positive orthant, let $\partial F_i/\partial N_j$ for $i,j = 1, 2, \ldots, m$ be continuous and uniformly bounded. This implies that there exists a constant matrix G such that

$$\partial F_i(\mathbf{N})/\partial N_i \leq G_{ii}, \quad i = 1, 2, \ldots, m, \tag{10}$$

$$|\partial F_i(\mathbf{N})/\partial N_j| \leq G_{ij}, \quad i \neq j, \tag{11}$$

at every point in the positive orthant. If the leading principal minors of $-G$ are positive, then Eq. (1) is globally stable.

Proof. Apply Taylor's Theorem to Eq. (6)

$$\dot{V}(\mathbf{N}) = \sum_{i=1}^{m} \sum_{j=1}^{m} d_i (N_i - N_i^*)(\partial F_i/\partial N_j)(N_j - N_j^*) \tag{12}$$

in which the partial derivatives are evaluated at a set of points between $\mathbf{N}$ and $\mathbf{N}^*$.

Equations (10) to (12) imply that

$$\dot{V}(\mathbf{N}) \leq \sum_{i=1}^{m} d_i G_{ii}(N_i - N_i^*)^2 + \sum_{i=1}^{m} \sum_{j \neq i}^{m} d_i G_{ij}|N_i - N_i^*||N_j - N_j^*|$$

$$= \tfrac{1}{2}\mathbf{y}^T(DG + G^T D)\mathbf{y},$$

where $\mathbf{y} = (|N_i - N_i^*|)$ and $D = \text{diag}(d_1, d_2, \ldots, d_m)$. Clearly if $DG + G^T D$ is negative definite then $\dot{V}(\mathbf{N})$ is negative for all $\mathbf{N}$ in the positive orthant such that $\mathbf{N} \neq \mathbf{N}^*$.

The diagonal elements of G are negative while the off-diagonal elements are positive. If all the leading principal minors of $-G$ are positive the matrix G is an M-matrix. For such a matrix there exists a positive diagonal matrix D such that $DG + G^T D$ is negative definite (Johnson, 1974). Hence Eq. (1) is globally stable.

Theorem 2 implies that an ecosystem model is robust if it is a collection of interacting stable subsystems such that the interactions between subsystems from the total systems point of view are weaker than the self-interactions of the subsystems. If a complex ecosystem model satisfies Eqs. (10) and (11), the condition that all the leading principal minors of $-G$ are positive provides a criterion for establishing that the interspecific interactions, from the total systems point of view, are weaker than the intraspecific interactions. The following example demonstrates that this is not a straightforward and intuitive concept.

Example 3. Let the matrices G_1 and G_2 be associated with two nonlinear models which satisfy Eqs. (10) and (11). Suppose each model contains three species. From Eq. (11) it is clear that the signs of the interspecific interactions can be positive or negative. It follows that the interspecific interactions could be any type of interactions. Let

$$G_1 = \begin{bmatrix} -1 & 0.6 & 0.3 \\ 0.5 & -1 & 0.7 \\ 0.7 & 0.4 & -1 \end{bmatrix}, \quad G_2 = \begin{bmatrix} -1 & 8 & 0.1 \\ 0.01 & -1 & 0.2 \\ 0.2 & 2 & -1 \end{bmatrix}.$$

From an inspection of the elements of G_1 and G_2, we see that some elements of G_2 are larger than those of G_1. But all the leading principal minors of $-G_2$ are positive. Hence from the total systems point, the interspecific interactions in a model with matrix G_2 are weaker than its intraspecific interactions. This is not the case in the model with G_1 because $\det(-G_1) < 0$.

We shall show that constant energy input to each population in a complex model, which satisfies Theorem 2 in the absence of energy input, is a stabilizing influence. If an ecosystem model does not satisfy Theorem 2, this type of energy input may be destabilizing. It is known in systems theory that a complex system can be destabilized by making some of its diagonal elements more negative (Willems, 1971).

Let $I_1, I_2, \ldots, I_m$ be nonnegative constants. In place of Eq. (1) let an ecosystem model be described by the set of equations

$$\dot{N}_i = N_i F_i(\mathbf{N}) + I_i, \quad i = 1, 2, \ldots, m. \tag{13}$$

Assume that this model has a feasible equilibrium at **N***. By assumption, Eq. (13) satisfies Eqs. (10) and (11) when $I_i = 0$ for $i = 1, 2, \ldots, m$. When $I_i \geq 0$ for $i = 1, 2, \ldots, m$, then

$$\partial[F_i + I_i/N_i]/\partial N_i \leq G_{ii} - I_i/N_i^2 \leq G_{ii}.$$

It follows that if Eq. (13) satisfies Theorem 2 when $I_i = 0$ for $i = 1, 2, \ldots, m$, it will also do so when $I_i \geq 0$ for $i = 1, 2, \ldots, m$. This establishes that constant energy input into an ecosystem model, which satisfies Theorem 2 in the absence of the energy input, is a stabilizing influence.

2.4. Exploited Complex Ecosystems

Ideally, in the exploitation of an ecosystem, optimal control theory should be employed to formulate optimal management policies (Clark, 1976; Goh, Chapter 15, this volume). The computational difficulties which are encountered in applying optimal control theory to a model ecosystem with more than a few state variables are horrendous. For this reason, in the exploitation of a complex ecosystem, it is more realistic for the time being to employ harvesting programs which satisfy the first and foremost requirement that they do not destabilize and destroy the ecosystem. Simple examples of counterintuitive consequences of the effects of enrichment and exploitation on an ecosystem have been examined by Rosenzweig (1971) and others.

A class of models are now described which remain stable under constant fraction harvestings of the component species.

Suppose the unexploited ecosystem is described by Eq. (1). Let E_i be the constant fraction harvesting program which is applied to the ith species in the ecosystem. Suppose $E_1, E_2, \ldots, E_m$ are nonnegative constants. A model of an exploited ecosystem is

$$\dot{N}_i = N_i[F_i(\mathbf{N}) - E_i], \qquad i = 1, 2, \ldots, m. \tag{14}$$

Clearly if Eq. (1) satisfies the conditions in Eqs. (10) and (11), then Eq. (14) also satisfies the same conditions. It follows that if Eq. (1) satisfies Theorem 2 and if Eq. (14) has a feasible equilibrium, then the latter also satisfies Theorem 2.

Thus an unexploited ecosystem which is globally stable and which satisfies Theorem 2 will remain globally stable for a wide range of constant fraction harvesting programs.

2.5. Connective Stability

Šiljak (1975) has developed a very interesting new concept of stability which is called *connective stability*. This concept describes a system's ability to remain stable in spite of structural changes in the interactions between components of the system. However, it is necessary to modify Šiljak's results on connective stability before it can be applied to ecological and economic models. The requirement that state variables in an ecosystem model have to be nonnegative restricts the set of admissible structural perturbations to those which do not cause the equilibrium of an ecosystem model to become nonfeasible.

Šiljak (1975) studied the connective stability of the model

$$\dot{x}_i = \sum_{j=1}^{m} A_{ij}(t, \mathbf{x}) x_j, \quad i = 1, 2, \ldots, m, \tag{15}$$

where (A_{ij}) is a matrix of nonlinear functions. This model tacitly assumes that structural perturbations which cause interactions between species to be connected or disconnected do not shift the equilibrium from the origin. This assumption is generally not satisfied in a nonlinear model of an ecosystem or an economic system.

Consider the prey–predator model,

$$\dot{N}_1 = N_1[1.8 - N_1 - 0.8 N_2], \quad \dot{N}_2 = N_2[-0.1 + 1.1 E_{21} N_1 - N_2]. \tag{16}$$

When $E_{21} = 1$, this model has a feasible equilibrium at $(1, 1)$. It follows that its linearized dynamics in the neighborhood of $(1, 1)$ are

$$\dot{x}_1 = -x_1 - 0.8 x_2, \quad \dot{x}_2 = 1.1 x_1 - x_2.$$

This linear system satisfies Šiljak's conditions for connective stability.

When $E_{21} = 0$, Eq. (16) reduces to the system,

$$\dot{N}_1 = N_1[1.8 - N_1 - 0.8 N_2], \quad \dot{N}_2 = N_2[-0.1 - N_2]. \tag{17}$$

This system has a nontrivial equilibrium at $(1.88, -0.1)$. It has three other equilibria at $(0, 0), (0, -0.1)$, and $(1.8, 0)$. It can be shown analytically that all solutions which initiate in the positive orthant tends to $(1.8, 0)$. This means that Eq. (17) does not describe a viable prey–predator system.

Šiljak's pioneering analysis of connective stability in nonlinear systems can be modified to apply to ecosystem and economic models. For simplicity the Lotka–Volterra models are employed in this discussion. From the proof of Theorem 2 and the subsequent analysis, it would be clear that this analysis of connective stability can be applied to a class of nonlinear models.

Consider the Lotka–Volterra model,

$$\dot{N}_i = N_i \left[b_i(s) + \sum_{j=1}^{m} E_{ij}(s) G_{ij} N_j \right], \quad i = 1, 2, \ldots, m, \tag{18}$$

where $b_i(s)$ and $E_{ij}(s)$ are functions of s. Let $E_{ii} = 1$ for $i = 1, 2, \ldots, m$ and

$-1 \leqslant E_{ij}(s) \leqslant 1$ for $i \neq j$. Suppose $G = (G_{ij})$ is a constant matrix such that $G_{ii} < 0$ for $i = 1, 2, \ldots, m$ and $G_{ij} \geqslant 0$ for $i \neq j$. Note that the assumption E_{ij} for $i \neq j$ can be positive or negative implies that Eq. (18) is a model for a community with any type of interspecific interactions.

The nontrivial equilibrium of Eq. (18) is

$$\mathbf{N}^*(s) = -C^{-1}\mathbf{b}, \tag{19}$$

where $\mathbf{b} = (b_i)$ and $C = (C_{ij}) = (E_{ij}G_{ij})$.

Consider two cases. First, the parameters b_i and E_{ij} are permitted to change impulsively but infrequently. Second, let δ be a small number and let $s = \delta t$. This means that $b_i(s)$ and $E_{ij}(s)$ are slowly varying functions of t. Therefore in both cases the question of whether a new equilibrium $\mathbf{N}^*(s)$, which is created by structural perturbations and environmental changes is feasible or not, is separated from the question of whether the newly created $\mathbf{N}^*(s)$ is stable or not.

Goh and Jennings (1977) showed that the subset of Lotka–Volterra models, each of which has a feasible equilibrium, is a small fraction in a set of randomly assembled Lotka–Volterra models. Moreover, this fraction decreases rapidly with the number of species in an ecosystem. These results imply that the requirement that the nontrivial equilibrium of a Lotka–Volterra model must be feasible imposes a very severe constraint on the values of its parameters. This in turn limits the class of structural perturbations which will not destroy the viability of a complex ecosystem.

Theorem 3. If under the above restricted class of environmental and structural perturbations the equilibrium $\mathbf{N}^* = -C^{-1}\mathbf{b}$ remains feasible and if all the leading principal minors of $-G$ are positive, then the model remains globally stable.

Proof. Equations (18) and (19) imply that

$$\dot{N}_i = N_i \sum_{j=1}^{m} E_{ij}(s) G_{ij}(N_j - N_j^*), \quad i = 1, 2, \ldots, m.$$

The function $V(N)$ in Eq. (5) is a suitable Liapunov function.

By assumption $E_{ii} = 1$ for $i = 1, 2, \ldots, m$ and $|E_{ij}| \leqslant 1$ if $j \neq i$. It follows that along solutions of the model

$$\dot{V}(\mathbf{N}) = \sum_{i=1}^{m} \sum_{j=1}^{m} d_i E_{ij} G_{ij} (N_i - N_i^*)(N_j - N_j^*)$$

$$\leqslant \sum_{i=1}^{m} d_i G_{ii}(N_i - N_i^*)^2 + \sum_{i=1}^{m} \sum_{j \neq i}^{m} d_i G_{ij}|N_i - N_i^*||N_j - N_j^*|$$

$$= \tfrac{1}{2} \mathbf{y}^T (DG + G^T D) \mathbf{y},$$

where $D = \text{diag}(d_1, d_2, \ldots, d_m)$ and $\mathbf{y} = (|N_i - N_i^*|)$. The diagonal elements of G are negative and its off-diagonal terms are nonnegative. It follows that if all the leading principal minors of $-G$ are positive, then there exists a positive diagonal matrix D such that $DG + G^T D$ is negative definite (Johnson, 1974). This implies that Eq. (18) is globally stable in spite of certain changes in the set of parameters $\{b_i, E_{jk}\}$.

Theorem 3 describes the conditions under which the model (Eq. 18) can track its equilibrium as the model varies because of certain environmental and structural perturbations. If a model satisfies Theorem 3, then each of its submodels with a feasible equilibrium will also satisfy the theorem. In other words a perturbation which causes the extinction of one or more species in a model which satisfies Theorem 3 will create a submodel which satisfies the same theorem.

From this analysis and the proof of Theorem 2, it is clear that this result on the stability of a model when it is subjected to a restricted class of environmental and structural perturbations can also be established in a class of ecosystem models which are more nonlinear than the Lotka–Volterra models.

3. NONVULNERABILITY

3.1. The Concept and Liapunov-Like Functions

In the real world, ecosystems are continually disturbed by unpredictable forces due to unpredictable changes in weather, migrating species, and diseases. Global stability in a model does not mean that the ecosystem can continue to exist in the face of continual and unpredictable disturbances. One way to establish an ecosystem's ability to remain viable in spite of continual and predictable disturbances is to use the theory of stochastic differential equations (Ludwig, 1975; May, 1974). Unfortunately in practice it is impossible to solve the Fokker–Planck equation for the full nonlinear model except in very simple cases. At present the most effective method to establish that an ecosystem model has the ability to persist in spite of large, continual, and unpredictable disturbances is by means of a Liapunov function (Goh, 1976). This method was previously applied to Lotka–Volterra models, but is also applicable to a general class of nonlinear models.

Let $u_i(t)$ be the magnitude of the resultant force of all the disturbances which act on the growth rate of the ith species. A model of an ecosystem which is subjected to density-independent disturbances is

$$\dot{N}_i = N_i F_i(\mathbf{N}) + u_i(t) N_i, \qquad i = 1, 2, \ldots, m. \tag{20}$$

Suppose $u_1(t), u_2(t), \ldots, u_m(t)$ are piecewise continuous functions which satisfy the *a priori* constraints

$$a_i \leq u_i(t) \leq b_i, \qquad i = 1, 2, \ldots, m, \tag{21}$$

where $\{a_i, b_i\}$ is a set of constraints. For convenience denote the admissible set of functions $\{u_i(t)\}$ by $\mathcal{U}$.

Let $\mathcal{S}(0)$ denote a set of acceptable states of the model ecosystem in Eq. (20). Let $\mathcal{Z}$ denote a set of undesirable states. The system is said to be vulnerable relative to the sets $\mathcal{U}, \mathcal{S}(0),$ and $\mathcal{Z}$ during the time interval $[0, t_1]$ if there exists an admissible vector $\mathbf{u}(t)$ which drives the system from an initial state in $\mathcal{S}(0)$ to the undesirable set $\mathcal{Z}$ during the time interval $[0, t_1]$. If there is no admissible disturbance $\mathbf{u}(t)$ which can drive the system from one of the states in $\mathcal{S}(0)$ to $\mathcal{Z}$ during the time interval $[0, t_1]$, then the system is said to be nonvulnerable relative to the sets $\mathcal{U}, \mathcal{S}(0),$ and $\mathcal{Z}$ during the time interval $[0, t_1]$.

At present the most effective method to establish nonvulnerability in a model ecosystem is to carry out a conservative analysis using Liapunov-like functions. Let $V(\mathbf{N})$ denote the function in Eq. (5). It is now called a Liapunov-like function because it is not used to establish asymptotic stability. It is convenient to define the sets $\mathcal{S}(0)$ and $\mathcal{Z}$ in terms of the level surfaces of $V(\mathbf{N})$. If V_S and V_Z are positive constants, let $\mathcal{S}(0) = \{\mathbf{N} | V(\mathbf{N}) \leq V_S\}$ and $\mathcal{Z} = \{\mathbf{N} | V(\mathbf{N}) \geq V_Z\}$. Usually, but not necessarily, the vector $\mathbf{N}^*$ in the function $V(\mathbf{N})$ is the feasible equilibrium of Eq. (20) when $\mathbf{u} = \mathbf{0}$.

Along solutions of Eq. (20),

$$\dot{V}(\mathbf{N}, \mathbf{u}) = (\mathbf{N} - \mathbf{N}^*)^T D F(\mathbf{N}) + (\mathbf{N} - \mathbf{N}^*)^T D \mathbf{u}, \tag{22}$$

where $D = \mathrm{diag}(d_1, d_2, \ldots, d_m)$. The following describes some sufficient conditions for the model (Eq. 20) to be nonvulnerable.

Theorem 4. The model ecosystem in Eq. (21) is nonvulnerable relative to the sets $\mathcal{U}, \mathcal{S}(0),$ and $\mathcal{Z}$ during the time interval $[0, t_1]$, where t_1 is any positive number, if there exists a positive number p such that (i) $V_S \leq p < V_Z$ and (ii) the global maximum of $\dot{V}(\mathbf{N}, \mathbf{u})$ for all $\mathbf{u} \in \mathcal{U}$ and $\mathbf{N} \in \{\mathbf{N} | V(\mathbf{N}) = p\}$ is negative.

Proof. The conditions in the theorem imply that $\dot{V}(\mathbf{N}, \mathbf{u})$ is negative on the hypersurface $\{\mathbf{N} | V(\mathbf{N}) = p\}$ for all admissible disturbances. It follows that all solutions of Eq. (20) which initiate in the set $\{\mathbf{N} | V(\mathbf{N}) \leq p\}$ remain in it indefinitely. Clearly the system cannot be driven from states in the set $\{\mathbf{N} | V(\mathbf{N}) \leq p\}$ to the set $\mathcal{Z}$ because $p < V_Z$. The set $\mathcal{S}(0)$ is a subset of the set $\{\mathbf{N} | V(\mathbf{N}) \leq p\}$. Hence the system is nonvulnerable relative to the sets $\mathcal{U}, \mathcal{S}(0),$ and $\mathcal{Z}$ during the time interval $[0, t_1]$, where t_1 is any positive number.

3.2. An Algebraic Method

The major difficulty in using Theorem 4 is due to the fact that global optimization algorithms are unreliable at the present time (Dixon and Szego, 1975). For this reason we shall now discuss a more conservative but reliable method. This is a generalization of the algebraic method which was developed for Lotka–Volterra models by Goh (1976).

Suppose there exists a constant matrix G such that the functions $F_1(\mathbf{N}), F_2(\mathbf{N}), \ldots, F_m(\mathbf{N})$ of Eq. (20) satisfy the inequalities in Eqs. (10) and (11). The diagonal elements of G are negative and its off-diagonal elements are nonnegative. It follows that if all the principal minors of $-G$ are positive then there exists a positive diagonal matrix D such that $DG + G^T D$ is negative definite. The elements of D are normalized by setting its smallest element equal to one. Let d^* be the largest element of D and let λ_1 be the smallest eigenvalue of $-(DG + G^T D)/2$. For simplicity we shall consider the special case in which the disturbance functions satisfy the constraints

$$-\xi \leq u_i(t) \leq \xi, \quad i = 1, 2, \ldots, m,$$

instead of Eq. (21).

Let r^* be the minimum distance from $\mathbf{N}^*$ to any of the axis hyperplanes. This means r^* is the smallest number in the set $\{N_1^*, N_2^*, \ldots, N_m^*\}$. For fixed values of λ_1, m, ξ, and d^*, let r_0 be the infimum of the values of r which satisfies the inequality

$$\lambda_1 > [d^* \xi m^{1/2}]/r. \tag{23}$$

Consider the hypersphere

$$(\mathbf{N} - \mathbf{N}^*)^T (\mathbf{N} - \mathbf{N}^*) = r^2, \tag{24}$$

where r satisfies the inequality $r_0 < r < r^*$. Let $\mathbf{N}^*$ be a point in $\mathscr{S}(0)$. Let p be the maximum value of $V(\mathbf{N})$ for all $\mathbf{N}$ on the hypersphere in Eq. (24). Let V_S and V_Z be positive numbers which satisfy the inequality $V_S \leq p < V_Z$. Let $\mathscr{S}(0) = \{\mathbf{N} | V(\mathbf{N}) \leq V_S\}$ and $\mathscr{Z} = \{\mathbf{N} | V(\mathbf{N}) \geq V_Z\}$.

Theorem 5. The ecosystem model in Eq. (20) is nonvulnerable relative to the sets $\mathscr{U}$, $\mathscr{S}(0)$, and $\mathscr{Z}$ during the time interval $[0, t_1]$, where t_1 is any positive number, if there exists a positive diagonal matrix D such that the inequalities in Eqs. (10), (11), and (23) are satisfied.

Proof. By assumption Eq. (20) satisfies the inequalities in Eqs. (10) and (11). It follows that along solutions of Eq. (20), we have

$$\dot{V}(\mathbf{N}, \mathbf{u}) = (\mathbf{N} - \mathbf{N}^*)^T DF(\mathbf{N}) + (\mathbf{N} - \mathbf{N}^*)^T D\mathbf{u}$$
$$\leq \tfrac{1}{2}\mathbf{y}^T (DG + G^T D)\mathbf{y} + (\mathbf{N} - \mathbf{N}^*)^T D\mathbf{u}, \quad (25)$$

where $\mathbf{y} = (|N_i - N_i^*|)$.

The product $(\mathbf{N} - \mathbf{N}^*)^T D\mathbf{u}$ is an inner product which is induced by the positive diagonal matrix D. Let $\mathbf{N}$ be a point on the hypersphere in Eq. (24). By the Cauchy–Schwarz inequality, we have

$$(\mathbf{N} - \mathbf{N}^*)^T D\mathbf{u} \leq [(\mathbf{N} - \mathbf{N}^*)^T D(\mathbf{N} - \mathbf{N}^*)]^{1/2} [\mathbf{u}^T D\mathbf{u}]^{1/2}$$
$$\leq d^* r (\mathbf{u}^T \mathbf{u})^{1/2} \leq d^* r \xi m^{1/2}. \quad (26)$$

Equations (25) and (26) imply that

$$\dot{V}(\mathbf{N}, \mathbf{u}) \leq -\lambda_1 r^2 + rd^* \xi m^{1/2}.$$

It follows that if Eq. (23) is satisfied, $\dot{V}(\mathbf{N}, \mathbf{u})$ is negative on the hypersphere in Eq. (24). Outside the hypersphere the quadratic term involving $(\mathbf{N} - \mathbf{N}^*)$ in Eq. (25) dominates the linear term in $(\mathbf{N} - \mathbf{N}^*)$. Hence, on and outside this hypersphere $\dot{V}(\mathbf{N}, \mathbf{u})$ is negative. Therefore $\dot{V}(\mathbf{N}, \mathbf{u})$ is negative on the hypersurface $\{\mathbf{N} | V(\mathbf{N}) = p\}$. This implies that a solution of Eq. (20) which begins in the set $\mathscr{S}(0)$ cannot be driven by disturbances to the set $\mathscr{Z}$. Hence the ecosystem model in Eq. (20) is nonvulnerable relative to the sets $\mathscr{U}$, $\mathscr{S}(0)$, $\mathscr{Z}$, and $[0, t_1]$, where t_1 is any positive number.

3.3. A Numerical Example

We shall demonstrate how Theorems 4 and 5 may be used to establish nonvulnerability in a prey–predator model.

Let the densities of a prey and its predator be N_1 and N_2, respectively. Suppose that the prey–predator system which is subjected to continual disturbances can be adequately described by the model,

$$\dot{N}_1 = N_1 [22 - 0.1 N_1 - 2.1 N_2] + u_1(t) N_1,$$
$$\dot{N}_2 = N_2 [-5 + 0.7 N_1 - 0.2 N_2] + u_2(t) N_2, \quad (27)$$

where $u_1(t)$ and $u_2(t)$ are piecewise continuous disturbances. We assume that there are adequate *a priori* estimates of the bounds on $u_1(t)$ and $u_2(t)$, namely,

$$-0.11 \leq u_i(t) \leq 0.11, \quad i = 1, 2.$$

If the system is not continually disturbed, we have $u_1(t) = 0 = u_2(t)$. In this case, Eq. (27) has a feasible equilibrium at $(N_1^*, N_2^*) = (10, 10)$. By trial and error we choose $d_1 = 1.0$ and $d_2 = 2.94$, the $V(\mathbf{N})$ function in Eq. (5) is a suitable Liapunov-like function. The $V(\mathbf{N})$ level surface which passes through the point $(8.37, 10)$ is $V(N) = 0.1493$. The global maximum of

$\dot{V}(\mathbf{N}, \mathbf{u})$ for all admissible disturbance functions, $u_1(t)$ and $u_2(t)$, and for all $\mathbf{N} \in \{\mathbf{N} | V(\mathbf{N}) \geq 0.1493, N_1 > 0, N_2 > 0\}$ is equal to -0.0035 when $N_1 = 9.511$, $N_2 = 10.428$, $u_1 = -0.11$, and $u_2 = 0.11$.

Let p be any positive number larger than 0.1493 and let V_S and V_Z be positive numbers such that $V_S \leq p < V_Z$. By definition $\mathscr{S}(0) = \{\mathbf{N} | V(\mathbf{N}) = V_S\}$ and $\mathscr{Z} = \{\mathbf{N} | V(\mathbf{N}) \geq V_Z\}$. Hence by Theorem 4 the model in Eq. (27) is nonvulnerable relative to the admissible set of disturbances, $\mathscr{S}(0)$ and $\mathscr{Z}$, during the time interval $[0, t_1]$ where t_1 is any positive number.

We can also use Theorem 5 to establish that Eq. (27) is nonvulnerable (Goh, 1976). By trial and error we choose $d_1 = 1$ and $d_2 = 3$; we get $\lambda_1 = 0.1$. This implies that we have to choose p which is greater than or equal to 5.79. Clearly in this example, Theorem 4 is more effective than Theorem 5. The main advantage of Theorem 5 compared with Theorem 4 is that it is applicable to a system which contains many species. The primitive state of global optimization techniques means that in practice we cannot be confident of computing the global maximum of $\dot{V}(\mathbf{N}, \mathbf{u})$ in a model with many species. The algebraic criterion in Theorem 5 can be verified readily in a model with many species.

4. SECTOR STABILITY

4.1. Introduction

Paine (1966) showed that the removal of the predator starfish *Pisaster ochraceus* from an area resulted in the reduction of a 15-species community to an 8-species community. Suppose a model of the dynamics of the 15-species community is given. It will be interesting if it can be established that the composition of the 8-species community is not sensitive to the population levels at the time when the starfish population was removed.

Simberloff (1976) carried out a large number of censuses of the communities or arboreal arthropods on 8 islands during a 3-year period. He found that members of a group of 22 species out of a total of 254 species were present at all censuses or were absent only once. This suggests that the persistence of some of the other species on an island is maintained by continual immigrations.

The stability analyses developed in the previous sections cannot be used to study effectively these ecological processes. They can only be used to examine the stability of a feasible equilibrium.

Goh and Jennings (1977) showed that the set of models with a feasible equilibrium is only a small fraction of a set of randomly assembled Lotka–Volterra models. This fraction is approximately equal to 2^{-m} where m is the number of species. This suggests that feasibility is a very severe constraint on a complex ecosystem model. On the other hand, a given Lotka–Volterra model of a large number of interacting populations, will usually have a large number of partially feasible equilibria. By definition, an equilibrium is partially feasible if some of its components are positive while the others are equal to zero.

It follows that in a complex ecosystem model it is of interest to know whether or not all of the trajectories of the model which initiate in the positive orthant will converge to a single, partially feasible equilibrium. This type of convergence property of a partially feasible equilibrium of a Lotka–Volterra model has been studied by Case and Casten (1979). In this section their results will be generalized so that they apply to a general class of nonlinear ecosystem models. This extension is important because in the study of a partially feasible equilibrium the main interest is on large perturbations of the initial state. A Lotka–Volterra model of a community cannot be expected to be valid in a large region of the state space.

4.2. A Class of Nonlinear Models

Suppose Eq. (1) has a partially feasible equilibrium at

$$\mathbf{N}^* = (N_1^*, N_2^*, \ldots, N_s^*, 0, 0, \ldots, 0), \tag{28}$$

where $N_i^* > 0$ for $i = 1, 2, \ldots, s$. Let $d_1, d_2, \ldots, d_m$ be positive constants. Consider the function,

$$V(\mathbf{N}) = \sum_{i=1}^{s} d_i [N_i - N_i^* - N_i^* \ln(N_i/N_i^*)] + \sum_{i=s+1}^{m} d_i |N_i|. \tag{29}$$

The level sets of this function are closed hypersurfaces. For $i = 1, 2, \ldots, s$ the function $V(\mathbf{N}) \to \infty$ as $N_i \to 0+$ or as $N_i \to \infty$ along any trajectory which remains in the positive orthant. For $i = s+1, s+2, \ldots, m$, the function $V(\mathbf{N}) \to \infty$ as $N_i \to \infty$.

Let the region

$$\mathscr{S} = \{\mathbf{N} | N_i > 0, \ i = 1, 2, \ldots, s, \ N_i \geq 0, i = s+1, s+2, \ldots, m\}.$$

By definition, the partially feasible equilibrium $\mathbf{N}^*$ in Eq. (28) is globally sector-stable if all of the trajectories of Eq. (1) which initiate in $\mathscr{S}$ remain in $\mathscr{S}$ for all finite time and tend to $\mathbf{N}^*$ as $t \to \infty$.

Theorem 6. The partially feasible equilibrium $\mathbf{N}^*$ in Eq. (28) is globally sector-stable if there exist positive constants $d_1, d_2, \ldots, d_m$ such that the function

$$W(\mathbf{N}) = \sum_{i=1}^{m} d_i(N_i - N_i^*)F_i(\mathbf{N}) \tag{30}$$

is negative definite in the set $\mathscr{S}$.

Proof. In the set $\mathscr{S}$ the total derivative of the function $V(\mathbf{N})$ in Eq. (29) along the trajectories of Eq. (1) is

$$\dot{V}(\mathbf{N}) = \sum_{i=1}^{s} d_i(N_i - N_i^*)F_i(\mathbf{N}) + \sum_{i=s+1}^{m} d_i N_i F_i(\mathbf{N}).$$

But $N_i^* = 0$ for $i = s+1, s+2, \ldots, m$. Therefore $\dot{V}(\mathbf{N})$ is equal to the function $W(\mathbf{N})$ in Eq. (30). It is interesting that the functions in Eqs. (6) and (30) are identical.

If the partial derivatives of the functions $\{F_i(\mathbf{N})\}$ are continuous in the nonnegative orthant Eq. (1) has locally a unique solution from every bounded point in the nonnegative orthant. This implies that a trajectory of Eq. (1) which initiates in the positive orthant does not intersect a coordinate axis hyperplane at a bounded point in finite time. By assumption $\dot{V}(\mathbf{N})$ is negative in the positive orthant. It follows that all of the trajectories of Eq. (1) which initiate in the positive orthant remain in it for all finite time and converge to $\mathbf{N}^*$ as $t \to \infty$. Similar conditions apply to all trajectories in $\mathscr{S}$. Therefore $\mathbf{N}^*$ is globally sector-stable.

Corollary 6.1. Suppose there exists a constant matrix G such that Eqs. (10) and (11) are satisfied in the set $\mathscr{S}$. If the leading principal minors of $-G$ are positive and

$$F_i(\mathbf{N}^*) \leq 0 \quad \text{for} \quad i = s+1, \, s+2, \ldots, m. \tag{31}$$

then the equilibrium $\mathbf{N}^*$ is globally sector-stable.

Proof. Apply Taylor's Theorem to Eq. (30)

$$\dot{V}(\mathbf{N}) = \sum_{i=1}^{m} \sum_{j=1}^{m} d_i(N_i - N_i^*)(\partial F_i/\partial N_j)(N_j - N_j^*) + \sum_{i=s+1}^{m} d_i N_i F_i(\mathbf{N}^*).$$

Equations (10) and (11) imply

$$\dot{V}(\mathbf{N}) \leq \tfrac{1}{2}\mathbf{y}^T(DG + G^T D)\mathbf{y} + \sum_{i=s+1}^{m} d_i N_i F_i(\mathbf{N}^*), \tag{32}$$

where $\mathbf{y} = (|N_i - N_i^*|)$ and $D = \mathrm{diag}(d_1, d_2, \ldots, d_m)$.

The arguments in the proof of Theorem 2 and Eq. (31) imply that $\dot{V}(\mathbf{N})$ (Eq. 32) is negative in the set $\mathscr{S}$. Hence $\mathbf{N}^*$ is globally sector-stable.

4.3. A Numerical Example

A model of three competing species is

$$\dot{N}_1 = N_1[11.7 - 4N_1 - 0.2N_2 - 0.1N_3],$$
$$\dot{N}_2 = N_2[1.2 - 0.8N_1 - N_2 - 0.2N_3], \qquad (33)$$
$$\dot{N}_3 = N_3[3 - 2N_1 - N_2 - 2N_3].$$

There are $2^3 = 8$ equilibria. They are at the points $(3, -1, -1)$, $(0, 1, 1)$, $(573/192, -57/48, 0)$, $(0, 6/5, 0)$, $(231/78, 0, -57/39)$, $(0, 0, 3/2)$, $(117/40, 0, 0)$, $(0, 0, 0)$. Clearly there are four partially feasible equilibria.

At the equilibrium $N^* = (117/40, 0, 0)$,

$$F_1(\mathbf{N}^*) = 0,$$
$$F_2(\mathbf{N}^*) = 1.2 - 0.8(117/40) = -1.14 < 0,$$
$$F_3(\mathbf{N}^*) = 3 - 2(117/40) = -2.85 < 0.$$

Using Eqs. (10) and (11) a matrix G is computed. All the leading principal minors of

$$-G = \begin{bmatrix} 4 & -0.2 & -0.1 \\ -0.8 & 1 & -0.2 \\ -2 & -1 & 2 \end{bmatrix}$$

are positive. It follows that there exists a positive diagonal matrix D such that $DG + G^T D$ is negative definite. Hence model (33) satisfies Corollary 6.1. It follows that all the trajectories of (33) which begin in the positive orthant will converge to the point $(117/40, 0, 0)$.

If $N_3 = 0$ the reduced system of equations in Eq. (33) satisfies Corollary 6.1. Hence all the trajectories of Eq. (33) which begin in the positive orthant of the (N_1, N_2)-space will remain in the (N_1, N_2)-space and converge to the point $(117/40, 0, 0)$. Similarly all the trajectories of Eq. (33) which begin in the positive orthant of the (N_1, N_3)-space or the N_1-space will converge to $(117/40, 0, 0)$.

If the species is absent (i.e., $N_1 = 0$), the N_2 and N_3 species have a feasible equilibrium at $(1, 1)$. This equilibrium is globally stable relative to the positive orthant of the (N_2, N_3)-space. This means that the N_2 and N_3 species can coexist in isolation. But if members of the N_1 species invade an area which is occupied by the N_2 and N_3 species, the N_1 species will drive the N_2 and N_3 species to extinction. Clearly this process models the replacement of some species by other species in a succession.

The three competing species can persist as a community if there are small and continual immigrations of N_2 and N_3 species. If the immigration rates of the N_2 and N_3 species are large the N_1 species may be driven to extinction (e.g., Christiansen and Fenchel, 1977).

The importance of Theorem 6 and Corollary 6.1 is that they can be used to carry out a complete analysis of the dynamical behavior in the nonnegative orthant of some ecosystem models. The theorems in Section 3 are not suitable for this purpose when a model has no feasible equilibrium.

5. CONCLUSION

This paper describes and establishes mathematical methods for studying the stability of ecosystem models when they are subjected to (a) large perturbations of the initial state, (b) certain finite changes in system parameters, (c) continual disturbances of the system dynamics, and (d) invasion and extinction of nonendemic species. The main conclusion is that a complex ecosystem model is robust relative to all types of perturbations if (a) it is a collection of subsystems each of which is self-regulating, and (b) from the total system point of view the interactions of the subsystems are weaker than the self-regulating interactions of the subsystems. It should be emphasized that we have not established that complex ecosystems in the real world have this type of system property.

REFERENCES

Arrow, K. J., Block, H. D., and Hurwicz, L. (1959). On the stability of competitive equilibrium. II. *Econometrica* 27, 82–109.

Case, T. D., and Casten, R. (1979). Global stability and multiple domains of attraction in ecological systems. *Am. Nat.* (in press).

Christiansen, F. B., and Fenchel, T. M. (1977). "Theories of Populations in Biological Communities." Springer-Verlag, Berlin and New York.

Clark, C. W. (1976). "Mathematical Bioeconomics: The Optimal Management of Renewable Resources." Wiley, New York.

Dixon, L. C. W., and Szego, G. P. (1975). "Towards Global Optimization." North-Holland Publ., Amsterdam.

Goh, B. S. (1976). Nonvulnerability of ecosystems in unpredictable environments. *Theor. Popul. Biol.* 10, 83–95.

Goh, B. S. (1977a). Global stability in many species systems. *Am. Nat.* 111, 135–143.

Goh, B. S. (1977b). Stability in a stock-recruitment model of an exploited fishery. *Math. Biosci.* 33, 359–372.

Goh, B. S., and Jennings, L. S. (1977). Feasibility and stability in randomly assembled Lotka–Volterra models. *Ecol. Modell.* 3, 63–71.

Holling, C. S. (1973). Resilience and stability of ecological systems. *Annu. Rev. Ecol. Syst.* 7, 1–23.

Jeffries, C. D. (1976). Stability of predation ecosystem models. *Ecology* 57, 1321–1325.

Johnson, C. R. (1974). Sufficient conditions for D-stability. *J. Econ. Theory* **9**, 53–62.
LaSalle, J. P., and Lefscheftz, S. (1961). "Stability by Liapunov's Direct Method with Applications." Academic Press, New York.
Liapunov, A. M. (1966). "Stability of Motion." Academic Press, New York.
Ludwig, D. (1975). Persistence of dynamical systems under random perturbations. *SIAM Rev.* **17**, 605–640.
May, R. M. (1974). "Stability and Complexity in Model Ecosystems," 2nd ed. Princeton Univ. Press, Princeton, New Jersey.
Odum, E. P. (1971). "Fundamentals of Ecology," 3rd ed., Saunders, Philadelphia, Pennsylvania.
Paine, R. T. (1966). Food web complexity and species diversity. *Am. Nat.* **100**, 65–75.
Rosenzweig, M. L. (1971). Paradox of enrichment: Destabilization of exploitation ecosystems in ecological time. *Science* **171**, 385–387.
Šiljak, D. D. (1975). Connective stability of competitive equilibrium. *Automatica* **11**, 389–400.
Simberloff, D. (1976). Experimental zoogeography of islands: Effects of island size. *Ecology* **57**, 629–648.
Takayama, A. (1974). "Mathematical Economics." Dryden Press, Hinsdale, Illinois.
Willems, J. L. (1970). "Stability Theory for Dynamical Systems." Nelson, London.
Willems, J. L. (1971). Optimum Liapunov functions and stability regions for multimachine power systems. *Proc. Inst. Elect. Eng.* **118**, 1631–1632.

Chapter **20**

STABILITY OF HOLISTIC ECOSYSTEM MODELS

Clark Jeffries

1. Introduction	489
1.1 Liapunov Functions	490
1.2 Stability of a Driven System	492
2. Holistic Ecosystem Modeling	494
2.1 Energy Accounting	494
2.2 Stability Theorem	495
2.3 Generalizations	498
3. Discussion	499
3.1 Acceptability of Conditions	499
3.2 Detritus Processing	500
3.3 Intercompartmental Competition	501
4. Conclusion	501
Appendix	502
References	503

1. INTRODUCTION

The basic model of this chapter, Eq. (7), amounts really to an energy accounting procedure. Each compartment of the model is characterized by a positive number, the energy level (in some fixed energy units such as calories) of the compartment. The model allows for two basic types of energy transfer between compartments: predation and detritus donation. Predation transfer depends on the levels of prey and predator compartments and time. Detritus donation terms depend upon the donor compartment, other compartments which have predation links with the donor compartment, and time. To elaborate, detritus donation via excretion would presumably depend on the donor compartment level and time. However, detritus donation as "carrion" (generalized to include by-products of herbivorous "predation") would presumably depend on "prey" compartment and "predator" compartment levels and time.

A "predation community" in this chapter means a set of compartments in a model interconnected directly or indirectly by "predation links," that is, energy transfer terms arising from predation interactions. A holistic ecosystem model should allow for energy transfer from photosynthesis-based predation communities to detritus-based predation communities (detritus compartments and decomposer compartments). That is, the system equations for the detritus-based community should depend on or be "driven" by the state vector of the photosynthesis-based community. The new result in this chapter is the description of such driving arrangements of two or more predation communities which comprise a hierarchy of energy states, a holistic ecosystem model.

1.1. Liapunov Functions

Mathematically this chapter explores how one stable dynamical system may drive another, and a straightforward extension is made to hierarchies of such systems. Stability of the whole system is derived using the Liapunov function technique.

Heuristically a Liapunov function (Willems, 1970, Chapter 2) is a natural, consistent method for quantifying how far the current state of an observed ecosystem is from a typical or predictable ecosystem development pattern. By comparing real ecosystems with standard ecosystem development patterns, field ecologists "know" about Liapunov functions, albeit perhaps not in mathematical language.

Here is a familiar Lotka–Volterra type model of interactions of a "prey" x_1 and "predator" x_2.

$$dx_1/dt = x_1(-0.1x_1 - 2x_2 + 2.1), \tag{1}$$
$$dx_2/dt = x_2(2x_1 - 0.1x_2 - 1.9). \tag{2}$$

The constant trajectory $(x_1(t), x_2(t)) = (1, 1)$ is a solution trajectory of this system and will be regarded as a candidate for attractor trajectory. A typical nonconstant trajectory for the system is shown in Fig. 1.

What the typical nonconstant trajectory looks like may be determined by phase plane techniques or brute force computer simulation. However, for the moment the reader is asked to accept my claim.

20. Stability of Holistic Ecosystem Models 491

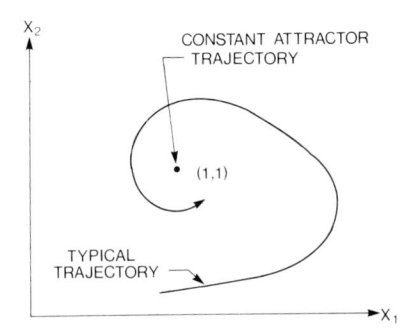

Figure 1. Two trajectories for the system given by Eqs. (1) and (2).

An analysis of Fig. 1 reveals that at certain instants the typical trajectory is actually heading away from the constant trajectory (1, 1). That is, if we draw concentric circles around (1, 1), then the typical trajectory will occasionally cross a circle toward the "outside." Thus the "populations" could actually be asymptotically approaching "equilibrium" in the systems sense but moving away from "equilibrium" in the geometric sense. We do not need Euclidean distance but Liapunov "distance" to measure how close (x_1, x_2) is to (1, 1).

From the calculus (Fleming, 1977) we know that concentric circles around (1, 1) are the *level sets* of the function $f(x_1, x_2) = (x_1 - 1)^2 + (x_2 - 1)^2$. For example, $[(x_1, x_2) | f(x_1, x_2) = 0.25]$ is a level set, a circle of radius 0.5 and center (1, 1). A level set is associated with each positive number. The level set of 0 is the point (1, 1).

Now it is possible to draw certain simple closed curves (*deformed* circles) concentrically around (1, 1), which *are* always crossed inwardly by typical trajectories. Some such curves are shown in Fig. 2.

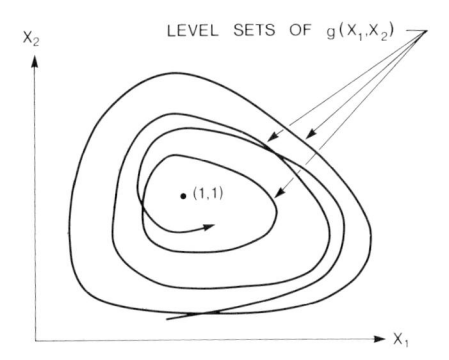

Figure 2. Curves arranged around (1, 1) which are always crossed "inwardly."

The curves in Fig. 2 are the level sets of the function $g(x_1, x_2) = x_1 - 1 - \ln x_1 + x_2 - 1 - \ln x_2$. To actually show that the level sets of g look something like the curves in Fig. 2 takes some analytic geometry and calculus. To actually show that a nonconstant trajectory of the system given by Eqs. (1) and (2) always crosses the special curves "inwardly" takes the calculus of composition functions, as follows. A typical trajectory may be written $\mathbf{x}(t) = (x_1(t), x_2(t))$, the two variables being functions of time. [The existence and, with specified initial values, uniqueness of solution trajectories follow from well-known theorems of dynamical systems theory (Rosen, 1970). Actually, only existence is needed in Liapunov theory generally. Moreover, existence is automatic with difference equation models used in simulations.] Thus along a trajectory g has the values $g[x_1(t), x_2(t)]$, a composition function. The derivative of g along a trajectory is

$$\dot{g} = dg/dt = -0.1[x_1(t) - 1]^2 - 0.1[x_2(t) - 1]^2. \tag{3}$$

Thus, $\dot{g}$ is always negative, except $\dot{g}(1, 1) = 0$. Therefore, a typical trajectory must either asymptotically approach $(1, 1)$ or must asymptotically approach some special level set. However, on any level set other than $(1, 1)$ and on a small neighborhood of any such level set, $\dot{g}$ would be negative and bounded away from zero. Asymptotic approach would require $\lim_{t \to +\infty} \dot{g}(t) = 0$. This contradiction implies that trajectories lying in the positive orthant (where g is defined) must asymptotically approach $(1, 1)$. In short, g is a Liapunov function for the system given by Eqs. (1) and (2) relative to the constant trajectory $(1, 1)$ (Willems, 1970).

The same ideas will be used in Section 2 to treat models with general time-dependent coefficients and time-dependent attractor trajectories.

1.2. Stability of a Driven System

A sequence of previous papers (Jeffries, 1974, 1975, 1976) has led to incorporation of general time dependence in the classical Lotka–Volterra model. This chapter gives a set of conditions for the stable "driving" of one such Lotka–Volterra predation community by another or generally a hierarchy of predation communities which comprise a holistic ecosystem model.

Here is an explicit illustration of the technique for proving stability of driving arrangements between dynamical systems (predation communities). Consider

$$dx_1/dt = -x_1, \tag{4}$$
$$dx_2/dt = -x_2 + x_1. \tag{5}$$

Here the "upstream" x_1 system obviously has as attractor trajectory $x_1(t)$

20. Stability of Holistic Ecosystem Models 493

$\equiv 0$. If the upstream system is "on" its attractor trajectory, then the downstream x_2 system is analogously attracted to $x_2(t) \equiv 0$. The stability question is: If the upstream system is on some nonconstant trajectory (say, following a perturbation), then will the driving arrangement still allow asymptotic approach of $x_2(t) \equiv 0$ (no longer a solution trajectory)? The answer is yes. The system given by Eqs. (4) and (5) is so simple that an explicit general solution exists, namely,

$$x_2(t) = x_2(0)e^{-t} + x_1(0)t e^{-t}, \qquad (6)$$

where $t_0 = 0$ is the initial time. [The general analysis of linear, time-dependent systems driven by other such linear systems may be studied in Willems (1970, pp. 105–106).]

One can see from Eq. (6) that $x_2(t) \equiv 0$ is asymptotically approached by solutions of Eq. (5). Now let us derive stability using a Liapunov-type approach for the sake of illustrating that approach.

Let $x_1 = x_1(0)e^{-t}$ be a fixed but arbitrary solution trajectory for Eq. (4). The function $\Lambda(x_2) = \tfrac{1}{2}x_2^2$ (a Liapunov function for Eq. (5) if $x_1(t) \equiv 0$) has derivative $\dot{\Lambda} = -x_2^2 + x_1 x_2$. Thus $\dot{\Lambda}$ is no longer negative at all points along all solution trajectories $x_2(t)$. However, choose a small positive

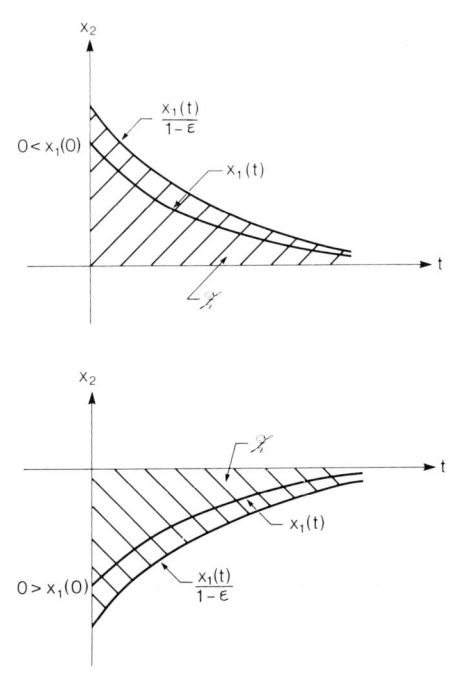

Figure 3. Two possible shapes for the region $\mathscr{X}$ of unknown $x_2(t)$ trajectory behavior.

constant ε ($0 < \varepsilon < 1$) and consider the zone $\mathscr{Z}$ defined by $\mathscr{Z} = [x_2, t | -x_2^2 + x_1(t)x_2 \geq -\varepsilon x_2^2]$. A little inequality chasing will verify that $-x_2^2 + x_1(t)x_2 \geq -\varepsilon x_2^2$ provided x_2 is between $x_1(t)/(1-\varepsilon)$ and 0. Figure 3 shows the two possible shapes of $\mathscr{Z}$. Which shape is used depends on the sign of $x_1(0)$.

Outside $\mathscr{Z}$ along any solution $x_2(t)$ we have $\dot{\Lambda} < -\varepsilon x_2^2$. By virtue of the existence and qualities of Λ any solution must asymptotically approach $x_2(t) \equiv 0$ while outside $\mathscr{Z}$. Inside $\mathscr{Z}$, Λ reveals nothing about solution trajectories. But the shape of $\mathscr{Z}$, the asymptotic squeezing of the boundary of $\mathscr{Z}$ with time, implies that while inside $\mathscr{Z}$ a solution trajectory must still be constrained ever more closely to 0.

The stability proof of the holistic model in this chapter will use an extension of the above technique, that is, a collapsing zone within which little is known of trajectory behavior and outside which lurks a nice Liapunov-type function.

2. HOLISTIC ECOSYSTEM MODELING

2.1. Energy Accounting

The holistic ecosystem model developed in this chapter uses the following notation.

(1) $[x_i]$ $i = 1, 2, \ldots, n$: levels of biomass energy (in calories, say) in n compartments corresponding to ecological groups of living organisms or detritus types.

(2) $[\hat{x}_i(t)]$ $i = 1, 2, \ldots, n$: predicted, typical, or frequently observed patterns of levels of biomass energy in the system compartments. Each $\hat{x}_i(t)$ is required to be positive and bounded as $0 < b_1 < \hat{x}_i(t) < b_2$.

(3) $[r_{ij}(t)x_i(t)x_j(t)]$ $i \neq j$: rate of predation interactions between x_i and x_j (so $r_{ij} = r_{ji} \geq 0$).

(4) $[e_{ij}(t)]$ $i \neq j$: rate of gain ($e_{ij} > 0$) or loss ($e_{ij} < 0$) to x_i from x_j per interaction. Thus if x_j preys upon x_i, the efficiency of predation transfer is $e_{ji}(t)/[-e_{ij}(t)]$.

(5) $[-a_i(t)x_i(t)x_i(t)]$: rate of energy loss from x_i due to intracompartmental predation, territorial conflicts, or other factors in animal compartments; or autolysis or other factors in detritus compartments.

(6) $[p_i(t)x_i(t) - a_i(t)x_i(t)x_i(t)]$: potential photosynthetic energy gain minus intracompartmental competition, that is, net photosynthetic gain for green plant compartments (possibly zero or negative for some t).

(7) $[-h_i(t)x_i(t)]$: energy losses to the environment such as heat lost in respiration, methane released, or peat deposited.

(8) $[d_i(\mathbf{x}, t)]$: rate of detritus input for a detritus compartment x_i.
(9) $[-l_i(t)x_i]$: rate of detritus loss or output from compartment x_i to other model compartments due to excretion, physical breakdown, or other nonpredation mechanisms.

All this leads to the model equation:

$$\frac{dx_i}{dt} = \sum_{j=1}^{n} x_i e_{ij}(t) r_{ij}(t) x_j - a_i(t) x_i^2 + p_i(t) x_i - h_i(t) x_i + d_i(\mathbf{x}, t) - l_i(t) x_i. \qquad (7)$$

There are two assumptions in this model: First, it makes sense to characterize an ecosystem by the energy content of its compartments (energy regarded as ecosystem currency); second, the parameter functions r_{ij}, e_{ij}, a_i, p_i, h_i, l_i all depend only on time, not the state vector $\mathbf{x}$. As indicated in Jeffries (1976), it is actually possible to retain local stability while allowing some general dependence on $\mathbf{x}$, depending on quantitative factors in a particular model. In Section 2.3 some allowable qualitative dependence of a_i upon x_i is developed, but for the present, dependence on t only is assumed.

The food web of an ecosystem model in the form of Eq. (7) may be described as follows. The food web is a signed directed graph with n numbered nodes corresponding to compartments. If $e_{ij} r_{ij} \neq 0$, or if detritus flows from x_j to x_i, then a directed edge is drawn from the jth node to the ith node and signed accordingly.

For purposes of studying stability below it is necessary to partition the food web of a model into predation communities, that is, maximal subsets of compartments interconnected by predation interactions. This leads to an idealized form for food webs, namely, tidy photosynthesis-based predation communities which donate energy to "downstream" detritus-based predation communities. In time these detritus-based predation communities donate energy to other detritus-based predation communities "further downstream" and so on. The detritus community structure is expected to be exceedingly intricate in detailed holistic ecosystem models.

At any rate, Fig. 4 shows a portion of a hypothetical boreal forest ecosystem model with some well-known energy pathways. Three predation communities are shown.

2.2. Stability Theorem

The main result of this chapter is the proof that certain conditions on the parameter functions in Eq. (7) are sufficient for the global (positive

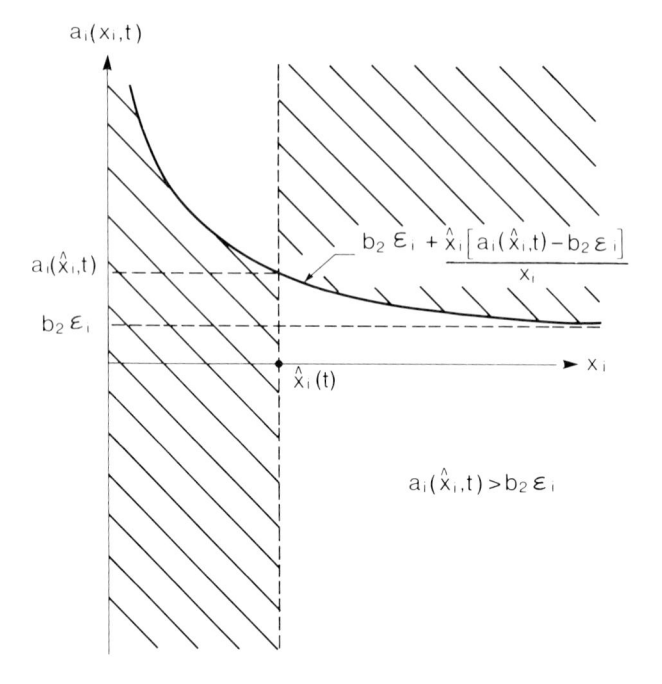

Figure 5. An allowable self-regulation function may lie in the cross-hatched region.

2.3. Generalizations

There are many approaches to relaxation of conditions (a)–(f) while retaining stability. Only one will be mentioned here.

Each self-regulation term a_i may actually depend upon x_i as well as t as follows. For the stability proof we really only need $[x_i \neq \hat{x}_i(t)]$

$$(x_i - \hat{x}_i)[-a_i(x_i, t)x_i + a_i(\hat{x}_i, t)\hat{x}_i] < -\varepsilon_i b_2(x_i - \hat{x}_i)^2$$

for some positive ε_i. This is equivalent to

$$a_i(x_i, t) > \varepsilon_i b_2 + \hat{x}_i[a_i(\hat{x}_i, t) - \varepsilon_i b_2]/x_i \quad \text{if} \quad x_i > \hat{x}_i,$$
$$a_i(x_i, t) < \varepsilon_i b_2 + \hat{x}_i[a_i(\hat{x}_i, t) - \varepsilon_i b_2]/x_i \quad \text{if} \quad x_i < \hat{x}_i.$$

Thus $a_i(x_i, t)$ is allowable for stability if $\varepsilon_i > 0$ exists such that for various times the graph of a_i lies in the cross-hatched regions of Figs. 5 or 6. I conjecture that almost any reasonable self-regulation function of x_i, t is allowable. In particular $-a_i(\hat{x}_i, t) > 0$ for some or all t is allowable, as may be seen in Fig. 6. The only inescapable condition is $-a_i(x_i, t) < 0$ for large x_i, that is, self-regulation in a compartment must occur at very high "population" levels.

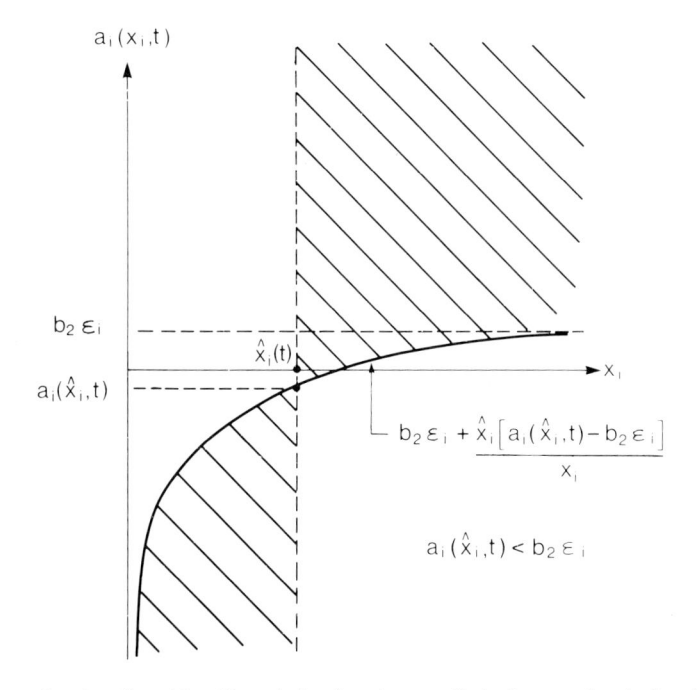

Figure 6. An allowable self-regulation function may lie in the cross-hatched region.

To indicate the direction of other generalizations the reader is invited to show that $h_i = h_i(x_i, t)$ [respectively $l_i = l_i(x_i, t)$] is allowable if $h_i(x_i, t) - h_i(\hat{x}_i, t)$ [respectively $l_i(x_i, t) - l_i(\hat{x}_i, t)$] always vanishes or has the same sign as $x_i - \hat{x}_i$. Still other directions may be surmised from the appendix of a paper of Jeffries (1976).

3. DISCUSSION

3.1. Acceptability of Conditions

Condition (a) would at first seem objectionable because some creatures derive energy from several very different sources. The arctic fox, for example, is said to consume berries, birds, small rodents, carrion, and polar bear dung. However, condition (a) would stand unless a significant energy source for some compartment in the predation community to which the fox belongs were energy which had previously been incorporated in a fox. [Condition (a) would not be violated just by intraspecific predation, a form of self-regulation.] For example, one might imagine gulls consuming fox carrion and foxes consuming gull eggs, gull juveniles, or gull carrion. However, the amount of energy actually recycled by such loops is probably small, so condition (a) might "almost" hold.

Alas, condition (b) precludes migration and certain other phenomena such as seasonal appearance (for example, emergence of adult dragonflies) and seasonal predation (for example, failure of large boreal forest carnivores to find small rodents in winter due to snow cover). However, $e_{ij}(t)$ will probably not vary much, and the complete arbitrariness of $r_{ij}(t)$ (except > 0 if not $\equiv 0$) may be often a valuable freedom in model building.

In the spirit of qualitative stability (Jeffries, 1974; Jeffries et al., 1977) condition (c) could be relaxed with some $a_i \equiv 0$, depending on the location of v_i in the food web. However, condition (c) itself requires a minimal degree of self-regulation in contrast to suggestions of others that stability should be associated with self-regulation dominating other interactions (Goh, Chapter 19, this volume; Harte, Chapter 18, this volume).

In practice $-e_{ij}(t)/e_{ji}(t)$ (predation efficiency) should not vary much. Hence, condition (d) should be associated with fairly constant predator to prey biomass ratios. In a model with all predation efficiencies at, say, 10%, condition (e) would require a neat hierarchy of trophic levels in each predation community, violated only by energy pathways of low importance.

Condition (f) precludes system information feedback via detritus donation. The continuity condition would only be violated by somewhat pathological models.

3.2. Detritus Processing

A model with only detritus compartments need only fulfill conditions (a), (c), and (f) to enjoy stability of $\hat{x}(t)$. The large 39 compartment model of Boling et al. (1975) of detritus processing in a woodland stream has only detritus compartments. If converted into difference equation form, the model features self-regulation in all compartments. Except for aggregation of whole individual leaves into masses of whole leaves, all flows in the model are consistent with conditions (a) and (f). (Each compartment is degenerately a "predation community.") An important feature of the Boling model is its extensive and strong time dependence (expressed as temperature, a function of time). Since condition (c) is fulfilled and (a) and (f) are "almost" fulfilled, it is not surprising that the observed annual trajectory is a stable attractor trajectory. In a similar model with a "bad" violation of condition (a), for example, with unrealistic transfer coefficients resulting in a net energy gain in the individual leaves–leaf aggregations loop, instability could well occur.

3.3. Intercompartmental Competition

The stability conditions (a)–(f) allow competition in the sense that two compartments may compete for energy from a third compartment provided the two do not directly interact. That is, the conditions allow competition provided one compartment does not influence the rate terms e_{ij}, r_{ij}, d_i, p_i which correspond to energy sources for a competitor.

However, competition between autotrophs for insolation generally does involve direct interference with photosynthesis terms p_i and so is not consistent with the above version of qualitative stability conditions. It may be shown that direct competition between autotrophs (dependence of p_i upon levels of various autotroph compartments) does not violate $\dot{\Lambda} < -\varepsilon \|\mathbf{x} - \hat{\mathbf{x}}\|^2$ in a region enclosing $\hat{\mathbf{x}}$ *provided* self-regulation dominates the influence of changing competitor levels in determining $p_i x_i - a_i x_i^2$ (May, 1976). Generally the actual quantitative expression of such conditions is messy and beyond direct ecological interpretation. Computer simulation may be a more efficient approach to studying stability than trying to analytically test $\dot{\Lambda}$.

4. CONCLUSION

Accounting for the flow of energy in ecosystems is a central theme of modern ecology. This chapter describes and develops a recipe for modeling ecosystems with predation interactions and detritus processing. The basic idea is that the mathematical conditions in the recipe are so general, yet the behavior of the model so "good," that nature might be expected to design and operate ecosystems more or less consistently with the conditions. In other words, this paper describes an archetypal holistic ecosystem model.

No attempt to model the flow of a specific nutrient (or pollutant) has been attempted. Modeling (as opposed to measuring) nutrient flow may be less cogent than modeling energy flow. Also, the nutrient content of a compartment should be proportional to energy content. The modeling of the *effects* of changing nutrient availability could be incorporated in the energy model provided nutrient availability were expressible as a function of time, so photosynthesis parameters and other parameters could still be expressed as functions of time.

To reiterate, the stability conditions are: Once energy leaves a predation community it may not return; for each predation pair, conversion efficiency (predator gain/prey loss) should be proportional to the current equilibrium prey/predator biomass ratio; and there should be equal net efficiencies for distinct energy pathways from one compartment to another within predation communities. Transfer rates of detritus energy by death, excretion, or physical breakdown to or among detritus compartments can be very general but must be consistent with one-way energy flow through the model. Considerable freedom in incorporating time dependence is allowable.

Last, the entire chapter may be placed in the context of stability theory with the following observations. The linear system $\dot{\mathbf{x}} = \mathbf{A}\mathbf{x}$ is asymptotically stable at $\mathbf{0}$ and if and only if symmetric positive definite matrices $\mathbf{M}_1$ and $\mathbf{M}_2$ exist satisfying $\mathbf{M}_1 \mathbf{A} + \mathbf{A}^* \mathbf{M}_1 = -\mathbf{M}_2$ (Willems, 1970, pp. 194–198). A symmetric matrix $\mathbf{M}$ is positive definite only if $M_{ii} > 0$ for all i and $M_{ii} M_{jj} > M_{ij}^2$ for all $i \neq j$. Roughly speaking, symmetric positive definite matrices are "not too far" from diagonal matrices (with positive entries). In the equation $\mathbf{M}_1 \mathbf{A} + \mathbf{A}^* \mathbf{M}_1 = -\mathbf{M}_2$ both $\mathbf{M}_1$ and $\mathbf{M}_2$ may be diagonal if and only if each $A_{ii} < 0$; for all $A_{ij} \neq 0$, $i \neq j$, we have $A_{ij} A_{ji} < 0$; and $\mathbf{A}$ has balanced loops, that is, for three or more distinct indices $i, j, k, \ldots, z$ we have $|A_{ij} A_{jk} \ldots A_{zi}| = |A_{iz} \ldots A_{kj} A_{ji}|$ (condition e). The stability of predation communities is a nonlinear, time-dependent extension of these observations. The main result is that coupling of such systems into hierarchies using a broad class of driving functions is still stable.

APPENDIX

Here is a proof of the theorem stated in Section 2.2.

Proof. Conditions (b), (d), and (e) suffice to guarantee the existence of positive constants (λ_i), $i = 1, 2, \ldots, n$, satisfying $\lambda_i e_{ij} r_{ij} \hat{x}_j = -\lambda_j e_{ji} r_{ji} \hat{x}_i$ for all $i \neq j$ at all times (Jeffries, 1974, 1975). A predation community upstream to all other predation communities, to which it is connected in the food web, is globally, uniformly, and asymptotically attracted to its portion of $\hat{\mathbf{x}}(t)$ by virtue of the Liapunov function of Eq. (8). For such a predation community with n_1 compartments, let,

$$\Lambda_1(\mathbf{x}, t) = \sum_{i=1}^{n_1} \lambda_i [x_i / \hat{x}_i(t) - 1 - \ln(x_i / \hat{x}_i(t))]. \tag{8}$$

Then along a solution trajectory $\mathbf{x}(t)$,

$$\dot{\Lambda}_1 = \sum_{i=1}^{n_1} -\frac{\lambda_i a_i (x_i - \hat{x}_i)^2}{\hat{x}_i}. \tag{9}$$

Since $b_1 < \hat{x}_i(t) < b_2$ and $-a_i < \varepsilon$ (condition (c)), the level sets of Λ_1 have bounded radii from $\hat{\mathbf{x}}(t)$ over time and on the level sets (except $\hat{\mathbf{x}}(t)$ itself) $\dot{\Lambda}_1$

is bounded below zero. These observations on Λ_1 may be translated into the claimed stability criteria.

The situation for a predation community directly downstream of the first predation community may be analyzed as follows. The analogous Λ_2 for the n_2 variables of the second predation community has as derivative along a solution trajectory,

$$\dot{\Lambda}_2 = \sum_{k=n_1+1}^{n_1+n_2} -\frac{\lambda_k a_k}{\hat{x}_k}(x_k - \hat{x}_k)^2 \qquad (10)$$
$$+ \sum_{k=n_1+1}^{n_1+n_2} \lambda_k \frac{(x_k - \hat{x}_k)}{\hat{x}_k^2 x_k}[\hat{x}_k d_k(\mathbf{x}, t) - x_k d_k(\hat{\mathbf{x}}, t)].$$

Each term in the second sum is positive only if $x_k/\hat{x}_k(t)$ is closer to 1 than $d_k(\mathbf{x}, t)/d_k(\hat{\mathbf{x}}, t)$. From condition (f), $d_k(\mathbf{x}, t)$ actually depends only on $(x_1, x_2, \ldots, x_{n_1}; t)$. Thus for a fixed but arbitrary "upstream" trajectory we have $\lim_{t \to \infty} d_k(\mathbf{x}, t)/d_k(\hat{\mathbf{x}}, t) = 1$, uniformly in t. Define a zone $\mathscr{Z}$ as

$$\mathscr{Z} = [(x_{n_1+1}, \ldots, x_{n_1+n_2}; t | m_1(t) < x_k/\hat{x}_k(t) < m_2(t), k = n_1+1, \ldots, n_1+n_2],$$

where $m_1(t) = \min_k [d_k(\mathbf{x}, t)/d_k(\hat{\mathbf{x}}, t)]$ and m_2 is the analogous maximum. The zone $\mathscr{Z}$ uniformly asymptotically collapses around "downstream" portion of $\hat{\mathbf{x}}(t)$. This is so because for any $\varepsilon > 0$ there exists $\delta > 0$ such that $\|\mathbf{x} - \hat{\mathbf{x}}\| < \delta$ implies $|d_k(\mathbf{x}, t)/d_k(\hat{\mathbf{x}}, t) - 1| < \varepsilon$, δ independent of t. Outside $\mathscr{Z}$ we have $\dot{\Lambda}_2$ less than a constant negative multiple of $\|\mathbf{x} - \hat{\mathbf{x}}\|^2$. The properties of Λ_2, $\dot{\Lambda}_2$, and $\mathscr{Z}$ may be translated into the stability criteria.

It follows by extension that any finite number of predation communities could be coupled consistently with condition (a) to form a huge model globally uniformly asymptotically attracted to $\hat{\mathbf{x}}$.

REFERENCES

Boling, R. H., Goodman, E. D., Van Sickle, J. V., Zimmer, J. O., Cummins, K. W., Petersen, R. C., and Reice, S. R. (1975). Toward a model of detritus processing in a woodland stream. *Ecology* **56**, 141–151.

Fleming, W. H. (1977). "Functions of Several Variables," p. 104. Springer-Verlag, Berlin and New York.

Jeffries, C. D. (1974). Qualitative stability and digraphs in model ecosystems. *Ecology* **55**, 1415–1419.

Jeffries, C. D. (1975). Stability of ecosystems with complex food webs. *Theor. Popul. Biol.* **7**, 149–155.

Jeffries, C. D. (1976). Stability of predation ecosystem models. *Ecology* **57**, 1321–1325.

Jeffries, C. D., Klee, V., and van den Driessche, P. (1977). When is a matrix sign stable? *Can. J. Math.* **29**, 315–326.

May, R. M. (1976). Models for two interacting populations. *In* "Theoretical Ecology, Principles and Applications" (R. M. May, ed.), pp. 58–59. Saunders, Philadelphia, Pennsylvania.
Rosen, R. (1970). "Dynamical System Theory in Biology," pp. 8–15. Wiley, New York.
Willems, J. L. (1970). "Stability Theory of Dynamical Systems." Wiley (Interscience), New York.

Index

A

Abstract object concept, *see also* Holon, 186
Abstraction, levels of, in aggregation problem, 40–41
Adaptability
 behavioral uncertainty in, 135
 defined, 133
 ecological variables and, 136–137
 environment and, 133–134
 evolution in time, 134
 genetic, 140–141
Adaptability theory
 construction of models and, 147
 dynamical independence of levels, 141–143
 relations among levels of organization, 139–140
Additive model, 63
Adequacy, of ecosystem models, 108–109
Aggregate, holon determinancy and, 196
Aggregation
 in ecosystem model, by digraph, 160–165
 of variables, in model design, 27–29
Aggregation model, 63–64
 random phase-space model, 39–40
Aggregation problem
 bias and, 67
 consequences of, 56
 discrete event model in, 19, 33–38
 discrete time system in, 33–38
 ecological modeling in, 18, 47
 error of, 64–69
 experimental frame and, 21–29
 formalism in, 20
 homomorphism and behavioral equivalance, 46

models and experimental frames, 20
 multifaceted system modeling, 45–47
 state-of-the-art, 56
Algae
 ice community, 199
 in river model, 283–285
 winter, as state variable, 362
Algorithm
 extended Kalman filter, 275
 for model sifting (GMDH), 332–335
 time delay hierarchical control, 430–434
Ammonia, as state variable, 313
Anaerobic digestion model, 271–273
Analysis
 of ecosystem model, 355–361
 of turnover times, in Wingra III model, 370
Artificial intelligence, 327
Artificial satellite control, 416
Andrews experimental forest, 317–320
Asymptotic stability, 454–455
Attribute
 model, 187
 of an object, 295
Average path length, in mathematical model, 378

B

Backward shift operator, 263
Base model, in aggregation problem, 3, 30–33
Base-lumped model pairs, in aggregation problem, 19
Behavior
 chaotic, 111
 definition of, 187

505

of generating system, 298
Behavioral uncertainty
 in adaptability theory, 135
Benthos, as state variable, 362
Bias
 in first order analysis, 92
 in model aggregation, 67
 in system identification, 282–283
Bilinear equation, in population model, 403
Biochemical oxygen demand (BOD), 283
 control of, in river, 437–444
 in river Cam, 284
 model of, 437
Biocide, 401
 application function, in model of, 406
Biological control
 adaptability and indifference, 132
 in ecosystem, 143–145
 evolutionary tendencies in, 134
 example of, 145
 fundamental mechanisms of, in ecosystem, 136
 levels of, 141–143
Bioregulation
 of ecosystem, 378
 in Lake Wingra model, 377–379
Black-box model
 defined, 263–265
 identification of, 270–271
 as input output mapping, 219
 parameter estimation, 268
"Black-box" ecological model as, 271–273
Black box modeling, advantages of, 264
Block diagram, 302
Blue-green algae, as state variable, 362

C

Calcium cycle model, 251
 identifiability characteristics, 253
Calculus of variations, and optimal control theory, 386
Callinassa major, 193, 231
Cam, river, 284
 biochemical oxygen demand and, 284
 dissolved oxygen data, 340
 pollution control in, 437–444
Canonical structure, of nonlinear system, 163

Carbon, dissolved organic, as state variable, 201
Carbon cycling, Ross Sea model and, 207
Carrying capacity, in logistic equation, 403
Catenary system, 366
Causal analysis of ecosystem, 192–199, 231–232
Causal bond
 dynamical systems and, 189
 for general systems, 187
 for input-output systems, 190–191
Causal chain, sequential dependence in, 193
Causal closure property, 193
Causal determination, 188
Causal holon, *see* Holon
Causal object, 185
Causality, 224
 holon, 187
 holon-environment reciprocity in, 191–192
 meanings of, 159
 transitive closure and, 193
Causal theory of environment, 185–192
Causal network
 causality and transitive closure in, 193
 cycling efficiency and, 207
 dynamic flow analysis and, 194
 feedback in, 189, 221
 flow analysis measures in, 207
 creaon inflow analysis and, 191, 206–207
 genon outflow analysis and, 191, 202–206
Causation, in hierarchies, 124
Cause, propagation of, in ecosystem, 186–187
Cause-effect
 functionality and, 224–225
 model, 185–189
Cephalopod, as state variable, 201
Chaos, as an emergert property, 112
Chlorophyll *a*, as state variable, 313
Choice of model, system identification and, 262
Closed-looped control, in hierarchical control, 445–448
Community, behavioral uncertainty of, 135
Community matrix, 457, 459–462
Compartment
 defined, 2
 function, in Lake Wingra model, 366
 role of, in system nutrient dynamics, 375

turnover times, analysis of, 368–371
variability of, in Wingra III model, 377
Compartment flux, defined, 367
Compartmental model
average path length, 378
competition within, 501
connectability, 244
controllability, 243, 245
defined, 239–241
as digraph, 495–496
generalities, 237–238
identifiability, 238
identification and parameter estimation, 238
observability, 243, 245
structural identifiability, 238, 241–243
transfer function matrix, 247
Compartmental structure, in ecosystem, 238
Compartmental system
generalities, 238
perturbation experiment, 238
Competition, intercompartmental, 501
Complement, in model development, 327
Complexity vs simplicity in ecosystem modeling, 46
Conifer, as state variable, 496
Connectability, input and output, 244
Connective stability, 475
Constants, evaluation of, in model, *see* parameter estimation
Constraint, in nonlinear programming, 388
Control, *see* Adaptability theory, Hierarchical control, Optimal control
in agroecosystem, 401, 416
artificial satellite, 416
biological, *see* Biological control
impulsive, 402, *see also* Impulsive control
in large scale systems, 420–426
on-line, 434
performance measure in, 388, 422
prediction optimization and, 338
problem statement, 385–388, 421
of river pollution, 420, 437–444, 449
Control matrix, 412, 449
Control problem, hierarchical methods and, 420
Control theory, 386, *see also* Optimal control theory
Control vector, 389
Controllability, 243

Coordination, in hierarchical systems, 420
Costate vector, in nonlinear programming, 390
Cost function, in control problem, 422
Coupling, system properties and, 356
system and environment, 192
Coupling matrix, among elements, 301
Coupling topology, effects on system behavior, 165
Creaon, as input environment
causation and, 196–197
defined, 190
in dynamic flow analysis, 200–208
in holon-environment reciprocity, 192
as input selector and receptor, 190
Creaon Inflow Analysis
in causal network, 195
in marine coprophagy model, 196–197
in Ross Sea model, 206–207
Criteria for model selection, 329–332
in self organization theory, hierarchy of, 334
Cross correlation function, 270
Curse of dimensionality, in optimization problem, 420
Cybernetics, 326
Cycling index, in Ross Sea pelagic model, 207

D

Data systems, in system identification, 293, 297
Dataless systems, 293, 295–297
Decomposer, as state variable, 496
Decomposistion of systems, in causal-effect theory, 221
Determinancy
holon, 187
principle of, 188
Determinism vs chaos, controversy, 109
Detritus, as state variable, 201, 362
Detritus processing, in ecosystem model, 500
Diatom, as state variable, 362
Differential system, 223
Digraph, 154–155
foodweb and, 495–496
partition and condensation, 160

predation community and, 495
reachability and, 155–160
stability and, 495
Dirac delta function, 402
Direct dependence, in causal sequence, 193
Directed generative system, in system identification, 300
Directed graph, see Digraph
Directed system, 294, 296–297
Discrete dynamical system, goal coordination method of control, 427
Discrete event model, in aggregation problem, 19, 33–38
Discrete time system, in aggregation problem, 33–38
Dissolved Oxygen (DO)
in aquatic environment, 284
control of, in river, 437–444
model of, 342, 437
Dual function, in control problem, 422
Duality of environment, 190–191
Dynamic ecological model, formulation of equations for, 30
Dynamic optimization, see also Hierarchical control
large scale, 420
Dynamical flow analysis, in causal network, 194
Dynamic programming, 386, 412
Dynamical system, causal bond and, 196

E

Ecological model, see also Ecosystem, Ecosystem model
aggregation problem in, 18–20, 55–57
analysis of, 355–361
BOD-DO river model, 284–286
development of, 20
input-output or mass balance approach in, 80–82
marine coprophagy, 192
multidimensional structure, 343
Ross Sea, 194
stability in, 143, 171–175
systems analysis of, 368–374
validation of, 269
Ecological system, see also Ecosystem
black box model of, 271–273

Ecological variable, in adaptability theory, 136–137
Ecology, hierarchial levels of organization, 125
systems approach in, 1
Eco-pressure, 115
Ecosystem
application of control in, 408–409
biological control in, 132, 143–145
bioregulation of, 378
boreal forest, 496
causal analysis, examples, 192–199, 231–232
causal closure in, 193
cause in, 185
as circular causal structure, 189
compartmental model, 239–241
compartmental structure of, 237–239
control theory and, 385–387
detritus processing, in stream, 500
evolution of, 375
exploited, model of, 475
Findley Lake, 85
heterotrophs' role in, 376
hierarchical organization, 119, 122–123, 137
hierarchy theory and, 126–127
holistic modeling of, 494–499
mass balance, 80
as multifaceted system, 18
perturbation response, 360
predator removal, effects on, 482
productivity, nutrient cycling and, 376
propagation of cause in, 186–187
properties of, 360
regulation of, with biocides, 401–402
response to stress, 453–455
"robustness," stability and, 469, 474
sampling design, 95
storage costs in, 376
stability properties of, 458
time domain analysis in, 222–224
as unit of investigation, 356
Wingra lake, 361
as whole, 184
Ecosystem analysis, theory in, 11–12
Ecosystem model
adequacy of, 108
evaluation of, 374–379
holon in, 187

as hypothesis about complexity, 19, 122
lumping of species populations, 56–57
state space representation in, 227–230
systems view in, 1–2
validation, 269
verification, 269
Eco-temperature, 114–115
Eigenvalue, in linear model, stability and, 455–457, 467
Emergent properties, chaos and, 109, 112
Environment
  adaptability theory and, 133–134
  cause and, 190–192
  concept of, 185–186
  defined, 185
  duality of, 190–191
  holon and, 187
Environmental problem, GMDH and, 339–350
Environmental uncertainty, in adaptability theory, 133, 135–136
Epistemological levels, hierarchy of, 292
Equilibrium state, in population model, 469
Error
  in hierarchical control, 424
  innovation, Kalman filter, 269
  in prediction, 337
  residual, 279
Error analysis, 81
  in aggregation problem, 64–69
  in Findley Lake model, 85–87
  in model development, 81–85
  in river flow, 83–85
  sampling design and, 95
  systems analysis and, 94
Error covariance, in Kalman filter, 278
Error sensitivity, in first order analysis, and parameter estimation, 100
Error types, in aggregation problem, 66
Estimation
  bias, 282–283
  combined state-parameter, 274
  instrumental variable, 268
  least squares, 268
  maximum likelihood, 268
  parameter, 267
  recursive, 271
Euphasia crystallorophias, 201
Euphasia superba, 201
Experimental design, 262

Findley Lake, sampling in, 95
first order analysis and, 94–101
system identification and, 262
Experimental frame, in aggregation problem, 21–29
Experimental investigation, in system identification, 305
Extended Kalman filter, 273–278

F

Feedback, in causal network, 189, 221
Feedback control, in hierarchical control, 444
Filter
  extended Kalman, 275
  Kalman, 274
  noise covariance, 278
  shaping, 264, 270
  stability, 281
Findley lake
  groundwater flow to, 85–86
  sampling design and, 95
  trophic status, 85
Findley lake model, error analysis of, 85–87
First order analysis, 82–92
  bias, 92
  Findley lake and, 85–87
  model development and, 81–85
  in river flow, 83–85
  sampling design and, 95
  systems analysis and, 94
Fish, as state variable, 201, 362
Fish management, control theory and, 391–393
Fish population, harvest of, 396
Fishery
  maximum sustainable yield, 397
  optimal harvest of, 393
Five variable model, lake Ontario, 315
Flow analysis, 194
Flow matrix, in causal analysis, 193–194
Flux rate, defined, 365
Food web, as digraph, 495–496
Forest, boreal, model of, 496
Forest ecosystem model, 317–320
Formalism of cause effect relationships, 213–233
Four variable model, lake Ontario, 313

510 Index

Fox, as state variable, 496
Framework, for systems problem solving, 291

G

Generative system, in system identification, 294, 297–300
Generative system identification, 305–307
Genon, as output environment
 causation and, 197–198
 defined, 191
 in dynamic flow analysis, 202–206
 in holon environment reciprocity, 191–192
Genon Outflow Analysis
 in causal network, 195
 in marine coprophagy model, 197–198
 in Ross Sea model, 202–205
GMDH, see Group method of data handling
Goal coordination method, discrete dynamical systems, 427
Gradient, of a function, 388, 423
Green algae, as state variable, 362
Group method of data handling, 326
 environmental problems and, 339–350
Groundwater, error estimate, 85–86

H

Hamiltonian function, in continuous time optimal control theory, 395
 in discrete time optimal control theory, 390
 in dual problems of hierarchical control, 429
Hare, as state variable, 496
Harvest of a fishery, optimal control theory and, 393, 396
Heterotrophs' role in ecosystem, 376
Hierarchical adaptability theory, 131–149, see also Adaptability, Adaptability theory
Hierarchical approach to optimization, problem formulation, 421
Hierarchical control
 closed-loop, 445–448
 feedback, linear quadratic problems and, 444–445
 goal coordination approach, 422–434
 interaction balance approach, 425
 interaction prediction approach, 434–437
 open loop, 434
 servomechanism case, 448–449
Hierarchical levels independence of, adaptability theory and, 141–143
Hierarchical model, stability of, 177
Hierarchical organization of nature, 119–125
Hierarchy
 defined, 120–121
 of epistemological levels, 292
 function, 123
 predation communities and, 492
 structure, 122
Hierarchy of morphisms, 40
Hierarchy theory and ecosystems, 126
Holistic ecosystem modeling, 494–499
Holon
 causal, 188
 creaon inflow analysis and, 190
 defined, 187
 genon outflow analysis and, 195
 in hierarchies, 124
 nonanticipation, 189
Holon coupling, 191–192
Holon determinancy, 188
Holon-environment reciprocity, 191–192
Homomorphism, defined, 40
Homomorphic image, in system identification, 293
Host-parasite interaction, chaos and, 112

I

Identifiability, defined, 238
Identification, see also System identification
 of internally descriptive model, 273–881
 of two dimensional model by GMDH, 348
 of generative system, 295, 305–307
 of model structure, 260
 of structure system, 295
 system, 261
Image system, in system identification, 292
Impulse response, in system identification, 262, 270
Impulsive control
 of agroecosystem, 402, 416
 optimal timing of biocide application, 408–409
 performance index, 404

Inference problem, in aggregation problem, 59–61
Inflow analysis
　in marine coprophagy model, 196
　in Ross Sea pelagic model, 206
Influence of inputs on states, 157, see also Creaon
Innovation, in Kalman filter, 279
Input
　connectability, in identification problem, 244
　environment, in causal analysis, 190
　reachability, 155–160
Input-output flow analysis, 194
Input-output function, in aggregation problem
Input-output mapping, 219
Input-output model, see Black box model
Input-output modeling formalism, 214
Input-output pair, identification of state and, 274
Input-output system, defined, 216
　causal bond and, 190–191
Input prewhitening, 270
Input signals, in model structure identification, 270
Input-state-output system, 228
Insect, as state variable, 496
Interaction balance approach, in control problem, 425
Interaction prediction approach, 434–437
　hierarchical solution to river pollution problem, 441–442
Interconnection matrix, digraph and, 154
Interconnection of systems, in causal-effect theory, 220–221
Internally descriptive model
　defined, 265
　identification of, 273–281
　parameter estimation of, 269
Invertebrate, as state variable, 200

J

Jacobian matrix, 173

K

Kalman filter, 275–278

Kalman gain, 277
Kronecker delta function, 269, 458

L

Lagrange multiplier, 423
Lake
　Findley, 85
　mass balance, 80
　Ontario, 312
　sampling design, 95
　Wingra, 361
Lake model, compartment turnover times, 369
Lake Wingra model, bioregulation, of, 377–379
Laplace transform, 247
Large-scale systems control see hierarchical control
Lattice, structure candidates and, 309
Leontief system analysis method, 194
Liapunov function, 470–471
Liapunov, method, stability and, 468–471
Linear black box model, 263–265
Linear model
　defined, 234
　dynamical equations, 240
　identifiability of, 241
　of Ross Sea pelagic ecosystem, 199
Linear-in-the-parameters, model, 266, 273
Linear system identification, 238
Linearity vs nonlinearity controversy, in ecological models, 112
Linearization, of nonlinear model, 275
Logistic equation, 111
　time varying, 403
Lotka-Volterra predator-prey model, 145
　generalized model, 472
　nonvulnerability in, 481
　stability of, 169–170
Lumped model, in aggregation problem, 3, 14, 33–38
Lumped parameters, 40

M

Macroscopic variables, 113

Index

Management, ecosystem and, 385–386
Management of a fishery, 391–393
Man-machine dialogue, 327
Mapping, as homomorphism, 40
Mapping, input-output, 219
Marine coprophagy model, 192
Mask, in system identification, 248
Mask selection, Andrews experimental forest model, 318
Mass balance
　in Findley lake, 85–92
　phosphorus, 86
　sampling design, and, 95
　in system identification, 280
Mass balance equation, 80
Mathematical model
　formulation of, 184
　system identification and, 260, 305
　theory and, 1–15
　stability of, see Stability
　variables in, 494–495
Mathematical systems theory, 214
Maximum principle, optimal control theory and, 386
Measurement, error covariance matrix, 277
Measurement-noise covariance, in Kalman filter, 277
Mechanistic model, see Internally descriptive model
Migration, in predator-prey model, 32
Model
　additivity in, 56
　additive, 63
　advection and diffusion, 344
　anaerobic digestion, 271–273
　analysis of, 355–361
　base, 3, 19, 30–33
　cause-effect, 185–189, 224
　constants, evaluation of, 273–286
　criteria for selection in self organization theory, 329–332
　defined, 1, 3, 216
　discrete event, 33
　of dynamical ecological system, 199
　fitting data in, 273–386
　formalism, 18–19
　hierarchical, 140–141
　identifiability, 238
　linear compartmental, 239
　Lotka-Volterra generalized, 472
　lumped, 33
　many-one correspondence in, 28
　nonparametric, 262
　parametric, 263
　purpose of, 184
　random phase space, 39–40
　vs. real system, 18
　self-organization, 326
　sifting (GMDH), 332–335
　simplification, 40
　uniqueness, 327
　validation, 269
　verification, 269
Model analysis, hypothesis in, 359–360
Model behavior, input signals and, see Input-Input-output
Model construction, adaptability theory and, 147
Model development, error analysis in, 81–85
Model example
　Andrews forest, 317
　BOD-DO, 437
　boreal forest, 496
　river Cam, 284
　exploited ecosystem, 425
　marine coprophagy, 192
　lake Ontario, 199
　lake Wingra, 361
　river water quality, 437
　Ross Sea pelagic ecosystem, 199
Model parameter estimation, 266–267
Model structure identification, 266–267
　internally descriptive model, 273–286
Modeling
　defined, 184
　holistic, of ecosystem, 494–499
　purpose of, 184
Modeling process, in ecological perspective, 107
Modeling steps
　in Klir's general systems theory, 291–294
　in Patten's ecosystem theory, 184
　in Zeigler's theory of modeling, 19–21
Multi-faceted systems modeling, 45–47

N

Nature, hierarchical organization of, 119–125

Index 513

Neighborhood stability, 468
Network, see Causal network
Neutral system, 294, 296–297
Nitrate, as state variable, 313
Nitrogen, recycling of, 368–381
Node, in digraph, 485
Noise, in Kalman filter, 278
Nonanticipation, 189, 225
Nonlinear programming, 388
Nonlinear stability analysis, 471–478
Nonlinear systems, canonical structure, 163
Nonparametric model, 262
Norwich sewage works, 272
Notation, 12
Numerical instability, in system identification, 282
Nutrient, as state variable, 201
 retention in a system, 366
Nutrient cycling, as system property, 356–357, 366

O

Object, 12, 186, see also Holon
 attribute, 295
Object causality, 188–189
Object interactions, in hierarchical theory, 123
Object system, as set of attributes, 292
Observability, 245
Ontario lake
 four variable model of, 313–315
 five variable model of, 315–316
 seven variable model of, 316–317
Open loop control, in hierarchical control, 434
Optimal control theory
 continuous time, 394
 discrete-time, 387
 fishery management and, 391–393
 hierarchical, see Hierarchical control
 nonlinear programming and, 387
 problem definition, 386, 389–390
Optimal impulsive control, 403
Optimization, hierarchical, see Hierarchical control
Organic matter, as state variable, 362
Organization of frames, in aggregation problem, 27

Outflow analysis
 in marine coprophagy model, 197
 in Ross sea pelagic model, 202–206
Output connectability, in identification problem, 244
Output environment, in causal analysis, 190
Output reachability, 156
Output system, 228
Oxygen, dissolved, see Dissolved oxygen

P

Parameter estimate, *a priori*, 280
Parameter estimation, 267
 of black box model, 268
 in group method of data handling, 335
 of Huffaker's universe model, 49–52
 of internally descriptive model, 269
 Kalman filter, 275–278
 recursive, 275–278
Parameter estimation, and error in first order analysis, 100
Parametric mapping, 220
Parametric model, 263
Partial order, as a logical relation, 27
Path matrix, digraph and, 156
Pattern recognition, in system identification, 271
Penguin, as state variable, 201
Perturbation, ecosystem and, 360, 453–455, 469
Phosphorus
 cycling of, in lake, 368–381
 mass balance in Findley lake, 86
 relative retention of, in model, 366–368
 sampling design, 95
 system turnover time of, 372–374
Phosphorus, soluble reactive, as state variable, 313
Phosphorus dynamics model, 254
 identifiability characteristics, 254–257
Phytoplankton, as state variable, 200
Pisaster ochraceus, 482
Pleurogamma antarticum, 201
Pool, defined, 356
Population, fish and optimal harvesting, 396
Population, as subsystem of ecosystems, 125
Population, as unit of functional interest, 56

Population dynamics, of blow flies, 111
Population model, of pest, 415
Predation community, in a model, 490
  digraph and, 495
  hierarchy of, 492
Predator-prey relationship, prey density and, 490-494
Predator removal, ecosystem effects on, 482
Prediction, criteria for model selection, in self organization theory, 330-337
Prediction optimization, automatic control and, 338
Prediction of predictions (GMDH), 337
Prediction problem, of Ukranian reservoir, 345

R

Random phase-space model, 39-40
Reachability, 155
Recycling of nutrient, in lake Wingra model, 366
Recycling effect, in ecosystem, 358
Reductionism, as an antithesis to holism, 123
Relative retention time, of a system, 366
Resilience, as system property, 360, 376, 455
Resistance, as system property, 360, 375, 455
Resource management, optimal control theory and, 387
River Cam, model of, 437
River water quality control, 420, 437-444
  multiple effluent inputs, 439-441
  simulation results, 442
  single effluent input, 439
Robust stability, 469, 474
Rocket trajectory, optimal control and, 386
Ross Sea pelagic ecosystem
  causal flow analysis, 202-209
  model description, 199-201

S

Sampling design, and first order analysis, 84
  in Findley lake, 95-97
Sampling variable, in system identification, 297

Sea, 199-201, *see also* Marine coprophagy model
Seal, as state variable, 201
Sector stability, 482-486
Self organization, principle of, 328
Seven variable model, lake Ontario, 316
Shrub, as state variable, 496
Signal flow graph, *see* Digraph
Similarity, definition, 2
SIMSCRIPT, computer language, 19
Singular control, 386
Source system, in system identification, 293
Space technology, application to ecological problems, 9, 386
Spectral analysis, 114
Spruce, as state variable, 496
Stability
  analysis of, 456-462
  concept of, 454, 468
  connective, 475
  criteria for, 454, 468
  defined, 455
  ecological significance of, 453, 468
  eigenvalues and, 455-457, 467
  in exploited ecosystem, 475
  hierarchical model, 177
  index of, 456
  Liapunov, 468-471
  in linear and nonlinear systems, 434
  local, 468
  in nonlinear models, 471
  robustness and, 469, 474
  sector, 482-486
  sufficient conditions for, 471-473
Stability, in adaptability theory, 143-145
Stability, and vulnerability, 171-174
Stable cycle, and chaotic behavior, 111
State
  in cause-effect relationship, 227-230
  defined, 227-228
  dynamical, 227
  holon determinancy and, 184
  properties, 228
  representation, 228
State space, 228
  equilibrium in, 469
State-parameter estimation problem, 274
State transition, as mapping, 229-230
State transition relation, in system identification, 298

Index 515

State vector, augmented, in Kalman filter, 274
Storage costs in ecosystem, 376
Stress, ecosystem response and, 453–455
Structural identifiability, 238
  of compartmental model, 246
  definitions related to, 242–243
  ecosystem examples and, 251–257
Structural perturbation, 167
Structure candidate, in system identification, 309
  Andrews experimental forest model, 38
Structure identification, 266–267
Structure identification problem, 295
Structure modeling, in system identification, 294
Structure system, in system identification, 294, 301–302
Structure system identification, 307–312
Submask, in system identification, 298
Symbolic language for computers, 35, 50
System, *see also* Model
  as bounded sector of reality, 18
  catenary, 366
  causal bond, and, 189
  causality and, 185, 224
  causal closure property, 193
  concept of, 1
  coupled or joined, 220–271
  couplings, 154, 186, 220
  defined, 1
  dynamic, 227
  emergent properties, 356
  identifiability, 238
  Klir's definition of, 242–294
  as mapping, 28, 219
  properties, 224–225
  relative retention time, 366
  resilience, 360, 455
  resistance to perturbation, 360, 455
  stability of, 454, *see also* Stability
  throughput, 207
  variability of, in Wingra III model, 377
  vulnerability of, 476
  Zadeh's and Mesarovich's definitions of, 215
System approach, 1
  deductive, 6–8, 214–215
  hierarchical, 10, 122–125, 137–143
  inductive, 6, 214, 291–294

System control, goal coordination method, 422
System cycling efficiency, 207
System flux, defined, 367
System identification
  bias and, 282–283
  choice of model and, 262
  defined, 6, 260
  in ecological model, 284, 312, 317
  experimental design and, 262
  experimental investigation and, 305
  mask, 298, 318
  mass balance and, 280
System, input-output, defined, 216
System noise covariance, in Kalman filter, 278
System nutrient dynamics, compartment role and, 375
System optimization, *see* Optimal control theory
System properties, as general features of all systems, 1
  coupling topology and, 356
  defined, 356
  material cycling and, 378
Systems analysis, error analysis and, 94
  of lake Wingra model, 368–374
Systems ecology, 185
Systems problem solving, conceptual framework, 241

T

Temperature, as state variable, 313
Terminology, 12
Throughflow, in a system, 378
Time delay representation, in black box model, 263
Time domain, defined, 222
Time series analysis, 262
Time systems, causal bond and, 222
Trajectory, time function, 23
Transfer function, 247
Transition matrix, 276
  of linear system, 226
Transitive closure, causality and, 193
Transitive closure matrix, 194
Transmittance, in causal network, 193–194

Turnover Time
  analysis of, in Wingra III model, 370
  data comparison, 374
  of compartments of Wingra III model, 369
  defined, 364–365
  ecosystem behavior and, 360
  of a system, 372–374

U

Ukranian SSR, 326
Uncertainty reduction, in system identification, 307
UNIVAC computer, 364
Universe, Huffaker's patch structured, 18–19

V

Validation, of model, 269
Variable
  explicit, 364
  implicit, 364
  internal, 243
Variables, aggregation of, 63–64

Verification, of model, 269
Vulnerability, and digraph, defined, 153
  of Lotka-Volterra model, 169
  of stability, 169
  of structure, 164

W

Water quality, *see* Dissolved oxygen
Whale, as state variable, 201
Wingra, lake, 361
Wisconsin, 361
Woodpecker, as state variable, 496

Y

Yield, maximum sustainable, of fish population, 397

Z

Zooplankton, as state variable, 200, 313, 362